国外优秀数学著作
原　版　系　列

伽罗瓦理论

Galois Theory (Fourth Edition)

（第4版）

（英文）

[英] 伊恩·斯图尔特（Ian Stewart）著

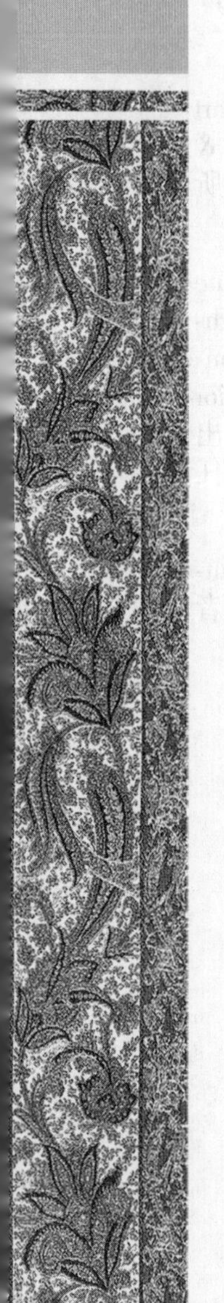

哈爾濱工業大學出版社
HARBIN INSTITUTE OF TECHNOLOGY PRESS

黑版贸审字 08-2019-191 号

Galois Theory: Fourth Edition/by Ian Stewart/ISBN:978 - 1 - 138 - 40170 - 9

图书在版编目(CIP)数据

伽罗瓦理论:第 4 版=Galois Theory: Fourth Edition: 英文/(英)伊恩·斯图尔特(Ian Stewart)著.—哈尔滨:哈尔滨工业大学出版社,2021.8
ISBN 978 - 7 - 5603 - 9643 - 9

Ⅰ.①伽… Ⅱ.①伊… Ⅲ.①伽罗瓦理论-英文
Ⅳ.①O153.4

中国版本图书馆 CIP 数据核字(2021)第 180278 号

策划编辑 刘培杰 杜莹雪
责任编辑 刘春雷 钱辰琛
封面设计 孙茵艾
出版发行 哈尔滨工业大学出版社
社 址 哈尔滨市南岗区复华四道街 10 号 邮编 150006
传 真 0451 - 86414749
网 址 http://hitpress.hit.edu.cn
印 刷 哈尔滨市颉升高印刷有限公司
开 本 787 mm×1 092 mm 1/16 印张 32 字数 510 千字
版 次 2021 年 8 月第 1 版 2021 年 8 月第 1 次印刷
书 号 ISBN 978 - 7 - 5603 - 9643 - 9
定 价 88.00 元

(如因印装质量问题影响阅读,我社负责调换)

Portrait of Évariste Galois, age 15.

Contents

Acknowledgements

The following illustrations are reproduced, with permission, from the sources listed.

Frontispiece and Figures 3–6, 22 from *Écrits et Mémoires Mathématiques d'Évariste Galois*, Robert Bourgne and J.-P. Azra, Gauthier-Villars, Paris 1962.

Figure 1 (left) from *Erwachende Wissenschaft 2: Die Anfänge der Astronomie*, B.L. van der Waerden, Birkhäuser, Basel 1968.

Figures 1 (right), 2 (right) from *The History of Mathematics: an Introduction*, David M. Burton, Allyn and Bacon, Boston 1985.

Figure 25 from *Carl Friedrich Gauss: Werke*, Vol. X, Georg Olms, Hildesheim and New York 1973.

The quotations in Chapter 25 are reproduced with permission from *The Mathematical Writings of Évariste Galois*, Peter M. Neumann, European Mathematical Society, Zürich 2011.

Preface to the First Edition

Galois theory is a showpiece of mathematical unification, bringing together several different branches of the subject and creating a powerful machine for the study of problems of considerable historical and mathematical importance. This book is an attempt to present the theory in such a light, and in a manner suitable for second- and third-year undergraduates.

The central theme is the application of the Galois group to the quintic equation. As well as the traditional approach by way of the 'general' polynomial equation I have included a direct approach which demonstrates the insolubility by radicals of a specific quintic polynomial with integer coefficients, which I feel is a more convincing result. Other topics covered are the problems of duplicating the cube, trisecting the angle, and squaring the circle; the construction of regular polygons; the solution of cubic and quartic equations; the structure of finite fields; and the 'Fundamental Theorem of Algebra'.

In order to make the treatment as self-contained as possible, and to bring together all the relevant material in a single volume, I have included several digressions. The most important of these is a proof of the transcendence of π, which all mathematicians should see at least once in their lives. There is a discussion of Fermat numbers, to emphasise that the problem of regular polygons, although reduced to a simple-looking question in number theory, is by no means completely solved. A construction for the regular 17-gon is given, on the grounds that such an unintuitive result requires more than just an existence proof.

Much of the motivation for the subject is historical, and I have taken the opportunity to weave historical comments into the body of the book where appropriate. There are two sections of purely historical matter: a short sketch of the history of polynomials, and a biography of Évariste Galois. The latter is culled from several sources, listed in the references.

I have tried to give plenty of examples in the text to illustrate the general theory, and have devoted one chapter to a detailed study of the Galois group of a particular field extension. There are nearly two hundred exercises, with twenty harder ones for the more advanced student.

Many people have helped, advised, or otherwise influenced me in writing this book, and I am suitably grateful to them. In particular my thanks are due to Rolph Schwarzenberger and David Tall, who read successive drafts of the manuscript; to Len Bulmer and the staff of the University of Warwick Library for locating documents relevant to the historical aspects of the subject; to Ronnie Brown for editorial guidance and much good advice; and to the referee who pointed out a multitude of sins of omission and commission on my part, whose name I fear will forever remain a mystery to me, owing to the system of secrecy without which referees would be in continual danger of violent retribution from indignant authors.

University of Warwick IAN STEWART
Coventry
April 1972

Preface to the Second Edition

It is sixteen years since the first edition of *Galois Theory* appeared. Classical Galois theory is not the kind of subject that undergoes tremendous revolutions, and a large part of the first edition remains intact in this, its successor. Nevertheless, a certain thinning at the temples and creaking of the joints have become apparent, and some rejuvenation is in order.

The main changes in this edition are the addition of an introductory overview and a chapter on the calculation of Galois groups. I have also included extra motivating examples and modified the exercises. Known misprints have been corrected, but since this edition has been completely reset there will no doubt be some new ones to tax the reader's ingenuity (and patience). The historical section has been modified in the light of new findings, and the publisher has kindly permitted me to do what I wanted to do in the first edition, namely, include photographs from Galois's manuscripts, and other historical illustrations. Some of the mathematical proofs have been changed to improve their clarity, and in a few cases their correctness. Some material that I now consider superfluous has been deleted. I have tried to preserve the informal style of the original, which for many people was the book's greatest virtue.

The new version has benefited from advice from several quarters. Lists of typographical and mathematical errors have been sent to me by Stephen Barber, Owen Brison, Bob Coates, Philip Higgins, David Holden, Frans Oort, Miles Reid, and C. F. Wright. The Open University used the first edition as the basis for course M333, and several members of its Mathematics Department have passed on to me the lessons that were learned as a result. I record for posterity my favourite example of OU wit, occasioned by a mistake in the index: '226: *Stéphanie D. xix.* Should refer to page xxi (the course of true love never does run smooth, nor does it get indexed correctly).'

I am grateful to them, and to their students, who acted as unwitting guinea-pigs: take heart, for your squeaks have not gone unheeded.

IAN STEWART

University of Warwick
Coventry
December 1988

Preface to the Third Edition

Galois Theory was the first textbook I ever wrote, although it was the third *book*, following a set of research-level lecture notes and a puzzle book for children. When I wrote it, I was an algebraist, and a closet Bourbakiste to boot; that is, I followed the fashion of the time which favoured generality and abstraction. For the uninitiated, 'Nicolas Bourbaki' is the pseudonym of a group of mathematicians—mostly French, mostly young—who tidied up the mathematics of the mid-20th Century in a lengthy series of books. Their guiding principle was never to prove a theorem if it could be deduced as a special case of a more general theorem. To study planar geometry, work in n dimensions and then 'let $n = 2$.'

Fashions change, and nowadays the presentation of mathematics has veered back towards specific examples and a preference for ideas that are more concrete, more down-to-Earth. Though what counts as 'concrete' today would have astonished the mathematicians of the 19th Century, to whom the general polynomial over the complex numbers was the height of abstraction, whereas to us it is *a single concrete example*.

As I write, *Galois Theory* has been in print for 30 years. With a lick of paint and a few running repairs, there is no great reason why it could not go on largely unchanged for another 30 years. 'If it ain't broke, don't fix it.' But I have convinced myself that psychologically it *is* broke, even if its logical mechanism is as bright and shiny as ever. In short: the time has come to bring the mathematical setting into line with the changes that have taken place in undergraduate education since 1973. For this reason, the story now starts with polynomials *over the complex numbers*, and the central quest is to understand when such polynomials have solutions that can be expressed by radicals—algebraic expressions involving nothing more sophisticated than nth roots.

Only after this tale is complete is any serious attempt made to generalise the theory to arbitrary fields, and to exploit the language and thought-patterns of rings, ideals, and modules. There is nothing wrong with abstraction and generality—they are still cornerstones of the mathematical enterprise. But 'abstract' is a verb as well as an adjective: general ideas should be abstracted *from* something, not conjured from thin air. Abstraction in this sense is highly non-Bourbakiste, best summed up by the counter-slogan 'let $2 = n$.' To do that we have to start with case 2, and fight our way through it using anything that comes to hand, however clumsy, *before* refining our methods into an elegant but ethereal technique which—without such preparation—lets us prove case n without having any idea of what the proof does, how it works, or where it came from.

It was with some trepidation that I undertook to fix my non-broke book. The process turned out to be rather like trying to reassemble a jigsaw puzzle to create a different picture. Many pieces had to be trimmed or dumped in the wastebasket, many new pieces had to be cut, discarded pieces had to be rescued and reinserted. Eventually order re-emerged from the chaos—or so I believe.

Along the way I made one change that may raise a few eyebrows. I have spent much of my career telling students that written mathematics should have punctuation as well as symbols. If a symbol or a formula would be followed by a comma if it were replaced by a word or phrase, then it should be followed by a comma—however strange the formula then looks.

I still think that punctuation is essential for formulas in the main body of the text. If the formula is $t^2 + 1$, say, then it should have its terminating comma. But I have come to the conclusion that eliminating visual junk from the printed page is more important than punctuatory pedantry, so that when the same formula is *displayed*, for example

$$t^2 + 1$$

then it looks silly if the comma is included, like this,

$$t^2 + 1,$$

and everything is much cleaner and less ambiguous without punctuation.

Purists will hate this, though many of them would not have noticed had I not pointed it out here. Until recently, I would have agreed. But I think it is time we accepted that the act of displaying a formula equips it with *implicit*—invisible—punctuation. This is the 21st Century, and typography has moved on.

Other things have also moved on, and instant gratification is one of them. Modern audiences want to see some payoff *today*, if not last week. So I have placed the more accessible applications, such as the 'Three Geometric Problems of Antiquity'—impossible geometric constructions—as early as possible. The price of doing this is that other material is necessarily delayed, and elegance is occasionally sacrificed for the sake of transparency.

I have preserved and slightly extended what was undoubtedly the most popular feature of the book, a wealth of historical anecdote and storytelling, with the romantic tale of Évariste Galois and his fatal duel as its centrepiece. 'Pistols at 25 paces!' *Bang!* Even though the tale has been over-romanticised by many writers, as Rothman (1982a, 1982b) has convincingly demonstrated, the true story retains elements of high drama. I have also added some of the more technical history, such as Vandermonde's analysis of 11th roots of unity, to aid motivation. I have rearranged the mathematics to put the concrete before the abstract, but I have not omitted anything of substance. I have invented new—or, at least, barely shop-soiled—proofs for old theorems when I felt that the traditional proofs were obscure or needlessly indirect. And I have revived some classical topics, such as the nontrivial expression of roots of unity by radicals, having felt for 30 years that $\sqrt[n]{1}$ is cheating.

The climax of the book remains the proof that the quintic equation cannot be solved by radicals. In fact, you will now be subjected to *four* proofs, of varying generality. There is a short, snappy proof that the 'general' polynomial equation of degree $n \geq 5$ cannot be solved by radicals *that are rational functions of the coefficients*. An optional section proving the Theorem on Natural Irrationalities, which was the big advance made by Abel in 1824, removes this restriction, and so provides the second proof. Lagrange came within a whisker of proving all of the above in 1770-1771, and Ruffini probably *did* prove it in 1799, but with the restriction to radicals that are rational functions of the coefficients. He seems to have thought that he had proved something stronger, which confused the issue. The proof given here has the merit of making the role of field automorphisms and the symmetric and alternating groups very clear, with hardly any fuss, and it could profitably be included in any elementary group theory course as an application of permutations and quotient groups. Proof 4 is a longer, abstract proof of the same fact, and this time the assumption that the radicals can be expressed as rational functions of the coefficients is irrelevant to the proof. In between is the third proof, which shows that a *specific* quintic equation, $x^5 - 6x + 3 = 0$, cannot be solved by radicals. This is the strongest statement of the four, and by far the most convincing; it takes full-blooded Galois Theory to prove it.

The sole remaining tasks in this preface are to thank Chapman and Hall/CRC Press for badgering me into preparing a revised edition and persisting for several years until I caved in, and for putting the whole book into LaTeX so that there was a faint chance that I might complete the task. And, as always, to thank careful readers, who for 30 years have sent in comments, lists of mistakes, and suggestions for new material. Two in particular deserve special mention. George Bergman suggested many improvements to the mathematical proofs, as well as pointing out typographical errors. Tom Brissenden sent a large file of English translations of documents related to Galois. Both have had a significant influence on this edition.

University of Warwick IAN STEWART
Coventry
April 2003

generally. There is a new, fairly gentle, proof that the general polynomial equation of degree $n \geq 5$ cannot be solved by radicals that are rational functions of the coefficients. An optional section proving the Theorem on Natural Irrationalities, which was the big advance made by Abel in 1824, removes this restriction, and so provides the second proof. Lagrange came within a whisker of proving all of this in about 1770–1771, and Ruffini probably did prove it in 1799, but with the restriction to radicals that are rational functions of the coefficients. He seems to have thought that he had proved an additional fact, which confused the issue. The proof given here has the merit of making the role of field automorphisms and the symmetric and alternating groups very clear, with hardly any fuss, and it could profitably be included in any elementary algebra course as an application of permutations and quotient groups. Proof 4, along Galois's lines, is a direct proof of the same fact, and this time the assumption that the radicals can be expressed as rational functions of the coefficients is irrelevant. The proof in between is the third proof, which shows that a specific quintic equation, $t^5 - 6t + 3 = 0$, cannot be solved by radicals. This is the strongest statement of the question, and by far the most convincing; it takes full-blooded Galois Theory to prove it.

The sole remaining task of this preface is to thank Chapman and Hall/CRC Press for badgering me into producing a revised edition and persisting for several years until I caved in, and for putting the whole book into LaTeX so that there was a faint chance that I might complete the task. And, as always, to thank the many readers, who for 30 years have sent in comments, lists of mistakes, and suggestions for new material. I would particularly like to give special mention to George Bergman, who sent detailed and many improvements to the mathematical content, as well as numerous typographical errors. [illegible] of English translations of documents related to Galois, which have had a significant influence on this edition.

University of Warwick IAN STEWART
Coventry
May 2003

Preface to the Fourth Edition

Another decade, another edition...

This time I have resisted the urge to tinker with the basic structure. I am grateful to George Bergman, David Derbes, Peter Mulligan, Gerry Myerson, Jean Pierre Ortolland, F. Javier Trigos-Arrieta, Hemza Yagoub, and Carlo Wood for numerous comments, corrections, and suggestions. This edition has greatly benefited from their advice. Known typographical errors have been corrected, though no doubt some ingenious new ones have been introduced. Material that needed updating, such as references, has been updated. Minor improvements to the exposition have been made throughout.

The main changes are as follows.

In Chapter 2, I have replaced the topological (winding number) proof of the Fundamental Theorem of Algebra by one that requires less sophisticated background: a simple and plausible result from point-set topology and estimates of a kind that will be familiar to anyone who has taken a first course in analysis.

Chapter 7 has been reformulated, identifying the Euclidean plane $\mathbb{R}^2$ with the complex plane $\mathbb{C}$. This makes it possible to talk of a point $x+iy=z\in\mathbb{C}$ being constructible by ruler and compass, instead of considering its coordinates x and y separately. The resulting theory is more elegant, some proofs are simpler, and attention focuses on the Pythagorean closure $\mathbb{Q}^{\mathrm{py}}$ of the rational numbers $\mathbb{Q}$, which consists precisely of the points that can be constructed from $\{0,1\}$. For consistency, similar but less extensive changes have been made in Chapter 20 on regular polygons. I have added a short section to Chapter 21 on constructions in which an angle-trisector is also permitted, since it is an intriguing and direct application of the methods developed.

Having read, and been impressed by, Peter Neumann's English translation of the publications and manuscripts of Évariste Galois (Neumann 2011), I have taken his warnings to heart and added a final historical Chapter 25. This takes a retrospective look at what Galois actually did, as compared to what many assume he did, and what is done in this book. It is all too easy to assume that today's presentation is merely a streamlined and generalised version of Galois's. However, the history of mathematics seldom follows what now seems the obvious path, and in this case it did not.

The issues are easier to discuss at the end of the book, when we have amassed the necessary terminology and understood the ideas required. The key question is the extent to which Galois relied on proving that the alternating group $\mathbb{A}_5$ is simple—or, at least, not soluble. The perhaps surprising answer is 'not at all'. His great contribution was to introduce the Galois correspondence, and to prove that (in our language) an equation is soluble by radicals if and only if its Galois group is soluble. He certainly knew that the group of the general quintic is the symmetric group $\mathbb{S}_5$, and that this is not soluble, but he did not emphasise that point. Instead, his main aim was to characterise equations (of prime degree) that *are* soluble by radicals. He did so by deducing the structure of the associated Galois group, which is clearly not the symmetric group since among other features it has smaller order. However, he did not point this out explicitly.

Neumann (2011) also discusses two myths: that Galois proved the alternating groups $\mathbb{A}_n$ are simple for $n \geq 5$, and that he proved that $\mathbb{A}_5$ is the smallest simple group aside from cyclic groups of prime order. As Neumann points out, there is absolutely no evidence for the first (and precious little to suggest that Galois cared about alternating groups). The sole evidence for the second is a casual statement that Galois made in his letter to his friend Auguste Chevalier, composed the night before the fatal duel. He states, enigmatically, that the smallest non-cyclic simple group has '5.4.3' elements. Neumann makes a very good case that here Galois is thinking not of $\mathbb{A}_5$ as such, but of the isomorphic group $\mathbb{PSL}(2,5)$. He definitely knew that $\mathbb{PSL}(2,5)$ is simple, but nothing in his extant works even hints at a proof that no non-cyclic simple group can have smaller order. The one issue on which I differ slightly from Neumann is whether Galois *could have* proved this. I believe it was possible, although I agree it is unlikely given the lack of supporting evidence. In justification, I have finished by giving a proof using only ideas that Galois could have' discovered and proved without difficulty. At the very least it shows that a proof is possible—and easier than we might expect—using only classical ideas and some bare-hands ingenuity.

University of Warwick IAN STEWART
Coventry
September 2014

Historical Introduction

Mathematics has a rich history, going back at least 5000 years. Very few subjects still make use of ideas that are as old as that, but in mathematics, important discoveries have lasting value. Most of the latest mathematical research makes use of theorems that were published last year, but it may also use results first discovered by Archimedes, or by some unknown Babylonian mathematician, astronomer, or priest. For example, ever since Archimedes proved (around 250 BC) that the volume of a sphere is what we would now write as $\frac{4}{3}\pi r^3$, that discovery has been available to any mathematician who is aware of the result, and whose research involves spheres. Although there are revolutions in mathematics, they are usually changes of viewpoint or philosophy; earlier *results* do not change—although the hypotheses needed to prove them may. In fact, there is a word in mathematics for previous results that are later changed: they are called 'mistakes'.

The history of Galois theory is unusually interesting. It certainly goes back to 1600 BC, where among the mud-brick buildings of exotic Babylon, some priest or mathematician worked out how to solve a quadratic equation, and they or their student inscribed it in cuneiform on a clay tablet. Some such tablets survive to this day, along with others ranging from tax accounts to observations of the motion of the planet Jupiter, Figure 1 (Left).

Adding to this rich historical brew, the problems that Galois theory solves, positively or negatively, have an intrinsic fascination—squaring the circle, duplicating the cube, trisecting the angle, constructing the regular 17-sided polygon, solving the quintic equation. If the hairs on your neck do not prickle at the very mention of these age-old puzzles, you need to have your mathematical sensitivities sharpened.

If those were not enough: Galois himself was a colourful and tragic figure—a youthful genius, one of the thirty or so greatest mathematicians who have ever lived, but also a political revolutionary during one of the most turbulent periods in the history of France. At the age of 20 he was killed in a duel, ostensibly over a woman and quite possibly with a close friend, and his work was virtually lost to the world. Only some smart thinking by Joseph Liouville, probably encouraged by Galois's brother Alfred, rescued it. Galois's story is one of the most memorable among the lives of the great mathematicians, even when the more excessive exaggerations and myths are excised.

Our tale therefore has two heroes: a mathematical one, the humble polynomial equation, and a human one, the tragic genius. We take them in turn.

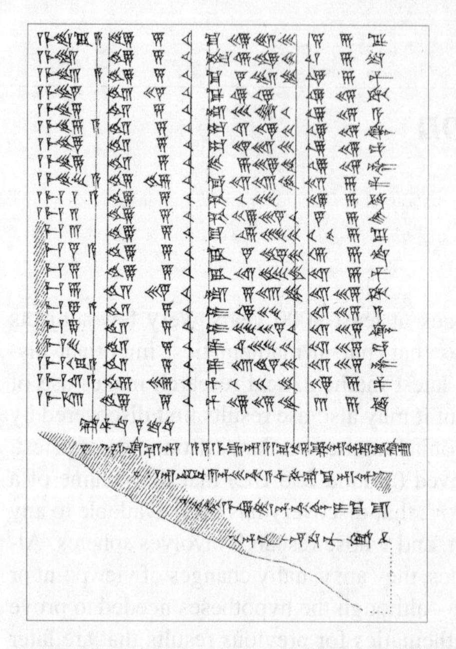

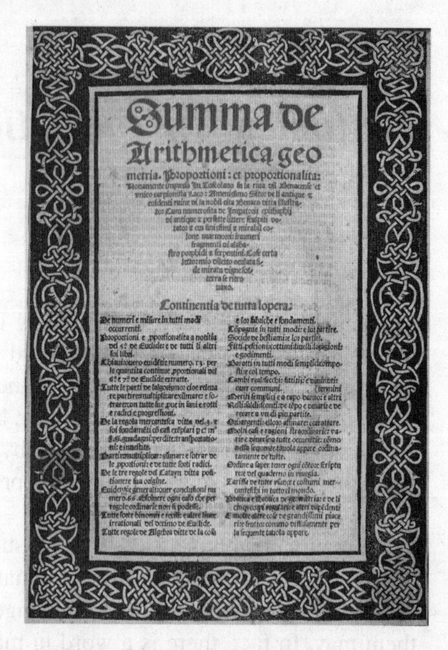
Summa de Arithmetica geo metria. Proportioni: et proportionalita:

Continentia de tutta lopera:

De numeri e misure in tutti modi occurrenti.

FIGURE 1: *Left*: A Babylonian clay tablet recording the motion of Jupiter. *Right*: A page from Pacioli's *Summa di Arithmetica*.

Polynomial Equations

A Babylonian clay tablet from about 1600 BC poses arithmetical problems that reduce to the solution of quadratic equations (Midonick 1965 page 48). The tablet also provides firm evidence that the Babylonians possessed general methods for solving quadratics, although they had no algebraic notation with which to express their solution. Babylonian notation for numbers was in base 60, so that (when transcribed into modern form) the symbols 7,4;3,11 denote the number $7 \times 60^2 + 4 \times 60 + 3 \times 60^{-1} + 11 \times 60^{-2} = 25440\frac{191}{3600}$. In 1930 the historian of science Otto Neugebauer announced that some of the most ancient Babylonian problem tablets contained methods for solving quadratics. For instance, one tablet contains this problem: find the side of a square given that the area minus the side is 14,30. Bearing in mind that $14,30 = 870$ in decimal notation, we can formulate this problem as the quadratic equation

$$x^2 - x = 870$$

The Babylonian solution reads:

> Take half of 1, which is 0;30, and multiply 0;30 by 0;30, which is 0;15. Add this to 14,30 to get 14,30;15. This is the square of 29;30. Now add 0;30 to 29;30. The result is 30, the side of the square.

Although this description applies to one specific equation, it is laid out so that similar reasoning can be applied in greater generality, and this was clearly the Babylonian scribe's intention. The method is the familiar procedure of completing the square, which nowadays leads to the usual formula for the solution of a quadratic. See Joseph (2000) for more on Babylonian mathematics.

The ancient Greeks in effect solved quadratics by geometric constructions, but there is no sign of an algebraic formulation until at least AD 100 (Bourbaki 1969 page 92). The Greeks also possessed methods for solving cubic equations, which involved the points of intersection of conics. Again, algebraic solutions of the cubic were unknown, and in 1494 Luca Pacioli ended his *Summa di Arithmetica* (Figure 1, right) with the remark that (in his archaic notation) the solution of the equations $x^3+mx=n$ and $x^3+n=mx$ was as impossible at the existing state of knowledge as squaring the circle.

This state of ignorance was soon to change as new knowledge from the Middle and Far East swept across Europe and the Christian Church's stranglehold on intellectual innovation began to weaken. The Renaissance mathematicians at Bologna discovered that the solution of the cubic can be reduced to that of three basic types: $x^3+px=q, x^3=px+q$, and $x^3+q=px$. They were forced to distinguish these cases because they did not recognise the existence of negative numbers. It is thought, on good authority (Bortolotti 1925), that Scipio del Ferro solved all three types; he certainly passed on his method for one type to a student, Antonio Fior. News of the solution leaked out, and others were encouraged to try their hand. Solutions for the cubic equation were rediscovered by Niccolo Fontana (nicknamed Tartaglia, 'The Stammerer'; Figure 2, left) in 1535.

One of the more charming customs of the period was the public mathematical contest, in which mathematicians engaged in mental duels using computational expertise as their weapons. Mathematics was a kind of performance art. Fontana demonstrated his methods in a public competition with Fior, but refused to reveal the details. Finally he was persuaded to tell them to the physician Girolamo Cardano, having first sworn him to secrecy. Cardano, the 'gambling scholar', was a mixture of genius and rogue, and when his *Ars Magna* (Figure 2, right) appeared in 1545, it contained a complete discussion of Fontana's solution. Although Cardano claimed motives of the highest order (see the modern translation of his *The Book of My Life*, 1931), and fully acknowledged Fontana as the discoverer, Fontana was justifiably annoyed. In the ensuing wrangle, the history of the discovery became public knowledge.

The *Ars Magna* also contained a method, due to Ludovico Ferrari, for solving the quartic equation by reducing it to a cubic. Ferrari was one of Cardano's students, so presumably he had given permission for his work to be published... or perhaps a student's permission was not needed. All the formulas discovered had one striking property, which can be illustrated by Fontana's solution $x^3+px=q$:

$$x=\sqrt[3]{\frac{q}{2}+\sqrt{\frac{p^3}{27}+\frac{q^2}{4}}}+\sqrt[3]{\frac{q}{2}-\sqrt{\frac{p^3}{27}+\frac{q^2}{4}}}$$

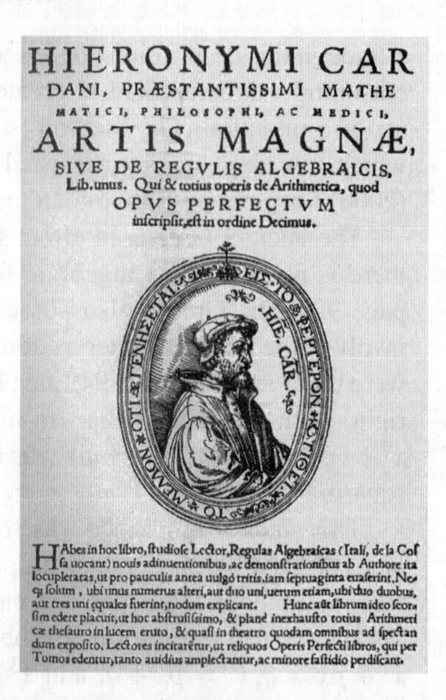

HIERONYMI CAR
DANI, PRÆSTANTISSIMI MATHE
MATICI, PHILOSOPHI, AC MEDICI,
ARTIS MAGNÆ,
SIVE DE REGVLIS ALGEBRAICIS,
Lib.unus. Qui & totius operis de Arithmetica, quod
OPVS PERFECTVM
inscripsit, est in ordine Decimus.

Habes in hoc libro, studiose Lector, Regulas Algebraicas (Itali, de la Cossa uocant) nouis adinuentionibus, ac demonstrationibus ab Authore ita locupletatas, ut pro pauculis antea uulgò tritis, iam septuaginta euaserint. Neq; solum, ubi unus numerus alteri, aut duo uni, uerum etiam, ubi duo duobus, aut tres uni æquales fuerint, nodum explicant. Hunc aūt librum ideo seorsim edere placuit, ut hoc abstrusissimo, & planè inexhausto totius Arithmeticæ thesauro in lucem eruto, & quasi in theatro quodam omnibus ad spectandum exposito, Lectores incitarentur, ut reliquos Operis Perfecti libros, qui per Tomos edentur, tanto auidius amplectantur, ac minore fastidio perdiscant.

FIGURE 2: *Left*: Niccolo Fontana (Tartaglia), who discovered how to solve cubic equations. *Right*: Title page of Girolamo Cardano's *Ars Magna*.

This expression, usually called Cardano's formula because he was the first to publish it, is built up from the coefficients p and q by repeated addition, subtraction, multiplication, division, and—crucially—extraction of roots. Such expressions became known as *radicals*.

Since all equations of degree ≤ 4 were now solved by radicals, it was natural to ask how to solve the quintic equation by radicals. Ehrenfried Walter von Tschirnhaus claimed a solution in 1683, but Gottfried Wilhelm Leibniz correctly pointed out that it was fallacious. Leonhard Euler failed to solve the quintic, but found new methods for the quartic, as did Etienne Bézout in 1765. Joseph-Louis Lagrange took a major step forward in his magnum opus *Réflexions sur la Résolution Algébrique des Équations* of 1770-1771, when he unified the separate tricks used for the equations of degree ≤ 4. He showed that they all depend on finding functions of the roots of the equation that are unchanged by certain permutations of those roots, and he showed that this approach *fails* when it is tried on the quintic. That did not prove that the quintic is insoluble by radicals, because other methods might succeed where this particular one did not. But the failure of such a general method was, to say the least, suspicious.

A realisation that the quintic might not be soluble by radicals was now dawning. In 1799 Paolo Ruffini published a two-volume book *Teoria Generale delle Equazioni* whose 516 pages constituted an attempt to prove the insolubility of the quintic. Tignol (1988) describes the history, saying that 'Ruffini's proof was received with scepticism in the mathematical community.' The main stumbling-block seems to have been the length and complexity of the proof; at any rate, no coherent criticisms emerged.

In 1810 Ruffini had another go, submitting a long paper about quintics to the French Academy; the paper was rejected on the grounds that the referees could not spare the time to check it. In 1813 he published yet another version of his impossibility proof. The paper appeared in an obscure journal, with several gaps in the proof (Bourbaki 1969 page 103). The most significant omission was to assume that all radicals involved must be based on rational functions of the roots (see Section 8.7). Nonetheless, Ruffini had made a big step forward, even though it was not appreciated at the time.

As far as the mathematical community of the period was concerned, the question was finally settled by Niels Henrik Abel in 1824, who proved conclusively that the general quintic equation is insoluble by radicals. In particular he filled in the big gap in Ruffini's work. But Abel's proof was unnecessarily lengthy and contained a minor error, which, fortunately, did not invalidate the method. In 1879 Leopold Kronecker published a simple, rigorous proof that tidied up Abel's ideas.

The 'general' quintic is therefore insoluble by radicals, but special quintic equations might still be soluble. Some are: see Section 1.4. Indeed, for all Abel's methods could prove, *every* particular quintic equation might be soluble, with a special formula for each equation. So a new problem now arose: to decide whether any particular equation can be solved by radicals. Abel was working on this question in 1829, just before he died of a lung condition that was probably tuberculosis.

In 1832 a young Frenchman, Évariste Galois, was killed in a duel. He had for some time sought recognition for his mathematical theories, submitting three memoirs to the Academy of Sciences in Paris. They were all rejected, and his work appeared to be lost to the mathematical world. Then, on 4 July 1843, Liouville addressed the Academy. He opened with these words:

> I hope to interest the Academy in announcing that among the papers of Évariste Galois I have found a solution, as precise as it is profound, of this beautiful problem: whether or not there exists a solution by radicals…

The Life of Galois

The most accessible account of Galois's troubled life, Bell (1965), is also one of the less reliable, and in particular it seriously distorts the events surrounding his death. The best sources I know are Rothman (1982a, 1982b). For Galois's papers and manuscripts, consult Bourgne and Azra (1962) for the French text and facsimiles of manuscripts and letters, and Neumann (2011) for English translation and parallel French text. Scans of the entire body of work can be found on the web at

www.bibliotheque-institutdefrance.fr/numerisation/

Évariste Galois (Figure 3) was born at Bourg-la-Reine near Paris on 25 October 1811. His father Nicolas-Gabriel Galois was a Republican (Kollros 1949)—that is, he favoured the abolition of the monarchy. He was head of the village liberal party, and after the return to the throne of Louis XVIII in 1814, Nicolas became town mayor. Évariste's mother Adelaide-Marie (*née* Demante) was the daughter of a jurisconsult—a legal expert who gives opinions about cases brought before them. She was a fluent reader of Latin, thanks to a solid education in religion and the classics.

For the first twelve years of his life, Galois was educated by his mother, who passed on to him a thorough grounding in the classics, and his childhood appears to have been a happy one. At the age of ten he was offered a place at the College of Reims, but his mother preferred to keep him at home. In October 1823 he entered a preparatory school, the College de Louis-le-Grand. There he got his first taste of revolutionary politics: during his first term the students rebelled and refused to chant in chapel. He also witnessed heavy-handed retribution, for a hundred of the students were expelled for their disobedience.

Galois performed well during his first two years at school, obtaining first prize in Latin, but then boredom set in. He was made to repeat the next year's classes, but predictably this just made things worse. During this period, probably as refuge from the tedium, Galois began to take a serious interest in mathematics. He came across a copy of Adrien-Marie Legendre's *Éléments de Géométrie*, a classic text which broke with the Euclidean tradition of school geometry. According to Bell (1965) Galois read it 'like a novel', and mastered it in one reading—but Bell is prone to exaggeration. Whatever the truth here, the school algebra texts certainly could not compete with Legendre's masterpiece as far as Galois was concerned, and he turned instead to the original memoirs of Lagrange and Abel. At the age of fifteen he was reading material intended only for professional mathematicians. But his classwork remained uninspired, and he seems to have lost all interest in it. His rhetoric teachers were particularly unimpressed by his attitude, and accused him of *affecting* ambition and originality, but even his own family considered him rather strange at that time.

Galois did make life very difficult for himself. For a start, he was was an untidy worker, as can be seen from some of his manuscripts (Bourgne and Azra 1962). Figures 4 and 5 are a sample. Worse, he tended to work in his head, committing only the results of his deliberations to paper. His mathematics teacher Vernier begged him to work systematically, no doubt so that ordinary mortals could follow his reasoning, but Galois ignored this advice. Without adequate preparation, and a year early, he took the competitive examination for entrance to the École Polytechnique. A pass would have ensured a successful mathematical career, for the Polytechnique was the breeding-ground of French mathematics. Of course, he failed. Two decades later Olry Terquem (editor of the journal *Nouvelles Annales des Mathématiques*) advanced the following explanation: 'A candidate of superior intelligence is lost with an examiner of inferior intelligence. Because they do not understand me, *I* am a barbarian...' To be fair to the examiner, communication skills are an important ingredient of success, as well as natural ability. We might counter Terquem with 'Because I do not take account of their inferior intelligence, *I* risk being misunderstood.' But Galois was too young and impetuous to see it that way.

In 1828 Galois enrolled in an advanced mathematics course offered by Louis-

FIGURE 3: Portrait of Évariste Galois drawn from memory by his brother Alfred, 1848.

Paul-Émile Richard, who recognised his ability and was very sympathetic towards him. He was of the opinion that Galois should be admitted to the Polytechnique without examination—probably because he recognised the dangerous combination of high talent and poor examination technique. If this opinion was ever communicated to the Polytechnique, it fell on deaf ears.

The following year saw the publication of Galois's first research paper (Galois 1897) on continued fractions; though competent, it held no hint of genius. Meanwhile, Galois had been making fundamental discoveries in the theory of polynomial equations, and he submitted some of his results to the Academy of Sciences. The referee was Augustin-Louis Cauchy, who had already published work on the behaviour of functions under permutation of the variables, a central theme in Galois's theory.

As Rothman (1982a) says, 'We now encounter a major myth.' Many sources state that Cauchy lost the manuscript, or even deliberately threw it away, either to conceal its contents or because he considered it worthless. But René Taton (1971) found a letter written by Cauchy in the archives of the Academy. Dated 18 January 1830, it reads in part:

> I was supposed to present today to the Academy first a report on the work of the young Galoi [spelling was not consistent in those days] and second a memoir on the analytic determination of primitive roots

> [by Cauchy]... Am indisposed at home. I regret not being able to attend today's session, and I would like you to schedule me for the following session for the two indicated subjects.

So Cauchy still had the manuscript in his possession, six months after Galois had submitted it. Moreover, he found the work sufficiently interesting to want to draw it to the Academy's attention. However, at the next session of the Academy, on 25 January, Cauchy presented only his own paper. What had happened to the paper by Galois?

Taton suggests that Cauchy was actually very impressed by Galois's researches, because he advised Galois to prepare a new (no doubt improved) version, and to submit it for the Grand Prize in Mathematics—the pinnacle of mathematical honour—which had a March 1 deadline. There is no direct evidence for this assertion, but the circumstantial evidence is quite convincing. We do know that Galois made such a submission in February. The following year the journal *Le Globe* published an appeal for Galois's aquittal during his trial for allegedly threatening the king's life (see below):

> Last year before March 1, M. Galois gave to the secretary of the Institute a memoir on the solution of numerical equations. This memoir should have been entered in the competition for the Grand Prize in Mathematics. It deserved the prize, for it could resolve some difficulties that Lagrange had failed to do. Cauchy had conferred the highest praise on the author about this subject. And what happened? The memoir is lost and the prize is given without the participation of the young savant.

Rothman points out that Cauchy fled France in September 1830, so the article is unlikely to have been based on Cauchy's own statements. *Le Globe* was a journal of the Saint-Simonian organisation, a neo-Christian socialist movement founded by the Comte de Sainte-Simone. When Galois left jail, his closest friend Auguste Chevalier invited him to join a Saint-Simonian commune founded by Prosper Enfantin. Chevalier was a very active member and an established journalist. It is plausible that Chevalier wrote the article, in which case the original source would have been Galois himself. If so, and if Galois was telling the truth, he knew that Cauchy had been impressed by the work.

The same year held two major disasters. On 2 July 1829 Galois's father committed suicide after a bitter political dispute in which the village priest forged Nicolas's signature on malicious epigrams aimed at his own relatives. It could not have happened at a worse time, for a few days later Galois again sat for entrance to the Polytechnique—his final chance. There is a legend (Bell 1965, Dupuy 1896) that he lost his temper and threw an eraser into the examiner's face, but according to Bertrand (1899) this tradition is false. Apparently the examiner, Dinet, asked Galois some questions about logarithms.

In one version of the story, Galois made some statements about logarithmic series, Dinet asked for proofs, and Galois refused on the grounds that the answer was completely obvious. A variant asserts that Dinet asked Galois to outline the theory of 'arithmetical logarithms'. Galois informed him, no doubt with characteristic bluntness, that there were no *arithmetical* logarithms. Dinet failed him.

Was Galois right, though? It depends on what Dinet had in mind. The phrase 'arithmetical logarithms' is not necessarily meaningless. In 1801 Carl Friedrich Gauss had published his epic *Disquisitiones Arithmeticae*, which laid the foundations of number theory for future generations of mathematicians. Ironically, Gauss had sent it to the French Academy in 1800, and it was rejected. In the *Disquisitiones* Gauss developed the notion of a primitive root modulo a prime. If g is a primitive root $(\mathrm{mod}\ p)$ then every nonzero element $m\ (\mathrm{mod}\ p)$ can be written as a power $m = g^{a(m)}$. Then $a(mn) = a(m) + a(n)$, so $a(m)$ is analogous to $\log m$. Gauss called $a(m)$ the *index* of m to base g, and Article 58 of his book begins by stating that 'Theorems pertaining to indices are completely analogous to those that refer to logarithms.' So if this is what Dinet was asking about, any properly prepared candidate should have recognised it, and known about it.

Because he had expected to be admitted to the Polytechnique, Galois had not studied for his final examinations. Now faced with the prospect of the École Normale, then called the École Preparatoire, which at that time was far less prestigious than the Polytechnique, he belatedly prepared for them. His performance in mathematics and physics was excellent, in literature less so; he obtained both the Bachelor of Science and Bachelor of Letters on 29 December 1829.

Possibly following Cauchy's recommendation, in February 1830 Galois presented a new version of his researches to the Academy of Sciences in competition for the Grand Prize in Mathematics. The manuscript reached the secretary Joseph Fourier, who took it home for perusal. But he died before reading it, and the manuscript could not be found among his papers. It may not have been Fourier who lost it, however; the Grand Prize committee had three other members: Legendre, Sylvestre-François Lacroix, and Louis Poinsot.

If the article in *Le Globe* is to be believed, no lesser a light than Cauchy had considered Galois's manuscript to have been worthy of the prize. The loss was probably an accident, but according to Dupuy (1896), Galois was convinced that the repeated losses of his papers were not just bad luck. He saw them as the inevitable effect of a society in which genius was condemned to an eternal denial of justice in favour of mediocrity, and he blamed the politically oppressive Bourbon regime. He may well have had a point, accident or not.

At that time, France was in political turmoil. King Charles X succeeded Louis XVIII in 1824. In 1827 the liberal opposition made electoral gains; in 1830 more elections were held, giving the opposition a majority. Charles, faced with abdication, attempted a *coup d'état*. On 25 July he issued his notorious *Ordonnances* suppressing the freedom of the press. The populace was in no mood to tolerate such repression, and revolted. The uprising lasted three days, after which as a compromise the Duke of Orléans, Louis-Philippe, was made king. During these three days, while the students of the Polytechnique were making history in the streets, Galois and his fellow students were locked in by Guigniault, Director of the École Normale. Galois was incensed, and subsequently wrote a blistering attack on the Director in the *Gazette des Écoles*, signing the letter with his full name. An excerpt (the letter was published in December) reveals the general tone:

> Gentlemen:
>
> The letter which M. Guignault placed in the Lycée yesterday, on the account of one of the articles in your journal, seemed to me most improper. I had thought that you would welcome eagerly any way of exposing this man.
>
> Here are the facts which can be vouched for by forty-six students.
>
> On the morning of July 28, when several students of the École Normale wanted to join in the struggle, M. Guigniault told them, twice, that he had the power to call the police to restore order in the school. The police on the 28th of July!
>
> The same day, M. Guigniault told us with his usual pedantry: 'There are many brave men fighting on both sides. If I were a soldier, I would not know what to decide. Which to sacrifice, liberty or LEGITIMACY?'
>
> There is the man who the next day covered his hat with an enormous tricolor cockade. There are our liberal doctrines!

The editor removed the signature, the Director was not amused, and Galois was expelled because of his 'anonymous' letter (Dalmas 1956).

Galois promptly joined the Artillery of the National Guard, a branch of the militia composed almost entirely of Republicans. On 21 December 1830 the Artillery of the National Guard, almost certainly including Galois, was stationed near the Louvre, awaiting the verdict of the trial of four ex-minsters. The public wanted these functionaries executed, and the Artillery was planning to rebel if they received only life sentences. Just before the verdict was announced, the Louvre was surrounded by the full National Guard, plus other troops who were far more trustworthy. When the verdict of a jail sentence was heralded by a cannon shot, the revolt failed to materialise. On 31 December, the king abolished the Artillery of the National Guard on the grounds that it constituted a serious security threat.

Galois was now faced with the urgent problem of making a living. On 13 January 1831 he tried to set up as a private teacher of mathematics, offering a course in advanced algebra. Forty students enrolled, but the class soon petered out, probably because Galois was too involved in politics. On 17 January he submitted a third version of his memoir to the Academy: *On the Conditions of Solubility of Equations by Radicals*. Cauchy was no longer in Paris, so Siméon Poisson and Lacroix were appointed referees. After two months Galois had heard no word from them. He wrote to the President of the Academy, asking what was happening. He received no reply.

During the spring of 1831, Galois's behaviour became more and more extreme, verging on the paranoid. On April 18 Sophie Germain, one of the few women mathematicians of the time, who studied with Gauss, wrote to Guillaume Libri about Galois's misfortunes: 'They say he will go completely mad, and I fear this is true.' See Henry (1879). Also in April, 19 members of the Artillery of the National Guard, arrested after the events at the Louvre, were put on trial charged with attempting to overthrow the government. The jury acquitted them, and on 9 May a celebratory banquet was held. About 200 Republicans were present, all extremely hostile to the government of Louis-Philippe. The proceedings became more and more riotous, and Galois was seen with a glass in one hand and a dagger in the other. His companions allegedly interpreted this as a threat to the king's life, applauded mightily, and ended up dancing and shouting in the street.

Next day, Galois was arrested. At his subsequent trial, he admitted everything, but claimed that the toast proposed was actually 'To Louis-Philippe, *if he turns traitor*,' and that the uproar had drowned the last phrase. But he also made it crystal clear that he expected Louis-Philippe to do just that. Nevertheless, the jury acquitted him, and he was freed on 15 June.

On 4 July he heard the fate of his memoir. Poisson declared it 'incomprehensible'. The report (reprinted in full in Taton, 1947) ended as follows:

> We have made every effort to understand Galois's proof. His reasoning is not sufficiently clear, sufficiently developed, for us to judge its correctness, and we can give no idea of it in this report. The author announces that the proposition which is the special object of this memoir is part of a general theory susceptible of many applications. Perhaps it will transpire that the different parts of a theory are mutually clarifying, are easier to grasp together rather than in isolation. We would then suggest that the author should publish the whole of his work in order to form a definitive opinion. But in the state which the part he has submitted to the Academy now is, we cannot propose to give it approval.

The report may well have been entirely fair. Tignol (1988) points out that Galois's entry 'did not yield any workable criterion to determine whether an equation is solvable by radicals.' The referees' report was explicit:

> [The memoir] does not contain, as [its] title promised, the condition of solubility of equations by radicals; indeed, assuming as true M. Galois's proposition, one could not derive from it any good way of deciding whether a given equation of prime degree is soluble or not by radicals, since one would first have to verify whether this equation is irreducible and next whether any of its roots can be expressed as a rational function of two others.

The final sentence here refers to a beautiful criterion for solubility by radicals of equations of prime degree that was the climax of Galois's memoir. It is indeed unclear how it can be applied to any specific equation. Tignol says that 'Galois's theory did not correspond to what was expected, it was too novel to be readily accepted.' What the referees wanted was some kind of condition on the *coefficients* that determined solubility; what Galois gave them was a condition on the *roots*. Tignol suggests that the referees' expectation was unreasonable; no simple criterion based on the coefficients has ever been found, nor is one remotely likely. But that was unclear at the time. See Chapter 25 for further discussion.

On 14 July, Bastille Day, Galois and his friend Ernest Duchâtelet were at the head of a Republican demonstration. Galois was wearing the uniform of the disbanded

Artillery and carrying a knife, several pistols, and a loaded rifle. It was illegal to wear the uniform, and even more so to be armed. Both men were arrested on the Pont-Neuf, and Galois was charged with the lesser offence of illegally wearing a uniform. They were sent to the jail at Sainte-Pélagie to await trial. While in jail, Duchâtelet drew a picture on the wall of his cell showing the king's head, labelled as such, lying next to a guillotine. This presumably did not help their cause. Duchâtelet was tried first; then it was Galois's turn. On 23 October he was tried and convicted, and his appeal was turned down on 3 December. By this time he had spent more than four months in jail. Now he was sentenced to six months there. He worked for a while on his mathematics (Figure 4 left); then in the cholera epidemic of 1832 he was transferred to a hospital. Soon he was put on parole.

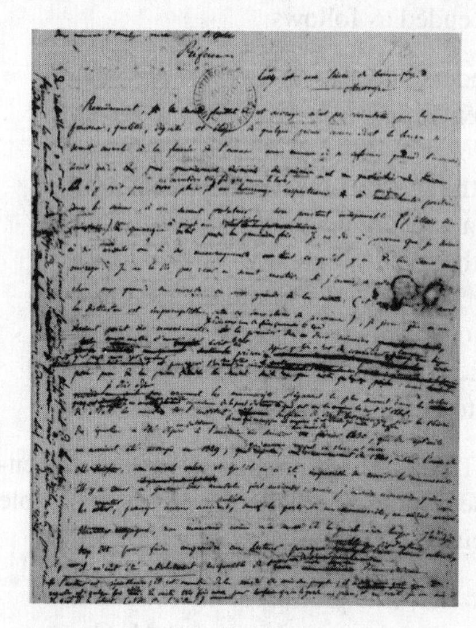

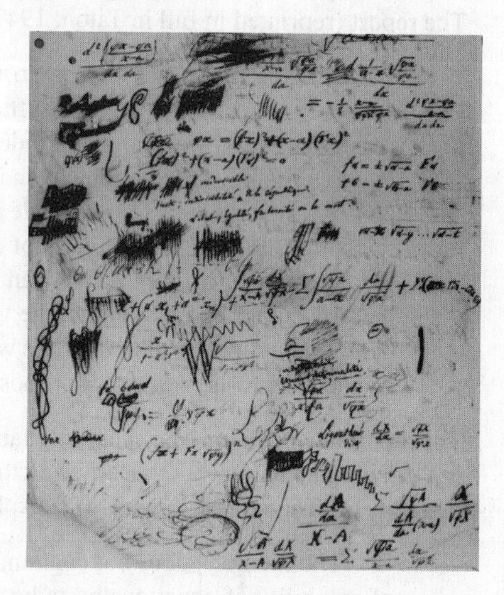

FIGURE 4: *Left*: First page of preface written by Galois when in jail. *Right*: Doodles left on the table before departing for the fatal duel. *'Une femme'*, with the second word scribbled out, can be seen near the lower left corner.

Along with his freedom he experienced his first and only love-affair, with a certain Mlle. 'Stéphanie D.' From this point on the history becomes very complicated and conjectural. Until recently, the lady's surname was unknown, adding to the romantic image of the *femme fatale*. The full name appears in one of Galois's manuscripts, but the surname has deliberately been scribbled over, no doubt by Galois. Some forensic work by Carlos Infantozzi (1968), deciphering the name that Galois had all but obliterated, led to the suggestion that the lady was Stéphanie-Felicie Poterin du Motel, the entirely respectable daughter of Jean-Louis Auguste Poterin du Motel. Jean-Louis was resident physician at the Sieur Faultrier, where Galois spent the last few months of his life. The identification is plausible, but it relies on extracting a sensible name from beneath Galois's scribbles, so naturally there is a some controversy about it.

In general, much mystery surrounds this interlude, which has a crucial bearing on subsequent events. Apparently Galois was rejected and took it very badly. On 25 May he wrote to Chevalier: 'How can I console myself when in one month I have exhausted the greatest source of happiness a man can have?' On the back of one of his papers he made fragmentary copies of two letters from Stéphanie (Tannery 1908, Bourgne and Azra 1962). One begins 'Please let us break up this affair' and continues '... and do not think about those things which did not exist and which never would have existed.' The other contains the sentences 'I have followed your advice and I have thought over what... has... happened... In any case, Sir, be assured there never would have been more. You're assuming wrongly and your regrets have no foundation.'

Not long afterwards, Galois was challenged to a duel, ostensibly because of his advances towards the young lady. Again, the circumstances are veiled in mystery, though Rothman (1982a, 1982b) has lifted a corner of the veil. One school of thought (Bell, 1965; Kollros, 1949) asserts that Galois's infatuation with Mlle. du Motel was used by his political opponents, who found it the perfect excuse to eliminate their enemy on a trumped-up 'affair of honour'. There are even suggestions that Galois was in effect assassinated by a police spy.

But in his *Mémoires*, Alexandre Dumas says that Galois was killed by Pescheux D'Herbinville, a fellow Republican, see Dumas (1967). Dumas described D'Herbinville as 'a charming young man who made silk-paper cartridges which he would tie up with silk ribbons.' The objects concerned seem to have been an early form of cracker, of the kind now familiar at Christmas. He was one of the 19 Republicans acquitted on charges of conspiring to overthrow the government, and something of a hero with the peasantry. D'Herbinville was certainly not a spy for the police: all such men were named in 1848 when Caussidière became chief of police. Dalmas (1956) cites evidence from the police report, suggesting that the other duellist was one of Galois's revolutionary comrades, and the duel was exactly what it appeared to be. This theory is largely borne out by Galois's own words on the matter (Bourgne and Azra, 1962):

> I beg patriots and my friends not to reproach me for dying otherwise than for my country. I die the victim of an infamous coquette. It is in a miserable brawl that my life is extinguished. Oh! why die for so trivial a thing, for something so despicable! ... Pardon for those who have killed me, they are of good faith.

Figure 4 right shows a doodle by Galois with the words 'Une femme' partially crossed out. It does appear that Stéphanie was at least a proximate cause of the duel, but very little else is clear.

On 29 May, the eve of the duel, Galois wrote a famous letter to his friend Auguste Chevalier, outlining his mathematical discoveries. This letter was eventually published by Chevalier in the *Revue Encyclopédique*. In it, Galois sketched the connection between groups and polynomial equations, stating that an equation is soluble by radicals provided its group is soluble. But he also mentioned many other ideas about elliptic functions and the integration of algebraic functions, and other things too cryptic to be identifiable.

The scrawled comment 'I have no time' in the margins (Figure 5) has given rise to another myth: that Galois spent the night before the duel frantically writing out his mathematical discoveries. However, that phrase has next to it '(Author's note)', which hardly fits such a picture; moreover, the letter was an explanatory accompaniment to Galois's rejected third manuscript, complete with a marginal note added by Poisson (Figure 6 left).

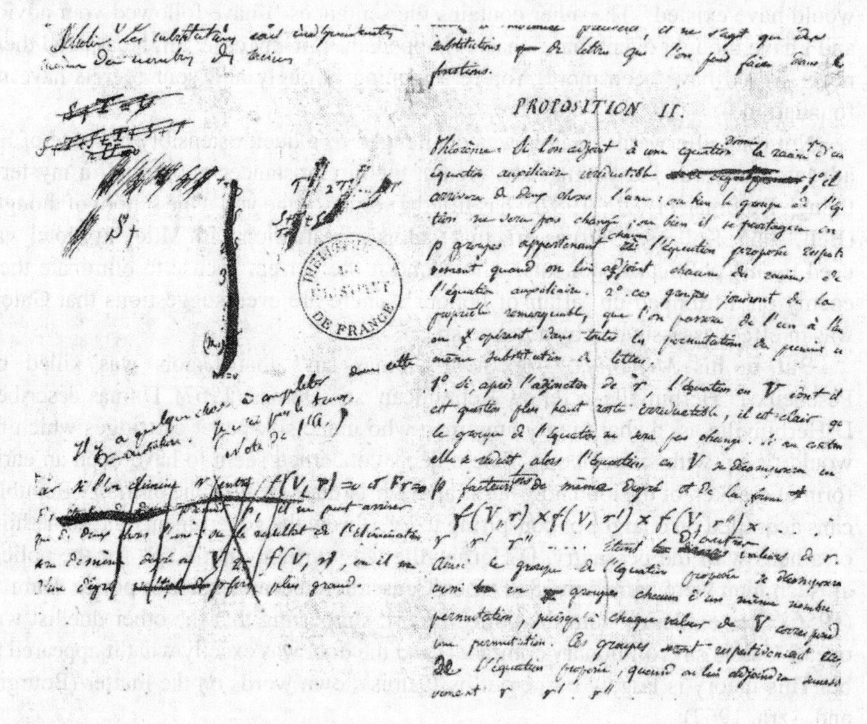
PROPOSITION II.

FIGURE 5: 'I have no time' (*je n' ai pas le temps*), above deleted paragraph in lower left corner. But consider the context.

The duel was with pistols. The post-mortem report (Dupuy 1896) states that they were fired at 25 paces, but the truth may have been even nastier. Dalmas reprints an article from the 4 June 1832 issue of *Le Precursor*, which reports:

> Paris, 1 June—A deplorable duel yesterday has deprived the exact sciences of a young man who gave the highest expectations, but whose celebrated precocity was lately overshadowed by his political activities. The young Évariste Galois... was fighting with one of his old friends, a young man like himself, like himself a member of the Society of Friends of the People, and who was known to have figured equally in a political trial. It is said that love was the cause of the combat. The pistol was the chosen weapon of the adversaries, but because of their old friendship they could not bear to look at one another and left the decision to blind fate. At point-blank range they were each armed with a pistol and fired.

> Only one pistol was charged. Galois was pierced through and through by a ball from his opponent; he was taken to the hospital Cochin where he died in about two hours. His age was 22. L.D., his adversary, is a bit younger.

Who was 'L.D.'? Does the initial 'D' refer to d'Herbinville? Perhaps. 'D' is acceptable because of the variable spelling of the period; the 'L' may have been a mistake. The article is unreliable on details: it gets the date of the duel wrong, and also the day Galois died and his age. So the initial might also be wrong. Rothman has another theory, and a more convincing one. The person who best fits the description here is not d'Herbinville, but Duchâtelet, who was arrested with Galois on the Pont-Neuf. Bourgne and Azra (1962) give his Christian name as 'Ernest', but that might be wrong, or again the 'L' may be wrong. To quote Rothman: 'we arrive at a very consistent and believable picture of two old friends falling in love with the same girl and deciding the outcome by a gruesome version of Russian roulette.'

This theory is also consistent with a final horrific twist to the tale. Galois was hit in the stomach, a particularly serious wound that was almost always fatal. If indeed the duel was at point-blank range, this is no great surprise. If at 25 paces, he was unlucky.

He did not die two hours later, as *Le Precursor* says, but a day later on 31 May, of peritonitis; he refused the office of a priest. On 2 June 1832 he was buried in the common ditch at the cemetery of Montparnasse.

His letter to Chevalier ended with these words (Figure 6 right):

> Ask Jacobi or Gauss publicly to give their opinion, not as to the truth, but as to the importance of these theorems. Later there will be, I hope, some people who will find it to their advantage to decipher all this mess...

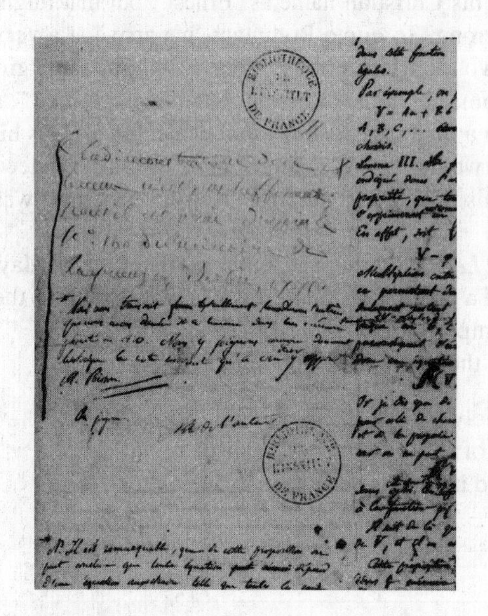

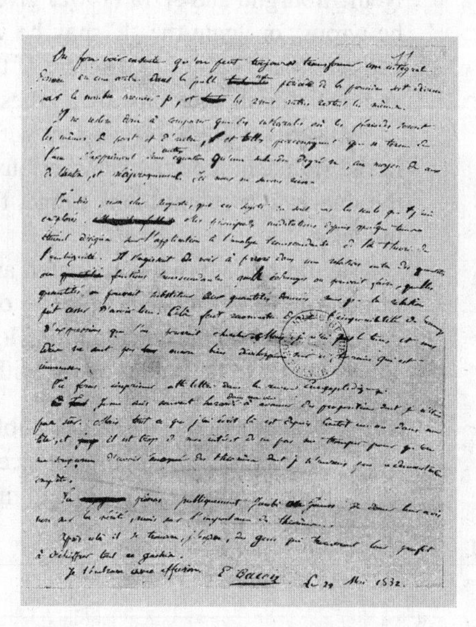

FIGURE 6: *Left*: Marginal comment by Poisson. *Right*: The final page written by Galois before the duel. 'To decipher all this mess' (*déchiffrer tout ce gâchis*, is the next to last line).

Chapter 1

Classical Algebra

In the first part of this book, Chapters 1-15, we present a (fairly) modern version of Galois's ideas in the same setting that he used, namely, the complex numbers. Later, from Chapter 16 onwards, we generalise the setting, but the complex numbers have the advantages of being familiar and concrete. By initially restricting ourselves to complex numbers, we can focus on the main ideas that Galois introduced, without getting too distracted by 'abstract nonsense'.

A warning is in order. The decision to work over the complex numbers has advantages in terms of accessibility of the material, but it sometimes makes the discussion seem clumsy by comparison with the elegance of an axiomatic approach. This is arguably a price worth paying, because this way we appreciate the abstract viewpoint when it makes its appearance, and we understand where it comes from. However, it also requires a certain amount of effort to verify that many of the proofs in the complex case go through unchanged to more general fields—and that some do not, and require modification.

We assume familiarity with the basic theory of real and complex numbers, but to set the scene, we recall some of the concepts involved. We begin with a brief discussion of complex numbers and introduce two important ideas. Both relate to subsets of the complex numbers that are closed under the usual arithmetic operations. A subring of the complex numbers is a subset closed under addition, subtraction, and mutliplication; a subfield is a subring that is also closed under division by any non-zero element. Both concepts were formalised by Richard Dedekind in 1871, though the ideas go back to Peter Gustav Lejeune-Dirichlet and Kronecker in the 1850s.

We then show that the historical sequence of extensions of the number system, from natural numbers to integers to rationals to reals to complex numbers, can with hindsight be interpreted as a quest to make more and more equations have solutions. We are thus led to the concept of a polynomial, which is central to Galois theory because it determines the type of equation that we wish to solve. And we appreciate that the existence of a solution depends on the kind of number that is permitted.

Throughout, we use the standard notation $\mathbb{N}, \mathbb{Z}, \mathbb{Q}, \mathbb{R}, \mathbb{C}$ for the natural numbers, integers, rationals, real numbers, and complex numbers. These systems sit inside each other:

$$\mathbb{N} \subseteq \mathbb{Z} \subseteq \mathbb{Q} \subseteq \mathbb{R} \subseteq \mathbb{C}$$

and each $\subseteq$ symbol hints at a lengthy historical process in which 'new numbers' were proposed for mathematical reasons—usually against serious resistance on the grounds that although their novelty was not in dispute, they were not numbers and therefore did not exist.

1.1 Complex Numbers

A *complex number* has the form

$$z = x + iy$$

where x, y are real numbers and $i^2 = -1$. Therefore $i = \sqrt{-1}$, in some sense. The easiest way to define what we mean by $\sqrt{-1}$ is to consider $\mathbb{C}$ as the set $\mathbb{R}^2$ of all pairs of real numbers (x, y), with algebraic operations

$$\begin{aligned} (x_1, y_1) + (x_2, y_2) &= (x_1 + x_2, y_1 + y_2) \\ (x_1, y_1)(x_2, y_2) &= (x_1x_2 - y_1y_2, x_1y_2 + x_2y_1) \end{aligned} \tag{1.1}$$

Then we identify $(x, 0)$ with the real number x to arrange that $\mathbb{R} \subseteq \mathbb{C}$, and define $i = (0, 1)$. In consequence, (x, y) becomes identified with $x + iy$. The formulas (1.1) imply that $i^2 = (0, 1)(0, 1) = (-1, 0)$ which is identified with the real number -1, so i is a 'square root of minus one'. Observe that $(0, 1)$ is not of the form $(x, 0)$, so i is not real, which is as it should be, since -1 has no real square root.

This approach seems to have first been published by the Irish mathematician William Rowan Hamilton in 1837, but in that year Gauss wrote to the geometer Wolfgang Bolyai that the same idea had occurred to him in 1831. This was probably true, because Gauss usually worked things out before anybody else did, but he set himself such high standards for publication that many of his more important ideas never saw print under his name. Moreover, Gauss was somewhat conservative, and shied away from anything potentially controversial.

Once we see that complex numbers are just pairs of real numbers, the previously mysterious status of the 'imaginary' number $\sqrt{-1}$ becomes much more prosaic. In fact, to the modern eye it is the 'real' numbers that are mysterious, because their rigorous definition involves analytic ideas such as sequences and convergence, which lead into deep philosophical waters and axiomatic set theory. In contrast, the step from $\mathbb{R}$ to $\mathbb{R}^2$ is essentially trivial—except for the peculiarities of human psychology.

1.2 Subfields and Subrings of the Complex Numbers

For the first half of this book, we keep everything as concrete as possible—but not more so, as Albert Einstein is supposed to have said about keeping things simple. Abstract algebra courses usually introduce (at least) three basic types of algebraic structure, defined by systems of axioms: groups, rings, and fields. Linear algebra adds a fourth: vector spaces. For the first half of this book, we steer clear of abstract rings and fields, but we do assume the basics of finite group theory and linear algebra.

Recall that a *group* is a set G equipped with an operation of 'multiplication' written $(g, h) \mapsto gh$. If $g, h \in G$ then $gh \in G$. The associative law $(gh)k = g(hk)$ holds for all $g,h,k \in G$. There is an identity $1 \in G$ such that $1g = g = g1$ for all $g \in G$. Finally, every $g \in G$ has an inverse $g^{-1} \in G$ such that $gg^{-1} = 1 = g^{-1}g$. The classic example here is the *symmetric group* $\mathbb{S}_n$, consisting of all permutations of the set $\{1,2,\ldots,n\}$ under the operation of composition. We assume familiarity with these axioms, and with subgroups, isomorphisms, homomorphisms, normal subgroups, and quotient groups.

Rings are sets equipped with operations of addition, subtraction, and multiplication; fields also have a notion of division. The formal definitions were supplied by Heinrich Weber in 1893. The axioms specify the formal properties assumed for these operations—for example, the commutative law $ab = ba$ for multiplication.

In the first part of this book, we do not assume familiarity with abstract rings and fields. Instead, we restrict attention to subrings and subfields of $\mathbb{C}$, or polynomials and rational functions over such subrings and subfields. Informally, we assume that the terms 'polynomial' and 'rational expression' (or 'rational function') are familiar, at least over $\mathbb{C}$, although for safety's sake we define them when the discussion becomes more formal, and redefine them when we make the whole theory more abstract in the second part of the book. There were no formal concepts of 'ring' or 'field' in Galois's day and linear algebra was in a rudimentary state. He had to invent groups for himself. So we are still permitting ourselves a more extensive conceptual toolkit than his.

Definition 1.1. A *subring* of $\mathbb{C}$ is a subset $R \subseteq \mathbb{C}$ such that $1 \in R$, and if $x,y \in R$ then $x+y$, $-x$, and $xy \in R$.

(The condition that $1 \in R$ is required here because we use 'ring' as an abbreviation for what is often called a 'ring-with-1' or 'unital ring'.)

A *subfield* of $\mathbb{C}$ is a subring $K \subseteq \mathbb{C}$ with the additional property that if $x \in K$ and $x \neq 0$ then $x^{-1} \in K$.

Here $x^{-1} = 1/x$ is the reciprocal. As usual we often write x/y for xy^{-1}.

It follows immediately that every subring of $\mathbb{C}$ contains $1+(-1) = 0$, and is closed under the algebraic operations of addition, subtraction, and multiplication. A subfield of $\mathbb{C}$ has all of these properties, and is also closed under division by any nonzero element. Because R and K in Definition 1.1 are subsets of $\mathbb{C}$, they inherit the usual rules for algebraic manipulation.

Examples 1.2. (1) The set of all $a+bi$, for $a,b \in \mathbb{Z}$, is a subring of $\mathbb{C}$, but not a subfield.

Since this is the first example we outline a proof. Let

$$R = \{a+bi : a,b \in \mathbb{Z}\}$$

Since $1 = 1+0i$, we have $1 \in R$. Let $x = a+bi, y = c+di \in R$. Then

$$\begin{aligned} x+y &= (a+c)+(b+d)i \in R \\ -x &= -a-bi \in R \\ xy &= (ac-bd)+(ad+bc)i \in R \end{aligned}$$

and the conditions for a subring are valid. However, $2 \in R$ but its reciprocal $2^{-1} = \frac{1}{2} \notin R$, so R is not a subfield.
(2) The set of all $a+bi$, for $a,b \in \mathbb{Q}$, is a subfield of $\mathbb{C}$.
Let
$$K = \{a+bi : a,b \in \mathbb{Q}\}$$
The proof is just like case (1), but now
$$(a+bi)^{-1} = \frac{a}{a^2+b^2} - \frac{b}{a^2+b^2} i \in K$$
so K is a subfield.
(3) The set of all polynomials in π, with integer coefficients, is a subring of $\mathbb{C}$, but not a subfield.
(4) The set of all polynomials in π, with rational coefficients, is a subring of $\mathbb{C}$. We can appeal to a result proved in Chapter 24 to show that this set is not a subfield. Suppose that $\pi^{-1} = f(\pi)$ where f is a polynomial over $\mathbb{Q}$. Then $\pi f(\pi) - 1 = 0$, so π satisfies a nontrivial polynomial equation with rational coefficients, contrary to Theorem 24.5 of Chapter 24.
(5) The set of all rational expressions in π with rational coefficients (that is, fractions $p(\pi)/q(\pi)$ where p,q are polynomials over $\mathbb{Q}$ and $q(\pi) \neq 0$) is a subfield of $\mathbb{C}$.
(6) The set $2\mathbb{Z}$ of all even integers is not a subring of $\mathbb{C}$, because (by our convention) it does not contain 1.
(7) The set of all $a+b\sqrt[3]{2}$, for $a,b \in \mathbb{Q}$, is not a subring of $\mathbb{C}$ because it is not closed under multiplication. However, it *is* closed under addition and subtraction.

Definition 1.3. Suppose that K and L are subfields of $\mathbb{C}$. An *isomorphism* between K and L is a map $\phi : K \to L$ that is one-to-one and onto, and satisfies the condition
$$\phi(x+y) = \phi(x) + \phi(y) \qquad \phi(xy) = \phi(x)\phi(y) \tag{1.2}$$
for all $x,y \in K$.

Proposition 1.4. *If $\phi : K \to L$ is an isomorphism, then:*
$$\begin{aligned} \phi(0) &= 0 \\ \phi(1) &= 1 \\ \phi(-x) &= -\phi(x) \\ \phi(x^{-1}) &= (\phi(x))^{-1} \end{aligned}$$

Proof. Since $0x = 0$ for all $x \in K$, we have $\phi(0)\phi(x) = \phi(0)$ for all $x \in K$. Let $x = \phi^{-1}(0)$, which exists since ϕ is one-to-one and onto. Then $\phi(0).0 = \phi(0)$, so $0 = \phi(0)$.

Since $1x = x$ for all $x \in K$, we have $\phi(1)\phi(x) = \phi(x)$ for all $x \in K$. Let $x = \phi^{-1}(1)$ to deduce that $\phi(1).1 = 1$, so $\phi(1) = 1$.

Since $x + (-x) = 0$ for all $x \in K$, we have $\phi(x) + \phi(-x) = \phi(0) = 0$. Therefore $\phi(-x) = -\phi(x)$.

Since $x.x^{-1} = 1$ for all $x \in K$, we have $\phi(x).\phi(x^{-1}) = \phi(1) = 1$. Therefore $\phi(x^{-1}) = (\phi(x))^{-1}$. □

If ϕ satisfies (1.2) and is one-to-one but not necessarily onto, it is a *monomorphism*. An isomorphism of K with itself is called an *automorphism* of K.

Throughout the book we make extensive use of the following terminology:

Definition 1.5. A *primitive* nth root of unity is an nth root of 1 that is not an mth root of 1 for any proper divisor m of n.

For example, i is a primitive fourth root of unity, and so is $-i$. Since $(-1)^4 = 1$, the number -1 is a fourth root of unity, but it is not a primitive fourth root of unity because $(-1)^2 = 1$.

Over $\mathbb{C}$ the standard choice for a primitive nth root of unity is

$$\zeta_n = e^{2\pi i/n}$$

We omit the subscript n when this causes no ambiguity.

The next result is standard, but we include a proof for completeness.

Proposition 1.6. *Let $\zeta = e^{2\pi i/n}$. Then $\zeta^k = e^{2k\pi i/n}$ is a primitive nth root of unity if and only if k is prime to n.*

Proof. We prove the equivalent statement: $\zeta^k = e^{2k\pi i/n}$ is not a primitive nth root of unity if and only if k is not prime to n.

Suppose that ζ^k is not a primitive nth root of unity. Then $(\zeta^k)^m = 1$ where m is a proper divisor of n. That is, $n = mr$ where $r > 1$. Therefore $\zeta^{km} = 1$, so $mr = n$ divides km. This implies that $r|k$, and since also $r|n$ we have $(n,k) \geq r > 1$, so k is not prime to n.

Conversely, suppose that k is not prime to n, and let $r > 1$ be a common divisor. Then $r|k$ and $n = mr$ where $m < n$. Now km is divisible by $mr = n$, so $(\zeta^k)^m = 1$. That is, ζ^k is not a primitive nth root of unity. □

Examples 1.7. (1) Complex conjugation $x + iy \mapsto x - iy$ is an automorphism of $\mathbb{C}$. Indeed, if we denote this map by α, then:

$$\begin{aligned}
\alpha((x+iy)+(u+iv)) &= \alpha((x+u)+i(y+v)) \\
&= (x+u)-i(y+v) \\
&= (x-iy)+(u-iv) \\
&= \alpha(x+iy)+\alpha(u+iv) \\
\alpha((x+iy)(u+iv)) &= \alpha((xu-yv)+i(xv+yu)) \\
&= xu-yv-i(xv+yu) \\
&= (x-iy)(u-iv) \\
&= \alpha(x+iy)\alpha(u+iv)
\end{aligned}$$

(2) Let K be the set of complex numbers of the form $p + q\sqrt{2}$, where $p, q \in \mathbb{Q}$. This is a subfield of $\mathbb{C}$ because

$$(p+q\sqrt{2})(p-q\sqrt{2}) = p^2 - 2q^2$$

so

$$(p+q\sqrt{2})^{-1} = \frac{p}{p^2-2q^2} - \frac{q}{p^2-2q^2}\sqrt{2}$$

if p and q are non-zero. The map $p+q\sqrt{2} \mapsto p-q\sqrt{2}$ is an automorphism of K.
(3) Let $\alpha = \sqrt[3]{2} \in \mathbb{R}$, and let

$$\omega = -\frac{1}{2} + i\frac{\sqrt{3}}{2}$$

be a primitive cube root of unity in $\mathbb{C}$. The set of all numbers $p+q\alpha+r\alpha^2$, for $p,q,r \in \mathbb{Q}$, is a subfield of $\mathbb{C}$, see Exercise 1.5. The map

$$p+q\alpha+r\alpha^2 \mapsto p+q\omega\alpha+r\omega^2\alpha^2$$

is a monomorphism onto its image, but not an automorphism, Exercise 1.6.

1.3 Solving Equations

A physicist friend of mine once complained that while every physicist knew what the big problems of physics were, his mathematical colleagues never seemed to be able to tell him what the big problems of mathematics were. It took me a while to realise that this doesn't mean that they didn't know, and even longer to articulate why. The reason, I claim, is that the big problems of physics, at any given moment, are very specific challenges: measure the speed of light, prove that the Higgs boson exists, find a theory to explain high-temperature superconductors. Mathematics has problems like that too; indeed, Galois tackled one of them—prove that the quintic cannot be solved by radicals. But the *big* problems of mathematics are more general, and less subject to fashion (or disappearance by virtue of being solved). They are things like 'find out how to solve equations like this one', 'find out what shape things like this are', or even 'find out how many of these gadgets can exist'. Mathematicians know this, but it is so deeply ingrained in their way of thinking that they seldom consciously recognise such questions as big problems. However, such problems have given rise to entire fields of mathematics—here, respectively, algebra, topology, and combinatorics. I mention this because it is the first of the above big problems that runs like an ancient river through the middle of the territory we are going to explore. *Find out how to solve equations.* Or, as often as not, prove that it cannot be done with specified methods.

What sort of equations? For Galois: polynomials. But let's work up to those in easy stages.

The usual reason for introducing a new kind of number is that the old ones are inadequate for solving some important problem. Most of the historical problems in this area can be formulated using equations—though it must be said that this is a modern interpretation and the ancient mathematicians did not think in quite those terms.

For example, the step from $\mathbb{N}$ to $\mathbb{Z}$ is needed because although some equations, such as

$$t + 2 = 7$$

can be solved for $t \in \mathbb{N}$, others, such as

$$t + 7 = 2$$

cannot. However, such equations *can* be solved in $\mathbb{Z}$, where $t = -5$ makes sense. (The symbol x is more traditional than t here, but it is convenient to standardise on t for the rest of the book, so we may as well start straight away.)

Similarly, the step from $\mathbb{Z}$ to $\mathbb{Q}$ (historically, it was initially from $\mathbb{N}$ to $\mathbb{Q}^+$, the positive rationals) makes it possible to solve the equation

$$2t = 7$$

because $t = \frac{7}{2}$ makes sense in $\mathbb{Q}$.

In general, an equation of the form

$$at + b = 0$$

where a, b are specific numbers and t is an unknown number, or 'variable', is called a *linear equation*. In a subfield of $\mathbb{C}$, any linear equation with $a \neq 0$ can be solved, with the unique solution $t = -b/a$.

The step from $\mathbb{Q}$ to $\mathbb{R}$ is related to a different kind of equation:

$$t^2 = 2$$

As the ancient Greeks understood (though in their own geometric manner—they did not possess algebraic notation and thought in a very different way from modern mathematicians), the 'solution' $t = \sqrt{2}$ is an *irrational* number—it is not in $\mathbb{Q}$. (See Exercise 1.2 for a proof, which may be different from the one you have seen before. It is essentially one of the old Greek proofs, translated into algebra. Paul Erdös used to talk of proofs being from 'The Book', by which he meant an alleged volume in the possession of the Almighty, in which only the very best mathematical proofs could be found. This Greek proof that the square root of 2 is irrational must surely be in The Book. An entirely different proof of a more general theorem is outlined in Exercise 1.3.)

Similarly, the step from $\mathbb{R}$ to $\mathbb{C}$ centres on the equation

$$t^2 = -1$$

which has no real solutions since the square of any real number is positive.

Equations of the form

$$at^2 + bt + c = 0$$

are called *quadratic equations*. The classic formula for their solutions (there can be 0, 1, or 2 of these) is of course

$$t = \frac{-b \pm \sqrt{b^2 - 4ac}}{2a}$$

and this gives all the solutions t provided the formula makes sense. For a start, we need $a \neq 0$. (If $a = 0$ then the equation is actually linear, so this restriction is not a problem.) Over the real numbers, the formula makes sense if $b^2 - 4ac \geq 0$, but not if $b^2 - 4ac < 0$. Over the complex numbers it makes sense for all a, b, c. Over the rationals, it makes sense only when $b^2 - 4ac$ is a perfect square—the square of a rational number.

1.4 Solution by Radicals

We begin by reviewing the state of the art regarding solutions of polynomial equations, as it was just before the time of Galois. We consider linear, quadratic, cubic, quartic, and quintic equations in turn. In the case of the quintic, we also describe some ideas that were discovered after Galois. Throughout, we make the default assumption of the period: the coefficients of the equation are complex numbers.

Linear Equations

Let $a, b \in \mathbb{C}$ with $a \neq 0$. The general *linear* equation is

$$at + b = 0$$

and the solution is clearly

$$t = -\frac{b}{a}$$

Quadratic Equations

Let $a, b, c \in \mathbb{C}$ with $a \neq 0$. The general *quadratic* equation is

$$at^2 + bt + c = 0$$

Dividing by a and renaming the coefficients, we can consider the equivalent equation

$$t^2 + at + b = 0$$

The standard way to solve this equation is to rewrite it in the form

$$\left(t + \frac{a}{2}\right)^2 = \frac{a^2}{4} - b$$

Taking square roots,

$$t + \frac{a}{2} = \pm\sqrt{\frac{a^2}{4} - b}$$

so that

$$t = -\frac{a}{2} \pm \sqrt{\frac{a^2}{4} - b}$$

which is the usual quadratic formula except for a change of notation. The process used here is called *completing the square*; as remarked in the Historical Introduction, it goes back to the Babylonians 3600 years ago.

Cubic Equations

Let $a, b, c \in \mathbb{C}$ with $a \neq 0$. The general *cubic* equation can be written in the form

$$t^3 + at^2 + bt + c = 0$$

where again we have divided by the leading coefficient to avoid unnecessary complications in the formulas.

The first step is to change the variable to make $a = 0$. This is achieved by setting $y = t + \frac{a}{3}$, so that $t = y - \frac{a}{3}$. Such a move is called a Tschirnhaus transformation, after the person who first made explicit and systematic use of it. The equation becomes

$$y^3 + py + q = 0 \tag{1.3}$$

where

$$p = \frac{-a^2 + 3b}{3}$$
$$q = \frac{2a^3 - 9ab + 27c}{27}$$

To find the solution(s) we try (rabbit out of hat) the substitution

$$y = \sqrt[3]{u} + \sqrt[3]{v}$$

Now

$$y^3 = u + v + 3\sqrt[3]{u}\sqrt[3]{v}(\sqrt[3]{u} + \sqrt[3]{v})$$

so that (1.3) becomes

$$(u + v + q) + (\sqrt[3]{u} + \sqrt[3]{v})(3\sqrt[3]{u}\sqrt[3]{v} + p) = 0$$

We now choose u and v to make *both* terms vanish:

$$u + v + q = 0 \tag{1.4}$$
$$3\sqrt[3]{u}\sqrt[3]{v} + p = 0 \tag{1.5}$$

which imply

$$u + v = -q \tag{1.6}$$
$$uv = -\frac{p^3}{27} \tag{1.7}$$

Multiply (1.6) by u and subtract (1.7) to get

$$u(u+v)-uv=-qu+\frac{p^3}{27}$$

which can be rearranged to give

$$u^2+qu-\frac{p^3}{27}=0$$

which is a quadratic.

The solution of quadratics now tells us that

$$u=-\frac{q}{2}\pm\sqrt{\frac{q^2}{4}+\frac{p^3}{27}}$$

Since $u+v=-q$ we have

$$v=-\frac{q}{2}\mp\sqrt{\frac{q^2}{4}+\frac{p^3}{27}}$$

Changing the sign of the square root just permutes u and v, so we can set the sign to $+$. Thus we find that

$$y=\sqrt[3]{-\frac{q}{2}+\sqrt{\frac{q^2}{4}+\frac{p^3}{27}}}+\sqrt[3]{-\frac{q}{2}-\sqrt{\frac{q^2}{4}+\frac{p^3}{27}}} \tag{1.8}$$

which (by virtue of publication, not discovery) is usually called *Cardano's formula*. (This version differs from the formula in the Historical Introduction because Cardano worked with $x^2+px=q$, so q changes sign.) Finally, remember that the solution t of the original equation is equal to $y-a/3$.

Peculiarities of Cardano's Formula

An old Chinese proverb says 'Be careful what you wish for: you might get it'. We have wished for a formula for the solution, and we've got one. It has its peculiarities.

First: recall that over $\mathbb{C}$ every nonzero complex number z has *three* cube roots. If one of them is α, then the other two are $\omega\alpha$ and $\omega^2\alpha$, where

$$\omega=-\frac{1}{2}+i\frac{\sqrt{3}}{2}$$

is a primitive cube root of 1. Then

$$\omega^2=-\frac{1}{2}-i\frac{\sqrt{3}}{2}$$

The expression for y therefore appears to lead to *nine* solutions, of the form

$$\begin{array}{ccc} \alpha+\beta & \alpha+\omega\beta & \alpha+\omega^2\beta \\ \omega\alpha+\beta & \omega\alpha+\omega\beta & \omega\alpha+\omega^2\beta \\ \omega^2\alpha+\beta & \omega^2\alpha+\omega\beta & \omega^2\alpha+\omega^2\beta \end{array}$$

where α, β are specific choices of the cube roots.

However, not all of these expressions are zeros. Equation (1.5) implies (1.7), but (1.7) implies (1.5) only when we make the correct choices of cube roots. If we choose α, β so that $3\alpha\beta + p = 0$, then the solutions are

$$\alpha + \beta \qquad \omega\alpha + \omega^2\beta \qquad \omega^2\alpha + \omega\beta$$

Another peculiarity emerges when we try to solve equations whose solutions we already know. For example,

$$y^3 + 3y - 36 = 0$$

has the solution $y = 3$. Here $p = 3, q = -36$, and Cardano's formula gives

$$y = \sqrt[3]{18 + \sqrt{325}} + \sqrt[3]{18 - \sqrt{325}}$$

which seems a far cry from 3. However, further algebra converts it to 3: see Exercise 1.4.

As Cardano observed in his book, it gets worse: if his formula is applied to

$$t^3 - 15t - 4 = 0 \tag{1.9}$$

it leads to

$$t = \sqrt[3]{2 + \sqrt{-121}} + \sqrt[3]{2 - \sqrt{-121}} \tag{1.10}$$

in contrast to the obvious solution $t = 4$. This is very curious even today, and must have seemed even more so in the Renaissance period.

Cardano had already encountered such baffling expressions when trying to solve the quadratic $t(10 - t) = 40$, with the apparently nonsensical solutions $5 + \sqrt{-15}$ and $5 - \sqrt{-15}$, but there it was possible to see the puzzling form of the 'solution' as expressing the fact that no solution exists. However, Cardano was bright enough to spot that if you ignore the question of what such expressions *mean*, and just manipulate them as if they are ordinary numbers, then they do indeed satisfy the equation. 'So,' Cardano commented, 'progresses arithmetic subtlety, the end of which is as refined as it is useless.'

However, this shed no light on why a cubic could possess a perfectly reasonable solution, but the formula (more properly, the equivalent numerical procedure) could not find it. Around 1560 Raphael Bombelli observed that $(2 \pm \sqrt{-1})^3 = 2 \pm \sqrt{-121}$, and recovered (see Exercise 1.7) the solution $t = 4$ of (1.9) from the formula (1.10), again assuming that such expressions can be manipulated just like ordinary numbers. But Bombelli, too, expressed scepticism that such manoeuvres had any sensible meaning. In 1629 Albert Girard argued that such expressions are valid as *formal* solutions of the equations, and should be included 'for the certitude of the general rules'. Girard was influential in making negative numbers acceptable, but he was way ahead of his time when it came to their square roots.

In fact, Cardano's formula is pretty much useless whenever the cubic has three *real* roots. This is called the 'irreducible case' of the cubic, and the traditional escape route is to use trigonometric functions, Exercise 1.8. All this rather baffled the Renaissance mathematicians, who did not even have effective algebraic notation, and were wary of negative numbers, let alone imaginary ones.

Using Galois theory, it is possible to prove that the cube roots of complex numbers that arise in the irreducible case of the cubic equation cannot be avoided. That is, there are no formulas in real radicals for the real and imaginary parts. See Van der Waerden (1953) volume 1 page 180, and Isaacs (1985).

Quartic Equations

An equation of the fourth degree

$$t^4+at^3+bt^2+ct+d=0$$

is called a *quartic* equation (an older term is *biquadratic*). To solve it, start by making the Tschirnhaus transformation $y=t+a/4$, to get

$$y^4+py^2+qy+r=0 \tag{1.11}$$

where

$$\begin{aligned} p &= b-\frac{3a^2}{8} \\ q &= c-\frac{ab}{2}+\frac{3a}{48} \\ r &= d-\frac{ac}{4}+\frac{a^2b}{16}-\frac{3a^4}{256} \end{aligned}$$

Rewrite this in the form

$$\left(y^2+\frac{p}{2}\right)^2=-qy-r+\frac{p^2}{4}$$

Introduce a new term u, and observe that

$$\begin{aligned} \left(y^2+\frac{p}{2}+u\right)^2 &= \left(y^2+\frac{p}{2}\right)^2+2\left(y^2+\frac{p}{2}\right)u+u^2 \\ &= -qy-r+\frac{p^2}{4}+2uy^2+pu+u^2 \end{aligned}$$

We choose u to make the right hand side a perfect square. If it is, it must be the square of $\sqrt{2u}y-\frac{q}{2\sqrt{2u}}$, and then we require

$$-r+\frac{p^2}{4}+pu+u^2=\frac{q^2}{8u}$$

Provided $u\neq 0$, this becomes

$$8u^3+8pu^2+(2p-8r)u-q^2=0 \tag{1.12}$$

which is a cubic in u. Solving by Cardano's method, we can find u. Now

$$\left(y^2+\frac{p}{2}+u\right)^2=\left(\sqrt{2u}y-\sqrt{2u}\right)^2$$

so

$$y^2+\frac{p}{2}+u=\pm\left(\sqrt{2u}y-\sqrt{2u}\right)$$

Finally, we can solve the above two quadratics to find y.

If $u=0$ we do not obtain (1.12), but if $u=0$ then $q=0$, so the quartic (1.11) is a quadratic in y^2, and can be solved using only square roots.

Equation (1.12) is called the *resolvent cubic* of (1.11). Explicit formulas for the roots can be obtained if required. Since they are complicated, we shall not give them here.

An alternative approach to the resolvent cubic, not requiring a preliminary Tschirnhaus transformation, is described in Exercise 1.13.

Quintic Equations

So far, we have a series of special tricks, different in each case. We can start to solve the general *quintic* equation

$$t^5+at^4+bt^3+ct^2+dt+e=0$$

in a similar way. A Tschirnhaus transformation $y=t+a/5$ reduces it to

$$y^5+py^3+qy^2+ry+s=0$$

However, all variations on the tricks that we used for the quadratic, cubic, and quartic equations grind to a halt.

In 1770–1771 Lagrange analysed all of the above special tricks, showing that they can all be 'explained' using general principles about symmetric functions of the roots. When he applied this method to the quintic, however, he found that it 'reduced' the problem to a sextic—an equation of degree 6. Instead of helping, the method made the problem *worse*. A fascinating description of these ideas, together with a method for solving quintics whenever they *are* soluble by radicals, can be found in a lecture by George Neville Watson, rescued from his unpublished papers and written up by Berndt, Spearman and Williams (2002). The same article contains a wealth of other information about the quintic, including a long list of historical and recent references. Because the formulas are messy and the story is lengthy, the most we can do here is give some flavour of what is involved.

Lagrange observed that all methods for solving polynomial equations by radicals involve constructing rational functions of the roots that take a small number of values when the roots α_j are permuted. Prominent among these is the expression

$$\delta=\prod_{1\leq j<k\leq n}(\alpha_j-\alpha_k) \tag{1.13}$$

where n is the degree. This takes just two values, $\pm\delta$: plus for even permutations and minus for odd ones. Therefore $\Delta = \delta^2$ (known as the *discriminant* because it is nonzero precisely when the roots are distinct, so it 'discriminates' among the roots) is a rational function of the coefficients. This gets us started, and it yields a complete solution for the quadratic, but for cubics upwards it does not help much unless we can find other expressions in the roots with similar properties under permutation.

Lagrange worked out what these expressions look like for the cubic and the quartic, and noticed a pattern. For example, if a cubic polynomial has roots $\alpha_1, \alpha_2, \alpha_3$ and ω is a primitive cube root of unity, then the expression

$$u = (\alpha_1 + \omega\alpha_2 + \omega^2\alpha_3)^3$$

takes exactly two distinct values. In fact, even permutations leave it unchanged, while odd permutations transform it to

$$v = (\alpha_1 + \omega^2\alpha_2 + \omega\alpha_3)^3$$

It follows that $u+v$ and uv are fixed by all permutations of the roots, and must therefore be expressible as rational functions of the coefficients. So $u+v = a, uv = b$ where a, b are rational functions of the coefficients. Therefore u and v are the solutions of the quadratic equation $t^2 - at + b = 0$, so they can be expressed using square roots. But now the further use of cube roots expresses $\alpha_1 + \omega\alpha_2 + \omega^2\alpha_3 = \sqrt[3]{u}$ and $\alpha_1 + \omega^2\alpha_2 + \omega\alpha_3 = \sqrt[3]{v}$ by radicals. Since we also know that $\alpha_1 + \alpha_2 + \alpha_3$ is minus the coefficient of t^2, we have three independent linear equations in the roots, which are easily solved.

Something very similar works for the quartic, with expressions like

$$(\alpha_1 - \alpha_2 + \alpha_3 - \alpha_4)^2$$

But when we try the same idea on the quintic, an obstacle appears. Suppose that the roots of the quintic are $\alpha_1, \alpha_2, \alpha_3, \alpha_4, \alpha_5$. Let ζ be a primitive fifth root of unity. Following Lagrange's lead, it is natural to consider

$$w = (\alpha_1 + \zeta\alpha_2 + \zeta^2\alpha_3 + \zeta^3\alpha_4 + \zeta^4\alpha_5)^5$$

There are 120 permutations of 5 roots, and they transform w into 24 distinct expressions. Therefore w is a root of a polynomial of degree 24—a big step in the wrong direction, since we started with a mere quintic.

The best that can be done is to use an expression derived by Arthur Cayley in 1861, based on an idea of Robert Harley in 1859. This expression is

$$x = (\alpha_1\alpha_2 + \alpha_2\alpha_3 + \alpha_3\alpha_4 + \alpha_4\alpha_5 + \alpha_5\alpha_1 - \alpha_1\alpha_3 - \alpha_2\alpha_4 - \alpha_3\alpha_5 - \alpha_4\alpha_1 - \alpha_5\alpha_2)^2$$

It turns out that x takes precisely 6 values when the variables are permuted in all 120 possible ways. Therefore x is a root of a sextic equation. The equation is very complicated and has no obvious roots; it is, perhaps, better than an equation of degree 24, but it is still no improvement on the original quintic. Except when the sextic happens, by accident, to have a root whose square is rational, in which case the quintic *is* soluble by radicals. Indeed, this is a necessary and sufficient condition for a quintic to be soluble by radicals, see Berndt, Spearman and Williams (2002). For instance, as they explain in detail, the equation

$$t^5 + 15t + 12 = 0$$

has the solution

$$t = \sqrt[5]{\frac{-75+21\sqrt{10}}{125}} + \sqrt[5]{\frac{-75-21\sqrt{10}}{125}} + \sqrt[5]{\frac{225+72\sqrt{10}}{125}} + \sqrt[5]{\frac{225-72\sqrt{10}}{125}}$$

with similar expressions for the other four roots.

Lagrange's general method, then, fails for the quintic. This does not prove that the general quintic is not soluble by radicals, because for all Lagrange or anyone else knew, there might be other methods that do *not* make the problem worse. But it does suggest that there is something very different about the quintic. Suspicion began to grow that *no* method would solve the quintic by radicals. Mathematicians stopped looking for such a solution, and started looking for an impossibility proof instead.

EXERCISES

1.1 Use (1.1) to prove that multiplication of complex numbers is commutative and associative. That is, if u, v, w are complex numbers, then $uv = vu$ and $(uv)w = u(vw)$.

1.2 Prove that $\sqrt{2}$ is irrational, as follows. Assume for a contradiction that there exist integers a, b, with $b \neq 0$, such that $(a/b)^2 = 2$.

1. Show that we may assume $a, b > 0$.
2. Observe that if such an expression exists, then there must be one in which b is as small as possible.
3. Show that
$$\left(\frac{2b-a}{a-b}\right)^2 = 2$$
4. Show that $2b - a > 0, a - b > 0$.
5. Show that $a - b < b$, a contradiction.

1.3 Prove that if $q \in \mathbb{Q}$ then $\sqrt{q}$ is rational if and only if q is a perfect square; that is, it can be written in the form $q = p_1^{a_1} \cdots p_n^{a_n}$ where the integers a_j, which may be positive or negative, are all even.

1.4* Prove without using Cardano's formula that

$$\sqrt[3]{18+\sqrt{325}}+\sqrt[3]{18-\sqrt{325}}=3$$

1.5 Let $\alpha = \sqrt[3]{2} \in \mathbb{R}$. Prove that the set of all numbers $p+q\alpha+r\alpha^2$, for $p,q,r \in \mathbb{Q}$, is a subfield of $\mathbb{C}$.

1.6 Let ω be a primitive cube root of unity in $\mathbb{C}$. With the notation of Exercise 1.5, show that the map

$$p+q\alpha+r\alpha^2 \mapsto p+q\omega\alpha+r\omega^2\alpha^2$$

is a monomorphism onto its image, but not an automorphism.

1.7 Use Bombelli's observation that $(2\pm\sqrt{-1})^3 = 2\pm\sqrt{-121}$ to show that (with one choice of values of the cube roots)

$$\sqrt[3]{2+\sqrt{-121}}+\sqrt[3]{2-\sqrt{-121}}=4$$

1.8 Use the identity $\cos 3\theta = 4\cos^3\theta - 3\cos\theta$ to solve the cubic equation $t^3 + pt + q = 0$ when $27q^2 + 4p^3 < 0$.

1.9 Find radical expressions for all three roots of $t^3 - 15t - 4 = 0$.

1.10 When $27q^2 + 4p^3 < 0$ it is possible to try to make sense of Cardano's formula by generalising Bombelli's observation; that is, to seek α, β such that

$$\left[\alpha \pm \beta\sqrt{\frac{q^2}{4}+\frac{p^3}{27}}\right]^3 = \frac{q}{2} \pm \sqrt{\frac{q^2}{4}+\frac{p^3}{27}}$$

Why is this usually pointless?

1.11* Let $P(n)$ be the number of ways to arrange n zeros and ones in a row, given that ones occur in groups of three or more. Show that

$$P(n) = 2P(n-1) - P(n-2) + P(n-4)$$

and deduce that as $n \to \infty$ the ratio $\frac{P(n+1)}{P(n)} \to x$, where $x > 0$ is real and $x^4 - 2x^3 + x^2 - 1 = 0$. Factorise this quartic as a product of two quadratics, and hence find x.

1.12* The largest square that fits inside an equilateral triangle can be placed in any of three symmetrically related positions. Eugenio Calabi noticed that there is exactly one other shape of triangle in which there are three equal largest squares, Figure 7. Prove that in this triangle the ratio x of the longest side the other two is a solution of the cubic equation $2x^3 - 2x^2 - 3x + 2 = 0$, and find an approximate value of x to three decimal places.

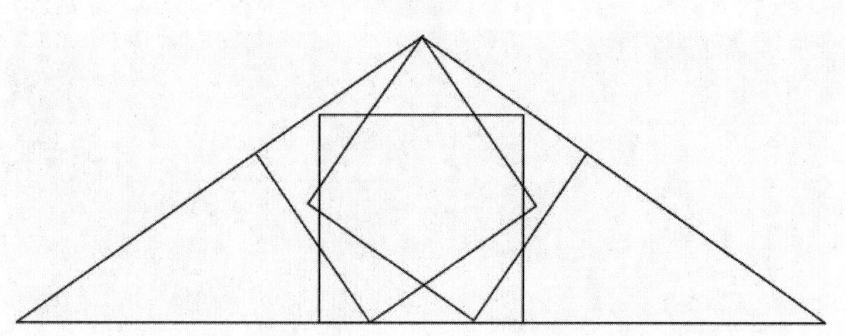

FIGURE 7: Calabi's triangle.

1.13 Investigate writing the general quartic $t^4+at^3+bt^2+ct+d$ in the form

$$(t^2+pt+q)^2-(rt+s)^2$$

which, being a difference of two squares, factorises into two quadratics

$$(t^2+pt+q+rt+s)(t^2+pt+q-rt-s)$$

and can thus be solved in radicals if p,q,r,s can be expressed in terms of the original coefficients a,b,c,d.

Show that doing this leads to a cubic equation.

1.14 Mark the following true or false.

(a) -1 has no square root.

(b) -1 has no real square root.

(c) -1 has two distinct square roots in $\mathbb{C}$.

(d) Every subring of $\mathbb{C}$ is a subfield of $\mathbb{C}$.

(e) Every subfield of $\mathbb{C}$ is a subring of $\mathbb{C}$.

(f) The set of all numbers $p+q\sqrt[7]{5}$ for $p,q\in\mathbb{Q}$ is a subring of $\mathbb{C}$.

(g) The set of all numbers $p+q\sqrt[7]{5}$ for $p,q\in\mathbb{C}$ is a subring of $\mathbb{C}$.

(h) Cardano's formula always gives a correct answer.

(i) Cardano's formula always gives a sensible answer.

(j) A quintic equation over $\mathbb{Q}$ can never be solved by radicals.

Chapter 2

The Fundamental Theorem of Algebra

At the time of Galois, the natural setting for most mathematical investigations was the complex number system. The real numbers were inadequate for many questions, because -1 has no real square root. The arithmetic, algebra, and—decisively—analysis of complex numbers were richer, more elegant, and more complete than the corresponding theories for real numbers.

In this chapter we establish one of the key properties of $\mathbb{C}$, known as the Fundamental Theorem of Algebra. This theorem asserts that every polynomial equation with coefficients in $\mathbb{C}$ has a solution in $\mathbb{C}$. This theorem is, of course, false over $\mathbb{R}$—consider the equation $t^2 + 1 = 0$. It was fundamental to classical algebra, but the name is somewhat archaic, and modern algebra bypasses $\mathbb{C}$ altogether, preferring greater generality. Because we find it convenient to work in the same setting as Galois, the theorem will be fundamental for us.

All rigorous proofs of the Fundamental Theorem of Algebra require quite a lot of background. Here, we give a proof that uses a few simple ideas from algebra and trigonometry, estimates of the kind that are familiar from any first course in analysis, and one simple basic result from point-set topology.

Later, we give an almost purely algebraic proof, but the price is the need for much more machinery: see Chapter 23. Ironically, that proof uses Galois theory to prove the Fundamental Theorem of Algebra, the exact opposite of what Galois did. The logic is not circular, because the proof in Chapter 23 rests on the abstract approach to Galois theory described in the second part of this book, which makes no use of the Fundamental Theorem of Algebra.

2.1 Polynomials

Linear, quadratic, cubic, quartic, and quintic equations are examples of a more general class: polynomial equations. These take the form

$$p(t) = 0$$

where $p(t)$ is a polynomial in t.

Mathematics is littered with polynomial equations, arising in a huge variety of contexts. As a sample, here are two from the literature. You don't need to think about them: just observe them like a butterfly-collector looking at a strange new specimen.

John Horton Conway came up with one of the strangest instances of a polynomial equation that I have ever encountered, in connection with the so-called *look and say sequence*. The sequence starts

$$1 \quad 11 \quad 21 \quad 1211 \quad 111221 \quad 312211 \quad 13112221 \quad \ldots$$

The rule of formation is most readily seen in verbal form. We start with '1', which can be read as 'one one', so the next term is 11. This reads 'two ones', leading to 21. Read this as 'one two, one one' and you see where 1211 comes from, and so on. If $L(n)$ is the length of the nth term in this sequence, approximately how big is $L(n)$? Conway (1985) proves that $L(n)$ satisfies a 72-term linear recurrence relation. Standard techniques from combinatorics then prove that for large n, the value of $L(n)$ is asymptotically proportional to α^n, where $\alpha = 1{\cdot}303577\ldots$ is the smallest real solution of the 71st degree polynomial equation

$$\begin{aligned}
&t^{71} - t^{69} - 2t^{68} - t^{67} + 2t^{66} + 2t^{65} - t^{63} - t^{62} - t^{61} - t^{60} + 2t^{58} \\
&+5t^{57} + 3t^{56} - 2t^{55} - 10t^{54} - 3t^{53} - 2t^{52} + 6t^{51} + 6t^{50} + t^{49} + 9t^{48} \\
&-3t^{47} - 7t^{46} - 8t^{45} - 8t^{44} + 10t^{43} + 6t^{42} + 8t^{41} - 5t^{40} - 12t^{39} \\
&+7t^{38} - 7t^{37} + 7t^{36} + t^{35} - 3t^{34} + 10t^{33} + t^{32} - 6t^{31} - 2t^{30} \\
&-10t^{29} - 3t^{28} + 2t^{27} + 9t^{26} - 3t^{25} + 14t^{24} - 8t^{23} - 7t^{21} + 9t^{20} \\
&+3t^{19} - 4t^{18} - 10t^{17} - 7t^{16} + 12t^{15} + 7t^{14} + 2t^{13} - 12t^{12} - 4t^{11} \\
&-2t^{10} + 5t^{9} + t^{7} - 7t^{6} + 7t^{5} - 4t^{4} + 12t^{3} - 6t^{2} + 3t - 6 = 0
\end{aligned} \tag{2.1}$$

The second example is from cosmology. Braden, Brown, Whiting, and York (1990) show that the entropy of a black hole is $\pi r_B^2 \alpha^2$, where α is a solution of the 7th degree equation

$$t^5(t-q^2)(t-1) + b^2(t^2-q^2)^2 = 0 \tag{2.2}$$

where b, q are expressions involving temperature and various fundamental physical constants such as the speed of light and Planck's constant.

With the importance of polynomial equations now established, we start to develop a coherent theory of their solutions. As the above examples illustrate, a polynomial is an algebraic expression involving the powers of a 'variable' or 'indeterminate' t. We are used to thinking of such a polynomial as the function that maps t to the value of the expression concerned, so that the first polynomial represents the function f such that $f(t) = t^2 - 2t + 6$. This 'function' viewpoint is familiar, and it causes no problems when we are thinking about polynomials with complex numbers as their coefficients. Later (Chapter 16) we will see that when more general fields are permitted, it is not such a good idea to think of a polynomial as a function. So it is worth setting up the concept of a polynomial so that it extends easily to the general context.

We therefore define a *polynomial over* $\mathbb{C}$ *in the indeterminate* t to be an expression

$$r_0 + r_1 t + \cdots + r_n t^n$$

where $r_0, \ldots, r_n \in \mathbb{C}$, $0 \leq n \in \mathbb{Z}$, and t is undefined. What, though, is an 'expression', logically speaking? For set-theoretic purity we can replace such an expression by the sequence $(r_0, \ldots, r_n)$. In fact, it is more convenient to use an infinite sequence $(r_0, r_1, \ldots)$ in which all entries $r_j = 0$ when $j > n$ for some finite n: see Exercise 2.2. In such a formalism, t is just a symbol for the sequence $\{0, 1, 0 \ldots\}$.

The elements $r_0, \ldots, r_n$ are the *coefficients* of the polynomial. In the usual way, terms $0t^m$ may be omitted or written as 0, and $1t^m$ can be replaced by t^m.

In practice we often write polynomials in descending order

$$r_n t^n + r_{n-1} t^{n-1} + \cdots + r_1 t + r_0$$

and from now on we make such changes without further comment.

Two polynomials are defined to be equal *if and only if* the corresponding coefficients are equal, with the understanding that powers of t not occurring in the polynomial may be taken to have zero coefficient. To define the sum and the product of two polynomials, write

$$\sum r_i t^i$$

instead of

$$r_0 + r_1 t + \cdots + r_n t^n$$

where the summation is considered as being over all integers $i \geq 0$, and r_k is defined to be 0 if $k \geq n$. Then, if

$$r = \sum r_i t^i \qquad s = \sum s_i t^i$$

we define

$$r + s = \sum (r_i + s_i) t^i \tag{2.3}$$

and

$$rs = \sum q_j t^j \quad \text{where} \quad q_j = \sum_{h+i=j} r_h s_i \tag{2.4}$$

It is now easy to check directly from these definitions that the set of all polynomials over $\mathbb{C}$ in the t obeys all of the usual algebraic laws (Exercise 2.3). We denote this set by $\mathbb{C}[t]$, and call it the *ring of polynomials over* $\mathbb{C}$ *in the indeterminate* t.

We can also define polynomials in several indeterminates $t_1, t_2, \ldots, t_n$, obtaining the ring of n-variable polynomials

$$\mathbb{C}[t_1, t_2, \ldots, t_n]$$

in an analogous way.

An element of $\mathbb{C}[t]$ will usually be denoted by a single letter, such as f, whenever it is clear which indeterminate is involved. If there is ambiguity, we write $f(t)$ to emphasise the role played by t. Although this looks like function notation, technically it is not. However, polynomials over $\mathbb{C}$ can be interpreted as functions, see Proposition 2.3 below.

Next, we introduce a simple but very useful concept, which quantifies how complicated a polynomial is.

Definition 2.1. If f is a polynomial over $\mathbb{C}$ and $f \neq 0$, then the *degree* of f is the highest power of t occurring in f with non-zero coefficient.

For example, t^2+1 has degree 2, and $723t^{1101}-9111t^{55}+43$ has degree 1101. The polynomial (2.1) has degree 71, and (2.2) has degree 7.

More generally, if $f=\sum r_it^i$ and $r_n \neq 0$ and $r_m=0$ for $m>n$, then f has degree n. We write ∂f for the degree of f. To deal with the case $f=0$ we adopt the convention that $\partial 0=-\infty$. This symbol is endowed with the following properties: $-\infty<n$ for any integer n, $-\infty+n=-\infty$, $-\infty\times n=-\infty$, $(-\infty)^2=-\infty$. We do *not* set $(-\infty)^2=+\infty$ because $0.0=0$.

The following result is immediate from this definition:

Proposition 2.2. *If f,g are polynomials over $\mathbb{C}$, then*

$$\partial(f+g)\leq \max(\partial f,\partial g) \qquad \partial(fg)=\partial f+\partial g$$

□

The inequality in the first line is due to the possibility of the highest terms 'cancelling', see Exercise 2.4.

The $f(t)$ notation makes f appear to be a function, with t as its 'independent variable', and in fact we can identify each polynomial f over $\mathbb{C}$ with the corresponding function. Specifically, each polynomial $f\in\mathbb{C}[t]$ can be considered as a function from $\mathbb{C}$ to $\mathbb{C}$, defined as follows: if $f=\sum r_it^i$ and $\alpha\in\mathbb{C}$, then α is mapped to $\sum r_i\alpha^i$. The next proposition proves that when the coefficients lie in $\mathbb{C}$, it causes no confusion if we use the same symbols f to denote a polynomial and the function associated with it.

Proposition 2.3. *Two polynomials f,g over $\mathbb{C}$ define the same function if and only if they are equal as polynomials; that is, they have the same coefficients.*

Proof. Equivalently, by taking the difference of the two polynomials, we must prove that if $f(t)$ is a polynomial over $\mathbb{C}$ and $f(t)=0$ for all t, then the coefficients of f are all 0. Let $P(n)$ be the statement: If a polynomial $f(t)$ over $\mathbb{C}$ has degree n, and $f(t)=0$ for all $t\in\mathbb{C}$, then $f=0$. We prove $P(n)$ for all n by induction on n.

Both $P(0)$ and $P(1)$ are obvious. Suppose that $P(n-1)$ is true. Write

$$f(t)=a_nt^n+\cdots+a_0$$

In particular, $f(0)=0$, so $a_0=0$ and

$$\begin{aligned} f(t) &= a_nt^n+\cdots+a_1t \\ &= t(a_nt^{n-1}+\cdots+a_1) \\ &= tg(t) \end{aligned}$$

where $g(t)=a_nt^{n-1}+\cdots+a_1$ has degree $n-1$. Now $g(t)$ vanishes for all $t\in\mathbb{C}$ except, perhaps, $t=0$. However, if $g(0)=a_1\neq 0$ then $g(t)\neq 0$ for t sufficiently small. (This follows by continuity of polynomial functions, but it can be proved directly by estimating the size of $g(\varepsilon)$ when ε is small.) Therefore $g(t)$ vanishes for all $t\in\mathbb{C}$. By induction, $g=0$. Therefore $f=0$, so $P(n)$ is true and the induction is complete. □

Proposition 2.3 implies that we can safely consider a polynomial over a subfield of $\mathbb{C}$ as either a formal algebraic expression or a function. It is easy to see that sums and products of polynomials agree with the corresponding sums and products of functions. Moreover, the same notational flexibility allows us to 'change the variable' in a polynomial. For example, if t, u are two indeterminates and $f(t) = \sum r_i t^i$, then we may define $f(u) = \sum r_i u^i$. It is also clear what is meant by such expressions as $f(t-3)$ or $f(t^2+1)$.

2.2 Fundamental Theorem of Algebra

In Section 1.3 we saw that the development of the complex numbers can be viewed as the culmination of a series of successive extensions of the natural number system. At each step, equations that cannot be solved within the existing number system become soluble in the new, extended system. For example, $\mathbb{C}$ arises from $\mathbb{R}$ by insisting that $t^2 = -1$ should have a solution.

The question then arises: why stop at $\mathbb{C}$? Why not find an equation that has no solutions over $\mathbb{C}$, and enlarge the number system still further to provide a solution?

The answer is that no such equation exists, at least if we limit ourselves to polynomials. *Every* polynomial equation over $\mathbb{C}$ has a solution in $\mathbb{C}$. This proposition was a matter of heated debate around 1700. In a paper of 1702, Leibniz disputes that it can be true, citing the example

$$x^4+a^4 = \left(x+a\sqrt{\sqrt{-1}}\right)\left(x-a\sqrt{\sqrt{-1}}\right)\left(x+a\sqrt{-\sqrt{-1}}\right)\left(x-a\sqrt{-\sqrt{-1}}\right)$$

and presumably thinking that $\sqrt{\sqrt{-1}}$ is not a complex number.

However, in 1676 Isaac Newton had already observed the factorisation into real quadratics:

$$x^4+a^4 = (x^2+a^2)^2 - 2a^2x^2 = (x^2+a^2+\sqrt{2}ax)(x^2+a^2-\sqrt{2}ax)$$

and Nicholas Bernoulli published the same formula in 1719. In effect, the resolution of the dispute rests on observing that $\sqrt{i} = \frac{1\pm i}{\sqrt{2}}$, which is in $\mathbb{C}$. In fact, every complex number has a complex square root:

$$\sqrt{a+bi} = \sqrt{\frac{a+\sqrt{a^2+b^2}}{2}} + i\sqrt{\frac{-a+\sqrt{a^2+b^2}}{2}} \tag{2.5}$$

(together with minus the same formula), as can be checked by squaring the right-hand side. Here the square root of a^2+b^2 is the positive one, and the signs of the other two square roots are chosen to make their product equal to b. Observe that

$$a+\sqrt{a^2+b^2} \geq 0 \qquad -a+\sqrt{a^2+b^2} \geq 0$$

because $a^2+b^2 \geq a^2$, so both of the main square roots on the right-hand side are real.

In 1742 Euler asserted, without proof, that every real polynomial can be decomposed into linear or quadratic factors with real coefficients; Bernoulli now erred the other way, citing

$$x^4-4x^3+2x^2+4x+4$$

with zeros $1+\sqrt{2+\sqrt{-3}}$, $1-\sqrt{2+\sqrt{-3}}$, $1+\sqrt{2-\sqrt{-3}}$, and $1-\sqrt{2-\sqrt{-3}}$. Euler responded, in a letter to his friend Christian Goldbach, that the four factors occur as two complex conjugate pairs, and that the product of such a pair of factors is a real quadratic. He showed this to be the case for Bernoulli's proposed counterexample. Goldbach suggested that $x^4+72x-20$ did not agree with Euler's assertion, and Euler pointed out a computational error, adding that he had proved the theorem for polynomials of degree ≤ 6. Euler and Jean Le Rond d'Alembert gave incomplete proofs for any degree; Lagrange claimed to have filled in the gaps in Euler's proof in 1772, but made the mistake of assuming that the roots existed, and using the laws of algebra to deduce that they must be complex numbers, without proving that the roots—whatever they were—must obey the laws of algebra. The first genuine proof was given by Gauss in his doctoral thesis of 1799. It involved the manipulation of complicated trigonometric series to derive a contradiction, and was far from transparent. The underlying idea can be reformulated in topological terms, involving the winding number of a curve about a point, see Hardy (1960) and Stewart (1977). Later Gauss gave three other proofs, all based on different ideas.

Other classical proofs use deep results in complex analysis, such as Liouville's Theorem: a bounded function analytic on the whole of the complex plane is constant. This depends on Cauchy's Integral Formula and takes most of a course in complex analysis to prove. See Titchmarsh (1960). An alternative approach uses Rouché's Theorem, Titchmarsh (1960) 3.44. Another proof uses the Maximum Modulus Theorem: if an analytic function is not constant, then the maximum value of its modulus on an arbitrary set occurs on the boundary of that set. A variant uses the Minimum Modulus Theorem (the minimum value of its modulus on an arbitrary set is either zero or occurs on the boundary of that set). See Stewart and Tall (1983) Theorems 10.14, 10.15. Euler's approach, which sets the real and imaginary parts of $p(z)$ to zero and proves that the resulting curves in the plane must intersect, can be made rigorous. William Kingdon Clifford gave a proof based on induction on the power of 2 that divides the degree n, which is most easily explained using Galois theory. We present this in Chapter 23, Corollary 23.13.

All of these proofs are quite sophisticated. But there's an easier way, using a few ideas from elementary point-set topology and estimates of the kind we encounter early on in any course on real analysis. It can be found on Wikipedia, and it deserves to be more widely known because it is simple and cuts straight to the heart of the issue. The necessary facts can be proved directly by elementary means, and would have been considered obvious before mathematicians started worrying about rigour in analysis around 1850. So Euler, Gauss, and other mathematicians of those periods could have discovered this proof.

We now state this property of the complex numbers formally, and explore some of its easier consequences. It is the aforementioned Fundamental Theorem of Algebra.

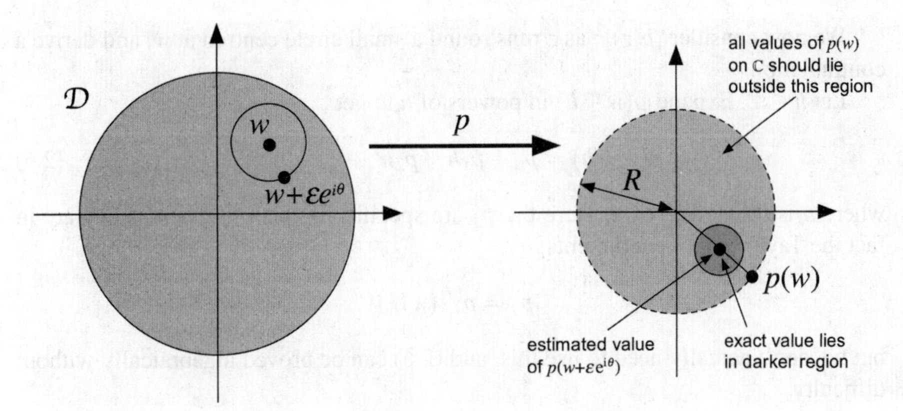

FIGURE 8: Idea of proof.

As we have observed, this is a good name if we are thinking of classical algebra, but not such a good name in the context of modern abstract algebra, which constructs suitable fields as it goes along and avoids explicit use of complex numbers.

Theorem 2.4 (Fundamental Theorem of Algebra). *If* $p(z)$ *is a non-constant polynomial over* $\mathbb{C}$*, then there exists* $z_0 \in \mathbb{C}$ *such that* $p(z_0) = 0$.

Such a number z is called a *root* of the equation $p(t) = 0$, or a *zero* of the polynomial p. For example, i is a root of the equation $t^2 + 1 = 0$ and a zero of $t^2 + 1$. Polynomial equations may have more than one root; indeed, $t^2 + 1 = 0$ has at least one other root, $-i$.

The idea behind the proof is illustrated in Figure 8, and can be summarised in a few lines. Assume for a contradiction that $p(z)$ is never zero. Then $|p(z)|^2$ has a nonzero minimum value and attains that minimum at some point $w \in \mathbb{C}$. Consider points v on a small circle centred at w, and use simple estimates to show that $|p(v)|^2$ must be less than $|p(w)|^2$ for some v. Contradiction.

Now for the details.

Proof of Theorem 2.4. Suppose for a contradiction that no such z_0 exists. For some $R > 0$ the set

$$\mathscr{D} = \{z : |p(z)|^2 \leq R\}$$

is non-empty. The map $\psi : \mathbb{C} \to \mathbb{R}^+$ defined by $\psi(z) = |p(z)|^2$ is continuous, so $\mathscr{D} = \psi^{-1}([0, R])$ is compact. For a subset of $\mathbb{C}$ this is equivalent to being closed and bounded. It follows that $|p(z)|^2$ attains its minimum value on $\mathscr{D}$. By the definition of $\mathscr{D}$ this is also its minimum value on $\mathbb{C}$.

Assume this minimum is attained at $w \in \mathbb{C}$. Then

$$|p(z)|^2 \geq |p(w)|^2$$

for all $z \in \mathbb{C}$, and by assumption $p(w) \neq 0$.

We now consider $|p(z)|^2$ as z runs round a small circle centred at w, and derive a contradiction.

Let $h \in \mathbb{C}$. Expand $p(w+h)$ in powers of h to get

$$p(w+h) = p_0 + p_1 h + p_2 h^2 + \cdots + p_n h^n \tag{2.6}$$

where n is the degree of p. Here the p_j are specific complex numbers. They are in fact the Taylor series coefficients

$$p_j = p^{(j)}(w)/j!$$

but we don't actually need to use this, and (2.6) can be proved algebraically without difficulty.

Clearly $p_0 = p(w)$, and we are assuming this is nonzero, so $p_0 \neq 0$. If $p_1 = p_2 = \cdots = p_n = 0$ then $p(z) = p_0$ is constant, contrary to hypothesis. So some $p_j \neq 0$. Let m be the smallest integer ≥ 1 from which $p_m \neq 0$. In (2.6) let $h = \varepsilon e^{i\theta}$ for small $\varepsilon > 0$. Then

$$p(w+\varepsilon e^{i\theta}) = p_0 + p_m \varepsilon^m e^{mi\theta} + O(\varepsilon^{m+1})$$

where $O(\varepsilon^n)$ indicates terms of order n or more in ε. Therefore

$$\begin{aligned} |p(w+\varepsilon e^{i\theta})|^2 &= |p_0 + p_m \varepsilon^m e^{mi\theta}|^2 + O(\varepsilon^{m+1}) \\ &= p_0\bar{p}_0 + \bar{p}_0 p_m \varepsilon^m e^{mi\theta} + p_0 \bar{p}_m \varepsilon^m e^{-mi\theta} + O(\varepsilon^{m+1}) \end{aligned}$$

Let $p_0\bar{p}_m = re^{i\phi}$ for $r \geq 0$. Since $p_0 \neq 0$ and $p_m \neq 0$ we have $r > 0$. Setting $h = 0$ we see that $p_0\bar{p}_0 = |p(w)|^2$. Now

$$\begin{aligned} |p(w+\varepsilon e^{i\theta})|^2 &= p_0\bar{p}_0 + re^{i\phi}\varepsilon^m e^{mi\theta} + re^{-i\phi}\varepsilon^m e^{-mi\theta} + O(\varepsilon^{m+1}) \\ &= |p(w)|^2 + 2\varepsilon^m r\cos(m\theta+\phi) + O(\varepsilon^{m+1}) \end{aligned}$$

Set $\theta = \frac{1}{m}(\phi - \pi)$, so that $\phi = \pi - m\theta$. Then $\cos(m\theta+\phi) = \cos(\pi) = -1$, and

$$|p(w+\varepsilon e^{i\theta})|^2 = |p(w)|^2 - 2\varepsilon^m r + O(\varepsilon^{m+1})$$

But $\varepsilon, r > 0$, so for sufficiently small ε we have

$$|p(w+\varepsilon e^{i\theta})|^2 < |p(w)|^2$$

contradicting the definition of w. Therefore there exists $z_0 \in \mathbb{C}$ such that $p(z_0) = 0$. □

2.3 Implications

The Fundamental Theorem of Algebra has some useful implications. Before proving the most basic of these, we first prove the Remainder Theorem.

Theorem 2.5 (Remainder Theorem). *Let $p(t) \in \mathbb{C}[t]$ with $\partial p \geq 1$, and let $\alpha \in \mathbb{C}$.*

(1) *There exist $q(t) \in \mathbb{C}[t]$ and $r \in \mathbb{C}$ such that $p(t) = (t-\alpha)q(t) + r$.*

(2) *The constant r satisfies $r = p(\alpha)$.*

Proof. Let $y = t - \alpha$ so that $t = y + \alpha$. Write $p(t) = p_n t^n + \cdots + p_0$ where $p_n \neq 0$ and $n \geq 1$. Then

$$p(t) = p_n(y+\alpha)^n + \cdots + p_0$$

Expand the powers of $y + \alpha$ by the binomial theorem, and collect terms to get

$$\begin{aligned} p(t) &= a_n y^n + \cdots + a_1 y + a_0 \qquad a_j \in \mathbb{C} \\ &= y(a_n y^{n-1} + \cdots + a_1) + a_0 \\ &= (t-\alpha)q(t) + r \end{aligned}$$

where

$$\begin{aligned} q(t) &= a_n(t-\alpha)^{n-1} + \cdots + a_2(t-\alpha) + a_1 0 \\ r &= a_0 \end{aligned}$$

Now substitute $t = \alpha$ in the identity $p(t) = (t-\alpha)q(t) + r$ to get

$$p(\alpha) = (\alpha - \alpha)q(\alpha) + r = 0.q(\alpha) + r = r$$

□

Corollary 2.6. *The complex number α is a zero of $p(t)$ if and only if $t - \alpha$ divides $p(t)$ in $\mathbb{C}[t]$.*

Proposition 2.7. *Let $p(t) \in \mathbb{C}[t]$ with $\partial p = n \geq 1$. Then there exist $\alpha_1, \ldots, \alpha_n \in \mathbb{C}$, and $0 \neq k \in \mathbb{C}$, such that*

$$p(t) = k(t-\alpha_1)\ldots(t-\alpha_n) \tag{2.7}$$

Proof. Use induction on n. The case $n = 1$ is obvious. If $n > 1$ we know, by the Fundamental Theorem of Algebra, that $p(t)$ has at least one zero in $\mathbb{C}$: call this zero α_n. By the Remainder Theorem, there exists $q(t) \in \mathbb{C}[t]$ such that

$$p(t) = (t-\alpha_n)q(t) \tag{2.8}$$

(note that the remainder $r = p(\alpha_n) = 0$). Then $\partial q = n - 1$, so by induction

$$q(t) = k(t-\alpha_1)\ldots(t-\alpha_{n-1}) \tag{2.9}$$

For suitable complex numbers $k, \alpha_1, \ldots, \alpha_{n-1}$. Substitute (2.9) in (2.8) and the induction step is complete. □

It follows immediately that the α_j are the *only* complex zeros of $p(t)$.

The zeros α_j need not be distinct. Collecting together those that are equal, we can rewrite (2.7) in the form

$$p(t) = k(t-\beta_1)^{m_1}\dots(t-\beta_l)^{m_l}$$

where $k = a_n$, the β_j are distinct, the m_j are integers ≥ 1, and $m_1 + \cdots + m_l = n$. We call m_j the *multiplicity* of the zero β_j of $p(t)$.

In particular, we have proved that every complex polynomial of degree n has precisely n complex zeros, counted according to multiplicity.

EXERCISES

2.1 Let $p(t) \in \mathbb{Q}[t]$. Show that $p(t)$ has a unique expression in the form

$$p(t) = (t-\alpha_1)\dots(t-\alpha_r)q(t)$$

(except for re-ordering the α_j) where $\alpha_j \in \mathbb{Q}$ for $1 \leq j \leq r$ and $q(t)$ has no zeros in $\mathbb{Q}[t]$. Prove that here, the α_j are precisely the zeros of $p(t)$ in $\mathbb{Q}$.

2.2 A formal definition of $\mathbb{C}[t]$ runs as follows. Consider the set S of all infinite sequences

$$(a_n)_{n\in\mathbb{N}} = (a_0, a_1, \dots, a_n, \dots)$$

where $a_n \in \mathbb{C}$ for all $n \in \mathbb{N}$, and such that $a_n = 0$ for all but a finite set of n. Define operations of addition and multiplication on S by the rules

$$\begin{aligned}(a_n)+(b_n) &= (u_n) \quad \text{where} \quad u_n = a_n + b_n \\ (a_n)(s_n) &= (v_n) \quad \text{where} \quad v_n = a_n b_0 + a_{n-1}b_1 + \cdots + a_0 b_n\end{aligned}$$

Prove that $\mathbb{C}[t]$, so defined, satisfies all of the usual laws of algebra for addition, subtraction, and multiplication. Define the map

$$\begin{aligned}\theta : \mathbb{C} &\to S \\ \theta(k) &= (k, 0, 0, 0, \dots)\end{aligned}$$

and prove that $\theta(\mathbb{C}) \subseteq S$ is isomorphic to $\mathbb{C}$.

Finally, prove that if we identify $a \in \mathbb{C}$ with $\theta(a) \in S$ and the 'indeterminate' t with $(0, 1, 0, 0, 0, \dots) \in S$, then $(a_n) = a_0 + \cdots + a_N t^N$, where N is chosen so that $a_n = 0$ for $n > N$. Thus we can define polynomials as sequences of complex numbers corresponding to the coefficients.

2.3 Using (2.3, 2.4), prove that polynomials over $\mathbb{C}$ obey the following algebraic laws:

$f+g = g+f$, $f+(g+h) = (f+g)+h$, $fg = gf$, $f(gh) = (fg)h$, and $f(g+h) = fg+fh$.

2.4 Show that $\partial(f+g)$ can be less than $\max(\partial f, \partial g)$, and indeed that $\partial(f+g)$ can be less than $\min(\partial f, \partial g)$.

2.5* If $z_1, z_2, \ldots, z_n$ are distinct complex numbers, show that the determinant

$$D = \begin{vmatrix} 1 & 1 & \cdots & 1 \\ z_1 & z_2 & \cdots & z_n \\ z_1^2 & z_2^2 & \cdots & z_n^2 \\ \vdots & \vdots & \ddots & \vdots \\ z_1^{n-1} & z_2^{n-1} & \cdots & z_n^{n-1} \end{vmatrix}$$

is non-zero.

(*Hint:* Consider the z_j as independent indeterminates over $\mathbb{C}$. Then D is a polynomial in the z_j, of total degree $0+1+2+\cdots+(n-1) = \frac{1}{2}n(n-1)$. Moreover, D vanishes whenever $z_j = z_k$, for $k \neq j$, since it then has two identical rows. Therefore D is divisible by $z_j - z_k$ for all $j \neq k$, hence it is divisible by $\prod_{j<k}(z_j - z_k)$. Now compare degrees.)

The determinant D is called a *Vandermonde determinant*, for obscure reasons (no such expression occurs in Alexandre-Theophile Vandermonde's published writings).

2.6 Use the Vandermonde determinant to prove that if a polynomial $f(t)$ vanishes for all $t \in \mathbb{C}$, then all coefficients of f are zero. (*Hint.* Substitute $t = 1, 2, 3, \ldots$ and solve the resulting system of linear equations for the coefficients.)

2.7 Prove, without using the Fundamental Theorem of Algebra, that every cubic polynomial over $\mathbb{R}$ can be expressed as a product of linear factors over $\mathbb{C}$.

2.8* Do the same for cubic polynomials over $\mathbb{C}$.

2.9 Mark the following true or false. Here f, g are polynomials over $\mathbb{C}$.

(a) $\partial(f-g) \geq \min(\partial f, \partial g)$.

(b) $\partial(f-g) \leq \min(\partial f, \partial g)$.

(c) $\partial(f-g) \leq \max(\partial f, \partial g)$.

(d) $\partial(f-g) \geq \max(\partial f, \partial g)$.

(e) Every polynomial over $\mathbb{C}$ has at least one zero in $\mathbb{C}$.

(f) Every polynomial over $\mathbb{C}$ of degree ≥ 1 has at least one zero in $\mathbb{R}$.

Chapter 3

Factorisation of Polynomials

Not only is there an algebra of polynomials: there is an arithmetic. That is, there are notions analogous to the integer-based concepts of divisibility, primes, prime factorisation, and highest common factors. These notions are essential for any serious understanding of polynomial equations, and we develop them in this chapter.

Mathematicians noticed early on that if f is a product gh of polynomials of smaller degree, then the solutions of $f(t) = 0$ are precisely those of $g(t) = 0$ together with those of $h(t) = 0$. For example, to solve the equation

$$t^3 - 6t^2 + 11t - 6 = 0$$

we can spot the factorisation $(t-1)(t-2)(t-3)$ and deduce that the roots are $t = 1, 2, 3$. From this simple idea emerged the arithmetic of polynomials—a systematic study of divisibility properties of polynomials with particular reference to analogies with the integers. In particular, there is an analogue for polynomials of the Euclidean Algorithm for finding the highest common factor of two integers.

In this chapter we define the relevant notions of divisibility and show that there are certain polynomials, the 'irreducible' ones, that play a similar role to prime numbers in the ring of integers. Every polynomial over a given subfield of $\mathbb{C}$ can be expressed as a product of irreducible polynomials over the same subfield, in an essentially unique way. We relate zeros of polynomials to the factorisation theory.

Throughout this chapter all polynomials are assumed to lie in $K[t]$, where K is a subfield of the complex numbers, or in $R[t]$, where R is a subring of the complex numbers. Some theorems are valid over R, while others are valid only over K: we will need both types.

3.1 The Euclidean Algorithm

In number theory, one of the key concepts is divisibility: an integer a is divisible by an integer b if there exists an integer c such that $a = bc$. For instance, 60 is divisible by 3 since $60 = 3.20$, but 60 is not divisible by 7. Divisibility properties of integers lead to such ideas as primes and factorisation. We wish to develop similar ideas for polynomials.

Many important results in the factorisation theory of polynomials derive from the observation that one polynomial may always be divided by another provided that a 'remainder' term is allowed. This is a generalisation of the Remainder Theorem, in which f is assumed to be linear.

Proposition 3.1 (Division Algorithm). *Let f and g be polynomials over K, and suppose that f is non-zero. Then there exist unique polynomials q and r over K, such that $g = fq + r$ and r has strictly smaller degree than f.*

Proof. Use induction on the degree of g. If $\partial g = -\infty$ then $g = 0$ and we may take $q = r = 0$. If $\partial g = 0$ then $g = k$ is an element of K. If also $\partial f = 0$ then f is an element of K, and we may take $q = k/f$ and $r = 0$. Otherwise $\partial f > 0$ and we may take $q = 0$ and $r = g$. This starts the induction.

Now assume that the result whenever the degree of g is less than n, and let $\partial g = n > 0$. If $\partial f > \partial g$, then we may as before take $q = 0$, $r = g$. Otherwise

$$f = a_m t^m + \cdots + a_0 \qquad g = b_n t^n + \cdots + b_0$$

where $a_m \neq 0 \neq b_n$ and $m \leq n$. Let

$$g_1 = b_n a_m^{-1} t^{n-m} f - g$$

Since the terms of highest degree cancel (which is the object of the exercise) we have $\partial g_1 < \partial g$. By induction there are polynomials q_1 and r_1 over K such that $g_1 = fq_1 + r_1$ and $\partial r_1 < \partial f$. Let

$$q = b_n a_m^{-1} t^{n-m} - q_1 \qquad r = -r_1$$

Then

$$fq + r = b_n a_m^{-1} t^{n-m} f - q_1 f - r_1 = g + g_1 - g_1 = g$$

so $g = fq + r$; clearly $\partial r < \partial f$ as required.

Finally we prove uniqueness. Suppose that

$$g = fq_1 + r_1 = fq_2 + r_2 \quad \text{where } \partial r_1, \partial r_2 < \partial f$$

Then $f(q_1 - q_2) = r_2 - r_1$. By Proposition 2.2, the polynomial on the left has higher degree than that on the right, unless both are zero. Since $f \neq 0$ we must have $q_1 = q_2$ and $r_1 = r_2$. Thus q and r are unique. □

With the above notation, q is called the *quotient* and r is called the *remainder* on dividing g by f. The inductive process we employed to find q and r is called the *Division Algorithm*.

Example 3.2. Divide $g(t) = t^4 - 7t^3 + 5t^2 + 4$ by $f = t^2 + 3$ and find the quotient and remainder.

Observe that

$$t^2(t^2 + 3) = t^4 + 3t^2$$

has the same leading coefficient as g. Then

$$g - t^2(t^2 + 3) = -7t^3 + 2t^2 + 4$$

which has the same leading coefficient as

$$-7t(t^2+3) = -7t^3 - 21t$$

Therefore

$$g - t^2(t^2+3) + 7t(t^2+3) = 2t^2 + 21t + 4$$

which has the same leading coefficient as

$$2(t^2+3) = 2t^2 + 6$$

Therefore

$$g - t^2(t^2+3) + 7t(t^2+3) - 2(t^2+3) = 21t - 2$$

So

$$g = (t^2+3)(t^2-7t+2) + (21t-2)$$

and the quotient $q(t) = t^2 - 7t + 2$, while the remainder $r(t) = 21t - 2$.

The next step is to introduce notions of divisibility for polynomials, and in particular the idea of 'highest common factor' which is crucial to the arithmetic of polynomials.

Definition 3.3. Let f and g be polynomials over K. We say that f *divides* g (or f is a *factor* of g, or g is a *multiple* of f) if there exists some polynomial h over K such that $g = fh$. The notation $f|g$ will mean that f divides g, while $f\nmid g$ will mean that f does not divide g.

Definition 3.4. A polynomial d over K is a *highest common factor* (hcf) of polynomials f and g over K if $d|f$ and $d|g$ and further, whenever $e|f$ and $e|g$, we have $e|d$.

Note that we have said *a* highest common factor rather than *the* highest common factor. This is because hcf's need not be unique. However, the next lemma shows that they are unique apart from constant factors.

Lemma 3.5. *If d is an hcf of the polynomials f and g over K, and if $0 \neq k \in K$, then kd is also an hcf for f and g.*

If d and e are two hcf's for f and g, then there exists a non-zero element $k \in K$ such that $e = kd$.

Proof. Clearly $kd|f$ and $kd|g$. If $e|f$ and $e|g$ then $e|d$ so that $e|kd$. Hence kd is an hcf.

If d and e are hcf's then by definition $e|d$ and $d|e$. Thus $e = kd$ for some polynomial k. Since $e|d$ the degree of e is less than or equal to the degree of d, so k must have degree ≤ 0. Therefore k is a constant, and so belongs to K. Since $0 \neq e = kd$, we must have $k \neq 0$. □

We shall prove that any two non-zero polynomials have an hcf by providing a method to calculate one. This method is a generalisation of the technique used by Euclid (*Elements* Book 7 Proposition 2) around 600 BC for calculating hcf's of integers, and is accordingly known as the *Euclidean Algorithm*.

Algorithm 3.6 (Euclidean Algorithm). ***Ingredients*** Two polynomials f and g over K, both non-zero.

Recipe For notational convenience let $f = r_{-1}, g = r_0$. Use the Division Algorithm to find successively polynomials q_j and r_i such that

$$\begin{array}{ll} r_{-1} = q_1 r_0 + r_1 & \partial r_1 < \partial r_0 \\ r_0 = q_2 r_1 + r_2 & \partial r_2 < \partial r_1 \\ r_1 = q_3 r_2 + r_3 & \partial r_3 < \partial r_2 \\ \dots & \\ r_i = q_{i+2} r_{i+1} + r_{i+2} & \partial r_{i+2} < \partial r_{i+1} \\ \dots & \end{array} \tag{3.1}$$

Since the degrees of the r_i decrease, we must eventually reach a point where the process stops; this can happen only if some $r_{s+2} = 0$. The last equation in the list then reads

$$r_s = q_{s+2} r_{s+1} \tag{3.2}$$

and it provides the answer we seek:

Theorem 3.7. *With the above notation, r_{s+1} is an hcf for f and g.*

Proof. First we show that r_{s+1} divides both f and g. We use descending induction to show that $r_{s+1}|r_i$ for all i. Clearly $r_{s+1}|r_{s+1}$. Equation (3.2) shows that $r_{s+1}|r_s$. Equation (3.1) implies that if $r_{s+1}|r_{i+2}$ and $r_{s+1}|r_{i+1}$ then $r_{s+1}|r_i$. Hence $r_{s+1}|r_i$ for all i; in particular $r_{s+1}|r_0 = g$ and $r_{s+1}|r_{-1} = f$.

Now suppose that $e|f$ and $e|g$. By (3.1) and induction, $e|r_i$ for all i. In particular, $e|r_{s+1}$. Therefore r_{s+1} is an hcf for f and g, as claimed. □

Example 3.8. Let $f = t^4 + 2t^3 + 2t^2 + 2t + 1$, $g = t^2 - 1$ over $\mathbb{Q}$. We compute an hcf as follows:

$$t^4 + 2t^3 + 2t^2 + 2t + 1 = (t^2 + 2t + 3)(t^2 - 1) + 4t + 4$$

$$t^2 - 1 = (4t + 4)(\frac{1}{4}t - \frac{1}{4})$$

Hence $4t + 4$ is an hcf. So is any rational multiple of it, in particular, $t + 1$.

We end this section by deducing from the Euclidean Algorithm an important property of the hcf of two polynomials.

Theorem 3.9. *Let f and g be non-zero polynomials over K, and let d be an hcf for f and g. Then there exist polynomials a and b over K such that*

$$d = af + bg$$

Proof. Since hcf's are unique up to constant factors we may assume that $d = r_{s+1}$ where equations (3.1) and (3.2) hold. We claim as induction hypothesis that there exist polynomials a_i and b_i such that

$$d = a_i r_i + b_i r_{i+1}$$

This is clearly true when $i = s+1$, for we may then take $a_i = 1$, $b_i = 0$. By (3.1)

$$r_{i+1} = r_{i-1} - q_{i+1}r_i$$

Hence by induction

$$d = a_i r_i + b_i(r_{i-1} - q_{i+1}r_i)$$

so that if we put

$$a_{i-1} = b_i \qquad b_{i-1} = a_i - b_i q_{i+1}$$

we have

$$d = a_{i-1}r_{i-1} + b_{i-1}r_i$$

Hence by descending induction

$$d = a_{-1}r_{-1} + b_{-1}r_0 = af + bg$$

where $a = a_{-1}$, $b = b_{-1}$. This completes the proof. □

The induction step above affords a practical method of calculating a and b in any particular case.

3.2 Irreducibility

Now we investigate the analogue, for polynomials, of prime numbers. The concept required is 'irreducibility'. In particular, we prove that every polynomial over a subring of $\mathbb{C}$ can be expressed as a product of irreducibles in an 'essentially' unique way.

An integer is prime if it cannot be expressed as a product of smaller integers. The analogue for polynomials is similar: we interpret 'smaller' as 'smaller degree'. So the following definition yields the polynomial analogue of a prime number.

Definition 3.10. A non-constant polynomial over a subring R of $\mathbb{C}$ is *reducible* if it is a product of two polynomials over R of smaller degree. Otherwise it is *irreducible*.

Examples 3.11. (1) All polynomials of degree 1 are irreducible, since they certainly cannot be expressed as a product of polynomials of smaller degree.
(2) The polynomial $t^2 - 2$ is irreducible over $\mathbb{Q}$. To show this we suppose, for a contradiction, that it is reducible. Then

$$t^2 - 2 = (at+b)(ct+d)$$

where $a, b, c, d, \in \mathbb{Q}$. Dividing out if necessary we may assume $a = c = 1$. Then $b + d = 0$ and $bd = -2$, so that $b^2 = 2$. But no rational number has its square equal to 2 (Exercise 1.2).

(3) However, t^2-2 is reducible over the larger subfield $\mathbb{R}$, for now

$$t^2-2=(t-\sqrt{2})(t+\sqrt{2})$$

This shows that an irreducible polynomial may become reducible over a larger subfield of $\mathbb{C}$.
(4) The polynomial $6t+3$ is irreducible in $\mathbb{Z}[t]$. Although it has factors

$$6t+3=3(2t+1)$$

the degree of $2t+1$ is the same as that of $6t+6$. So this factorisation does not count.
(5) The constant polynomial 6 is irreducible in $\mathbb{Z}[t]$. Again, $6=2\cdot 3$ does not count.

Any reducible polynomial can be written as the product of two polynomials of smaller degree. If either of these is reducible it too can be split up into factors of smaller degree ... and so on. This process must terminate since the degrees cannot decrease indefinitely. This is the idea behind the proof of:

Theorem 3.12. *Any non-zero polynomial over a subring R of $\mathbb{C}$ is a product of irreducible polynomials over R.*

Proof. Let g be any non-zero polynomial over R. We proceed by induction on the degree of g. If $\partial g=0$ or 1 then g is automatically irreducible. If $\partial g>1$, then either g is irreducible or $g=hk$ where $\partial h,\partial k<\partial g$. By induction, h and k are products of irreducible polynomials, whence g is such a product. The theorem follows by induction. □

Example 3.13. We can use Theorem 3.12 to prove irreducibility in some cases, especially for cubic polynomials over $\mathbb{Z}$. For instance, let $R=\mathbb{Z}$. The polynomial

$$f(t)=t^3-5t+1$$

is irreducible. If not, then it must have a linear factor $t-\alpha$ over $\mathbb{Z}$, and then $\alpha\in\mathbb{Z}$ and $f(\alpha)=0$. Moreover, there must exist $\beta,\gamma\in\mathbb{Z}$ such that

$$\begin{aligned} f(t) &= (t-\alpha)(t^2+\beta t+\gamma) \\ &= t^3+(\beta-\alpha)t^2+(\gamma-\alpha\beta)t-\alpha\gamma \end{aligned}$$

so in particular $\alpha\gamma=-1$. Therefore $\alpha=\pm1$. But $f(1)=-3\neq0$ and $f(-1)=5\neq0$. Therefore no such factor exists.

Irreducible polynomials are analogous to prime numbers. The importance of prime numbers in $\mathbb{Z}$ stems in part from the possibility of factorising every integer into primes, but even more so from the *uniqueness* (up to order) of the prime factors. Likewise the importance of irreducible polynomials depends upon a uniqueness theorem. Uniqueness of factorisation is not obvious, see Stewart and Tall (2002) Chapter 4. In certain cases it is possible to express every element as a product of irreducible elements, without this expression being in any way unique. We shall heed the warning and prove the uniqueness of factorisation for polynomials. To avoid technical issues like those in Examples 3.1(4,5), we restrict attention to polynomials over a sub*field* K of $\mathbb{C}$. It is possible to prove more general theorems by introducing the idea of a 'unique factorisation domain', see Fraleigh (1989) Chapter 6.

For convenience we make the following:

Definition 3.14. If f and g are polynomials over a subfield K of $\mathbb{C}$ with hcf equal to 1, we say that f and g are *coprime*, or *f is prime to g*. (The common phrase 'coprime to' is wrong. The prefix 'co' and the 'to' say the same thing, so it is redundant to use both.)

The key to unique factorisation is a statement analogous to an important property of primes in $\mathbb{Z}$, and is used in the same way:

Lemma 3.15. *Let K be a subfield of $\mathbb{C}$, f an irreducible polynomial over K, and g,h polynomials over K. If f divides gh, then either f divides g or f divides h.*

Proof. Suppose that $f \nmid g$. We claim that f and g are coprime. For if d is an hcf for f and g, then since f is irreducible and $d|f$, either $d = kf$ for some $k \in K$, or $d = k \in K$. In the first case $f|g$, contrary to hypothesis. In the second case, 1 is also an hcf for f and g, so they are coprime. By Theorem 3.9, there exist polynomials a and b over K such that

$$1 = af + bg$$

Then

$$h = haf + hbg$$

Now $f|haf$, and $f|hbg$ since $f|gh$. Hence $f|h$. This completes the proof. □

We may now prove the uniqueness theorem.

Theorem 3.16. *For any subfield K of $\mathbb{C}$, factorisation of polynomials over K into irreducible polynomials is unique up to constant factors and the order in which the factors are written.*

Proof. Suppose that $f = f_1 \cdots f_r = g_1 \cdots g_s$ where f is a polynomial over K and $f_1, \ldots, f_r, g_1, \ldots, g_s$ are irreducible polynomials over K. If all the f_i are constant then $f \in K$, so all the g_j are constant. Otherwise we may assume that no f_i is constant, by dividing out all of the constant terms. Then $f_1 | g_1 \ldots g_s$. By an obvious induction based on Lemma 3.15, $f_1 | g_j$ for some j. We can choose notation so that $j = 1$, and then $f_1 | g_1$. Since f_1 and g_1 are irreducible and f_1 is not a constant, we must have $f_1 = k_1 g_1$ for some constant k_1. Similarly $f_2 = k_2 g_2, \ldots, f_r = k_r g_r$ where $k_2, \ldots, k_r$ are constant. The remaining $g_l (l > r)$ must also be constant, or else the degree of the right-hand side would be too large. The theorem is proved. □

3.3 Gauss's Lemma

It is in general very difficult to decide—without using computer algebra, at any rate—whether a given polynomial is irreducible. As an example, think about

$$t^{16}+t^{15}+t^{14}+t^{13}+t^{12}+t^{11}+t^{10}+t^{9}+t^{8}+t^{7}+t^{6}+t^{5}+t^{4}+t^{3}+t^{2}+t+1 \quad (3.3)$$

This is not an idle example: we shall be considering precisely this polynomial in Chapter 20, in connection with the regular 17-gon, and its irreducibility (or not) will be crucial.

To test for irreducibility by trying all possible factors is usually futile. Indeed, at first sight there are infinitely many potential factors to try, although with suitable short cuts the possibilities can be reduced to a finite—usually unfeasibly large—number. In principle the resulting method can be applied to polynomials over $\mathbb{Q}$, for example: see van der Waerden (1953), Garling (1960). But the method is not really practicable.

Instead, we have to invent a few useful tricks. In the next two sections we describe two of them: Eisenstein's Criterion and reduction modulo a prime. Both tricks apply in the first instance to polynomials over $\mathbb{Z}$. However, we now prove that irreducibility over $\mathbb{Z}$ is equivalent to irreducibility over $\mathbb{Q}$. This extremely useful result was proved by Gauss, and we use it repeatedly.

Lemma 3.17 (Gauss's Lemma). *Let f be a polynomial over $\mathbb{Z}$ that is irreducible over $\mathbb{Z}$. Then f, considered as a polynomial over $\mathbb{Q}$, is also irreducible over $\mathbb{Q}$.*

Proof. The point of this lemma is that when we extend the subring of coefficients from $\mathbb{Z}$ to $\mathbb{Q}$, there are hosts of new polynomials which, perhaps, might be factors of f. We show that in fact they are not. For a contradiction, suppose that f is irreducible over $\mathbb{Z}$ but reducible over $\mathbb{Q}$, so that $f = gh$ where g and h are polynomials over $\mathbb{Q}$, of smaller degree, and seek a contradiction. Multiplying through by the product of the denominators of the coefficients of g and h, we can rewrite this equation in the form $nf = g'h'$, where $n \in \mathbb{Z}$ and g', h' are polynomials over $\mathbb{Z}$. We now show that we can cancel out the prime factors of n one by one, without going outside $\mathbb{Z}[t]$.

Suppose that p is a prime factor of n. We claim that if

$$g' = g_0 + g_1t + \cdots + g_rt^r \qquad h' = h_0 + h_1t + \cdots + h_st^s$$

then either p divides all the coefficients g_i, or else p divides all the coefficients h_j. If not, there must be smallest values i and j such that $p \nmid g_i$ and $p \nmid h_j$. However, p divides the coefficient of t^{i+j} in $g'h'$, which is

$$h_0g_{i+j} + h_1g_{i+j-1} + \cdots + h_jg_i + \cdots + h_{i+j}g_0$$

and by the choice of i and j, the prime p divides every term of this expression except perhaps h_jg_i. But p divides the whole expression, so $p|h_jg_i$. However, $p \nmid h_j$ and $p \nmid g_i$, a contradiction. This establishes the claim.

Without loss of generality, we may assume that p divides every coefficient g_i. Then $g' = pg''$ where g'' is a polynomial over $\mathbb{Z}$ of the same degree as g' (or g). Let $n = pn_1$. Then $pn_1 f = pg''h'$, so that $n_1 f = g''h'$. Proceeding in this way we can remove all the prime factors of n, arriving at an equation $f = \bar{g}\bar{h}$. Here $\bar{g}$ and $\bar{h}$ are polynomials over $\mathbb{Z}$, which are rational multiples of the original g and h, so $\partial\bar{g} = \partial g$ and $\partial\bar{h} = \partial h$. But this contradicts the irreducibility of f over $\mathbb{Z}$, so the lemma is proved. □

Corollary 3.18. *Let $f \in \mathbb{Z}[t]$ and suppose that over $\mathbb{Q}[t]$ there is a factorisation into irreducibles:*

$$f = g_1 \dots g_s$$

Then there exist $a_i \in \mathbb{Q}$ such that $a_i g_i \in \mathbb{Z}[t]$ and $a_1 \dots a_s = 1$. Furthermore,

$$f = (a_1 g_1) \dots (a_s g_s)$$

is a factorisation of f into irreducibles in $\mathbb{Z}[t]$.

Proof. Factorise f into irreducibles over $\mathbb{Z}[t]$, obtaining $f = h_1 \dots h_r$. By Gauss's Lemma, each h_j is irreducible over $\mathbb{Q}$. By uniqueness of factorisation in $\mathbb{Q}[t]$, we must have $r = s$ and $h_j = a_j g_j$ for $a_j \in \mathbb{Q}$. Clearly $a_1 \dots a_s = 1$. The Corollary is now proved. □

3.4 Eisenstein's Criterion

No, not 'Einstein'. Ferdinand Gotthold Eisenstein was a student of Gauss, and greatly impressed his tutor. We can apply the tutor's lemma to prove the student's criterion for irreducibility:

Theorem 3.19 (Eisenstein's Criterion). *Let*

$$f(t) = a_0 + a_1 t + \cdots + a_n t^n$$

be a polynomial over $\mathbb{Z}$. Suppose that there is a prime q such that

(1) $q \nmid a_n$

(2) $q | a_i \ (i = 0, \dots, n-1)$

(3) $q^2 \nmid a_0$

Then f is irreducible over $\mathbb{Q}$.

Proof. By Gauss's Lemma it is sufficient to show that f is irreducible over $\mathbb{Z}$. Suppose for a contradiction that $f = gh$, where

$$g = b_0 + b_1 t + \cdots + b_r t^r \qquad h = c_0 + c_1 t + \cdots + c_s t^s$$

are polynomials of smaller degree over $\mathbb{Z}$. Then $r \geq 1, s \geq 1$, and $r+s=n$. Now $b_0c_0 = a_0$ so by (2) $q|b_0$ or $q|c_0$. By (3) q cannot divide both b_0 and c_0, so without loss of generality we can assume $q|b_0$, $q \nmid c_0$. If all b_j are divisible by q, then a_n is divisible by q, contrary to (1). Let b_j be the first coefficient of g not divisible by q. Then

$$a_j = b_jc_0 + \cdots + b_0c_j$$

where $j < n$. This implies that q divides c_0, since q divides $a_j, b_0, \ldots, b_{j-1}$, but not b_j. This is a contradiction. Hence f is irreducible. □

Example 3.20. Consider

$$f(t) = \tfrac{2}{9}t^5 + \tfrac{5}{3}t^4 + t^3 + \tfrac{1}{3} \text{ over } \mathbb{Q}$$

This is irreducible over $\mathbb{Q}$ if and only if

$$9f(t) = 2t^5 + 15t^4 + 9t^3 + 3$$

is irreducible over $\mathbb{Q}$. Eisenstein's criterion now applies with $q = 3$, showing that f is irreducible.

We now turn to the polynomial (3.3). This provides an instructive example that leads to a useful general result. In preparation, we prove a standard number-theoretic property of binomial coefficients:

Lemma 3.21. *If p is prime, the binomial coefficient*

$$\binom{p}{r}$$

is divisible by p if $1 \leq r \leq p-1$.

Proof. The binomial coefficient is an integer, and

$$\binom{p}{r} = \frac{p!}{r!(p-r)!}$$

The factor p in the numerator cannot cancel with any factor in the denominator unless $r = 0$ or $r = p$. □

We then have:

Lemma 3.22. *If p is a prime then the polynomial*

$$f(t) = 1 + t + \cdots + t^{p-1}$$

is irreducible over $\mathbb{Q}$.

Proof. Note that $f(t) = (t^p - 1)/(t - 1)$. Put $t = 1 + u$ where u is a new indeterminate. Then $f(t)$ is irreducible over $\mathbb{Q}$ if and only if $f(1+u)$ is irreducible. But

$$\begin{aligned} f(1+u) &= \frac{(1+u)^p - 1}{u} \\ &= u^{p-1} + ph(u) \end{aligned}$$

where h is a polynomial in u over $\mathbb{Z}$ with constant term 1, by Lemma 3.21. By Eisenstein's Criterion, Theorem 3.19, $f(1+u)$ is irreducible over $\mathbb{Q}$. Hence $f(t)$ is irreducible over $\mathbb{Q}$. □

Setting $p = 17$ shows that the polynomial (3.3) is irreducible over $\mathbb{Q}$.

3.5 Reduction Modulo p

A second trick to prove irreducibility of polynomials in $\mathbb{Z}[t]$ involves 'reducing' the polynomial modulo a prime integer p.

Recall that if $n \in \mathbb{Z}$, two integers a, b are *congruent modulo n*, written

$$a \equiv b \pmod{n}$$

if $a - b$ is divisible by n. The number n is the *modulus*, and 'modulo' is Latin for 'to the modulus'. Congruence modulo n is an equivalence relation, and the set of equivalence classes is denoted by $\mathbb{Z}_n$. Arithmetic in $\mathbb{Z}_n$ is just like arithmetic in $\mathbb{Z}$, except that $n \equiv 0$.

The test for irreducibility that we now wish to discuss is most easily explained by an example. The idea is this. There is a natural map $\mathbb{Z} \to \mathbb{Z}_n$ in which each $m \in \mathbb{Z}$ maps to its congruence class modulo n. The natural map extends in an obvious way to a map $\mathbb{Z}[t] \to \mathbb{Z}_n[t]$. Now a reducible polynomial over $\mathbb{Z}$ is a product gh of polynomials of lower degree, and this factorisation is preserved by the map. Provided n does not divide the highest coefficient of the given polynomial, the image is reducible over $\mathbb{Z}_n$. So if the image of a polynomial is irreducible over $\mathbb{Z}_n$, then the original polynomial must be irreducible over $\mathbb{Z}$. (The corresponding statement for reducible polynomials is in general false: consider $t^2 - 2 \in \mathbb{Z}[t]$ when $p = 2$.) Since $\mathbb{Z}_n$ is finite, there are only finitely many possibilities to check when deciding irreducibility.

In practice, the trick is to choose the right value for n.

Example 3.23. Consider

$$f(t) = t^4 + 15t^3 + 7 \text{ over } \mathbb{Z}$$

Over $\mathbb{Z}_5$ this becomes $t^4 + 2$. If this is reducible over $\mathbb{Z}_5$, then either it has a factor of degree 1, or it is a product of two factors of degree 2. The first possibility gives rise to an element $x \in \mathbb{Z}_5$ such that $x^4 + 2 = 0$. No such element exists (there are only five

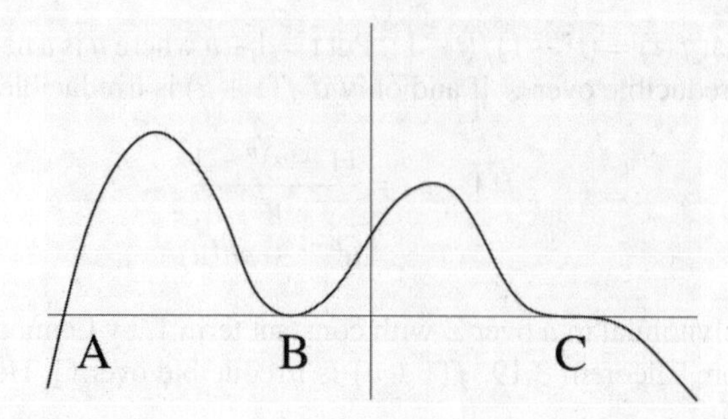

FIGURE 9: Multiple zeros of a (real) polynomial. The multiplicity is 1 at (A), 2 at (B), and 3 at (C).

elements to check) so this case is ruled out. In the remaining case we have, without loss of generality,

$$t^4+2=(t^2+at+b)(t^2+ct+d)$$

Therefore $a+c=0, ac+b+d=0, ad+bc=0, bd=2$. Combining $ad+bc=0$ with $a+c=0$ we get $a(b-d)=0$. So either $a=0$ or $b=d$.

If $a=0$ then $c=0$, so $b+d=0, bd=2$. That is, $b^2=-2=3$ in $\mathbb{Z}_5$. But this is not possible.

If $b=d$ then $b^2=2$, also impossible in $\mathbb{Z}_5$.

Hence t^4+2 is irreducible over $\mathbb{Z}_5$, and therefore the original $f(t)$ is irreducible over $\mathbb{Z}$, hence over $\mathbb{Q}$.

Notice that if instead we try to work in $\mathbb{Z}_3$, then $f(t)$ becomes t^4+1, which equals $(t^2+t-1)(t^2-t-1)$ and so is reducible. Thus working (mod 3) fails to prove irreducibility.

3.6 Zeros of Polynomials

We have already studied the zeros of a polynomial over $\mathbb{C}$. It will be useful to employ similar terminology for polynomials over a subring R of $\mathbb{C}$, because then we can keep track of where the zeros lie. We begin with a formal definition.

Definition 3.24. Let R be a subring of $\mathbb{C}$, and let f be a polynomial over R. An element $\alpha \in R$ such that $f(\alpha)=0$ is a *zero of f in R*.

To illustrate some basic phenomena associated with zeros, we consider polynomials over the real numbers. In this case, we can draw the graph $y=f(x)$ (in standard terminology, with $x \in \mathbb{R}$ in place of t). The graph might, for example, resemble Figure 9.

The zeros of f are the values of x at which the curve crosses the x-axis. Consider the three zeros marked A, B, C in the diagram. At A the curve cuts straight through the axis; at B it 'bounces' off it; at C it 'slides' through horizontally. These phenomena are generally distinguished by saying that B and C are 'multiple zeros' of $f(t)$. The single zero B must be thought of as two equal zeros (or more) and C as three (or more).

But if they are equal, how can there be two of them? The answer is the concept of 'multiplicity' of a zero, introduced in Section 2.3. We now reformulate this concept *without* using the Fundamental Theorem of Algebra, which in this context is the proverbial nut-cracking sledgehammer. The key is to look at linear factors of f.

Lemma 3.25. *Let f be a polynomial over the subfield K of $\mathbb{C}$. An element $\alpha \in K$ is a zero of f if and only if $(t-\alpha)|f(t)$ in $K[t]$.*

Proof. We know that $(t-\alpha)|f(t)$ in $\mathbb{C}[t]$ by Theorem 2.5, but we want slightly more. If $(t-\alpha)|f(t)$ in $K[t]$, then $f(t)=(t-\alpha)g(t)$ for some polynomial g over K, so that $f(\alpha)=(\alpha-\alpha)g(\alpha)=0$.

Conversely, suppose $f(\alpha)=0$. By the Division Algorithm, there exist polynomials $q,r \in K[t]$ such that

$$f(t)=(t-\alpha)q(t)+r(t)$$

where $\partial r < 1$. Thus $r(t)=r \in K$. Substituting α for t,

$$0=f(\alpha)=(\alpha-\alpha)q(\alpha)+r$$

so $r=0$. Hence $(t-\alpha)|f(t) \in K[t]$ as required. □

We can now say what we mean by a multiple zero, without appealing to the Fundamental Theorem of Algebra.

Definition 3.26. Let f be a polynomial over the subfield K of $\mathbb{C}$. An element $\alpha \in K$ is a *simple zero* of f if $(t-\alpha)|f(t)$ but $(t-\alpha)^2 \nmid f(t)$. The element α is a zero of f of *multiplicity* m if $(t-\alpha)^m|f(t)$ but $(t-\alpha)^{m+1} \nmid f(t)$. Zeros of multiplicity greater than 1 are *repeated* or *multiple zeros*.

For example, t^3-3t+2 over $\mathbb{Q}$ has zeros at $\alpha=1,-2$. It factorises as $(t-1)^2(t+2)$. Hence -2 is a simple zero, while 1 is a zero of multiplicity 2.

When $K=\mathbb{R}$ and we draw a graph, as in Figure 9, points like A are the simple zeros; points like B are zeros of even multiplicity; and points like C are zeros of odd multiplicity >1. For subfields of $\mathbb{C}$ other than $\mathbb{R}$ (except perhaps $\mathbb{Q}$, or other subfields of $\mathbb{R}$) a graph has no evident meaning, but the simple geometric picture for $\mathbb{R}$ is often helpful.

Lemma 3.27. *Let f be a non-zero polynomial over the subfield K of $\mathbb{C}$, and let its distinct zeros be $\alpha_1,\dots,\alpha_r$ with multiplicities $m_1,\dots,m_r$ respectively. Then*

$$f(t)=(t-\alpha_1)^{m_1}\dots(t-\alpha_r)^{m_r}g(t) \tag{3.4}$$

where g has no zeros in K.

Conversely, if (3.4) *holds and g has no zeros in K, then the zeros of f in K are $\alpha_1,\dots,\alpha_r$, with multiplicities $m_1,\dots,m_r$ respectively.*

Proof. For any $\alpha \in K$ the polynomial $t - \alpha$ is irreducible. Hence for distinct $\alpha, \beta \in K$ the polynomials $t - \alpha$ and $t - \beta$ are coprime in $K[t]$. By uniqueness of factorisation (Theorem 3.16) equation (3.4) must hold. Moreover, g cannot have any zeros in K, or else f would have extra zeros or zeros of larger multiplicity.

The converse follows easily from uniqueness of factorisation, Theorem 3.12 and Theorem 3.16. □

From this lemma we deduce a famous theorem:

Theorem 3.28. *The number of zeros of a nonzero polynomial over a subfield of* $\mathbb{C}$, *counted according to multiplicity, is less than or equal to its degree.*

Proof. In equation (3.4) we must have $m_1 + \cdots + m_r \leq \partial f$. □

EXERCISES

3.1 For the following pairs of polynomials f and g over $\mathbb{Q}$, find the quotient and remainder on dividing g by f.

(a) $g = t^7 - t^3 + 5,\ f = t^3 + 7$

(b) $g = t^2 + 1,\ f = t^2$

(c) $g = 4t^3 - 17t^2 + t - 3,\ f = 2t + 5$

(d) $g = t^4 - 1,\ f = t^2 + 1$

(e) $g = t^4 - 1,\ f = 3t^2 + 3t$

3.2 Find hcf's for these pairs of polynomials, and check that your results are common factors of f and g.

3.3 Express these hcf's in the form $af + bg$.

3.4 Decide the irreducibility or otherwise of the following polynomials:

(a) $t^4 + 1$ over $\mathbb{R}$.

(b) $t^4 + 1$ over $\mathbb{Q}$.

(c) $t^7 + 11t^3 - 33t + 22$ over $\mathbb{Q}$.

(d) $t^4 + t^3 + t^2 + t + 1$ over $\mathbb{Q}$.

(e) $t^3 - 7t^2 + 3t + 3$ over $\mathbb{Q}$.

3.5 Decide the irreducibility or otherwise of the following polynomials:

(a) $t^4 + t^3 + t^2 + t + 1$ over $\mathbb{Q}$. (*Hint:* Substitute $t + 1$ in place of t and appeal to Eisenstein's Criterion.)

(b) $t^5 + t^4 + t^3 + t^2 + t + 1$ over $\mathbb{Q}$.

(c) $t^6+t^5+t^4+t^3+t^2+t+1$ over $\mathbb{Q}$.

3.6 In each of the above cases, factorise the polynomial into irreducibles.

3.7 Say that a polynomial f over a subfield K of $\mathbb{C}$ is *prime* if whenever $f|gh$ either $f|g$ or $f|h$. Show that a polynomial $f \neq 0$ is prime if and only if it is irreducible.

3.8 Find the zeros of the following polynomials; first over $\mathbb{Q}$, then $\mathbb{R}$, then $\mathbb{C}$.

(a) t^3+1

(b) $t^3-6t^2+11t-6$

(c) t^5+t+1

(d) t^2+1

(e) $t^4+t^3+t^2+t+1$

(f) t^4-6t^2+11

3.9 Mark the following true or false. (Here 'polynomial' means 'polynomial over $\mathbb{C}$'.)

(a) Every polynomial of degree n has n distinct zeros.

(b) Every polynomial of degree n has at most n distinct zeros.

(c) Every polynomial of degree n has at least n distinct zeros.

(d) If f,g are non-zero polynomials and f divides g, then $\partial f < \partial g$.

(e) If f,g are non-zero polynomials and f divides g, then $\partial f \leq \partial g$.

(f) Every polynomial of degree 1 is irreducible.

(g) Every irreducible polynomial has prime degree.

(h) If a polynomial f has integer coefficients and is irreducible over $\mathbb{Z}$, then it is irreducible over $\mathbb{Q}$.

(i) If a polynomial f has integer coefficients and is irreducible over $\mathbb{Z}$, then it is irreducible over $\mathbb{R}$.

(j) If a polynomial f has integer coefficients and is irreducible over $\mathbb{R}$, then it is irreducible over $\mathbb{Z}$.

Chapter 4

Field Extensions

Galois's original theory was couched in terms of polynomials over the complex field. The modern approach is a consequence of the methods used, starting around 1890 and flourishing in the 1920s and 1930s, to generalise the theory to arbitrary fields. From this viewpoint the central object of study ceases to be a polynomial, and becomes instead a 'field extension' related to a polynomial. Every polynomial f over a field K defines another field L containing K (or at any rate a subfield isomorphic to K). There are conceptual advantages in setting up the theory from this point of view. In this chapter we define field extensions (always working inside $\mathbb{C}$) and explain the link with polynomials.

4.1 Field Extensions

Suppose that we wish to study the quartic polynomial

$$f(t) = t^4 - 4t^2 - 5$$

over $\mathbb{Q}$. Its irreducible factorisation over $\mathbb{Q}$ is

$$f(t) = (t^2 + 1)(t^2 - 5)$$

so the zeros of f in $\mathbb{C}$ are $\pm i$ and $\pm\sqrt{5}$. There is a natural subfield L of $\mathbb{C}$ associated with these zeros; in fact, it is the unique smallest subfield that contains them. We claim that L consists of all complex numbers of the form

$$p + qi + r\sqrt{5} + si\sqrt{5} \qquad (p, q, r, s \in \mathbb{Q})$$

Clearly L must contain every such element, and it is not hard to see that sums and products of such elements have the same form. It is harder to see that inverses of (non-zero) such elements also have the same form, but it is true: we postpone the proof to Example 4.8. Thus the study of a polynomial over $\mathbb{Q}$ leads us to consider a subfield L of $\mathbb{C}$ that contains $\mathbb{Q}$. In the same way the study of a polynomial over an arbitrary subfield K of $\mathbb{C}$ will lead to a subfield L of $\mathbb{C}$ that contains K. We shall call L an 'extension' of K. For technical reasons this definition is too restrictive; we wish to allow cases where L contains a subfield isomorphic to K, but not necessarily equal to it.

Definition 4.1. A *field extension* is a monomorphism $\iota : K \to L$, where K and L are subfields of $\mathbb{C}$. We say that K is the *small* field and L is the *large* field.

Notice that with a strict set-theoretic definition of function, the map ι determines both K and L. See Definition 1.3 for the definition of 'monomorphism'. We often think of a field extension as being a pair of fields (K,L), when it is clear which monomorphism is intended.

Examples 4.2. 1. The inclusion maps $\iota_1 : \mathbb{Q} \to \mathbb{R}$, $\iota_2 : \mathbb{R} \to \mathbb{C}$, and $\iota_3 : \mathbb{Q} \to \mathbb{C}$ are all field extensions.

2. Let K be the set of all real numbers of the form $p + q\sqrt{2}$, where $p,q \in \mathbb{Q}$. Then K is a subfield of $\mathbb{C}$ by Example 1.7. The inclusion map $\iota : \mathbb{Q} \to K$ is a field extension.

If $\iota : K \to L$ is a field extension, then we can usually identify K with its image $\iota(K)$, so that ι can be thought of as an inclusion map and K can be thought of as a subfield of L. Under these circumstances we use the notation

$$L : K$$

for the extension, and say that L is an *extension of* K. In future we shall identify K and $\iota(K)$ whenever this is legitimate.

The next concept is one which pervades much of abstract algebra:

Definition 4.3. Let X be a subset of $\mathbb{C}$. Then the subfield of $\mathbb{C}$ *generated by* X is the intersection of all subfields of $\mathbb{C}$ that contain X.

It is easy to see that this definition is equivalent to either of the following:

1. The (unique) smallest subfield of $\mathbb{C}$ that contains X.

2. The set of all elements of $\mathbb{C}$ that can be obtained from elements of X by a finite sequence of field operations, provided $X \neq \{0\}$ or $\emptyset$.

Proposition 4.4. *Every subfield of* $\mathbb{C}$ *contains* $\mathbb{Q}$.

Proof. Let $K \subseteq \mathbb{C}$ be a subfield. Then $0, 1 \in K$ by definition, so inductively we find that $1 + \ldots + 1 = n$ lies in K for every integer $n > 0$. Now K is closed under additive inverses, so $-n$ also lies in K, proving that $\mathbb{Z} \subseteq K$. Finally, if $p,q \in \mathbb{Z}$ and $q \neq 0$, closure under products and multiplicative inverses shows that $pq^{-1} \in K$. Therefore $\mathbb{Q} \subseteq K$ as claimed. □

Corollary 4.5. *Let* X *be a subset of* $\mathbb{C}$. *Then the subfield of* $\mathbb{C}$ *generated by* X *contains* $\mathbb{Q}$.

Because of Corollary 4.5, we use the notation

$$\mathbb{Q}(X)$$

for the subfield of $\mathbb{C}$ generated by X.

Example 4.6. We find the subfield K of $\mathbb{C}$ generated by $X = \{1, i\}$. By Proposition 4.4, K must contain $\mathbb{Q}$. Since K is closed under the arithmetical operations, it must contain all complex numbers of the form $p + qi$, where $p, q \in \mathbb{Q}$. Let M be the set of all such numbers. We claim that M is a subfield of $\mathbb{C}$. Clearly M is closed under sums, differences, and products. Further

$$(p+qi)^{-1} = \frac{p}{p^2+q^2} - \frac{q}{p^2+q^2}i$$

so that every non-zero element of M has a multiplicative inverse in M. Hence M is a subfield, and contains X. Since K is the smallest subfield containing X, we have $K \subseteq M$. But $M \subseteq K$ by definition. Hence $K = M$, and we have found a description of the subfield generated by X.

In the case of a field extension $L : K$ we are mainly interested in subfields lying between K and L. This means that we can restrict attention to subsets X that contain K; equivalently, to sets of the form $K \cup Y$ where $Y \subseteq L$.

Definition 4.7. If $L : K$ is a field extension and Y is a subset of L, then the subfield of $\mathbb{C}$ generated by $K \cup Y$ is written $K(Y)$ and is said to be obtained from K by *adjoining* Y.

Clearly $K(Y) \subseteq L$ since L is a subfield of $\mathbb{C}$. Notice that $K(Y)$ is in general considerably larger than $K \cup Y$.

This notation is open to all sorts of useful abuses. If Y has a single element y we write $K(y)$ instead of $K(\{y\})$, and in the same spirit $K(y_1, \ldots, y_n)$ will replace $K(\{y_1, \ldots, y_n\})$.

Example 4.8. Let $K = \mathbb{Q}$ and let $Y = \{i, \sqrt{5}\}$. Then $K(Y)$ must contain K and Y. It also contains the product $i\sqrt{5}$. Since $K \supseteq \mathbb{Q}$, the subfield $K(Y)$ must contain all elements

$$\alpha = p + qi + r\sqrt{5} + si\sqrt{5} \qquad (p, q, r, s \in \mathbb{Q}).$$

Let $L \subseteq \mathbb{C}$ be the set of all such α. If we prove that L is a subfield of $\mathbb{C}$, then it follows that $K(Y) = L$. Moreover, it is easy to check that L is a subring of $\mathbb{C}$, hence L is a subfield of $\mathbb{C}$ if and only if for $\alpha \neq 0$ we can find an inverse $\alpha^{-1} \in L$. If fact, we shall prove that if $(p, q, r, s) \neq (0, 0, 0, 0)$ then $\alpha \neq 0$, and then

$$(p + qi + r\sqrt{5} + si\sqrt{5})^{-1} \in L$$

First, suppose that $p + qi + r\sqrt{5} + si\sqrt{5} = 0$. Then

$$p + r\sqrt{5} = -i(q + s\sqrt{5})$$

Now both $p + r\sqrt{5}$ and $-(q + s\sqrt{5})$ are real, but i is imaginary. Therefore $p + r\sqrt{5} = 0$ and $q + s\sqrt{5} = 0$. If $r \neq 0$ then $\sqrt{5} = -p/r \in \mathbb{Q}$, but $\sqrt{5}$ is irrational. Therefore $r = 0$, whence $p = 0$. Similarly, $q = s = 0$.

Now we prove the existence of α^{-1} in two stages. Let M be the subset of L containing all $p + qi$ $(p, q \in \mathbb{Q})$. Then we can write

$$\alpha = x + y\sqrt{5}$$

where $x = p + iq$ and $y = r + is \in M$. Let

$$\beta = p + qi - r\sqrt{5} - si\sqrt{5} = x - y\sqrt{5} \in L$$

Then

$$\alpha\beta = (x + y\sqrt{5})(x - y\sqrt{5}) = x^2 - 5y^2 = z$$

say, where $z \in M$. Since $\alpha \neq 0$ and $\beta \neq 0$ we have $z \neq 0$, so $\alpha^{-1} = \beta z^{-1}$. Now write $z = u + vi$ $(u, v \in \mathbb{Q})$ and consider $w = u - vi$. Since $zw = u^2 + v^2 \in \mathbb{Q}$ we have

$$z^{-1} = (u^2 + v^2)^{-1} w \in M$$

so $\alpha^{-1} = \beta z^{-1} \in L$.

Alternatively, we can obtain an explicit formula by working out the expression

$$(p + qi + r\sqrt{5} + si\sqrt{5})(p - qi + r\sqrt{5} - si\sqrt{5}) \\ \times (p + qi - r\sqrt{5} - si\sqrt{5})(p - qi - r\sqrt{5} + si\sqrt{5})$$

and showing that it belongs to $\mathbb{Q}$, and then dividing out by

$$(p + qi + r\sqrt{5} + si\sqrt{5})$$

See Exercise 4.6.

Examples 4.9. (1) The subfield $\mathbb{R}(i)$ of $\mathbb{C}$ must contain all elements $x + iy$ where $x, y \in \mathbb{R}$. But those elements comprise the whole of $\mathbb{C}$. Therefore $\mathbb{C} = \mathbb{R}(i)$.
(2) The subfield P of $\mathbb{R}$ consisting of all numbers $p + q\sqrt{2}$ where $p, q \in \mathbb{Q}$ is easily seen to equal $\mathbb{Q}(\sqrt{2})$.
(3) It is not always true that a subfield of the form $K(\alpha)$ consists of all elements of the form $j + k\alpha$ where $j, k \in K$. It certainly contains all such elements, but they need not form a subfield.

For example, in $\mathbb{R} : \mathbb{Q}$ let α be the real cube root of 2, and consider $\mathbb{Q}(\alpha)$. As well as α, the subfield $\mathbb{Q}(\alpha)$ must contain α^2. We show that $\alpha^2 \neq j + k\alpha$ for $j, k \in \mathbb{Q}$. For a contradiction, suppose that $\alpha^2 = j + k\alpha$. Then $2 = \alpha^3 = j\alpha + k\alpha^2 = jk + (j + k^2)\alpha$. Therefore $(j + k^2)\alpha = 2 - jk$. Since α is irrational, $(j + k^2) = 0 = 2 - jk$. Eliminating j, we find that $k^3 = 2$, contrary to $k \in \mathbb{Q}$.

In fact, $\mathbb{Q}(\alpha)$ is precisely the set of all elements of $\mathbb{R}$ of the form $p + q\alpha + r\alpha^2$, where $p, q, r \in \mathbb{Q}$. To show this, we prove that the set of such elements is a subfield. The only (minor) difficulty is finding a multiplicative inverse: see Exercise 4.7.

4.2 Rational Expressions

We can perform the operations of addition, subtraction, and multiplication in the polynomial ring $\mathbb{C}[t]$, but (usually) not division. For example, $\mathbb{C}[t]$ does not contain an inverse t^{-1} for t, see Exercise 4.8.

However, we can enlarge $\mathbb{C}[t]$ to provide inverses in a natural way. We have seen that we can think of polynomials $f(t) \in \mathbb{C}[t]$ as functions from $\mathbb{C}$ to itself. Similarly, we can think of fractions $p(t)/q(t) \in \mathbb{C}(t)$ as functions. These are called *rational functions* of the complex variable t, and their formal statements in terms of polynomials are *rational expressions* in the indeterminate t. However, there is now a technical difficulty. The domain of such a function is not the whole of $\mathbb{C}$: all of the zeros of $q(t)$ have to be removed, or else we are trying to divide by zero. Complex analysts often work in the Riemann sphere $\mathbb{C} \cup \{\infty\}$, and cheerfully let $1/\infty = 0$, but care must be exercised if this is done; the civilised way to proceed is to remove all the potential troublemakers. So we take the domain of $p(t)/q(t)$ to be

$$\{z \in \mathbb{C} : q(z) \neq 0\}$$

As we have seen, any complex polynomial q has only finitely many zeros, so the domain here is 'almost all' of $\mathbb{C}$. We have to be careful, but we shouldn't get into much trouble provided we are.

In the same manner we can also construct the set

$$\mathbb{C}(t_1, \ldots, t_n)$$

of all rational functions in n variables (rational expressions in n indeterminates). One use of such functions is to specify the subfield generated by a given set X. It is straightforward to prove that $\mathbb{Q}(X)$ consists of all rational expressions

$$\frac{p(\alpha_1, \ldots, \alpha_n)}{q(\beta_1, \ldots, \beta_n)}$$

for all n, where $p, q \in \mathbb{Q}[t_1, \ldots, t_n]$, the α_j and β_j belong to X, and $q(\beta_1, \ldots, \beta_n) \neq 0$. See Exercise 4.9.

It is also possible to define such expressions without using functions. See 'field of fractions' in Chapter 16, immediately after Corollary 16.18. This approach is necessary in the more abstract development of the subject.

4.3 Simple Extensions

The basic building-blocks for field extensions are those obtained by adjoining one element:

Definition 4.10. A *simple extension* is a field extension $L : K$ such that $L = K(\alpha)$ for some $\alpha \in L$.

Examples 4.11. (1) As the notation shows, the extensions in Examples 4.9 are all simple.

(2) *Beware:* An extension may be simple without appearing to be. Consider $L =$

$\mathbb{Q}(i,-i,$
$\sqrt{5},-\sqrt{5})$. As written, it appears to require the adjunction of four new elements. Clearly just two, i and $\sqrt{5}$, suffice. But we claim that in fact only one element is needed, because $L = L'$ where $L' = \mathbb{Q}(i+\sqrt{5})$, which is obviously simple. To prove this, it is enough to show that $i \in L'$ and $\sqrt{5} \in L'$, because these imply that $L \subseteq L'$ and $L' \subseteq L$, so $L = L'$. Now L' contains

$$(i+\sqrt{5})^2 = -1+2i\sqrt{5}+5 = 4+2i\sqrt{5}$$

Thus it also contains

$$(i+\sqrt{5})(4+2i\sqrt{5}) = 14i-2\sqrt{5}$$

Therefore it contains

$$14i-2\sqrt{5}+2(i+\sqrt{5}) = 16i$$

so it contains i. But then it also contains $(i+\sqrt{5})-i = \sqrt{5}$. Therefore $L = L'$ as claimed, and the extension $\mathbb{Q}(i,-i,\sqrt{5},-\sqrt{5}) : \mathbb{Q}$ is in fact simple.
(3) On the other hand, $\mathbb{R} : \mathbb{Q}$ is not a simple extension (Exercise 4.5).

Our aim in the next chapter will be to classify all possible simple extensions. We end this chapter by formulating the concept of isomorphism of extensions. In Chapter 5 we will develop techniques for constructing all possible simple extensions up to isomorphism.

Definition 4.12. An *isomorphism* between two field extensions $\iota : K \to \hat{K}, j : L \to \hat{L}$ is a pair (λ, μ) of field isomorphisms $\lambda : K \to L, \mu : \hat{K} \to \hat{L}$, such that for all $k \in K$

$$j(\lambda(k)) = \mu(\iota(k))$$

Another, more pictorial, way of putting this is to say that the diagram

$$\begin{array}{ccc} K & \xrightarrow{\iota} & \hat{K} \\ \lambda \downarrow & & \downarrow \mu \\ L & \xrightarrow{j} & \hat{L} \end{array}$$

commutes; that is, the two paths from K to $\hat{L}$ compose to give the same map.

The reason for setting up the definition like this is that as well as the field structure being preserved by isomorphism, the embedding of the small field in the large one is also preserved.

Various identifications may be made. If we identify K and $\iota(K)$, and L and $j(L)$, then ι and j are inclusions, and the commutativity condition now becomes

$$\mu|_K = \lambda$$

where $\mu|_K$ denotes the restriction of μ to K. If we further identify K and L then λ becomes the identity, and so $\mu|_K$ is the identity. In what follows we shall attempt to use these 'identified' conditions wherever possible. But on a few occasions (notably Theorem 9.6) we shall need the full generality of the first definition.

EXERCISES

4.1 Prove that isomorphism of field extensions is an equivalence relation.

4.2 Find the subfields of $\mathbb{C}$ generated by:

(a) $\{0,1\}$
(b) $\{0\}$
(c) $\{0,1,i\}$
(d) $\{i,\sqrt{2}\}$
(e) $\{\sqrt{2},\sqrt{3}\}$
(f) $\mathbb{R}$
(g) $\mathbb{R}\cup\{i\}$

4.3 Describe the subfields of $\mathbb{C}$ of the form

(a) $\mathbb{Q}(\sqrt{2})$
(b) $\mathbb{Q}(i)$
(c) $\mathbb{Q}(\alpha)$ where α is the real cube root of 2
(d) $\mathbb{Q}(\sqrt{5},\sqrt{7})$
(e) $\mathbb{Q}(i\sqrt{11})$
(f) $\mathbb{Q}(\mathrm{e}^2+1)$
(g) $\mathbb{Q}(\sqrt[3]{\pi})$

4.4 This exercise illustrates a technique that we will tacitly assume in several subsequent exercises and examples.

Prove that $1,\sqrt{2},\sqrt{3},\sqrt{6}$ are linearly independent over $\mathbb{Q}$.

(*Hint:* Suppose that $p+q\sqrt{2}+r\sqrt{3}+s\sqrt{6}=0$ with $p,q,r,s\in\mathbb{Q}$. We may suppose that $r\neq 0$ or $s\neq 0$ (why?). If so, then we can write $\sqrt{3}$ in the form

$$\sqrt{3}=\frac{a+b\sqrt{2}}{c+d\sqrt{2}}=e+f\sqrt{2}$$

where $a,b,c,d,e,f\in\mathbb{Q}$. Square both sides and obtain a contradiction.)

4.5 Show that $\mathbb{R}$ is not a simple extension of $\mathbb{Q}$ as follows:

(a) $\mathbb{Q}$ is countable.

(b) Any simple extension of a countable field is countable.

(c) $\mathbb{R}$ is not countable.

4.6 Find a formula for the inverse of $p+qi+r\sqrt{5}+si\sqrt{5}$, where $p,q,r,s\in\mathbb{Q}$.

4.7 Find a formula for the inverse of $p+q\alpha+r\alpha^2$, where $p,q,r\in\mathbb{Q}$ and $\alpha=\sqrt[3]{2}$.

4.8 Prove that t has no multiplicative inverse in $\mathbb{C}[t]$.

4.9 Prove that $\mathbb{Q}(X)$ consists of all rational expressions

$$\frac{p(\alpha_1,\dots,\alpha_n)}{q(\beta_1,\dots,\beta_n)}$$

for all n, where $p,q\in\mathbb{Q}[t_1,\dots,t_n]$, the α_j and β_j belong to X, and $q(\beta_1,\dots,\beta_n)\neq 0$.

4.10 Mark the following true or false.

(a) If X is the empty set then $\mathbb{Q}(X)=\mathbb{Q}$.

(b) If X is a subset of $\mathbb{Q}$ then $\mathbb{Q}(X)=\mathbb{Q}$.

(c) If X contains an irrational number, then $\mathbb{Q}(X)\neq\mathbb{Q}$.

(d) $\mathbb{Q}(\sqrt{2})=\mathbb{Q}$.

(e) $\mathbb{Q}(\sqrt{2})=\mathbb{R}$.

(f) $\mathbb{R}(\sqrt{2})=\mathbb{R}$.

(g) Every subfield of $\mathbb{C}$ contains $\mathbb{Q}$.

(h) Every subfield of $\mathbb{C}$ contains $\mathbb{R}$.

(i) If $\alpha\neq\beta$ and both are irrational, then $\mathbb{Q}(\alpha,\beta)$ is not a simple extension of $\mathbb{Q}$.

Chapter 5

Simple Extensions

The basic building block of field theory is the simple field extension. Here *one* new element α is adjoined to a given subfield K of $\mathbb{C}$, along with all rational expressions in that element over K. Any finitely generated extension can be obtained by a sequence of simple extensions, so the structure of a simple extension provides vital information about all of the extensions that we shall encounter.

We first classify simple extensions into two very different kinds: transcendental and algebraic. If the new element α satisfies a polynomial equation over K, then the extension is algebraic; if not, it is transcendental. Up to isomorphism, K has exactly one simple transcendental extension. For most fields K there are many more possibilities for simple algebraic extensions; they are classified by the irreducible polynomials m over K.

The structure of simple algebraic extensions can be described in terms of the polynomial ring $K[t]$, with operations being performed 'modulo m'. In Chapter 16 we generalise this construction using the notion of an ideal.

5.1 Algebraic and Transcendental Extensions

Recall that a simple extension of a subfield K of $\mathbb{C}$ takes the form $K(\alpha)$ where in nontrivial cases $\alpha \notin K$. We classify the possible simple extensions for any K. There are two distinct types:

Definition 5.1. Let K be a subfield of $\mathbb{C}$ and let $\alpha \in \mathbb{C}$. Then α is *algebraic* over K if there exists a non-zero polynomial p over K such that $p(\alpha) = 0$. Otherwise, α is *transcendental* over K.

We shorten 'algebraic over $\mathbb{Q}$' to 'algebraic', and 'transcendental over $\mathbb{Q}$' to 'transcendental'.

Examples 5.2. (1) The number $\alpha = \sqrt{2}$ is algebraic, because $\alpha^2 - 2 = 0$.
(2) The number $\alpha = \sqrt[3]{2}$ is algebraic, because $\alpha^3 - 2 = 0$.
(3) The number $\pi = 3 \cdot 14159\ldots$ is transcendental. We postpone a proof to Chapter 24. In Chapter 7 we use the transcendence of π to prove the impossibility of 'squaring the circle'.
(4) The number $\alpha = \sqrt{\pi}$ is algebraic over $\mathbb{Q}(\pi)$, because $\alpha^2 - \pi = 0$.

(5) However, $\alpha = \sqrt{\pi}$ is transcendental over $\mathbb{Q}$. To see why, suppose that $p(\sqrt{\pi}) = 0$ where $0 \neq p(t) \in \mathbb{Q}[t]$. Separating out terms of odd and even degree, we can write this as $a(\pi) + b(\pi)\sqrt{\pi} = 0$, so $a(\pi) = -b(\pi)\sqrt{\pi}$ and $a^2(\pi) = \pi b^2(\pi)$. Thus $f(\pi) = 0$, where

$$f(t) = a^2(t) - tb^2(t) \in \mathbb{Q}[t]$$

Now $\partial(a^2)$ is even, and $\partial(tb^2)$ is odd, so the difference $f(t)$ is nonzero. But this implies that π is algebraic, a contradiction.

In the next few sections we classify all possible simple extensions and find ways to construct them. The transcendental case is very straightforward: if $K(t)$ is the set of rational functions of the indeterminate t over K, then $K(t) : K$ is the unique simple transcendental extension of K up to isomorphism. If $K(\alpha) : K$ is algebraic, the possibilities are richer, but tractable. We show that there is a unique monic irreducible polynomial m over K such that $m(\alpha) = 0$, and that m determines the extension uniquely up to isomorphism.

We begin by constructing a simple transcendental extension of any subfield.

Theorem 5.3. *The set of rational expressions $K(t)$ is a simple transcendental extension of the subfield K of $\mathbb{C}$.*

Proof. Clearly $K(t) : K$ is a simple extension, generated by t. If p is a polynomial over K such that $p(t) = 0$ then $p = 0$ by definition of $K(t)$, so the extension is transcendental. □

5.2 The Minimal Polynomial

The construction of simple algebraic extensions is a much more delicate issue. It is controlled by a polynomial associated with the generator α of $K(\alpha) : K$, called the 'minimal polynomial'. (An alternative name often encountered is 'minimum polynomial'.) To define it we first set up a technical definition.

Definition 5.4. A polynomial $f(t) = a_0 + a_1 t + \cdots + a_n t^n$ over a subfield K of $\mathbb{C}$ is *monic* if $a_n = 1$.

Clearly every polynomial is a constant multiple of some monic polynomial, and for a non-zero polynomial this monic polynomial is unique. Further, the product of two monic polynomials is again monic.

Now suppose that $K(\alpha) : K$ is a simple algebraic extension. There is a polynomial p over K such that $p(\alpha) = 0$. We may suppose that p is monic. Therefore there exists at least one monic polynomial *of smallest degree* that has α as a zero. We claim that p is unique. To see why, suppose that p, q are two such. then $p(\alpha) - q(\alpha) = 0$, so if $p \neq q$ then some constant multiple of $p - q$ is a monic polynomial with α as a zero, contrary to the definition. Hence there is a unique monic polynomial p of smallest degree such that $p(\alpha) = 0$. We give this a name:

Definition 5.5. Let $L : K$ be a field extension, and suppose that $\alpha \in L$ is algebraic over K. Then the *minimal polynomial* of α over K is the unique monic polynomial m over K of smallest degree such that $m(\alpha) = 0$.

For example, $i \in \mathbb{C}$ is algebraic over $\mathbb{R}$. If we let $m(t) = t^2 + 1$ then $m(i) = 0$. Clearly m is monic. The only monic polynomials over $\mathbb{R}$ of smaller degree are those of the form $t + r$, where $r \in \mathbb{R}$, or the constant polynomial 1. But i cannot be a zero of any of these, or else we would have $i \in \mathbb{R}$. Hence the minimal polynomial of i over $\mathbb{R}$ is $t^2 + 1$.

It is natural to ask which polynomials can be minimal. The next lemma provides information on this question.

Lemma 5.6. *If α is an algebraic element over the subfield K of $\mathbb{C}$, then the minimal polynomial of α over K is irreducible over K. It divides every polynomial of which α is a zero.*

Proof. Suppose that the minimal polynomial m of α over K is reducible, so that $m = fg$ where f and g are of smaller degree. We may assume f and g are monic. Since $m(\alpha) = 0$ we have $f(\alpha)g(\alpha) = 0$, so either $f(\alpha) = 0$ or $g(\alpha) = 0$. But this contradicts the definition of m. Hence m is irreducible over K.

Now suppose that p is a polynomial over K such that $p(\alpha) = 0$. By the Division Algorithm, there exist polynomials q and r over K such that $p = mq + r$ and $\partial r < \partial m$. Then $0 = p(\alpha) = 0 + r(\alpha)$. If $r \neq 0$ then a suitable constant multiple of r is monic, which contradicts the definition of m. Therefore $r = 0$, so m divides p. □

Conversely, if K is a subfield of $\mathbb{C}$, then it is easy to show that any irreducible polynomial over K can be the mimimum polynomial of an algebraic element over K:

Theorem 5.7. *If K is any subfield of $\mathbb{C}$ and m is any irreducible monic polynomial over K, then there exists $\alpha \in \mathbb{C}$, algebraic over K, such that α has minimal polynomial m over K.*

Proof. Let α be any zero of m in $\mathbb{C}$. Then $m(\alpha) = 0$, so the minimal polynomial f of α over K divides m. But m is irreducible over K and both f and m are monic; therefore $f = m$. □

5.3 Simple Algebraic Extensions

Next, we describe the structure of the field extension $K(\alpha) : K$ when α has minimal polynomial m over K. We proceed by analogy with a basic concept of number theory. Recall from Section 3.5 that for any positive integer n it is possible to perform arithmetic *modulo n*, and that integers a, b are *congruent modulo n*, written

$$a \equiv b \pmod{n}$$

if $a-b$ is divisible by n. In the same way, given a polynomial $m \in K[t]$, we can calculate with polynomials *modulo m*. We say that polynomials $a,b \in K[t]$ are *congruent modulo m*, written

$$a \equiv b \pmod{m}$$

if $a(t)-b(t)$ is divisible by $m(t)$ in $K[t]$.

Lemma 5.8. *Suppose that* $a_1 \equiv a_2 \pmod{m}$ *and* $b_1 \equiv b_2 \pmod{m}$. *Then* $a_1+b_1 \equiv a_2+b_2 \pmod{m}$, *and* $a_1b_1 \equiv a_2b_2 \pmod{m}$.

Proof. We know that $a_1-a_2 = am$ and $b_1-b_2 = bm$ for polynomials $a,b \in K[t]$. Now

$$(a_1+b_1)-(a_2+b_2) = (a_1-a_2)+(b_1-b_2) = (a-b)m$$

which proves the first statement. For the product, we need a slightly more elaborate argument:

$$\begin{aligned} a_1b_1-a_2b_2 &= a_1b_1-a_1b_2+a_1b_2-a_2b_2 \\ &= a_1(b_1-b_2)+b_2(a_1-a_2) \\ &= (a_1b+b_2a)m \end{aligned}$$

□

Lemma 5.9. *Every polynomial* $a \in K[t]$ *is congruent modulo m to a unique polynomial of degree* $< \partial m$.

Proof. Divide a by m with remainder, so that $a = qm+r$ where $q,r \in K[t]$ and $\partial r < \partial m$. Then $a-r = qm$, so $a \equiv r \pmod{m}$. To prove uniqueness, suppose that $r \equiv s \pmod{m}$ where $\partial r, \partial s < \partial m$. Then $r-s$ is divisible by m but has smaller degree than m. Therefore $r-s=0$, so $r=s$, proving uniqueness. □

We call r the *reduced form* of a modulo m. Lemma 5.9 shows that we can calculate with polynomials modulo m in terms of their reduced forms. Indeed, the reduced form of $a+b$ is the reduced form of a plus the reduced form of b, while the reduced form of ab is the remainder, after dividing by m, of the product of the reduced form of a and the reduced form of b.

Slightly more abstractly, we can work with equivalence classes. The relation $\equiv \pmod{m}$ is an equivalence relation on $K[t]$, so it partitions $K[t]$ into equivalence classes. We write $[a]$ for the equivalence class of $a \in K[t]$. Clearly

$$[a] = \{f \in K[t] : m|(a-f)\}$$

The sum and product of $[a]$ and $[b]$ can be defined as:

$$[a]+[b] = [a+b] \qquad [a][b] = [ab]$$

It is straightforward to show that these operations are well-defined; that is, they do not depend on the choice of elements from equivalence classes. Each equivalence class contains a unique polynomial of degree less than ∂m, namely, the reduced form of a. Therefore algebraic computations with equivalence classes are the same as computations with reduced forms, and both are the same as computations in $K[t]$ with the added convention that $m(t)$ is identified with 0. In particular, the classes $[0]$ and $[1]$ are additive and multiplicative identities respectively.

We write

$$K[t]/\langle m\rangle$$

for the set of equivalence classes of $K[t]$ modulo m. Readers who know about ideals in rings will see at once that $K[t]/\langle m\rangle$ is a thin disguise for the quotient ring of $K[t]$ by the ideal generated by m, and the equivalence classes are cosets of that ideal, but at this stage of the book these concepts are more abstract than we really need.

A key result is:

Theorem 5.10. *Every nonzero element of $K[t]/\langle m\rangle$ has a multiplicative inverse in $K[t]/\langle m\rangle$ if and only if m is irreducible in $K[t]$.*

Proof. If m is reducible then $m = ab$ where $\partial a, \partial b < \partial m$. Then $[a][b] = [ab] = [m] = [0]$. Suppose that $[a]$ has an inverse $[c]$, so that $[c][a] = [1]$. Then $[0] = [c][0] = [c][a][b] = [1][b] = [b]$, so m divides b. Since $\partial b < \partial m$ we must have $b = 0$, so $m = 0$, contradiction.

If m is irreducible, let $a \in K[t]$ with $[a] \neq [0]$; that is, $m \nmid a$. Therefore a is prime to m, so their highest common factor is 1. By Theorem 3.9, there exist $h, k \in K[t]$ such that $ha + km = 1$. Then $[h][a] + [k][m] = [1]$, but $[m] = [0]$ so $[1] = [h][a] + [k][m] = [h][a] + [k][0] = [h][a] + [0] = [h][a]$. Thus $[h]$ is the required inverse. □

Again, in abstract terminology, what we have proved is that $K[t]/\langle m\rangle$ is a field if and only if m is irreducible in $K[t]$. See Chapter 17 for a full explanation and generalisations.

5.4 Classifying Simple Extensions

We now demonstrate that the above methods suffice for the construction of all possible simple extensions (up to isomorphism). Again transcendental extensions are easily dealt with.

Theorem 5.11. *Every simple transcendental extension $K(\alpha) : K$ is isomorphic to the extension $K(t) : K$ of rational expressions in an indeterminate t over K. The isomorphism $K(t) \to K(\alpha)$ can be chosen to map t to α, and to be the identity on K.*

Proof. Define a map $\phi : K(t) \to K(\alpha)$ by

$$\phi(f(t)/g(t)) = f(\alpha)/g(\alpha)$$

If $g \neq 0$ then $g(\alpha) \neq 0$ (since α is transcendental) so this definition makes sense. It is clearly a homomorphism, and a simple calculation shows that it is a monomorphism.

It is clearly onto, and so is an isomorphism. Further, $\phi|_K$ is the identity, so that ϕ defines an isomorphism of extensions. Finally, $\phi(t) = \alpha$. □

The classification for simple algebraic extensions is just as straightforward, but more interesting:

Theorem 5.12. *Let $K(\alpha) : K$ be a simple algebraic extension, and let the minimal polynomial of α over K be m. Then $K(\alpha) : K$ is isomorphic to $K[t]/\langle m\rangle : K$. The isomorphism $K[t]/\langle m\rangle \to K(\alpha)$ can be chosen to map t to α (and to be the identity on K).*

Proof. The isomorphism is defined by $[p(t)] \mapsto p(\alpha)$, where $[p(t)]$ is the equivalence class of $p(t)$ (mod m). This map is well-defined because $p(\alpha) = 0$ if and only if $m|p$. It is clearly a field monomorphism. It maps t to α, and its restriction to K is the identity. □

Corollary 5.13. *Suppose $K(\alpha) : K$ and $K(\beta) : K$ are simple algebraic extensions, such that α and β have the same minimal polynomial m over K. Then the two extensions are isomorphic, and the isomorphism of the large fields can be taken to map α to β (and to be the identity on K).*

Proof. Both extensions are isomorphic to $K[t]/\langle m\rangle$. The isomorphisms concerned map t to α and t to β respectively. Call them ι, j respectively. Then $j\iota^{-1}$ is an isomorphism from $K(\alpha)$ to $K(\beta)$ that is the identity on K and maps α to β. □

Lemma 5.14. *Let $K(\alpha) : K$ be a simple algebraic extension, let the minimal polynomial of α over K be m, and let $\partial m = n$. Then $\{1, \alpha, \ldots, \alpha^{n-1}\}$ is a basis for $K(\alpha)$ over K.*

Proof. The theorem is a restatement of Lemma 5.9. □

For certain later applications we need a slightly stronger version of Theorem 5.12, to cover extensions of isomorphic (rather than identical) fields. Before we can state the more general theorem we need the following:

Definition 5.15. Let $\iota : K \to L$ be a field monomorphism. Then there is a map $\hat{\iota} : K[t] \to L[t]$, defined by

$$\hat{\iota}(k_0 + k_1 t + \cdots + k_n t^n) = \iota(k_0) + \iota(k_1)t + \cdots + \iota(k_n)t^n$$

$(k_0, \ldots, k_n \in K)$. It is easy to prove that $\hat{\iota}$ is a monomorphism. If ι is an isomorphism, then so is $\hat{\iota}$.

The hat is unnecessary, once the statement is clear, and it may be dispensed with. So in future we use the same symbol ι for the map between subfields of $\mathbb{C}$ and for its extension to polynomial rings. This should not cause confusion since $\hat{\iota}(k) = \iota(k)$ for any $k \in K$.

Theorem 5.16. *Suppose that K and L are subfields of $\mathbb{C}$ and $\iota : K \to L$ is an isomorphism. Let $K(\alpha), L(\beta)$ be simple algebraic extensions of K and L respectively, such that α has minimal polynomial $m_\alpha(t)$ over K and β has minimal polynomial $m_\beta(t)$ over L. Suppose further that $m_\beta(t) = \iota(m_\alpha(t))$. Then there exists an isomorphism $j : K(\alpha) \to L(\beta)$ such that $j|_K = \iota$ and $j(\alpha) = \beta$.*

Proof. We can summarise the hypotheses in the diagram

$$\begin{array}{ccc} K & \to & K(\alpha) \\ \iota\downarrow & & \downarrow j \\ L & \to & L(\beta) \end{array}$$

where j is yet to be determined. Using the reduced form, every element of $K(\alpha)$ is of the form $p(\alpha)$ for a polynomial p over K of degree $< \partial m_\alpha$. Define $j(p(\alpha)) = (\iota(p))(\beta)$ where $\iota(\rho)$ is defined as above. Everything else follows easily from Theorem 5.12. □

The point of this theorem is that the given map ι can be extended to a map j between the larger fields. Such *extension theorems*, saying that under suitable conditions maps between sub-objects can be extended to maps between objects, constitute important weapons in the mathematician's armoury. Using them we can extend our knowledge from small structures to large ones in a sequence of simple steps.

Theorem 5.16 implies that under the given hypotheses the extensions $K(\alpha) : K$ and $L(\beta) : L$ are isomorphic. This allows us to identify K with L and $K(\alpha)$ with $L(\beta)$, via the maps ι and j.

Theorems 5.7 and 5.12 together give a complete characterisation of simple algebraic extensions in terms of polynomials. To each extension corresponds an irreducible monic polynomial, and given the small field and this polynomial, we can reconstruct the extension.

EXERCISES

5.1 Is the extension $\mathbb{Q}(\sqrt{5}, \sqrt{7})$ simple? If so, why? If not, why not?

5.2 Find the minimal polynomials over the small field of the following elements in the following extensions:

(a) i in $\mathbb{C} : \mathbb{Q}$

(b) i in $\mathbb{C} : \mathbb{R}$

(c) $\sqrt{2}$ in $\mathbb{R} : \mathbb{Q}$

(d) $(\sqrt{5}+1)/2$ in $\mathbb{C} : \mathbb{Q}$

(e) $(i\sqrt{3}-1)/2$ in $\mathbb{C} : \mathbb{Q}$

5.3 Show that if α has minimal polynomial $t^2 - 2$ over $\mathbb{Q}$ and β has minimal polynomial $t^2 - 4t + 2$ over $\mathbb{Q}$, then the extensions $\mathbb{Q}(\alpha) : \mathbb{Q}$ and $\mathbb{Q}(\beta) : \mathbb{Q}$ are isomorphic.

5.4 For which of the following $m(t)$ and K do there exist extensions $K(\alpha)$ of K for which α has minimal polynomial $m(t)$?

(a) $m(t) = t^2 - 4, K = \mathbb{R}$

(b) $m(t) = t^2 - 3, K = \mathbb{R}$

(c) $m(t) = t^2 - 3, K = \mathbb{Q}$

(d) $m(t) = t^7 - 3t^6 + 4t^3 - t - 1, K = \mathbb{R}$

5.5 Let K be any subfield of $\mathbb{C}$ and let $m(t)$ be a quadratic polynomial over K ($\partial m = 2$). Show that all zeros of $m(t)$ lie in an extension $K(\alpha)$ of K where $\alpha^2 = k \in K$. Thus allowing 'square roots' $\sqrt{k}$ enables us to solve all quadratic equations over K.

5.6 Construct extensions $\mathbb{Q}(\alpha) : \mathbb{Q}$ where α has the following minimal polynomial over $\mathbb{Q}$:

(a) $t^2 - 5$

(b) $t^4 + t^3 + t^2 + t + 1$

(c) $t^3 + 2$

5.7 Is $\mathbb{Q}(\sqrt{2}, \sqrt{3}, \sqrt{5}) : \mathbb{Q}$ a simple extension?

5.8 Suppose that $m(t)$ is irreducible over K, and α has minimal polynomial $m(t)$ over K. Does $m(t)$ necessarily factorise over $K(\alpha)$ into linear (degree 1) polynomials? (*Hint:* Try $K = \mathbb{Q}, \alpha =$ the real cube root of 2.)

5.9 Mark the following true or false.

(a) Every field has non-trivial extensions.

(b) Every field has non-trivial algebraic extensions.

(c) Every simple extension is algebraic.

(d) Every extension is simple.

(e) All simple algebraic extensions of a given subfield of $\mathbb{C}$ are isomorphic.

(f) All simple transcendental extensions of a given subfield of $\mathbb{C}$ are isomorphic.

(g) Every minimal polynomial is monic.

(h) Monic polynomials are always irreducible.

(i) Every polynomial is a constant multiple of an irreducible polynomial.

Chapter 6

The Degree of an Extension

A technique which has become very useful in mathematics is that of associating with a given structure a different one, of a type better understood. In this chapter we exploit the technique by associating with any field extension a vector space. This places at our disposal the machinery of linear algebra—a very successful algebraic theory—and with its aid we can make considerable progress. The machinery is sufficiently powerful to solve three notorious problems which remained unanswered for over two thousand years. We shall discuss these problems in the next chapter, and devote the present chapter to developing the theory.

6.1 Definition of the Degree

It is not hard to define a vector space structure on a field extension. It already has one! More precisely:

Theorem 6.1. *If $L:K$ is a field extension, then the operations*

$$(\lambda, u) \mapsto \lambda u \qquad (\lambda \in K, u \in L)$$
$$(u, v) \mapsto u+v \qquad (u, v \in L)$$

define on L the structure of a vector space over K.

Proof. The set L is a vector space over K if the two operations just defined satisfy the following axioms:

(1) $u+v = v+u$ for all $u, v \in L$.

(2) $(u+v)+w = u+(v+w)$ for all $u, v, w \in L$.

(3) There exists $0 \in L$ such that $0+u = u$ for all $u \in L$.

(4) For any $u \in L$ there exists $-u \in L$ such that $u+(-u) = 0$.

(5) If $\lambda \in K, u, v \in L$, then $\lambda(u+v) = \lambda u + \lambda v$.

(6) If 1 is the multiplicative identity of K, then $1u = u$ for all $u \in L$.

(7) If $\lambda, \mu \in K$, then $\lambda(\mu u) = (\lambda\mu)u$ for all $u \in L$.

Each of these statements follows immediately because K and L are subfields of $\mathbb{C}$ and $K \subseteq L$. □

We know that a vector space V over a subfield K of $\mathbb{C}$ (indeed over *any* field, but we're not supposed to know about those yet) is uniquely determined, up to isomorphism, by its dimension. The dimension is the number of elements in a basis—a subset of vectors that spans V and is linearly independent over K. The following definition is the traditional terminology in the context of field extensions:

Definition 6.2. The *degree* $[L : K]$ of a field extension $L : K$ is the dimension of L considered as a vector space over K.

Examples 6.3. (1) The complex numbers $\mathbb{C}$ are two-dimensional over the real numbers $\mathbb{R}$, because a basis is $\{1, i\}$. Hence $[\mathbb{C} : \mathbb{R}] = 2$.
(2) The extension $\mathbb{Q}(i, \sqrt{5}) : \mathbb{Q}$ has degree 4. The elements $\{1, \sqrt{5}, i, i\sqrt{5}\}$ form a basis for $\mathbb{Q}(i, \sqrt{5})$ over $\mathbb{Q}$, by Example 4.8.

Isomorphic field extensions obviously have the same degree.

6.2 The Tower Law

The next theorem lets us calculate the degree of a complicated extension if we know the degrees of certain simpler ones.

Theorem 6.4 (Short Tower Law). *If K, L, M are subfields of $\mathbb{C}$ and $K \subseteq L \subseteq M$, then*

$$[M : K] = [M : L][L : K]$$

Note: For those who are happy with infinite cardinals this formula needs no extra explanation; the product on the right is just multiplication of cardinals. For those who are not, the formula needs interpretation if any of the degrees involved is infinite. This interpretation is the obvious one: if either $[M : L]$ or $[L : K] = \infty$ then $[M : K] = \infty$; and if $[M : K] = \infty$ then either $[M : L] = \infty$ or $[L : K] = \infty$.

Proof. Let $(x_i)_{i\in I}$ be a basis for L as vector space over K and let $(y_j)_{j\in J}$ be a basis for M over L. For all $i \in I$ and $j \in J$ we have $x_i \in L, y_j \in M$. We shall show that $(x_i y_j)_{i\in I, j\in J}$ is a basis for M over K (where $x_i y_j$ is the product in the subfield M). Since dimensions are cardinalities of bases, the theorem follows.

First, we prove linear independence. Suppose that some finite linear combination of the putative basis elements is zero; that is,

$$\sum_{i,j} k_{ij} x_i y_j = 0 \qquad (k_{ij} \in K)$$

We can rearrange this as

$$\sum_j \left(\sum_i k_{ij} x_i \right) y_j = 0$$

Since the coefficients $\sum_i k_{ij} x_i$ lie in L and the y_j are linearly independent over L,

$$\sum_i k_{ij} x_i = 0$$

Repeating the argument inside L we find that $k_{ij} = 0$ for all $i \in I$, $j \in J$. So the elements $x_i y_j$ are linearly independent over K.

Finally we show that the $x_i y_j$ span M over K. Any element $x \in M$ can be written

$$x = \sum_j \lambda_j y_j$$

for suitable $\lambda_j \in L$, since the y_j span M over L. Similarly for any $j \in J$

$$\lambda_j = \sum_i \lambda_{ij} x_i$$

for $\lambda_{ij} \in K$. Putting the pieces together,

$$x = \sum_{i,j} \lambda_{ij} x_i y_j$$

as required. □

Example 6.5. Suppose we wish to find $[\mathbb{Q}(\sqrt{2}, \sqrt{3}) : \mathbb{Q}]$. It is easy to see that $\{1, \sqrt{2}\}$ is a basis for $\mathbb{Q}(\sqrt{2})$ over $\mathbb{Q}$. For let $\alpha \in \mathbb{Q}(\sqrt{2})$. Then $\alpha = p + q\sqrt{2}$ where $p, q \in \mathbb{Q}$, proving that $\{1, \sqrt{2}\}$ spans $\mathbb{Q}(\sqrt{2})$ over $\mathbb{Q}$. It remains to show that 1 and $\sqrt{2}$ are linearly independent over $\mathbb{Q}$. Suppose that $p + q\sqrt{2} = 0$, where $p, q \in \mathbb{Q}$. If $q \neq 0$ then $\sqrt{2} = p/q$, which is impossible since $\sqrt{2}$ is irrational. Therefore $q = 0$. But this implies $p = 0$.

In much the same way we can show that $\{1, \sqrt{3}\}$ is a basis for $\mathbb{Q}(\sqrt{2}, \sqrt{3})$ over $\mathbb{Q}(\sqrt{2})$. Every element of $\mathbb{Q}(\sqrt{2}, \sqrt{3})$ can be written as $p + q\sqrt{2} + r\sqrt{3} + s\sqrt{6}$ where $p, q, r, s \in \mathbb{Q}$. Rewriting this as

$$(p + q\sqrt{2}) + (r + s\sqrt{2})\sqrt{3}$$

we see that $\{1, \sqrt{3}\}$ spans $\mathbb{Q}(\sqrt{2}, \sqrt{3})$ over $\mathbb{Q}(\sqrt{2})$. To prove linear independence we argue much as above: if

$$(p + q\sqrt{2}) + (r + s\sqrt{2})\sqrt{3} = 0$$

then either $(r + s\sqrt{2}) = 0$, whence also $(p + q\sqrt{2}) = 0$, or else

$$\sqrt{3} = (p + q\sqrt{2})/(r + s\sqrt{2}) \in \mathbb{Q}(\sqrt{2})$$

Therefore $\sqrt{3} = a + b\sqrt{2}$ where $a, b \in \mathbb{Q}$. Squaring, we find that $ab\sqrt{2}$ is rational, which is possible only if either $a=0$ or $b=0$. But then $\sqrt{3}=a$ or $\sqrt{3}=b\sqrt{2}$, both of which are absurd. Then $(p+q\sqrt{2})=(r+s\sqrt{2})=0$ and we have proved that $\{1,\sqrt{3}\}$ is a basis. Hence

$$\begin{aligned}[\mathbb{Q}(\sqrt{2},\sqrt{3}):\mathbb{Q}] &= [\mathbb{Q}(\sqrt{2},\sqrt{3}):\mathbb{Q}(\sqrt{2})][\mathbb{Q}(\sqrt{2}):\mathbb{Q}] \\ &= 2\times 2=4\end{aligned}$$

The theorem even furnishes a basis for $\mathbb{Q}(\sqrt{2},\sqrt{3})$ over $\mathbb{Q}$: form all possible pairs of products from the two bases $\{1,\sqrt{2}\}$ and $\{1,\sqrt{3}\}$, to get the 'combined' basis $\{1,\sqrt{2},\sqrt{3},\sqrt{6}\}$.

By induction on n we easily parlay the Short Tower Law into a useful generalisation:

Corollary 6.6 (Tower Law). *If $K_0 \subseteq K_1 \subseteq \cdots \subseteq K_n$ are subfields of $\mathbb{C}$, then*

$$[K_n:K_0]=[K_n:K_{n-1}][K_{n-1}:K_{n-2}]\cdots[K_1:K_0]$$

□

In order to use the Tower Law we have to get started. The degree of a simple extension is fairly easy to find:

Proposition 6.7. *Let $K(\alpha):K$ be a simple extension. If it is transcendental then $[K(\alpha):K]=\infty$. If it is algebraic then $[K(\alpha):K]=\partial m$, where m is the minimal polynomial of α over K.*

Proof. For the transcendental case it suffices to note that the elements $1,\alpha,\alpha^2,\ldots$ are linearly independent over K. For the algebraic case, we appeal to Lemma 5.14. □

For example, we know that $\mathbb{C}=\mathbb{R}(i)$ where i has minimal polynomial t^2+1, of degree 2, Hence $[\mathbb{C}:\mathbb{R}]=2$, which agrees with our previous remarks.

Example 6.8. We now illustrate a technique that we shall use, without explicit reference, whenever we discuss extensions of the form $\mathbb{Q}(\sqrt{\alpha_1},\ldots,\sqrt{\alpha_n}):\mathbb{Q}$ with rational α_j. The technique can be used to prove a general theorem about such extensions, see Exercise 6.15. The question we tackle is: find $[\mathbb{Q}(\sqrt{2},\sqrt{3},\sqrt{5}):\mathbb{Q}]$.

By the Tower Law,

$$\begin{aligned}&[\mathbb{Q}(\sqrt{2},\sqrt{3},\sqrt{5}):\mathbb{Q}] \\ &=[\mathbb{Q}(\sqrt{2},\sqrt{3},\sqrt{5}):\mathbb{Q}(\sqrt{2},\sqrt{3})][\mathbb{Q}(\sqrt{2},\sqrt{3}):\mathbb{Q}(\sqrt{2})][\mathbb{Q}(\sqrt{2}):\mathbb{Q}]\end{aligned}$$

It is 'obvious' that each factor equals 2, but it takes some effort to prove it. As a cautionary remark: the degree $[\mathbb{Q}(\sqrt{6},\sqrt{10},\sqrt{15}):\mathbb{Q}]$ is 4, not 8 (Exercise 6.14).
(a) Certainly $[\mathbb{Q}(\sqrt{2}):\mathbb{Q}]=2$.
(b) If $\sqrt{3}\notin\mathbb{Q}(\sqrt{2})$ then $[\mathbb{Q}(\sqrt{2},\sqrt{3}):\mathbb{Q}(\sqrt{2})]=2$. So suppose $\sqrt{3}\in\mathbb{Q}(\sqrt{2})$, implying that

$$\sqrt{3}=p+q\sqrt{2} \qquad p,q\in\mathbb{Q}$$

We argue as in Example 6.5. Squaring,

$$3 = (p^2 + 2q^2) + 2pq\sqrt{2}$$

so

$$p^2 + 2q^2 = 3 \qquad pq = 0$$

If $p = 0$ then $2q^2 = 3$, which is impossible by Exercise 1.3. If $q = 0$ then $p^2 = 3$, which is impossible for the same reason. Therefore $\sqrt{3} \notin \mathbb{Q}(\sqrt{2})$, and $[\mathbb{Q}(\sqrt{2}, \sqrt{3}) : \mathbb{Q}(\sqrt{2})] = 2$.

(c) Finally, we claim that $\sqrt{5} \notin \mathbb{Q}(\sqrt{2}, \sqrt{3})$. Here we need a new idea. Suppose

$$\sqrt{5} = p + q\sqrt{2} + r\sqrt{3} + s\sqrt{6} \qquad p, q, r, s \in \mathbb{Q}$$

Squaring:

$$5 = p^2 + 2q^2 + 3r^2 + 6s^2 + (2pq + 6rs)\sqrt{2} + (2pr + 4qs)\sqrt{3} + (2ps + 2qr)\sqrt{6}$$

whence

$$\begin{aligned} p^2 + 2q^2 + 3r^2 + 6s^2 &= 5 \\ pq + 3rs &= 0 \\ pr + 2qs &= 0 \\ ps + qr &= 0 \end{aligned} \tag{6.1}$$

The new idea is to observe that if (p, q, r, s) satisfies (6.1), then so do $(p, q, -r, -s)$, $(p, -q, r, -s)$, and $(p, -q, -r, s)$. Therefore

$$\begin{aligned} p + q\sqrt{2} + r\sqrt{3} + s\sqrt{6} &= \sqrt{5} \\ p + q\sqrt{2} - r\sqrt{3} - s\sqrt{6} &= \pm\sqrt{5} \\ p - q\sqrt{2} + r\sqrt{3} - s\sqrt{6} &= \pm\sqrt{5} \\ p - q\sqrt{2} - r\sqrt{3} + s\sqrt{6} &= \pm\sqrt{5} \end{aligned}$$

Adding the first two equations, we get $p + q\sqrt{2} = 0$ or $p + q\sqrt{2} = \sqrt{5}$. The first implies that $p = q = 0$. The second implies that $p^2 + 2q^2 + 2pq\sqrt{2} = 5$, which is easily seen to be impossible. Adding the first and third, $r\sqrt{3} = 0$ or $r\sqrt{3} = \sqrt{5}$, so $r = 0$. Finally, $s = 0$ since $s\sqrt{6} = \sqrt{5}$ is impossible by Exercise 1.3.

Having proved the claim, we immediately deduce that

$$[\mathbb{Q}(\sqrt{2}, \sqrt{3}, \sqrt{5}) : \mathbb{Q}(\sqrt{2}, \sqrt{3})] = 2$$

which implies that $[\mathbb{Q}(\sqrt{2}, \sqrt{3}, \sqrt{5}) : \mathbb{Q}] = 8$.

Linear algebra is at its most powerful when dealing with finite-dimensional vector spaces. Accordingly we shall concentrate on field extensions that give rise to such vector spaces.

Definition 6.9. A *finite extension* is one whose degree is finite.

Proposition 6.7 implies that any simple algebraic extension is finite. The converse is not true, but certain partial results are: see Exercise 6.16. In order to state what is true we need:

Definition 6.10. *An extension $L:K$ is algebraic if every element of L is algebraic over K.*

Algebraic extensions need not be finite, see Exercise 6.11, but every finite extension is algebraic. More generally:

Lemma 6.11. *An extension $L:K$ is finite if and only if $L=K(\alpha,\ldots,\alpha_r)$ where r is finite and each α_i is algebraic over K.*

Proof. Induction using Theorem 6.4 and Proposition 6.7 shows that any extension of the form $K(\alpha_1,\ldots,\alpha_s):K$ for algebraic α_j is finite.

Conversely, let $L:K$ be a finite extension. Then there is a basis $\{\alpha_1,\ldots,\alpha_s\}$ for L over K, whence $L=K(\alpha_1,\ldots,\alpha_s)$. Each α_j is clearly algebraic. □

EXERCISES

6.1. Find the degrees of the following extensions:

(a) $\mathbb{C}:\mathbb{Q}$

(b) $\mathbb{R}(\sqrt{5}):\mathbb{R}$

(c) $\mathbb{Q}(\alpha):\mathbb{Q}$ where α is the real cube root of 2

(d) $\mathbb{Q}(3,\sqrt{5},\sqrt{11}):\mathbb{Q}$

(e) $\mathbb{Q}(\sqrt{6}):\mathbb{Q}$

(f) $\mathbb{Q}(\alpha):\mathbb{Q}$ where $\alpha^7=3$

6.2. Show that every element of $\mathbb{Q}(\sqrt{5},\sqrt{7})$ can be expressed uniquely in the form

$$p+q\sqrt{5}+r\sqrt{7}+s\sqrt{35}$$

where $p,q,r,s\in\mathbb{Q}$. Calculate explicitly the inverse of such an element.

6.3. If $[L:K]$ is a prime number show that the only fields M such that $K\subseteq M\subseteq L$ are K and L themselves.

6.4. If $[L:K]=1$ show that $K=L$.

6.5. Write out in detail the inductive proof of Corollary 6.6.

6.6. Let $L:K$ be an extension. Show that multiplication by a fixed element of L is a linear transformation of L considered as a vector space over K. When is this linear transformation nonsingular?

6.7. Let $L : K$ be a finite extension, and let p be an irreducible polynomial over K. Show that if ∂p does not divide $[L : K]$, then p has no zeros in L.

6.8. If $L : K$ is algebraic and $M : L$ is algebraic, is $M : K$ algebraic? Note that you may not assume the extensions are finite.

6.9. Prove that $\mathbb{Q}(\sqrt{3}, \sqrt{5}) = \mathbb{Q}(\sqrt{3} + \sqrt{5})$. Try to generalise your result.

6.10* Prove that the square roots of all prime numbers are linearly independent over $\mathbb{Q}$. Deduce that algebraic extensions need not be finite.

6.11 Find a basis for $\mathbb{Q}(\sqrt{(1+\sqrt{3})})$ over $\mathbb{Q}$ and hence find the degree of $\mathbb{Q}(\sqrt{(1+\sqrt{3})}) : \mathbb{Q}$. (*Hint:* You will need to prove that $1+\sqrt{3}$ is not a square in $\mathbb{Q}(\sqrt{3})$.)

6.12 If $[L : K]$ is prime, show that L is a simple extension of K.

6.13 Show that $[\mathbb{Q}(\sqrt{6}, \sqrt{10}, \sqrt{15}) : \mathbb{Q}] = 4$, not 8.

6.14* Let K be a subfield of $\mathbb{C}$ and let $a_1, \ldots, a_n$ be elements of K such that any product $a_{j_1} \cdots a_{j_k}$, with distinct indices j_l, is not a square in K. Let $\alpha_j = \sqrt{a_j}$ for $1 \leq j \leq n$. Prove that $[K(\alpha_1, \ldots, \alpha_n) : K] = 2^n$.

If $K = \mathbb{Q}$, how can we verify the hypotheses on the a_j by looking at their prime factorisations?

6.15* Let $L : K$ be an algebraic extension and suppose that K is an infinite field. Prove that $L : K$ is simple if and only if there are only finitely many fields M such that $K \subseteq M \subseteq L$, as follows.

(a) Assume only finitely many M exist. Use Lemma 6.11 to show that $L : K$ is finite.

(b) Assume $L = K(\alpha_1, \alpha_2)$. For each $\beta \in K$ let $J_\beta = K(\alpha_1 + \beta\alpha_2)$. Only finitely many distinct J_β can occur: hence show that $L = J_\beta$ for some β.

(c) Use induction to prove the general case.

(d) For the converse, let $L = K(\alpha)$ be simple algebraic, with $K \subseteq M \subseteq L$. Let m be the minimal polynomial of α over K, and let m_M be the minimal polynomial of α over M. Show that $m_M | m$ in $L[t]$. Prove that m_M determines M uniquely, and that only finitely many m_M can occur.

6.16 Mark the following true or false.

(a) Extensions of the same degree are isomorphic.

(b) Isomorphic extensions have the same degree.

(c) Every algebraic extension is finite.

(d) Every transcendental extension is not finite.

(e) Every element of $\mathbb{C}$ is algebraic over $\mathbb{R}$.

(f) Every extension of $\mathbb{R}$ that is a subfield of $\mathbb{C}$ is finite.

(g) Every algebraic extension of $\mathbb{Q}$ is finite.

Chapter 7

Ruler-and-Compass Constructions

Already we are in a position to see some payoff. The degree of a field extension is a surprisingly powerful tool. Even before we get into Galois theory proper, we can apply the degree to a warm-up problem—indeed, several. The problems come from classical Greek geometry, and we will do something much more interesting and difficult than solving them. We will prove that no solutions exist, subject to certain technical conditions on the permitted methods.

According to Plato the only 'perfect' geometric figures are the straight line and the circle. In the most widely known parts of ancient Greek geometry, this belief had the effect of restricting the (conceptual) instruments available for performing geometric constructions to two: the ruler and the compass. The ruler, furthermore, was a single unmarked straight edge.

Strictly, the term should be '*pair* of compasses', for the same reason we call a single cutting instrument a pair of scissors. However, 'compass' is shorter, and there is no serious danger of confusion with the navigational instrument that tells you which way is north. So 'compass' it is.

With these instruments alone it is possible to perform a wide range of constructions, as Euclid systematically set out in his *Elements* somewhere around 300 BC. This series of books opens with 23 definitions of basic objects ranging from points to parallels, five axioms (called 'postulates' in the translation by Sir Thomas Heath), and five 'common notions' about equality and inequality. The first three axioms state that certain constructions may be performed:

(1) To draw a straight line from any point to any point.

(2) To produce a finite straight line continuously in a straight line.

(3) To describe a circle with any centre and any distance.

The first two model the use of a ruler (or straightedge); the third models the use of a compass.

Definition 7.1. A *ruler-and-compass construction* in the sense of Euclid is a finite sequence of operations of the above three types.

Note the restriction to finite constructions. Infinite constructions can sometimes make theoretical sense, and are more powerful: see Exercise 7.12. They provide arbitrarily good approximations if we stop after a finite number of steps.

Later Greek geometry introduced other 'drawing instruments', such as conic sections and a curve called the quadratrix. But long-standing tradition associates Euclid with geometric constructions carried out using an unmarked ruler and a compass. The *Elements* includes ruler-and-compass constructions to bisect a line or an angle, to divide a line into any specified number of equal parts, and to draw a regular pentagon.

However, there are many geometric problems that clearly 'should' have solutions, but for which the tools of ruler and compasses are inadequate. In particular, there are three famous constructions which the Greeks could not perform using these tools: *Duplicating the Cube, Trisecting the Angle*, and *Squaring the Circle*. These ask respectively for a cube twice the volume of a given cube, an angle one-third the size of a given angle, and a square of area equal to a given circle.

It seems likely that Euclid would have included such constructions if he knew any, and it is a measure of his mathematical taste that he did not present fallacious constructions that are approximately correct but not exact. The Greeks were ingenious enough to find exact constructions if they existed, unless they had to be extraordinarily complicated. (The construction of a regular 17-gon is an example of a complicated construction that they missed: see Chapter 19.) We now know why they failed to find ruler-and-compass constructions for the three classical problems: they don't exist. But the Greeks lacked the algebraic techniques needed to prove that.

The impossibility of trisecting an arbitrary angle using ruler and compass was not proved until 1798 when Gauss was writing his *Disquisitiones Arithmeticae*, published in 1801. Discussing his construction of the regular 17-gon, he states without proof that such constructions do not exist for the 9-gon, 25-gon, and other numbers that are not a power of 2 times a product of distinct Fermat primes—those of the form $2^{2^n}+1$. He also writes that he can 'prove in all rigour that these higher-degree equations [involved in the construction] cannot be avoided in any way', but adds 'the limits of the present work exclude this demonstration here.' Constructing the regular 9-gon is clearly equivalent to trisecting $\frac{2\pi}{3}$, so Gauss's claim disposes of trisections. He did not publish a proof; the first person to do so was Pierre Wantzel in 1837.

This result does not imply that an angle one third the size of a given one does not exist, or that practical constructions with very small errors cannot be devised; it tells us that the specified instruments are inadequate to find it *exactly*. Wantzel also proved that it is impossible to duplicate the cube with ruler and compass. Squaring the circle had to wait even longer for an impossibility proof.

In this chapter we mention approximate constructions, which are entirely acceptable for practical work. We make some brief historical remarks to point out that the Greeks could solve the three classical problems using 'instruments' that went beyond just ruler and compass. We identify the Euclidean plane $\mathbb{R}^2$ with the complex plane $\mathbb{C}$, which lets us avoid considering the two coordinates of a point separately and greatly simplifies the discussion. We formalise the concept of ruler-and-compass construction by defining the notion of a constructible point in $\mathbb{C}$. We introduce a series of specific constructions that correspond to field operations $(+,-,\times,/)$ and square roots in $\mathbb{C}$. We characterise constructible points in terms of the 'Pythagorean closure' $\mathbb{Q}^{\text{py}}$ of $\mathbb{Q}$, and deduce a simple algebraic criterion for a point to be constructible. By applying this criterion, we prove that the three classical problems cannot be solved by ruler-and-compass construction. We also prove that there is no such construction for a regular heptagon (7-sided polygon).

7.1 Approximate Constructions and More General Instruments

For the technical drawing expert we emphasise that we are discussing *exact* constructions. There are many approximate constructions for trisecting the angle, for instance, but no exact methods. Dudley (1987) is a fascinating collection of approximate methods that were thought by their inventors to be exact. Figure 10 is a typical example. To trisect angle BOA, draw line BE parallel to OA. Mark off AC and CD equal to OA, draw arc DE with centre C and radius CD. Drop a perpendicular EF to OD and draw arc FT centre O radius OF to meet BE at T. Then angle AOT approximately trisects angle BOA. See Exercise 7.10.

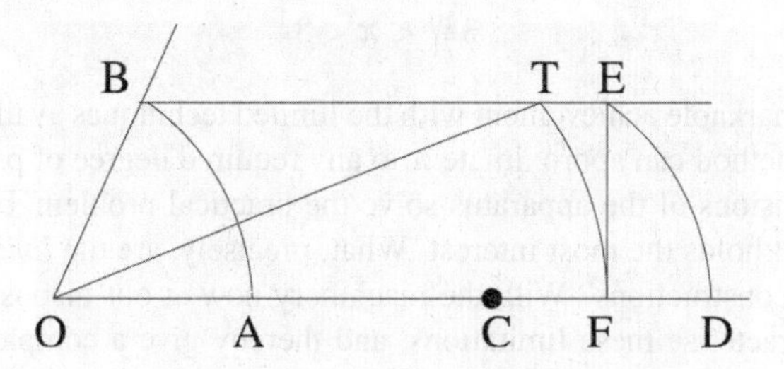

FIGURE 10: Close—but no banana.

The Greeks were well aware that by going outside the Platonic constraints, all three classical problems can be solved. Archimedes and others knew that angles can be trisected using a *marked* ruler, as in Figure 11. The ruler has marked on it two points distance r apart. Given $\angle AOB = \theta$ draw a circle centre O with radius r, cutting OA at X, OB at Y. Place the ruler with its edge through X and one mark on the line OY at D; slide it until the other marked point lies on the circle at E. Then $\angle EDO = \theta/3$. For a proof, see Exercise 7.3. Exercise 7.14 shows how to duplicate the cube using a marked ruler.

Setting your compasses up against the ruler so that the pivot point and the pencil effectively constitute such marks also provides a trisection, but again this goes beyond the precise concept of a 'ruler-and-compass construction'. Many other uses of 'exotic' instruments are catalogued in Dudley (1987), which examines the history of trisection attempts. Euclid may have limited himself to an unmarked ruler (plus compasses) because it made his axiomatic treatment more convincing. It is not entirely clear what conditions should apply to a marked ruler—the distance between the marks causes difficulties. Presumably it ought to be constructible, for example.

The Greeks solved all three problems using conic sections, or more recondite curves such as the conchoid of Nichomedes or the quadratrix (Klein 1962, Coolidge

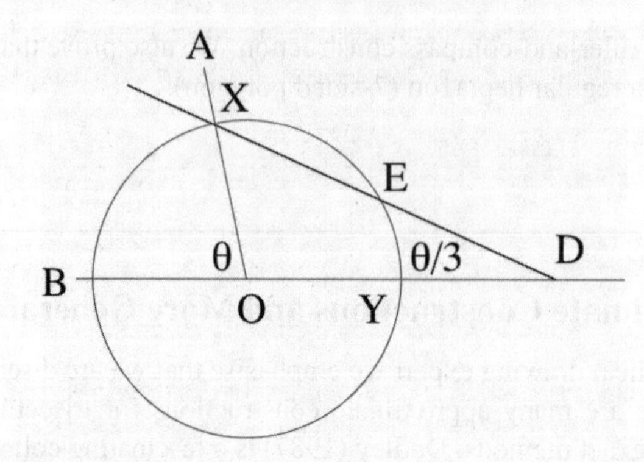

FIGURE 11: Trisecting an angle with a marked ruler.

1963). Archimedes tackled the problem of Squaring the Circle in a characteristically ingenious manner, and proved a result which would now be written

$$3\tfrac{10}{71} < \pi < 3\tfrac{1}{7}$$

This was a remarkable achievement with the limited techniques available, and refinements of his method can approximate π to any required degree of precision.

Such extensions of the apparatus solve the practical problem, but it is the theoretical one that holds the most interest. What, precisely, are the *limitations* on ruler-and-compass constructions? With the machinery now at our disposal it is relatively simple to characterise these limitations, and thereby give a complete answer to all three problems. We use coordinate geometry to express problems in algebraic terms, and apply the theory of field extensions to the algebraic questions that arise.

7.2 Constructions in $\mathbb{C}$

We begin by formalising the notion of a ruler-and-compass construction. Assume that initially we are given two distinct points in the plane. Equivalently, by Euclid's Axiom 1, we can begin with the line segment that joins them. These points let us choose an origin and set a scale. So we can identify the Euclidean plane $\mathbb{R}^2$ with $\mathbb{C}$, and assume that these two points are 0 and 1.

Euclid dealt with finite line segments (his condition (1) above) but could make them as long as he pleased by extending the line (condition (2)). We find it more convenient to work with infinitely long lines (modelling an infinitely long ruler), which in effect combines Euclid's conditions into just one: the possibility of drawing the (infinitely long) line that passes through two given points. From now on, 'line' is always used in this sense.

If $z_1, z_2 \in \mathbb{C}$ and $0 \le r \in \mathbb{R}$, define

$$\begin{aligned} L(z_1, z_2) &= \text{the line joining } z_1 \text{ to } z_2 \quad (z_1 \neq z_2) \\ C(z_1, r) &= \text{the circle centre } z_1 \text{ with radius } r > 0 \end{aligned}$$

We now define constructible points, lines, and circles recursively:

Definition 7.2. For each $n \in \mathbb{N}$ define sets $\mathscr{P}_n, \mathscr{L}_n$, and $\mathscr{C}_n$ of *n-constructible points, lines*, and *circles*, by:

$$\begin{aligned} \mathscr{P}_0 &= \{0, 1\} \\ \mathscr{L}_0 &= \emptyset \\ \mathscr{C}_0 &= \emptyset \\ \mathscr{L}_{n+1} &= \{L(z_1, z_2) : z_1, z_2 \in \mathscr{P}_n\} \\ \mathscr{C}_{n+1} &= \{C(z_1, |z_2 - z_3|) : z_1, z_2, z_3 \in \mathscr{P}_n\} \\ \mathscr{P}_{n+1} &= \{z \in \mathbb{C} : z \text{ lies on two distinct lines in } \mathscr{L}_{n+1}\} \cup \\ &\quad\ \{z \in \mathbb{C} : z \text{ lies on a line in } \mathscr{L}_{n+1} \text{ and a circle in } \mathscr{C}_{n+1}\} \cup \\ &\quad\ \{z \in \mathbb{C} : z \text{ lies on two distinct circles in } \mathscr{L}_{n+1}\} \end{aligned}$$

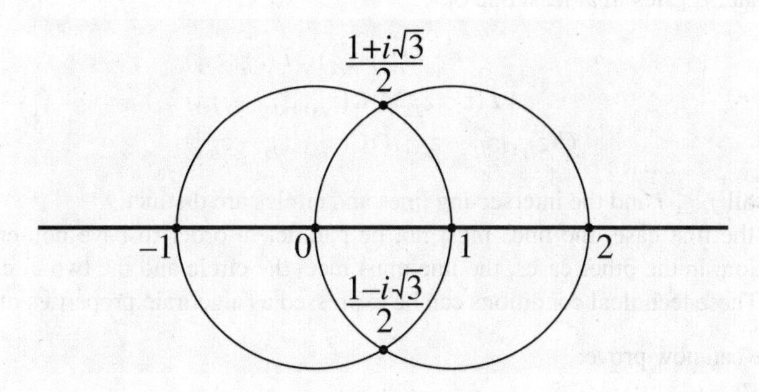

FIGURE 12: The set $\mathscr{P}_1$.

Figure 12 shows that

$$\mathscr{P}_1 = \{-1, 0, 1, 2, \frac{1 \pm i\sqrt{3}}{2}\}$$

Lemma 7.3. *For all $n \in \mathbb{N}$,*

$$\mathscr{P}_n \subseteq \mathscr{P}_{n+1} \qquad \mathscr{L}_n \subseteq \mathscr{L}_{n+1} \qquad \mathscr{C}_n \subseteq \mathscr{C}_{n+1}$$

and each is a finite set.

Proof. The inclusions are clear. Let p_n be the number of points in $\mathscr{P}_n$, l_n the number of lines in $\mathscr{L}_n$, and c_n the number of circles in $\mathscr{C}_n$. Then

$$\begin{aligned}|\mathscr{L}_{n+1}| &\leq \tfrac{1}{2}p_n(p_n+1)\\ |\mathscr{C}_{n+1}| &\leq p_n\tfrac{1}{2}p_n(p_n+1)\\ |\mathscr{P}_{n+1}| &\leq \tfrac{1}{2}l_{n+1}(l_{n+1}+1)+2l_nc_n+c_{n+1}(c_{n+1}+1)\end{aligned}$$

bearing in mind that a line or circle meets a distinct circle in ≤ 2 points. By induction, all three sets are finite for all n. □

We formalise a Euclidean ruler-and-compass construction using these sets. The intuitive idea is that starting from 0 and 1, such a construction generates a finite sequence of points by drawing a line through two previously constructed points, or a circle whose centre is one previously constructed point and whose radius is the distance between two previously constructed points, and then defining a new point using intersections of these.

Definition 7.4. A point $z \in \mathbb{C}$ is *constructible* if there is a finite sequence of points

$$z_0 = 0, z_1 = 1, z_2, z_3, \ldots z_k = z \tag{7.1}$$

such that z_{j+1} lies in at least one of:

$$\begin{gathered}L(z_{j_1}, z_{j_2}) \cap L(z_{j_3}, z_{j_4})\\ L(z_{j_1}, z_{j_2}) \cap C(z_{j_3}, |z_{j_4} - z_{j_5}|)\\ C(z_{j_1}, |z_{j_2} - z_{j_3}|) \cap C(z_{j_4}, |z_{j_5} - z_{j_6}|)\end{gathered}$$

where all $j_i \leq j$ and the intersecting lines and circles are distinct.

In the first case, the lines must not be parallel in order to have non-empty intersection; in the other cases, the line must meet the circle and the two circles must meet. These technical conditions can be expressed as algebraic properties of the z_j.

We can now prove:

Theorem 7.5. *A point $z \in \mathbb{C}$ is constructible if and only if $z \in \mathscr{P}_n$ for some $n \in \mathbb{N}$.*

Proof. Let $z \in \mathbb{C}$ be constructible, using the sequence (7.1). Inductively, it is clear that $z = z_k \in \mathscr{P}_k$.

Conversely, let $z \in \mathscr{P}_k$. Then we can find a sequence $z_j \in \mathscr{P}_j$, where $0 \leq j \leq k$, satisfying (7.1). □

To characterise constructible points, we need:

Definition 7.6. The *Pythagorean closure* $\mathbb{Q}^{\text{py}}$ of $\mathbb{Q}$ is the smallest subfield $K \subseteq \mathbb{C}$ with the property:

$$z \in K \implies \pm\sqrt{z} \in K \tag{7.2}$$

The Pythagorean closure of $\mathbb{Q}$ exists because every subfield of $\mathbb{C}$ contains $\mathbb{Q}$, so $\mathbb{Q}^{\text{py}}$ is the intersection of all subfields of $\mathbb{C}$ satisfying (7.2).

The main theorem of this section is:

Theorem 7.7. *A point* $z \in \mathbb{C}$ *is constructible if and only if* $z \in \mathbb{Q}^{\text{py}}$. *Equivalently,*

$$\bigcup_{n=0}^{\infty} \mathscr{P}_n = \mathbb{Q}^{\text{py}} \tag{7.3}$$

Pre-proof Discussion.

We can summarise the main idea succinctly. Coordinate geometry in $\mathbb{C}$ shows that each step in a ruler-and-compass construction leads to points that can be expressed using rational functions of the previously constructed points together with the square root of a rational function of those points. Conversely, all rational functions of given points can be constructed, and so can square roots of given points. Therefore anything that can be constructed lies in $\mathbb{Q}^{\text{py}}$, and anything in $\mathbb{Q}^{\text{py}}$ can be constructed.

The details require some algebraic computations in $\mathbb{C}$ and some basic Euclidean geometry. We prove Theorem 7.7 in two stages. In this section we show that

(A) $\mathscr{P}_n \subseteq \mathbb{Q}^{\text{py}}$ for all $n \in \mathbb{N}$.

In the next section, after describing some basic constructions for arithmetical operations and square roots, we complete the proof by establishing

(B) If $z \in \mathbb{Q}^{\text{py}}$ then $z \in \mathscr{P}_n$ for some $n \in \mathbb{N}$.

Equation (7.3) is an immediate consequence of (A) and (B).

Proof of Part (A). Part (A) follows by coordinate geometry in $\mathbb{C} \equiv \mathbb{R}^2$. The details are tedious, but we give them for completeness. Use induction on n. Since $\mathscr{P}_0 = \{0,1\} \subseteq \mathbb{Q}$, we have $\mathscr{P}_0 \in \bar{z}$. Suppose inductively that $\mathscr{P}_n \subseteq \mathbb{Q}^{\text{py}}$, and let $z \in \mathscr{P}_{n+1}$. We have to prove that $z \in \mathbb{Q}^{\text{py}}$.

There are three cases: line meets line, line meets circle, circle meets circle.

Case 1: Line meets line. Here $\{z\} = L(z_1,z_2) \cap L(z_3,z_4)$ where the $z_j \in \mathscr{P}_n \subseteq \mathbb{Q}^{\text{py}}$ (induction hypothesis) and the lines are distinct. Therefore there exist real α, β such that

$$z = \alpha z_1 + (1-\alpha) z_2$$
$$z = \beta z_3 + (1-\beta) z_4$$

Therefore

$$\alpha = \frac{\beta(z_3 - z_4) + z_4 - z_2}{z_1 - z_2}$$

Since $\alpha, \beta \in \mathbb{R}$, we also have

$$\alpha = \frac{\beta(\bar{z}_3 - \bar{z}_4) + \bar{z}_4 - \bar{z}_2}{\bar{z}_1 - \bar{z}_2}$$

where the bar is complex conjugate. These two equations have a unique solution for

α, β because we are assuming that the lines meet at a unique point z, and the solution is:

$$\alpha = \frac{z_2(\bar{z}_4 - \bar{z}_3) + \bar{z}_2(z_3 - z_4) - z_3\bar{z}_4 + z_4\bar{z}_3}{(z_1 - z_2)(\bar{z}_3 - \bar{z}_4) + (z_4 - z_3)(\bar{z}_1 - \bar{z}_2)}$$

$$\beta = \frac{z_3(\bar{z}_1 - \bar{z}_2) + \bar{z}_3(z_2 - z_1) - z_2\bar{z}_1 + z_1\bar{z}_2}{(z_4 - z_3)(\bar{z}_2 - \bar{z}_1) + (z_1 - z_2)(\bar{z}_4 - \bar{z}_3)}$$

so $\alpha, \beta \in \mathbb{Q}^{\text{py}}$. Then $z = \alpha z_1 + (1 - \alpha) z_2 \in \mathbb{Q}^{\text{py}}$.

Case 2: Line meets circle. Here $z \in L(z_1, z_2) \cap C(z_3, |z_4 - z_5|)$ where the $z_j \in \mathscr{P}_n \subseteq \mathbb{Q}^{\text{py}}$ (induction hypothesis). Let $r = |z_4 - z_5|$. There exist $\alpha, \theta \in \mathbb{R}$ such that

$$z = \alpha z_1 + (1 - \alpha z_2)$$
$$z = z_3 + re^{i\theta}$$

Therefore

$$\alpha(z_1 - z_2) + z_2 = z_3 + re^{i\theta}$$
$$\alpha(\bar{z}_1 - \bar{z}_2) + \bar{z}_2 = \bar{z}_3 + re^{-i\theta}$$

where we take the complex conjugate to get the second equation. We can eliminate θ to get

$$(\alpha(z_1 - z_2) + z_2 - z_3)(\alpha(\bar{z}_1 - \bar{z}_2) + \bar{z}_2 - \bar{z}_3) = re^{i\theta}.re^{-i\theta} = r^2 = (z_4 - z_5)(\bar{z}_4 - \bar{z}_5)$$

which is a quadratic equation for α with coefficients in $\mathbb{Q}^{\text{py}}$. Since the quadratic formula involves only rational functions of the coefficients and a square root, $\alpha \in \mathbb{Q}^{\text{py}}$. Therefore $z \in \mathbb{Q}^{\text{py}}$.

Case 3: Circle meets circle. Here $z \in C(z_1, |z_2 - z_3|) \cap C(z_4, |z_5 - z_6|)$ where the $z_j \in \mathscr{P}_n \subseteq \mathbb{Q}^{\text{py}}$ (induction hypothesis). Let $r = |z_2 - z_3|, s = |z_5 - z_6|$. There exist $\theta, \phi \in \mathbb{R}$ such that

$$z = z_1 + re^{i\theta}$$
$$z = z_4 + se^{i\phi}$$

Take conjugates and eliminate θ, ϕ as above to get

$$(z - z_1)(\bar{z} - \bar{z}_1) = r^2$$
$$(z - z_4)(\bar{z} - \bar{z}_4) = s^2$$

Solving for z and $\bar{z}$ (left as an exercise) we find that z satisifies a quadratic equation with coefficients in $\mathbb{Q}^{\text{py}}$. Therefore $z \in \mathbb{Q}^{\text{py}}$. □

7.3 Specific Constructions

To prove the converse (B) above we first discuss constructions that implement algebraic operations and square roots in $\mathbb{C}$. The next lemma begins the process of assembling useful constructions and bounding the number of steps they require.

Lemma 7.8. (1) *A line can be bisected using a 2-step construction.*

(2) *An angle can be bisected using a 2-step construction.*

(3) *An angle can be copied (so that its vertex is a given point and one leg lies along a given line through that point) using a 3-step construction.*

(4) *A perpendicular to a given line at a given point can be constructed using a 2-step construction.*

Proof. See Figure 13 for diagrams.
(1) Let the line be $L[z,w]$.
Draw circles $C[z,|z-w|]$ and $C[w,|z-w|]$. These meet at two points u,v.
The midpoint p of $L[z,w]$ is its intersection with $L[u,v]$.
(2) Let θ be the angle between $L[a,b]$ and $L[a,c]$.
Draw $C[a,1]$ meeting $L[a,b]$ at p and $L[a,c]$ at q.
Draw $C[p,1]$ and $C[q,1]$ meeting at s,t. Then $L[a,s]$ (or $L[a,t]$) bisects θ.
(3) Let θ be the angle between $L[a,b]$ and $L[a,c]$.
Suppose $p,q \in \mathbb{C}$ are given, and we wish to construct angle θ at p with one side $L[p,q]$.
Let $C[a,1]$ meet $L[a,b]$ at d and $L[a,c]$ at e.
Let $L[p,1]$ meet $L[p,q]$ at s.
Let $C[s,|d-e|]$ meet $C[p,1]$ at t as shown. Then the angle between $L[p,t]$ and $L[p,q]$ is θ for the appropriate choice of t.
(4) Let a lie on a line L. Let the circle $C[a,1]$ meet L at b,c.
Let $C[b,|b-c|]$ meet $C[c,|b-c|]$ at p,q.
Then $L[p,q]$ is the required perpendicular. □

The next lemma continues the process of collecting useful constructions.

Lemma 7.9. (1) *A parallel to a given line through a given point not on that line can be constructed using a 3-step construction.*

(2) *A triangle similar to a given triangle, with one edge prescribed, can be constructed using a 7-step construction.*

Proof. See Figure 14 for diagrams.
(1) Let the line be $L[a,b]$ and let $p \in \mathbb{C}$ be a point that does not lie on the line. Using Lemma 7.8(3), copy the angle between $L[a,b]$ and $L[a,p]$ to vertex p, with one leg lying along $L[a,p]$ produced. The other leg is then parallel to $L[a,b]$.
(2) Let the vertices of the first triangle be a,b,c. Suppose two vertices p,q of the required similar triangle are given, such that the similarity maps a to p and b to q.

Using Lemma 7.8(3), copy angles θ,ϕ at a,b to locations p,q, with one leg of each lying along $L[p,q]$. Then the other legs meet at s, which is the third vertex of the similar triangle required. □

We can now prove the existence of constructions that produce useful algebra results:

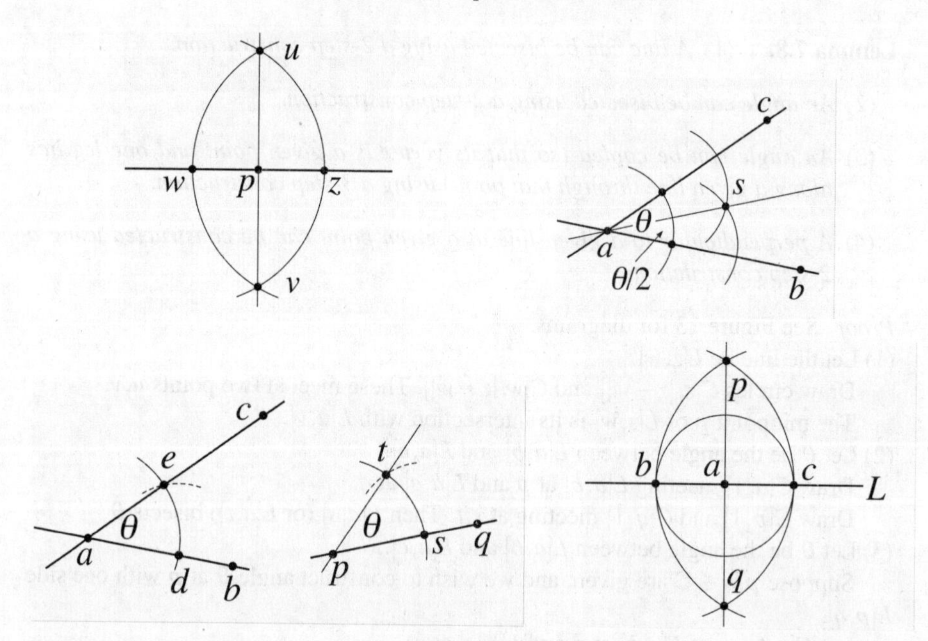

FIGURE 13: Four basic constructions. Top left: Bisecting a line. Top right: Bisecting an angle. Bottom left: Copying an angle. Bottom right: Constructing a perpendicular.

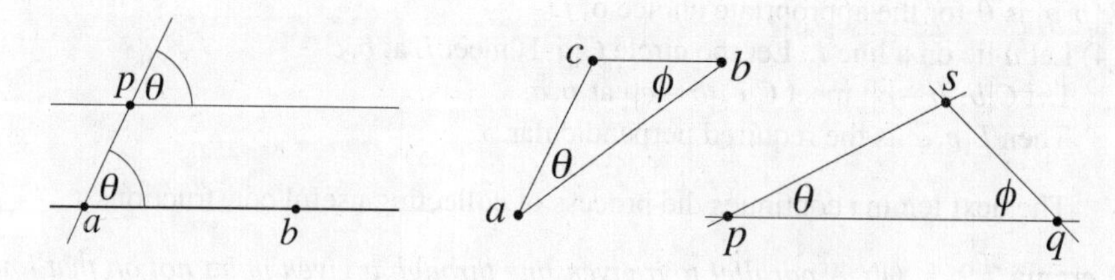

FIGURE 14: Left: Constructing a parallel. Right: Constructing a similar triangle.

Theorem 7.10. *Let $z, w \in \mathbb{C}$. Then, assuming z and w are already constructed:*

(1) *$z+w$ can be constructed using a 7-step construction.*

(2) *$-z$ can be constructed using a 1-step construction.*

(3) *zw can be constructed using a 7-step construction.*

(4) *$1/z$ can be constructed using an 8-step construction.*

(5) *$\pm\sqrt{z}$ can be constructed using an 8-step construction.*

Proof. See Figure 15 for diagrams.
(1) If z, w are not collinear with 0, complete the parallelogram with vertices $0, z, w$. The remaining vertex is $z+w$.

If z, w are collinear with 0, circle $C[z, |w|]$ meets $L[0, z]$ in two points, $z + w$ and $z - w$.
(2) The circle $C[0, |z|]$ meets the line $L[0, z]$ at z and at $-z$.
(3) Consider the triangle T with vertices $0, 1, z$. Construct point p so that the triangle with vertices $0, w, p$ is similar to T.

We claim that $p = zw$. By similarity $|p|/|w| = |z|/1$, so $|p| = |z||w|$. Further, $\arg(p) = \arg z + \arg w$, where arg denotes the argument. Therefore $p = zw$.
(4) Let $C[0, 1]$ meet $L[0, z]$ at p (with 0 lying between z and p). Then $|p| = 1$.

Construct a triangle with vertices $0, p, q$ similar to $0, z, 1$. Then $|q|/1 = |p|/|z| = 1/|z|$, so $|q = 1/|z|$.

Let $C[0, q]$ meet $L[p, z]$ at s, on the same side of the origin as p. Then $|s| = 1/|z|$ and $\arg(s) = \pi + \arg(z)$, so $p = 1/z$.
(5) Let $z = e^{i\theta}$. Then $\sqrt{z} = e^{i\theta/2}, e^{i(\pi+\theta/2)}$. So we have to bisect θ and construct $\sqrt{r} \in \mathbb{R}^+$.

Use $C[0, 1]$ to construct -1.

Bisect $L[-1, r]$ to get $a = (r-1)/2$.

Construct the perpendicular P to $L[0, 1]$ at 0.

Let circle $C[a, |r - a|]$ meet P at s. Then the intersecting chords theorem (or a short calculation with coordinates) implies that $s.s = 1.r$, so $s = \sqrt{r}$.

Construct line L through 0 bisecting the angle between $L[0, r]$ and $L[0, z]$.

This meets the circle $C[0, |s|]$ at $\pm\sqrt{z}$. For the other square root use (2) above. □

Next we characterise the elements of $\mathbb{Q}^{\text{py}}$ in terms of field extensions.

Theorem 7.11. *A complex number α is an element of $\mathbb{Q}^{\text{py}}$ if and only if there is a tower of field extensions*

$$\mathbb{Q} = K_0 \subseteq K_1 \subseteq \ldots \subseteq K_n = \mathbb{Q}(\alpha)$$

such that

$$[K_{j+1} : K_j] = 2$$

for $0 \leq j \leq n-1$.

Proof. First, suppose such a tower exists. We prove by induction on j that $K_j \subseteq \mathbb{Q}^{\text{py}}$. This is clear for $j = 0$. Now, K_{j+1} is an extension of K_j of degree 2, so $K_{j+1} = K_j(\beta)$ where the minimum polynomial of β over K_j is quadratic. Since quadratics can be solved by extracting square roots, $\beta \in \mathbb{Q}^{\text{py}}$, so $K_{j+1} \subseteq \mathbb{Q}^{\text{py}}$. Therefore $\alpha \in \mathbb{Q}^{\text{py}}$.

Next, suppose that $\alpha \in \mathbb{Q}^{\text{py}}$. We prove that such a tower exists. By the definition of $\mathbb{Q}^{\text{py}}$ there is a tower

$$\mathbb{Q} = L_0 \subseteq L_1 \subseteq \ldots \subseteq L_n \supseteq \mathbb{Q}(\alpha)$$

such that $[L_{j+1} : L_j] = 2$ for $0 \leq j \leq n-1$. Define

$$M_j = L_j \cap \mathbb{Q}(\alpha)$$

Consider the L_j and M_j as vector spaces over $\mathbb{Q}$, and note that they are finite-dimensional. We have $\dim L_{j+1} = 2 \dim L_j$ for all relevant j. Therefore either $M_{j+1} =$

M_j or $\dim M_{j+1} = 2\dim M_j$. Delete M_{j+1} if it equals M_j and renumber the resulting M_j as $K_0, K_1, \ldots, K_n$, with $K_0 = \mathbb{Q}$. Clearly $K_n = \mathbb{Q}(\alpha)$. □

From this we immediately deduce a simple *necessary* condition for a point to be constructible:

Theorem 7.12. *If α is constructible then $[\mathbb{Q}(\alpha) : \mathbb{Q}]$ is a power of 2.* □

Now we are ready for the:

Proof. Proof of Part (B) To complete the proof, we must prove (B). If $z \in \mathbb{Q}^{\mathrm{py}}$ then there is a finite sequence of points $z_0 = 0, z_1 = 1, \ldots z_k = z$ such that $z_{l+1} \in \mathbb{Q}(z_0, \ldots, z_l, \alpha)$ where $\alpha^2 \in \mathbb{Q}(z_0, \ldots, z_l)$. Inductively, z_l is constructible by Theorem 7.10, so z_{l+1} is constructible. □

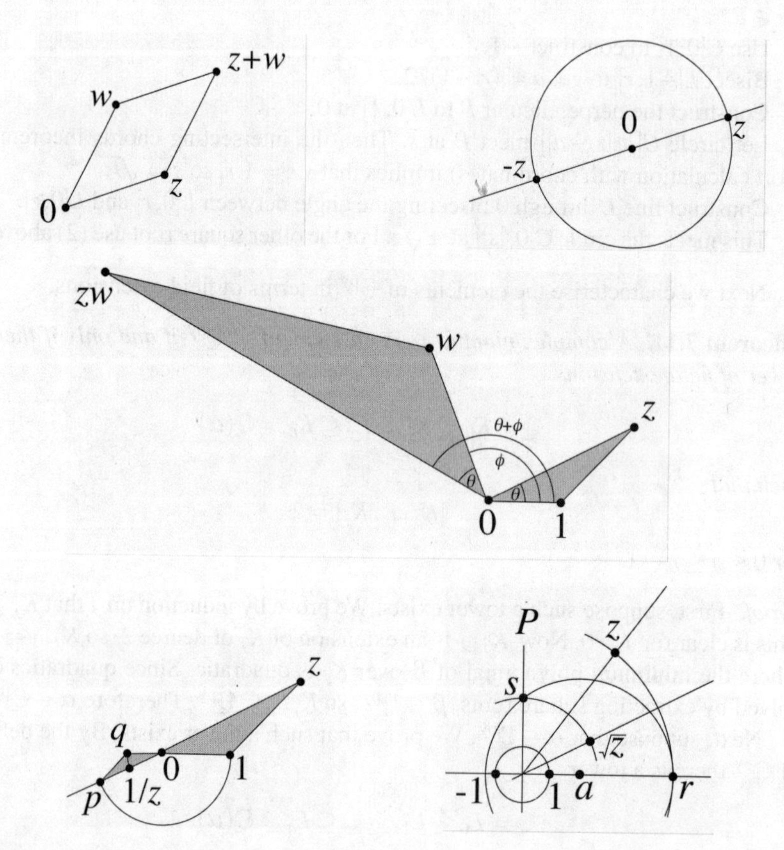

FIGURE 15: Constructions for five operations. Top left: $z+w$. Top right: $-z$. Middle: zw. Bottom left: $1/z$. Bottom right: $\pm\sqrt{z}$.

7.4 Impossibility Proofs

We now apply the above theory to prove that there do not exist ruler-and-compass constructions that solve the three classical problems mentioned in the introduction to this chapter.

We first prove the impossibility of Duplicating the Cube, where the method is especially straightforward.

Theorem 7.13. *The cube cannot be duplicated by ruler and compass construction.*

Proof. Duplicating the cube is equivalent to constructing $\alpha = \sqrt[3]{2}$. Suppose for a contradiction that $\alpha \in \mathbb{Q}^{\mathrm{py}}$, and let m be its minimum polynomial over $\mathbb{Q}$. By Theorem 7.12, $\partial m = 2^k$ for some k.

However, since $\alpha^3 = 2$, the minimum polynomial of α divides $x^3 - 2$. But this is irreducible over $\mathbb{Q}$. If not, it would have a linear factor $x - a$ with $a \in \mathbb{Q}$, and then $a^3 = 2$, so $a = \alpha$. But α is irrational.Therefore $\partial m = 3$, which is not a power of 2, contradicting Theorem 7.12. □

Some angles can be trisected, for example $\pi/2$. However, the required construction should work for any angle, so to prove impossibility it is enough to exhibit one specific angle that cannot be trisected. We prove:

Theorem 7.14. *There exists an angle that cannot be trisected by ruler-and-compass construction.*

Proof. We prove something more specific: the angle $\frac{2\pi}{3}$ cannot be trisected. We know that $\omega = e^{2\pi i/3} \in \mathbb{Q}^{\mathrm{py}}$, since $\omega = \frac{-1+i\sqrt{3}}{2}$. Suppose for a contradiction that such a construction exists. Then $\zeta = e^{2\pi i/9} \in \mathbb{Q}^{\mathrm{py}}$. Therefore $\alpha = \zeta + \zeta^{-1} \in \mathbb{Q}^{\mathrm{py}}$, so its minimum polynomial m over $\mathbb{Q}$ has degree $\partial m = 2^k$ for some k. Now $\zeta^3 = \omega$ and $\omega^2 + \omega + 1 = 0$, so $\zeta^6 + \zeta^3 + 1 = 0$. Therefore $\zeta^6 + \zeta^3 = -1$. But

$$\begin{aligned}\alpha^3 &= (\zeta + \zeta^{-1})^3 \\ &= \zeta^3 + 3\zeta + 3\zeta^{-1} + \zeta^{-3} \\ &= \zeta^3 + 3\zeta + 3\zeta^{-1} + \zeta^6 \\ &= 3\alpha - 1\end{aligned}$$

Therefore m divides $x^3 - 3x + 1$. But this is irreducible over $\mathbb{Q}$ by Gauss's lemma, so $m = x^3 - 3x + 1$ and $\partial m = 3$, contradicting Theorem 7.12. □

This is the place for a word of warning to would-be trisectors, who are often aware of Wantzel's impossibility proof but somehow imagine that they can succeed despite it (Dudley 1987). If you claim a trisection of a general angle using ruler and compasses according to our standing conventions (such as 'unmarked ruler') then you are in particular claiming a trisection of $\pi/3$ using those instruments. The above proof shows that you are therefore claiming that 3 is a power of 2; in particular, since $3 \neq 1$, you are claiming that 3 *is an even number*.

Do you *really* want to go down in history as believing you have proved this?

The final problem of antiquity is more difficult:

Theorem 7.15. *The circle cannot be squared using ruler-and-compass constructions.*

Proof. Such a construction is equivalent to constructing the point $(0, \sqrt{\pi})$ from the initial set of points $P_0 = \{(0,0),(1,0)\}$. From this we can easily construct $(0,\pi)$. So if such a construction exists, then $[\mathbb{Q}(\pi):\mathbb{Q}]$ is a power of 2, and in particular π is algebraic over $\mathbb{Q}$. On the other hand, a famous theorem of Ferdinand Lindemann asserts that π is *not* algebraic over $\mathbb{Q}$. The theorem follows. □

We prove Lindemann's theorem in Chapter 24. We could give the proof now, but it involves ideas off the main track of the book, and has therefore been placed in the Chapter 24. If you are willing to take the result on trust, you can skip the proof.

As a bonus, and to set the scene for Chapter 19 on regular polygons, we dispose of another construction that the ancients might well have wondered about. They knew constructions for regular polygons with 3, 4, 5, sides, and it is easy to double these to get 6, 8, 10, 12, 16, 20, and so on. The impossibility of trisecting $2\pi/3$ also proves that a regular 9-gon (enneagon) cannot be constructed with ruler and compass. But the first 'missing' case is the regular 7-gon (heptagon). Our methods easily prove this impossible, too:

Theorem 7.16. *The regular 7-gon (heptagon) cannot be constructed with ruler and compass.*

Proof. Constructing the regular heptagon is equivalent to proving that

$$\zeta = e^{2\pi i/7} \in \mathbb{Q}^{py}$$

and this complex 7th root of unity satisfies the polynomial equation

$$\zeta^6+\zeta^5+\zeta^4+\zeta^3+\zeta^2+\zeta+1=0$$

because $\zeta^7-1=0$ and the polynomial t^7-1 factorises as

$$t^7-1=(t-1)(t^6+t^5+t^4+t^3+t^2+t+1)$$

Since 7 is prime, Lemma 3.22, implies that $t^6+t^5+t^4+t^3+t^2+t+1$ is irreducible. Its degree is 6, which is not a power of 2, so the regular 7-gon is not constructible.

There is an alternative approach in this case, which does not appeal to Eisenstein's Criterion. Rewrite the above equation as

$$\zeta^3+\zeta^2+\zeta+1+\zeta^{-1}+\zeta^{-2}+\zeta^{-3}=0$$

Now $\zeta \in \mathbb{Q}^{py}$ if and only if $\alpha = \zeta+\zeta^{-1} \in \mathbb{Q}^{py}$, as above. Observe that

$$\begin{aligned}\alpha^3 &= \zeta^3+3\zeta+3\zeta^{-1}+\zeta^{-3}\\ \alpha^2 &= \zeta^2+2+\zeta^{-2}\end{aligned}$$

so

$$\alpha^3 + \alpha^2 - 3\alpha - 1 = 0$$

The polynomial $x^3 + x^2 - 3x - 1$ is irreducible by Gauss's Lemma, Lemma 3.17, so the degree of the minimum polynomial of α over $\mathbb{Q}$ is 3. Therefore $\alpha \notin \mathbb{Q}^{\text{py}}$. $\square$

7.5 Construction From a Given Set of Points

There is a 'relative' version of the theory of this chapter, in which we start not with $\{0,1\}$ but some finite subset $P \subseteq \mathbb{C}$, satisfying some simple technical conditions. This set-up is more appropriate for discussing constructions such as '*given* an angle, bisect it', without assuming that the original angle is itself constructible. In this context, Definition 7.4 is modified to:

Definition 7.17. Let P be a finite subset of $\mathbb{C}$ containing at least two distinct elements, with $0,1 \in P$ (to identify the plane with $\mathbb{C}$). For each $n \in \mathbb{N}$ define sets $\mathscr{P}_n, \mathscr{L}_n$, and $\mathscr{C}_n$ of *points, lines*, and *circles* that are *n-constructible from P* by:

$$\begin{aligned}
\mathscr{P}_0 &= P \\
\mathscr{L}_0 &= \emptyset \\
\mathscr{C}_0 &= \emptyset \\
\mathscr{L}_{n+1} &= \{L(z_1,z_2) : z_1,z_2 \in \mathscr{P}_n\} \\
\mathscr{C}_{n+1} &= \{C(z_1,|z_2 - z_3|) : z_1,z_2,z_3 \in \mathscr{P}_n\} \\
\mathscr{P}_{n+1} &= \{z \in \mathbb{C} : z \text{ lies on two distinct lines in } \mathscr{L}_{n+1}\} \cup \\
&\quad\ \{z \in \mathbb{C} : z \text{ lies on a line in } \mathscr{L}_{n+1} \text{ and a circle in } \mathscr{C}_{n+1}\} \cup \\
&\quad\ \{z \in \mathbb{C} : z \text{ lies on two distinct circles in } \mathscr{L}_{n+1}\}
\end{aligned}$$

A point is *constructible from P* if it is *n-constructible from P* for some *n*.

The entire theory then goes through, with essentially the same proofs, except that the ground field $\mathbb{Q}$ must be replaced by $\mathbb{Q}(P)$ throughout. The constructible points are precisely those in $\mathbb{Q}(P)^{\text{py}}$, defined in the obvious way, and they are characterised by the existence of a tower of subfields of $\mathbb{C}$ starting from $\mathbb{Q}(P)$ such that each successive extension has degree 2. More precisely, Theorem 7.11 becomes

Theorem 7.18. *A complex number α is an element of $\mathbb{Q}(P)^{\text{py}}$ if and only if there is a tower of field extensions*

$$\mathbb{Q}(P) = K_0 \subseteq K_1 \subseteq \ldots \subseteq K_n = \mathbb{Q}(\alpha)$$

such that

$$[K_{j+1} : K_j] = 2$$

for $0 \leq j \leq n-1$.

The proof is the same.

EXERCISES

7.1 Express in the language of this chapter methods of constructing, by ruler and compasses:

(a) The perpendicular bisector of a line.

(b) The points trisecting a line.

(c) Division of a line into n equal parts.

(d) The tangent to a circle at a given point.

(e) Common tangents to two circles.

7.2 Estimate the degrees of the field extensions corresponding to the constructions in Exercise 7.1, by giving reasonably good upper bounds.

7.3 Prove using Euclidean geometry that the 'marked ruler' construction of Figure 11 does indeed trisect the given angle AOB.

7.4 Can the angle $2\pi/5$ be trisected using ruler and compasses?

7.5 Show that it is impossible to construct a regular 9-gon using ruler and compasses.

7.6 By considering a formula for $\cos 5\theta$ find a construction for the regular pentagon.

7.7 Prove that the angle θ can be trisected by ruler and compasses if and only if the polynomial

$$4t^3 - 3t - \cos\theta$$

is reducible over $\mathbb{Q}(\cos\theta)$.

7.8 Verify the following approximate construction for π due to Ramanujan (1962, p. 35), see Figure 16. Let AB be the diameter of a circle centre O. Bisect AO at M, trisect OB at T. Draw TP perpendicular to AB meeting the circle at P. Draw BQ = PT, and join AQ. Draw OS, TR parallel to BQ. Draw AD = AS, and AC = RS tangential to the circle at A. Join BC, BD, CD. Make BE = BM. Draw EX parallel to CD. Then the square on BX has approximately the same area as the circle.
(You will need to know that π is approximately $\frac{355}{113}$. This approximation is first found in the works of the Chinese astronomer Zu Chongzhi in about AD 450.)

7.9 Prove that the construction in Figure 10 is correct if and only if the identity

$$\sin\frac{\theta}{3} = \frac{\sin\theta}{2+\cos\theta}$$

holds. Disprove the identity and estimate the error in the construction.

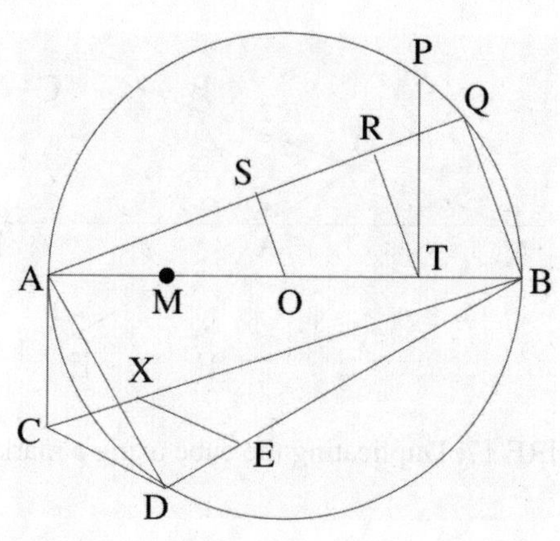

FIGURE 16: Srinivasa Ramanujan's approximate squaring of the circle.

7.10 Show that the 'compasses' operation can be replaced by 'draw a circle centre P_0 and passing through some point other than P_0' without altering the set of constructible points.

7.11 Find a construction with infinitely many steps that trisects any given angle θ, in the sense that the angle ϕ_n obtained by stopping the construction after n steps converges to $\phi = \theta/3$ when n tends to infinity. (*Hint*: consider the infinite series

$$\frac{1}{4}+\frac{1}{16}+\frac{1}{64}+\cdots$$

which converges to $\frac{1}{3}$.)

7.12 A race of alien creatures living in n-dimensional hyperspace $\mathbb{R}^n$ wishes to duplicate the hypercube by ruler-and-compass construction. For which n can they succeed?

7.13 Figure 17 shows a regular hexagon of side AB = 1 and some related lines. If XY = 1, show that YB = $\sqrt[3]{2}$. Deduce that the cube can be duplicated using a marked ruler.

7.14 Since the angles $\frac{\theta}{3}, \frac{\theta}{3}+\frac{2\pi}{3}, \frac{\theta}{3}+\frac{4\pi}{3}$ are all distinct, but equal θ when multiplied by 3, it can be argued that every angle has three distinct trisections. Show that Archimedes's construction with a marked ruler (Figure 11) can find them all.

7.15 Prove that the regular 11-gon cannot be constructed with ruler and compass. [*Hint*: Let $\zeta = e^{2\pi i/11}$ and mimic the proof for a heptagon.]

7.16 Prove that the regular 13-gon cannot be constructed with ruler and compass. [*Hint*: Let $\zeta = e^{2\pi i/13}$ and mimic the proof for a heptagon.]

7.17 The regular 15-gon and 16-gon can be constructed with ruler and compass. So the next regular polygon to consider is the 17-gon.

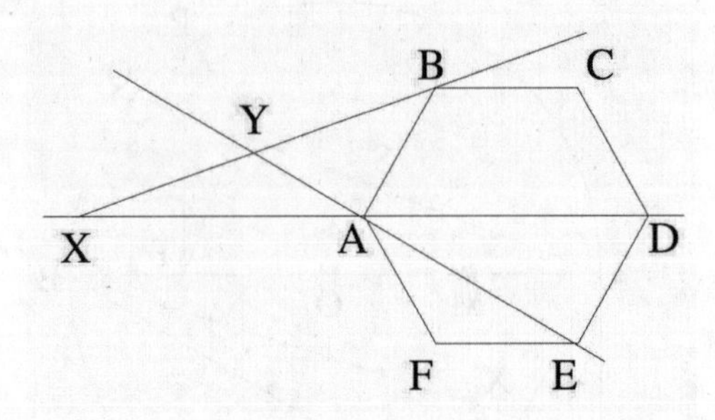

FIGURE 17: Duplicating the cube using a marked ruler.

Why does the method used in the previous questions *fail* for the 17-gon?

7.18* Prove that an angle (which you must specify and which must itself be constructible) cannot be divided into five equal pieces with ruler and compass. [*Hint*: Do not start with $2\pi/3$ or $\pi/2$, both of which *can* be divided into five equal pieces with ruler and compass (why?).]

7.19 If $\alpha \in \mathbb{Q}$, prove that the angle θ such that $\tan\theta = \alpha$ is constructible.

7.20* Let θ be such that $\tan\theta = a/b$ where $a, b \in \mathbb{Z}$ are coprime and $b \neq 0$. Prove the following:

(a) If $a+b$ is odd, then θ can be trisected using ruler and compass if and only if a^2+b^2 is a perfect cube.

(b) If $a+b$ is even, then θ can be trisected using ruler and compass if and only if $(a^2+b^2)/2$ is a perfect cube.

(c) The angles $\tan^{-1} 2/11$ and $\tan^{-1} 9/13$ can be trisected using ruler and compass.

[*Hint*: Use the fact that the ring of Gaussian integers $\mathbb{Z}[i] = \{p+iq : p, q \in \mathbb{Z}\}$ has the property of unique prime factorisation, together with the standard formula for $\tan 3\theta$ in terms of $\tan\theta$.]

This Exercise is based on Chang and Gordon (2014).

7.21 Mark the following true or false.

(a) There exist ruler-and-compass constructions trisecting the angle to an arbitrary degree of approximation.

(b) Such constructions are sufficient for practical purposes but insufficient for mathematical ones.

(c) A point is constructible if it lies in a subfield of $\mathbb{C}$ whose degree over $\mathbb{Q}$ is a power of 2.

(d) The angle π cannot be trisected using ruler and compass.

(e) A line of length π cannot be constructed using ruler and compass.

(f) It is impossible to triplicate the cube (that is, construct one with three times the volume of a given cube) by ruler and compass.

(g) The real number π is transcendental over $\mathbb{Q}$.

(h) The real number π is transcendental over $\mathbb{R}$.

(i) If α cannot be constructed by ruler and compass, then α is transcendental over $\mathbb{Q}$.

Chapter 8

The Idea Behind Galois Theory

Having satisfied ourselves that field extensions are good for something, we can focus on the main theme of this book: the elusive quintic, and Galois's deep insights into the solubility of equations by radicals. We start by outlining the main theorem that we wish to prove, and the steps required to prove it. We also explain where it came from.

We have already associated a vector space to each field extension. For some problems this is too coarse an instrument; it measures the size of the extension, but not its shape, so to speak. Galois went deeper into the structure. To any polynomial $p \in \mathbb{C}[t]$, he associated a group of permutations, now called the *Galois group* of p in his honour. Complicated questions about the polynomial can sometimes be reduced to much simpler questions about the group—especially when it comes to solution by radicals. What makes his work so astonishing is that in his day the group concept existed only in rudimentary form. Others had investigated ideas that we now interpret as early examples of groups, but Galois was arguably the first to recogne the concept in sufficient generality, and to understand its importance.

We introduce the main ideas in a very simple context—a quartic polynomial equation whose roots are obvious. We show that the reason for the roots being obvious can be stated in terms of the *symmetries* of the polynomial—in an appropriate sense—and that any polynomial equation with those symmetries will also have 'obvious' roots.

With a little extra effort, we then subvert the entire reason for the existence of this book, by proving that the 'general' polynomial equation of the nth degree cannot be solved by radicals—of a particular, special kind—when $n \geq 5$. This is a spectacular application of the Galois group, but in a very limited context: it corresponds roughly to what Ruffini proved (or came close to proving) in 1813. By stealing one further idea from Abel, we can even remove Ruffini's assumption, and prove that there is no general radical expression in the coefficients of a quintic, or any polynomial of degree ≥ 5, that determines a zero.

We could stop there. But Galois went much further: his methods are not only more elegant, they give much stronger results. The material in this chapter provides a sprinbgboard, from which we can launch into the full beauty of the theory.

8.1 A First Look at Galois Theory

Galois theory is a fascinating mixture of classical and modern mathematics, and it takes a certain amount of effort to get used to its thought patterns. This section is intended to give a quick survey of the basic principles of the subject, and explain how the abstract treatment has developed from Galois's original ideas.

The aim of Galois theory is to study the solutions of polynomial equations

$$f(t) = t^n + a_{n-1}t^{n-1} + \cdots + a_0 = 0$$

and, in particular, to distinguish those that can be solved by a 'formula' from those that cannot. By a formula we mean a *radical expression*: anything that can be built up from the coefficients a_j by the operations of addition, subtraction, multiplication, and division, and also—the essential ingredient—by nth roots, $n = 2, 3, 4, \ldots$.

In Chapter 1 we saw that polynomial equations over $\mathbb{C}$ of degree 1, 2, ,3 or 4 can be solved by radicals. The central objective of this book is a proof that the quintic equation is different. It cannot, in general, be solved by radicals. Along the way we come to appreciate the deep, general reason *why* quadratics, cubics, and quartics *can* be solved using radicals.

In modern terms, Galois's main idea is to look at the symmetries of the polynomial $f(t)$. These form a group, its *Galois group,* and the solution of the polynomial equation is reflected in various properties of the Galois group.

8.2 Galois Groups According to Galois

Galois had to invent the concept of a group, quite aside from sorting out how it relates to the solution of equations. Not surprisingly, his approach was relatively concrete by today's standards, but by those of his time it was highly abstract. Indeed Galois is one of the founders of modern abstract algebra. So to understand the modern approach, it helps to take a look at something rather closer to what Galois had in mind.

As an example, consider the polynomial equation

$$f(t) = t^4 - 4t^2 - 5 = 0$$

which we encountered in Chapter 4. As we saw, this factorises as

$$(t^2 + 1)(t^2 - 5) = 0$$

so there are four roots $t = i, -i, \sqrt{5}, -\sqrt{5}$. These form two natural pairs: i and $-i$ go together, and so do $\sqrt{5}$ and $-\sqrt{5}$. Indeed, itis impossible to distinguish i from

$-i$, or $\sqrt{5}$ from $-\sqrt{5}$, by algebraic means, in the following sense. Write down any polynomial equation, with rational coefficients, that is satisfied by some selection from the four roots. If we let

$$\alpha = i \qquad \beta = -i \qquad \gamma = \sqrt{5} \qquad \delta = -\sqrt{5}$$

then such equations include

$$\alpha^2+1=0 \qquad \alpha+\beta=0 \qquad \delta^2-5=0 \qquad \gamma+\delta=0 \qquad \alpha\gamma-\beta\delta=0$$

and so on. There are infinitely many valid equations of this kind. On the other hand, infinitely many other algebraic equations, such as $\alpha+\gamma=0$, are manifestly false.

Experiment suggests that if we take any valid equation connecting α, β, γ, and δ, and interchange α and β, we again get a valid equation. The same is true if we interchange γ and δ. For example, the above equations lead by this process to

$$\begin{array}{cccc} \beta^2+1=0 & \beta+\alpha=0 & \gamma^2-5=0 & \delta+\gamma=0 \\ \beta\gamma-\alpha\delta=0 & \alpha\delta-\beta\gamma=0 & \beta\delta-\alpha\gamma=0 & \end{array}$$

and all of these are valid. In contrast, if we interchange α and γ, we obtain equations such as

$$\gamma^2+1=0 \qquad \gamma+\beta=0 \qquad \alpha+\delta=0$$

which are false. Exercise 8.1 outlines a simple proof that these operations preserve all valid equations connecting α, β, γ, and δ.

The operations that we are using here are *permutations* of the zeros $\alpha, \beta, \gamma, \delta$. In fact, in the usual permutation notation, the interchange of α and β is

$$R = \begin{pmatrix} \alpha & \beta & \gamma & \delta \\ \beta & \alpha & \gamma & \delta \end{pmatrix} \tag{8.1}$$

and that of γ and δ is

$$S = \begin{pmatrix} \alpha & \beta & \gamma & \delta \\ \alpha & \beta & \delta & \gamma \end{pmatrix} \tag{8.2}$$

These are elements of the symmetric group $\mathbb{S}_4$ on four symbols, which includes all 24 possible permutations of $\alpha, \beta, \gamma, \delta$.

If these two permutations turn valid equations into valid equations, then so must the permutation obtained by performing them both in turn, which is

$$T = \begin{pmatrix} \alpha & \beta & \gamma & \delta \\ \beta & \alpha & \delta & \gamma \end{pmatrix}$$

Are there any other permutations that preserve all the valid equations? Yes, of course, the identity

$$I = \begin{pmatrix} \alpha & \beta & \gamma & \delta \\ \alpha & \beta & \gamma & \delta \end{pmatrix}$$

It can be checked that only these four permutations preserve valid equations: the other 20 all turn some valid equation into a false one. For example, if α, δ are fixed and β, γ are swapped, the value equation $\alpha+\beta=0$ becomes the invalid equation $\alpha+\gamma=0$.

It is a general fact, and an easy one to prove, that the invertible transformations of a mathematical object that preserve some feature of its structure always form a group. We call this the *symmetry group* of the object. This terminology is especially common when the object is a geometrical figure and the transformations are rigid motions, but the same idea applies more widely. And indeed these four permutations do form a group, which we denote by G.

What Galois realised is that the structure of this group to some extent controls how we should set about solving the equation.

He did not use today's notation for permutations, and this led to potential confusion. To him, a *permutation* of, say, $\{1,2,3,4\}$, was an ordered list, such as 2413. Given a second list, say 3214, he then considered the *substitution* that changes 2413 to 3214; that is, the map $2 \mapsto 3, 4 \mapsto 2, 1 \mapsto 1, 3 \mapsto 4$. Nowadays we would write this as

$$\begin{pmatrix} 2 & 4 & 1 & 3 \\ 3 & 2 & 1 & 4 \end{pmatrix}$$

or, reordering the top row,

$$\begin{pmatrix} 1 & 2 & 3 & 4 \\ 1 & 3 & 4 & 2 \end{pmatrix}$$

but Galois did not even have the $\mapsto$ notation or associated concepts, so he had to write the substitution as 1342. His use of similar notation for both permutations and substitutions takes some getting used to, and probably did not make life easier for the people asked to referee his papers. Today's definition of 'function' or 'map' dates from about 1950; it certainly helps to clarify the ideas.

To see why permutations/substitutions of the roots matter, consider the subgroup $H=\{I,R\}$ of G. Certain expressions in α, β, γ, δ are fixed by the permutations in this group. For example, if we apply R to $\alpha^2+\beta^2-5\gamma\delta^2$, then we obtain $\beta^2+\alpha^2-5\gamma\delta^2$, which is clearly the same. In fact an expression is fixed by R if and only if it is symmetric in α and β.

It is not hard to show that any polynomial in α, β, γ, δ that is symmetric in α and β can be rewritten as a polynomial in $\alpha+\beta, \alpha\beta$, γ, and δ. For example, the above expression can be written as $(\alpha+\beta)^2-2\alpha\beta-5\gamma\delta^2$. But we know that $\alpha=i, \beta=-i$, so that $\alpha+\beta=0$ and $\alpha\beta=1$. Hence the expression reduces to $-2-5\gamma\delta^2$. Now α and β have been eliminated altogether.

8.3 How to Use the Galois Group

Pretend for a moment that we don't know the explicit zeros $i, -i, \sqrt{5}, -\sqrt{5}$, but that we do know the Galois group G. In fact, consider any quartic polynomial $g(t)$

with the same Galois group as our example $f(t)$ above; that way we cannot possibly know the zeros explicitly. Let them be $\alpha, \beta, \gamma, \delta$. Consider three subfields of $\mathbb{C}$ related to $\alpha, \beta, \gamma, \delta$, namely

$$\mathbb{Q} \subseteq \mathbb{Q}(\gamma, \delta) \subseteq \mathbb{Q}(\alpha, \beta, \gamma, \delta)$$

Let $H = \{I, R\} \subseteq G$. Assume that we also know the following two facts:

(1) The numbers fixed by H are precisely those in $\mathbb{Q}(\gamma, \delta)$.

(2) The numbers fixed by G are precisely those in $\mathbb{Q}$.

Then we can work out how to solve the quartic equation $g(t) = 0$, as follows.

The numbers $\alpha + \beta$ and $\alpha\beta$ are obviously both fixed by H. By fact (1) they lie in $\mathbb{Q}(\gamma, \delta)$. But since

$$(t-\alpha)(t-\beta) = t^2 - (\alpha+\beta)t + \alpha\beta$$

this means that α and β satisfy a quadratic equation whose coefficients are in $\mathbb{Q}(\gamma, \delta)$. That is, we can use the formula for solving a quadratic to express α, β in terms of rational functions of γ and δ, together with nothing worse than square roots. Thus we obtain α and β as radical expressions in γ and δ.

But we can repeat the trick to find γ and δ. The numbers $\gamma + \delta$ and $\gamma\delta$ are fixed by the whole of G: they are clearly fixed by R, and also by S, and these generate G. Therefore $\gamma + \delta$ and $\gamma\delta$ belong to $\mathbb{Q}$ by fact (2) above. Therefore γ and δ satisfy a quadratic equation over $\mathbb{Q}$, so they are given by radical expressions in rational numbers. Plugging these into the formulas for α and γ we find that all four zeros are radical expressions in rational numbers.

We have not found the formulas explicitly. But we have shown that certain information about the Galois group necessarily implies that they exist. Given more information, we can finish the job completely.

This example illustrates that the subgroup structure of the Galois group G is closely related to the possibility of solving the equation $g(t) = 0$. Galois discovered that this relationship is very deep and detailed. For example, the proof that an equation of the fifth degree cannot be solved by a formula boils down to this: *the quintic has the wrong sort of Galois group.* Galois's surviving papers do not make this proof explicit, probably because he considered the insolubility of the quintic to be a known theorem, but it is an easy deduction from results that he does state: see Chapter 25.

We present a simplified version of this argument, in a restricted setting, in Section 8.7. In Section 8.8 we remove this technical restriction using Abel's classical methods.

8.4 The Abstract Setting

The modern approach follows Galois closely in principle, but differs in several respects in practice. The permutations of $\alpha, \beta, \gamma, \delta$ that preserve all algebraic relations between them turns out to be the symmetry group of the subfield $\mathbb{Q}(\alpha,\beta,\gamma,\delta)$ of $\mathbb{C}$ generated by the zeros of g, or more precisely its *automorphism group*, which is a fancy name for the same thing.

Moreover, we wish to consider polynomials not just with integer or rational coefficients, but coefficients that lie in a subfield K of $\mathbb{C}$ (or, later, any field). The zeros of a polynomial $f(t)$ with coefficients in K determine another field L which contains K, but may well be larger. Thus the primary object of consideration is a pair of fields $K \subset L$, or in a slight generalisation, a field extension $L : K$. Thus when Galois talks of polynomials, the modern approach talks of field extensions. And the Galois group of the polynomial becomes the group of K-automorphisms of L, that is, of bijections $\theta : L \to L$ such that for all $x, y \in L$ and $k \in K$

$$\begin{aligned} \theta(x+y) &= \theta(x) + \theta(y) \\ \theta(xy) &= \theta(x)\theta(y) \\ \theta(k) &= k \end{aligned}$$

Thus the bulk of the theory is described in terms of field extensions and their groups of K-automorphisms. This point of view was introduced in 1894 by Dedekind, who also gave axiomatic definitions of subrings and subfields of $\mathbb{C}$.

The method used above to solve $g(t) = 0$ relies crucially on knowing the conditions (1) and (2) at the start of Section 8.3. But can we lay hands on that kind of information if we do not already know the zeros of g? The answer is that we can—though not easily—provided we make a general study of the automorphism groups of field extensions, their subgroups, and the subfields fixed by those subgroups. This study leads to the *Galois correspondence* between subgroups of the Galois group and subfields M of L that contain K. Chapters 9-11 set up the Galois correspondence and prove its key properties, and the main theorem is stated and proved in Chapter 12. Chapter 13 studies one example in detail to drive the ideas home. Chapters 15 and 18 derive the spectacular consequences for the quintic. Then, starting in Chapter 16, we generalise the Galois correspondence to arbitrary fields, and develop the resulting theory in several directions.

8.5 Polynomials and Extensions

In this section we define the Galois group of a field extension $L : K$. We begin by defining a special kind of automorphism.

Definition 8.1. Let $L : K$ be a field extension, so that K is a subfield of the subfield L of $\mathbb{C}$. A *K-automorphism* of L is an automorphism α of L such that

$$\alpha(k) = k \quad \text{for all } k \in K \tag{8.3}$$

We say that α *fixes* $k \in K$ if (8.3) holds.

Effectively condition (8.3) makes α an automorphism of the *extension* $L:K$, rather than an automorphism of the large field L alone. The idea of considering automorphisms of a mathematical object relative to a sub-object is a useful general method; it falls within the scope of the famous 1872 'Erlangen Programme' of Felix Klein. Klein's idea was to consider every 'geometry' as the theory of invariants of an associated transformation group. Thus Euclidean geometry is the study of invariants of the group of distance-preserving transformations of the plane; projective geometry arises if we allow projective transformations; topology comes from the group of all continuous maps possessing continuous inverses (called 'homeomorphisms' or 'topological transformations'). According to this interpretation any field extension is a geometry, and we are simply studying the geometrical figures.

The pivot upon which the whole theory turns is a result which is not in itself hard to prove. As Lewis Carroll said in *The Hunting of the Snark,* it is a 'maxim tremendous but trite'.

Theorem 8.2. *If $L:K$ is a field extension, then the set of all K-automorphisms of L forms a group under composition of maps.*

Proof. Suppose that α and β are K-automorphisms of L. Then $\alpha\beta$ is clearly an automorphism; further if $k \in K$ then $\alpha\beta(k) = \alpha(k) = k$, so that $\alpha\beta$ is a K-automorphism. The identity map on L is obviously a K-automorphism. Finally, α^{-1} is an automorphism of L, and for any $k \in K$ we have

$$k = \alpha^{-1}\alpha(k) = \alpha^{-1}(k)$$

so that α^{-1} is a K-automorphism. Composition of maps is associative, so the set of all K-automorphisms of L is a group. □

Definition 8.3. The *Galois group* $\Gamma(L:K)$ of a field extension $L:K$ is the group of all K-automorphisms of L under the operation of composition of maps.

Examples 8.4. (1) The extension $\mathbb{C}:\mathbb{R}$. Suppose that α is an $\mathbb{R}$-automorphism of $\mathbb{C}$. Let $j = \alpha(i)$ where $i = \sqrt{-1}$. Then

$$j^2 = (\alpha(i))^2 = \alpha(i^2) = \alpha(-1) = -1$$

since $\alpha(r) = r$ for all $r \in \mathbb{R}$. Hence either $j = i$ or $j = -i$. Now for any $x, y \in \mathbb{R}$

$$\alpha(x+iy) = \alpha(x) + \alpha(i)\alpha(y) = x + jy$$

Thus we have two candidates for $\mathbb{R}$-automorphisms:

$$\alpha_1 : x+iy \mapsto x+iy$$
$$\alpha_2 : x+iy \mapsto x-iy$$

Obviously α_1 is the identity, and thus is an $\mathbb{R}$-automorphism of $\mathbb{C}$. The map α_2 is complex conjugation, and is an automorphism by Example 1.7(1). Moreover,

$$\alpha_2(x+0i) = x - 0i = x$$

so α_2 is an $\mathbb{R}$-automorphism. Obviously $\alpha_2^2 = \alpha_1$, so the Galois group $\Gamma(\mathbb{C}:\mathbb{R})$ is a cyclic group of order 2.
(2) Let c be the real cube root of 2, and consider $\mathbb{Q}(c):\mathbb{Q}$. If α is a $\mathbb{Q}$-automorphism of $\mathbb{Q}(c)$, then

$$(\alpha(c))^3 = \alpha(c^3) = \alpha(2) = 2$$

Since $\mathbb{Q}(c) \subseteq \mathbb{R}$ we must have $\alpha(c) = c$. Hence α is the identity map, and $\Gamma(\mathbb{Q}(c):\mathbb{Q})$ has order 1.
(3) Let the field extension be $\mathbb{Q}(\sqrt{2},\sqrt{3},\sqrt{5}):\mathbb{Q}$, as in Example 6.8. The analysis presented in that example shows that t^2-5 is irreducible over $\mathbb{Q}(\sqrt{2},\sqrt{3})$. Similarly, t^2-2 is irreducible over $\mathbb{Q}(\sqrt{3},\sqrt{5})$ and t^2-3 is irreducible over $\mathbb{Q}(\sqrt{2},\sqrt{5})$. Thus there are three $\mathbb{Q}$-automorphisms of $\mathbb{Q}(\sqrt{2},\sqrt{3},\sqrt{5})$, defined by

$$\begin{array}{llll} \rho_2: & \sqrt{2}\mapsto-\sqrt{2} & \sqrt{3}\mapsto\sqrt{3} & \sqrt{5}\mapsto\sqrt{5} \\ \rho_3: & \sqrt{2}\mapsto\sqrt{2} & \sqrt{3}\mapsto-\sqrt{3} & \sqrt{5}\mapsto\sqrt{5} \\ \rho_5: & \sqrt{2}\mapsto\sqrt{2} & \sqrt{3}\mapsto\sqrt{3} & \sqrt{5}\mapsto-\sqrt{5} \end{array}$$

It is easy to see that these maps commute, and hence generate the group $\mathbb{Z}_2\times\mathbb{Z}_2\times\mathbb{Z}_2$ Moreover, any $\mathbb{Q}$-automorphism of $\mathbb{Q}(\sqrt{2},\sqrt{3},\sqrt{5})$ must map $\sqrt{2}\mapsto\pm\sqrt{2}$, $\sqrt{3}\mapsto\pm\sqrt{3}$, and $\sqrt{5}\mapsto\pm\sqrt{5}$ by considering minimal polynomials. All combinations of signs occur in the group $\mathbb{Z}_2\times\mathbb{Z}_2\times\mathbb{Z}_2$, so this must be the Galois group.

8.6 The Galois Correspondence

Although it is easy to prove that the set of all K-automorphisms of a field extension $L:K$ forms a group, that fact alone does not significantly advance the subject. To be of any use, the Galois group must reflect aspects of the structure of $L:K$. Galois made the discovery (which he expressed in terms of polynomials) that, under certain extra hypotheses, there is a one-to-one correspondence between:

(1) Subgroups of the Galois group of $L:K$.

(2) Subfields M of L such that $K\subseteq M$.

As it happens, this correspondence *reverses* inclusion relations: larger subfields correspond to smaller groups. First, we explain how the correspondence is set up.

If $L:K$ is a field extension, we call any field M such that $K\subseteq M\subseteq L$ an *intermediate* field. To each intermediate field M we associate the group $M^*=\Gamma(L:M)$ of all M-automorphisms of L. Thus K^* is the whole Galois group, and $L^*=1$ (the group consisting of just the identity map on L). Clearly if $M\subseteq N$ then $M^*\supseteq N^*$, because any automorphism of L that fixes the elements of N certainly fixes the elements of M. This is what we mean by 'reverses inclusions'.

Conversely, to each subgroup H of $\Gamma(L:K)$ we associate the set $H^\dagger$ of all elements $x\in L$ such that $\alpha(x)=x$ for all $\alpha\in H$. In fact, this set is an intermediate field:

Lemma 8.5. *If H is a subgroup of $\Gamma(L:K)$, then $H^\dagger$ is a subfield of L containing K.*

Proof. Let $x, y \in H^\dagger$, and $\alpha \in H$. Then

$$\alpha(x+y) = \alpha(x) + \alpha(y) = x + y$$

so $x+y \in H^\dagger$. Similarly $H^\dagger$ is closed under subtraction, multiplication, and division (by nonzero elements), so $H^\dagger$ is a subfield of L. Since $\alpha \in \Gamma(L:K)$ we have $\alpha(k) = k$ for all $k \in K$, so $K \subseteq H^\dagger$. □

Definition 8.6. With the above notation, $H^\dagger$ is the *fixed field* of H.

It is easy to see that like $*$, the map $\dagger$ reverses inclusions: if $H \subseteq G$ then $H^\dagger \supseteq G^\dagger$. It is also easy to verify that if M is an intermediate field and H is a subgroup of the Galois group, then

$$\begin{aligned} M &\subseteq M^{*\dagger} \\ H &\subseteq H^{\dagger *} \end{aligned} \tag{8.4}$$

Indeed, every element of M is fixed by every automorphism that fixes all of M, and every element of H fixes those elements that are fixed by all of H. Example 8.4(2) shows that these inclusions are not always equalities, for there

$$\mathbb{Q}^{*\dagger} = \mathbb{Q}(c) \neq \mathbb{Q}$$

If we let $\mathscr{F}$ denote the set of intermediate fields, and $\mathscr{G}$ the set of subgroups of the Galois group, then we have defined two maps

$$\begin{aligned} * &: \mathscr{F} \to \mathscr{G} \\ \dagger &: \mathscr{G} \to \mathscr{F} \end{aligned}$$

which reverse inclusions and satisfy equation (8.4). These two maps constitute the *Galois correspondence* between $\mathscr{F}$ and $\mathscr{G}$. Galois's results can be interpreted as giving conditions under which * and † are mutual inverses, setting up a bijection between $\mathscr{F}$ and $\mathscr{G}$. The extra conditions needed are called *separability* (which is automatic over $\mathbb{C}$) and *normality*. We discuss them in Chapter 9.

Example 8.7. The polynomial equation

$$f(t) = t^4 - 4t^2 - 5 = 0$$

was discussed in Section 8.2. Its roots are $\alpha = i$, $\beta = -i$, $\gamma = \sqrt{5}$, $\delta = -\sqrt{5}$. The associated field extension is $L : \mathbb{Q}$ where $L = \mathbb{Q}(i, \sqrt{5})$, which we discussed in Example 4.8. There are four $\mathbb{Q}$-automorphisms of L, namely I, R, S, T where I is the identity, and in cycle notation $R = (\alpha\beta), S = (\gamma\delta)$, and $T = (\alpha\beta)(\gamma\delta)$. Recall that a *cycle* $(a_1 \dots a_k) \in \mathbb{S}_n$ is the permutation σ such that $\sigma(a_j) = a_{j+1}$ when $1 \leq j \leq k-1$, $\sigma(a_k) = a_1$, and $\sigma(a) = a$ when $a \notin \{a_1, \dots, a_k\}$. Every element of $\mathbb{S}_n$ is a product of disjoint cycles, which commute, and this expression is unique except for the order in which the cycles are composed.

In fact I,R,S,T are all possible $\mathbb{Q}$-automorphisms of L, because any $\mathbb{Q}$-automorphism must send i to $\pm i$ and $\sqrt{5}$ to $\pm\sqrt{5}$. Therefore the Galois group is

$$G = \{I,R,S,T\}$$

The proper subgroups of G are

$$1 \quad \{I,R\} \quad \{I,S\} \quad \{I,T\}$$

where $1 = \{I\}$. It is easy to check that the corresponding fixed fields are respectively

$$L \quad \mathbb{Q}(\sqrt{5}) \quad \mathbb{Q}(i) \quad \mathbb{Q}(i\sqrt{5})$$

Extensive but routine calculations (Exercise 8.2) show that these, together with K, are the only subfields of L. So in this case the Galois correspondence is bijective.

8.7 Diet Galois

To provide further motivation, we now pursue a modernised version of Lagrange's train of thought in his memoir of 1770-1771, which paved the way for Galois. Indeed we will follow a line of argument that is very close to the work of Ruffini and Abel, and prove that the general quintic is not soluble by radicals. Why, then, does the rest of this book exist? Because 'general' has a paradoxically special meaning in this context, and we have to place a very strong restriction on the kind of radical that is permitted. A major feature of Galois theory is that it does not assume this restriction. However, quadratics, cubics, and quartics *are* soluble by these restricted types of radical, so the discission here does have some intrinsic merit. It could profitably be included as an application in a first course of group theory, or a digression in a course on rings and fields.

We have already encountered the symmetric group $\mathbb{S}_n$, which comprises all permutations of the set $\{1,2,\ldots,n\}$. Its order is $n!$. When $n \geq 2, \mathbb{S}_n$ has a subgroup of index 2 (that is, of order $n!/2$); namely, the *alternating group* $\mathbb{A}_n$,F which consists of all products of an even number of transpositions (ab). The elements of $\mathbb{A}_n$ are the *even permutations*. The group $\mathbb{A}_n$ is a normal subgroup of $\mathbb{S}_n$. It is well known that $\mathbb{A}_n$ is generated by all 3-cycles (abc): see Exercise 8.7. The group $\mathbb{A}_5$ holds the secret of the quintic, as we now explain.

Introduce the polynomial ring $\mathbb{C}[t_1,\ldots,t_n]$ in n indeterminates. Let its field of fractions be $\mathbb{C}(t_1,\ldots,t_n)$, consisting of rational expressions in the t_j. Consider the polynomial

$$F(t) = (t-t_1)\ldots(t-t_n)$$

over $\mathbb{C}(t_1,\ldots,t_n)$, whose zeros are $t_1,\ldots,t_n$. Expanding and using induction, we see that

$$F(t) = t^n - s_1t^{n-1} + s_2t^{n-2} + \cdots + (-1)^n s_n \tag{8.5}$$

where the s_j are the *elementary symmetric polynomials*

$$s_1 = t_1 + \cdots + t_n$$
$$s_2 = t_1t_2 + t_1t_3 + \cdots + t_{n-1}t_n$$
$$\cdots$$
$$s_n = t_1 \dots t_n$$

Here s_r is the sum of all products of r distinct t_j.

The symmetric group $\mathbb{S}_n$ acts as symmetries of $\mathbb{C}(t_1, \dots, t_n)$:

$$\sigma f(t_1, \dots, t_n) = f(t_{\sigma(1)}, \dots, t_{\sigma(n)})$$

for $f \in \mathbb{C}(t_1, \dots, t_n)$. The fixed field K of $\mathbb{S}_n$ consists, by definition, of all symmetric rational functions in the t_j, which is known to be generated over $\mathbb{C}$ by the n elementary symmetric polynomials in the t_j. That is, $K = \mathbb{C}(s_1, \dots, s_n)$. Moreover, the s_j satisfy no nontrivial polynomial relation: they are independent. There is a classical proof of these facts based on induction, using 'symmetrised monomials'

$$t_1^{a_1} t_2^{a_2} \cdots t_n^{a_n} + \text{all permutations thereof}$$

and the so-called 'lexicographic ordering' of the list of exponents $a_1, \dots, a_n$. See Exercise 8.5. A more modern but less constructive proof is given in Chapter 18.

Assuming that the s_j generate the fixed field, we consider the extension

$$\mathbb{C}(t_1, \dots, t_n) : \mathbb{C}(s_1, \dots, s_n)$$

We know that in $\mathbb{C}(t_1, \dots, t_n)$ the polynomial $F(t)$ in (8.5) factorises completely as

$$F(t) = (t - t_1) \dots (t - t_n)$$

Since the s_j are independent indeterminates, $F(t)$ is traditionally called the *general* polynomial of degree n. The reason for this name is that this polynomial has a universal property. If we can solve $F(t) = 0$ by radicals, then we can solve any *specific* complex polynomial equation of degree n by radicals. Just substitute specific numbers for the coefficients s_j. The converse, however, is not obvious. We might be able to solve every specific complex polynomial equation of degree n by radicals, but using a different formula each time. Then we would not be able to deduce a radical expression to solve $F(t) = 0$. So the adjective 'general' is somewhat misleading; 'generic' would be better, and is sometimes used.

The next definition is not standard, but its name is justified because it reflects the assumptions made by Ruffini in his attempted proof that the quintic is insoluble.

Definition 8.8. The general polynomial equation $F(t) = 0$ is *soluble by Ruffini radicals* if there exists a finite tower of subfields

$$\mathbb{C}(s_1, \dots, s_n) = K_0 \subseteq K_1 \subseteq \cdots \subseteq K_r = \mathbb{C}(t_1, \dots, t_n) \tag{8.6}$$

such that for $j = 1, \dots, r$,

$$K_j = K_{j-1}(\alpha_j) \quad \text{and} \quad \alpha_j^{n_j} \in K_j \quad \text{for} \quad n_j \geq 2,\ n_j \in \mathbb{N}$$

The aim of this definition is to exclude possibilities like the $\sqrt{-121}$ in Cardano's solution (1.10) of the quartic equation $t^4 - 15t - 4 = 0$, which does not lie in the field generated by the roots, but is used to express them by radicals.

Ruffini tacitly assumed that if $F(t) = 0$ is soluble by radicals, then those radicals are all expressible as rational functions of the roots $t_1, \dots, t_n$. Indeed, this was the situation studied by his predecessor Lagrange in his deep but inconclusive researches on the quintic. So Lagrange and Ruffini considered only solubility by Ruffini radicals. However, this is a strong assumption. It is conceivable that a solution by radicals might exist, for which some of the α_j constructed along the way do *not* lie in $\mathbb{C}(t_1, \dots, t_n)$, but in some extension of $\mathbb{C}(t_1, \dots, t_n)$. For example, $\sqrt[5]{s_1}$ might be useful. (It *is* useful to solve $t^5 - s_1 = 0$, for instance, but the solutions of this equation do not belong to $\mathbb{C}(t_1, \dots, t_n)$.) However, the more we think about this possibility, the less likely it seems. Abel thought about it very hard, and *proved* that if $F(t) = 0$ is soluble by radicals, then those radicals are all expressible in terms of rational functions of the roots—they are Ruffini radicals after all. This step, historically called 'Abel's Theorem', is more commonly referred to as the 'Theorem on Natural Irrationalities'. From today's perspective, it is the main difficulty in the impossibility proof. So, following Lagrange and Ruffini, we start by defining the main difficulty away. In compensation, we gain excellent motivation for the remainder of this book.

For completeness, we prove the Theorem on Natural Irrationalities in Section 8.8, using classical (pre-Galois) methods. As preparation for all of the above, we need:

Proposition 8.9. *If there is a finite tower of subfields* (8.6)*, then it can be refined (if necessary increasing its length) to make all n_j prime.*

Proof. For fixed j write $n_j = p_1 \dots p_k$ where the p_l are prime. Let $\beta_l = \alpha_j^{p_{l+1} \cdots p_k}$, for $0 \leq l \leq k$. Then $\beta_0 \in K_j$ and $\beta_l^{p_l} \in K_j(\beta_{l-1})$, and the rest is easy. □

For the remainder of this chapter we assume that this refinement has been performed, and write p_j for n_j as a reminder. With this preliminary step completed, we will prove:

Theorem 8.10. *The general polynomial equation $F(t) = 0$ is insoluble by Ruffini radicals if $n \geq 5$.*

All we need is a simple group-theoretic lemma.

Lemma 8.11. (1) *The symmetric group $\mathbb{S}_n$ has a cyclic quotient group of prime order p if and only if $p = 2$ and $n \geq 2$, in which case the kernel is the alternating group $\mathbb{A}_n$.*
(2) *The alternating group $\mathbb{A}_n$ has a cyclic quotient group of prime order p if and only if $p = 3$ and $n = 3, 4$.*

Proof. (1) We may assume $n \geq 3$ since there is nothing to prove when $n = 1, 2$. Suppose that N is a normal subgroup of $\mathbb{S}_n$ and $\mathbb{S}_n / N \cong \mathbb{Z}_p$. Then $\mathbb{S}_n / N$ is abelian, so N contains every *commutator* $ghg^{-1}h^{-1}$ for $g, h \in \mathbb{S}_n$. To see why, let $\bar{g}$ denote the image of $g \in \mathbb{S}_n$ in the quotient group $\mathbb{S}_n / N$. Since $\mathbb{S}_n / N$ is abelian, $\bar{g}\bar{h}\bar{g}^{-1}\bar{h}^{-1} = \bar{1}$ in $\mathbb{S}_n / N$; that is, $ghg^{-1}h^{-1} \in N$.

Let g, h be 2-cycles of the form $g = (ab), h = (ac)$ where a, b, c are distinct. Then

$$ghg^{-1}h^{-1} = (bca)$$

is a 3-cycle, and all possible 3-cycles can be obtained in this way. Therefore N contains all 3-cycles. But the 3-cycles generate $\mathbb{A}_n$, so $N \supseteq \mathbb{A}_n$. Therefore $p = 2$ since $|\mathbb{S}_n/\mathbb{A}_n| = 2$.

(2) Suppose that N is a normal subgroup of $\mathbb{A}_n$ and $\mathbb{A}_n/N \cong \mathbb{Z}_p$. Again, N contains every commutator. If $n = 2$ then $\mathbb{A}_n$ is trivial. When $n = 3$ we know that $\mathbb{A}_n \cong \mathbb{Z}_3$.

Suppose first that $n = 4$. Consider the commutator $ghg^{-1}h^{-1}$ where $g = (abc), h = (abd)$ for a, b, c, d distinct. Computation shows that

$$ghg^{-1}h^{-1} = (ab)(cd)$$

so N must contain $(12)(34), (13)(24)$, and $(14)(23)$. It also contains the identity. But these four elements form a group $\mathbb{V}$. Thus $\mathbb{V} \subseteq N$. Since $\mathbb{V}$ is a normal subgroup of $\mathbb{A}_4$ and $\mathbb{A}_4/\mathbb{V} \cong \mathbb{Z}_3$, we are done.

The symbol $\mathbb{V}$ comes from Klein's term *Vierergruppe*, or 'fours-group'. Nowadays it is usually called the *Klein four-group*.

Finally, assume that $n \geq 5$. The same argument shows that N contains all permutations of the form $(ab)(cd)$. If a, b, c, d, e are all distinct (which is why the case $n = 4$ is special) then

$$(ab)(cd) \cdot (ab)(ce) = (ced)$$

so N contains all 3-cycles. But the 3-cycles generate $\mathbb{A}_n$, so this case cannot occur. □

As our final preparatory step, we recall the expression (1.13)

$$\delta = \prod_{j<k}^{n} (t_j - t_k)$$

It is not a symmetric polynomial in the t_j, but its square $\Delta = \delta^2$ is, because

$$\Delta = (-1)^{n(n-1)/2} \prod_{j \neq k}^{n} (t_j - t_k)$$

The expression Δ, mentioned in passing in Section 1.4, is called the *discriminant* of $F(t)$. If $\sigma \in \mathbb{S}_n$, then the action of σ sends δ to $\pm\delta$. The even permutations (those in $\mathbb{A}_n$) fix δ, and the odd ones map δ to $-\delta$. Indeed, this is a standard way to define odd and even permutations.

We are now ready for the:

Proof of Theorem 8.10

Assume that $F(t) = 0$ is soluble by Ruffini radicals, with a tower (8.6) of subfields K_j in which all $n_j = p_j$ are prime. Let $K = \mathbb{C}(s_1, \ldots, s_n)$ and $L = \mathbb{C}(t_1, \ldots, t_n)$. Consider the first step in the tower,

$$K \subseteq K_1 \subseteq L$$

where $K_1 = K(\alpha_1), \alpha_1^p \in K, \alpha_1 \notin K$, and $p = p_1$ is prime.

Since $\alpha_1 \in L$ we can act on it by $\mathbb{S}_n$, and since every $\sigma \in \mathbb{S}_n$ fixes K we have

$$(\sigma(\alpha_1))^p = \alpha_1^p$$

Therefore $\sigma(\alpha_1) = \zeta^{j(\sigma)}\alpha_1$, for ζ a primitive pth root of unity and $j(\sigma)$ an integer between 0 and $p-1$. The set of all pth roots of unity in $\mathbb{C}$ is a group under multiplication, and this group is cyclic, isomorphic to $\mathbb{Z}_p$. Indeed $\zeta^a\zeta^b = \zeta^{a+b}$ where $a+b$ is taken modulo p.

Clearly the map

$$\begin{aligned} j : \mathbb{S}_n &\to \mathbb{Z}_p \\ \sigma &\mapsto j(\sigma) \end{aligned}$$

is a group homomorphism. Since $\alpha_1 \notin K$, some $\sigma(\alpha_1) \neq \alpha_1$, so j is nontrivial. Since $\mathbb{Z}_p$ has prime order, hence no nontrivial proper subgroups, j must be onto. Therefore $\mathbb{S}_n$ has a homomorphic image that is cyclic of order p. By Lemma 8.11, $p = 2$ and the kernel is $\mathbb{A}_n$. Therefore α_1 is fixed by $\mathbb{A}_n$.

We claim that this implies that $\alpha_1 \in K(\delta)$. Since $p = 2$, the relation $\alpha_1^p \in K$ becomes $\alpha_1^2 \in K$, so α_1 is a zero of $t^2 - \alpha_1^2 \in K[t]$. The images of α_1 under $\mathbb{S}_n$ must all be zeros of this, namely $\pm\alpha_1$. Now α_1 is fixed by $\mathbb{A}_n$ but not by $\mathbb{S}_n$, so some permutation $\sigma \in \mathbb{S}_n \setminus \mathbb{A}_n$ satisifes $\sigma(\alpha_1) = -\alpha_1$. Then $\delta\alpha_1$ is fixed by both $\mathbb{A}_n$ and σ, hence by $\mathbb{S}_n$. So $\delta\alpha_1 \in K$ and $\alpha_1 \in K(\delta)$.

If $n = 2$ we are finished. Otherwise consider the second step in the tower

$$K(\delta) \subseteq K_2 = K(\delta)(\alpha_2)$$

By a similar argument, α_2 defines a group homomorphism $j : \mathbb{A}_n \to \mathbb{Z}_p$, which again must be onto. By Lemma 8.11, $p = 3$ and $n = 3, 4$. In particular, no tower of Ruffini radicals exists when $n \geq 5$. □

It is plausible that any tower of radicals that leads from $\mathbb{C}(s_1, \ldots, s_n)$ to a subfield containing $\mathbb{C}(t_1, \ldots, t_n)$ must give rise to a tower of Ruffini radicals. However, it is not at all clear how to prove this, and in fact, this is where the main difficulty of the problem really lies, once the role of permutations is understood. Ruffini appeared not to notice that this needed proof. Abel tackled the obstacle head on.

Galois worked his way round it, by way of the Galois group—an extremely elegant solution. The actual details of his work differ considerably from the modern presentation, see Neumann (2011), both notationally and strategically. However, the underlying idea of studying what we now interpret as the symmetry group of the polynomial, and deriving properties related to solubility by radicals, is central to Galois's approach. His method also went much further: it applies not just to the general polynomial $F(t)$, but to any polynomial whatsoever. And it provides necessary *and sufficient* conditions for solutions by radicals to exist.

Exercises 8.9-8.11 provide enough hints for you to show that when $n = 2, 3, 4$ the equation $F(t) = 0$, where F is defined by (8.5), *can* be solved by Ruffini radicals. Therefore, despite the special nature of Ruffini radicals, we see that the quintic equation differs (radically) from the quadratic, cubic, and quartic equations. We also appreciate the significant role of group theory and symmetries of the roots of a polynomial for the existence—or not—of a solution by radicals. This will serve us in good stead when the going gets tougher.

8.8 Natural Irrationalities

With a little more effort we can go the whole hog. Abel's proof contains one further idea, which lets us delete the word 'Ruffini' from Theorem 8.10. This section is an optional extra, and nothing later depends on it. We continue to work with the general polynomial, so throughout this section $L = \mathbb{C}(t_1, \ldots, t_n)$ and $K = \mathbb{C}(s_1, \ldots, s_n)$, where the s_j are the elementary symmetric polynomials in the t_j.

To delete 'Ruffini' we need:

Definition 8.12. An extension $L : K$ in $\mathbb{C}$ is *radical* if $L = K(\alpha_1, \ldots, \alpha_m)$ where for each $j = 1, \ldots, m$ there exists an integer n_j such that

$$\alpha_j^{n_j} \in K(\alpha_1, \ldots, \alpha_{j-1}) \qquad (j \geq 2)$$

The elements α_j form a *radical sequence* for $L : K$. The *radical degree* of the radical α_j is n_j.

The essential point is:

Theorem 8.13. *If the general polynomial equation $F(t) = 0$ can be solved by radicals, then it can be solved by Ruffini radicals.*

Corollary 8.14. *The general polynomial equation $F(t) = 0$ is insoluble by radicals if $n \geq 5$.*

To prove the above, all we need is the so-called 'Theorem on Natural Irrationalities', which states that extraneous radicals like $\sqrt[5]{s_1}$ cannot help in the solution of $F(t) = 0$. More precisely:

Theorem 8.15 (Natural Irrationalities). *If L contains an element x that lies in some radical extension R of K, then there exists a radical extension R' of K with $x \in R'$ and $R' \subseteq L$.*

Once we have proved Theorem 8.15, any solution of $F(t) = 0$ by radicals can be converted into one by Ruffini radicals. Theorem 8.13 and Corollary 8.14 are then immediate.

It remains to prove Theorem 8.15. A proof using Galois theory is straightforward, see Exercise 15.11. With what we know at the moment, we have to work a little harder—but, following Abel's strategic insights, not much harder. We need several lemmas, and a technical definition.

Definition 8.16. *Let $R:K$ be a radical extension. The* height *of $R:K$ is the smallest integer h such that there exist elements $\alpha_1,\dots,\alpha_h \in R$ and primes $p_1,\dots,p_h$ such that $R = K(\alpha_1,\dots,\alpha_h)$ and*

$$\alpha_j^{p_j} \in K(\alpha_1,\dots,\alpha_{j-1}) \qquad 1 \leq j \leq h$$

where when $j=1$ we interpret $K(\alpha_1,\dots,\alpha_{j-1})$ as K.

Proposition 8.9 shows that the height of every radical extension is defined.

We prove Theorem 8.15 by induction on the height of a radical extension R that contains x. The key step is extensions of height 1, and this is where all the work is put in.

Lemma 8.17. *Let M be a subfield of L such that $K \subseteq M$, and let $a \in M$, where a is not a pth power in M. Then*

(1) *a^k is not a pth power in M for $k = 1,2,\dots,p-1$.*

(2) *The polynomial $m(t) = t^p - a$ is irreducible over M.*

Proof. (1) Since k is prime to p there exist integers q,l such that $qp+lk=1$. If $a^k = b^p$ with $b \in M$, then

$$(a^q b^l)^p = a^{qp} b^{lp} = a^{qp} a^{kl} = a$$

contrary to a not being a pth power in M.

(2) Assume for a contradiction that $t^p - a$ is reducible over M. Suppose that $P(t)$ is a monic irreducible factor of $m(t) = t^p - a$ over M. For $0 \leq j \leq p-1$ let $P_j(t) = P(\zeta^j t)$, where $\zeta \in \mathbb{C} \subseteq K \subseteq M$ is a primitive pth root of unity. Then $P_0 = P$, and P_j is irreducible for all j, for if $P(\zeta^j t) = g(t)h(t)$ then $P(t) = g(\zeta^{-j}t)h(\zeta^{-j}t)$. Moreover, $m(\zeta^j t) = (\zeta^j t)^p - a = t^p - a = m(t)$, so P_j divides m for all $j = 0,\dots,p-1$ by Lemma 5.6.

We claim that P_k and P_j are coprime whenever $0 \leq j < k \leq p-1$. If not, by irreducibility

$$P_j(t) = cP_k(t) \qquad c \in M$$

Let

$$P(t) = p_0 + p_1 t + \cdots + p_{r-1}t^{r-1} + t^r$$

where $r \leq p$. By irreducibility, $p_0 \neq 0$. Then

$$\begin{aligned} P_j(t) &= p_0 + p_1\zeta^j t + \cdots + p_{r-1}\zeta^{j(r-1)}t^{r-1} + \zeta^{jr}t^r \\ P_k(t) &= p_0 + p_1\zeta^k t + \cdots + p_{r-1}\zeta^{k(r-1)}t^{r-1} + \zeta^{kr}t^r \end{aligned}$$

so $c = \zeta^{(j-k)r}$ from the coefficient of t^r. But then $p_0 = \zeta^{(j-k)r}p_0$. Since $p_0 \neq 0$, we must have $\zeta^{(j-k)r} = 1$, so $r = p$. But this implies that $\partial P = \partial m$, so m is irreducible over M.

Thus we may assume that the P_j are pairwise coprime. We know that $P_j | m$ for all j, so

$$P_0 P_1 \dots P_{p-1} \,|\, m$$

Since $\partial p = r$, it follows that $pr \leq p$, so $r = 1$. Thus P is linear, so there exists $b \in M$ such that $(t-b)|m(t)$. But this implies that $b^p = a$, contradicting the assumption that a is not a pth power. Thus $t^p - a$ is irreducible. □

Now suppose that R is a radical extension of height 1 over M. Then $R = M(\alpha)$ where $\alpha^p \in M$, $\alpha \notin M$. Therefore every $x \in R \setminus M$ is uniquely expressible as

$$x = x_0 + x_1\alpha + x_2\alpha^2 + \cdots x_{p-1}\alpha^{p-1} \tag{8.7}$$

where the $x_j \in M$.This follows since $[M(\alpha):M] = p$ by irreducibility of m. We want to put x into a more convenient form, and for this we need the following result:

Lemma 8.18. *Let $L \subseteq M$ be fields, and let p be a prime such that L contains a primitive pth root of unity ζ. Suppose that $\alpha, x_0, \ldots, x_{p-1} \in M$ with $\alpha \neq 0$, and L contains all of the elements*

$$X_r = x_0 + (\zeta^r\alpha)x_1 + (\zeta^r\alpha)^2 x_2 + \cdots + (\zeta^r\alpha)^{p-1}x_{p-1} \tag{8.8}$$

for $0 \leq r \leq p-1$. Then each of the elements $x_0, \alpha x_1, \alpha^2 x_2, \ldots, \alpha^{p-1}x_{p-1}$ also lies in L. Hence, if $x_1 = 1$, then α and each x_j $(0 \leq j \leq p-1)$ lies in L.

Proof. For any m with $0 \leq m \leq p-1$, consider the sum

$$X_0 + \zeta^{-m}X_1 + \zeta^{-2m}X_2 + \cdots + \zeta^{-(p-1)m}X_{p-1}$$

Since $1 + \zeta + \zeta^2 + \cdots + \zeta^{p-1} = 0$, all terms vanish except for those in which the power of ζ is zero. These terms sum to $p\alpha^m x_m$. Therefore $p\alpha^m x_m \in L$, so $\alpha^m x_m \in L$.

If $x_1 = 1$ then the case $m = 1$ shows that $\alpha \in L$, so now $x_m \in L$ for all m with $0 \leq m \leq p-1$. □

We can also prove:

Lemma 8.19. *With the above notation, for a given $x \in R$, there exist $\beta \in M(\alpha)$ and $b \in M$ with $b = \beta^p$, such that b is not the pth power of an element of M, and*

$$x = y_0 + \beta + y_2\beta^2 + \cdots y_{p-1}\beta^{p-1}$$

where the $y_j \in M$.

Proof. We know that $x \notin M$, so in (8.7) some $x_s \neq 0$ for $1 \leq s \leq p-1$. Let $\beta = x_s\alpha^s$, and let $b = \beta^p$. Then $b = x_s^p\alpha^{sp} = x_s^p a^s$, and if b is a pth power of an element of M then a^s is a pth power of an element of M, contrary to Lemma 8.17(2). Therefore b is not the pth power of an element of M.

Now s is prime to p, and the additive group $\mathbb{Z}_p$ is cyclic of prime order p, so s generates $\mathbb{Z}_p$. Therefore, up to multiplication by nonzero elements of M, the powers β^j of β run through the powers of α precisely once as j runs from 0 to $p-1$. Since $\beta^0 = 1, \beta^1 = x_s\alpha^s$, we have

$$x = y_0 + \beta + y_2\beta^2 + \cdots + y_{p-1}\beta^{p-1}$$

for suitable $y_j \in M$, where in fact $y_0 = x_0$. □

Lemma 8.20. *Let $q \in L$. Then the minimal polynomial of q over K splits into linear factors over L.*

Proof. The element q is a rational expression $q(t_1, \ldots, t_n) \in \mathbb{C}(t_1, \ldots, t_n)$. The polynomial

$$f_q(t) = \prod_{\sigma \in \mathbb{S}_n} (t - q(t_{\sigma(1)}, \ldots, t_{\sigma(n)}))$$

has q as a zero. Symmetry under $\mathbb{S}_n$ implies that $f_q(t) \in K[t]$. The minimal polynomial m_q of q over K divides f_q, and f_q is a product of linear factors; therefore m_q is the product of some subset of those linear factors. □

We are now ready for the climax of Diet Galois:

Proof of Theorem 8.15. We prove the theorem by induction on the height h of R.

If $h = 0$ then the theorem is obvious.

Suppose that $h \geq 1$. Then $R = R_1(\alpha)$ where R_1 is a radical extension of K of height $h-1$, and $\alpha^p \in R_1$, $\alpha \notin R_1$, with p prime. Let $\alpha^p = a \in R_1$.

By Lemma 8.19 we may assume without loss of generality that

$$x = x_0 + \alpha + x_2\alpha^2 + \cdots + x_{p-1}\alpha^{p-1}$$

where the $x_j \in R_1$. (Replace α by β as in the lemma, and then change notation back to α.) The mimimum polynomial $m(t)$ of x over K splits into linear factors in L by Lemma 8.20. In particular, x is a zero of $m(t)$, while all zeros of $m(t)$ lie in L.

Take the equation $m(\alpha) = 0$, write x as above in terms of powers of α with coefficients in R_1, and consider the result as an equation satisfied by α. The equation has the form $f(\alpha) = 0$ where $f(t) \in R_1[t]$. Therefore $f(t)$ is divisible by the minimal polynomial of α, which is $t^p - a$. Hence all the roots of that equation, namely $\zeta^r\alpha$ for $0 \leq r \leq p-1$, are also roots of $f(t)$. Therefore all the elements X_r in (8.8) are roots of $m(t)$, so they lie in L. Lemma 8.18 now shows that $\alpha, x_0, x_2, \ldots x_{p-1} \in L$.

Also, $\alpha^p, x_0, x_2, \ldots x_{p-1} \in R_1$. The height of R_1 is $h-1$, so by induction, each of these elements lies in some radical extension of K that is contained in L. The subfield J generated by all of these radical extensions is clearly radical (Exercise 8.12), and contains $\alpha^p, x_0, x_2, \ldots x_{p-1}$. Then $x \in J(\alpha) \subseteq L$, and $J(\alpha)$ is radical. This completes the induction step, and with it, the proof. □

So much for the general quintic. We have used virtually everything that led up to Galois theory, but instead of thinking of a group of automorphisms of a field extension, we have used a group of permutations of the roots of a polynomial. Indeed, we have used only the group $\mathbb{S}_n$, which permutes the roots t_j of the general polynomial $F(t)$. It would be possible to stop here, with a splendid application of group theory to the insolubility of the 'general' quintic. But for Galois, and for us, there is much more to do. The general quintic is not general *enough*, and it would be nice to find out why the various tricks used above actually *work*. At the moment, they seem to be fortunate accidents. In fact, they conceal an elegant theory (which, in particular, makes the Theorem on Natural Irrationalities entirely obvious; so much so that we can ignore it altogether). That theory is, of course, Galois theory. Now motivated up to the hilt, we can start to develop it in earnest.

EXERCISES

8.1 Prove that in Section 8.2, the permutations R and S of equations (8.1, 8.2) preserve every valid polynomial equation over Q relating α, β, γ, and δ. (*Hint*: The permutation R has the same effect as complex conjugation. For the permutation S, observe that any polynomial equation in $\alpha, \beta, \gamma, \delta$ can be expressed as

$$p\gamma + q\delta = 0$$

where $p, q \in \mathbb{Q}(i)$. Substitute $\gamma = \sqrt{5}, \delta = -\sqrt{5}$ to derive a condition on p and q. Show that this condition also implies that the equation holds if we change the values so that $\gamma = -\sqrt{5}, \delta = \sqrt{5}$.)

8.2 Show that the only subfields of $\mathbb{Q}(i, \sqrt{5})$ are $\mathbb{Q}$, $\mathbb{Q}(i)$, $\mathbb{Q}(\sqrt{5})$, $\mathbb{Q}(i\sqrt{5})$, and $\mathbb{Q}(i, \sqrt{5})$.

8.3 Express the following in terms of elementary symmetric polynomials of α, β, γ.

(a) $\alpha^2 + \beta^2 + \gamma^2$

(b) $\alpha^3 + \beta^3 + \gamma^3$

(c) $\alpha^2\beta + \alpha^2\gamma + \beta^2\alpha + \beta^2\gamma + \gamma^2\alpha + \gamma^2\beta$

(d) $(\alpha - \beta)^2 + (\beta - \gamma)^2 + (\gamma - \alpha)^2$

8.4 Prove that every symmetric polynomial $p(x, y) \in \mathbb{Q}[x, y]$ can be written as a polynomial in xy and $x + y$, as follows. If p contains a term ax^iy^j, with $i \neq j \in \mathbb{N}$ and $a \in \mathbb{Q}$, show that it must also contain the term ax^jy^i. Use this to write p as a sum of terms of the form $a(x^iy^j + x^jy^i)$ or ax^iy^i. Observe that

$$\begin{aligned} x^iy^j + x^jy^i &= x^iy^i(x^{j-i} + y^{j-i}) \quad \text{if } i < j \\ x^iy^i &= (xy)^i \\ (x^i + y^i) &= (x + y)(x^{i-1} + y^{i-1}) - xy(x^{i-2} + y^{i-2}). \end{aligned}$$

Hence show that p is a sum of terms that are polynomials in $x + y, xy$.

8.5* This exercise generalises Exercise 8.3 to n variables. Suppose that $p(t_1, \ldots, t_n) \in K[t_1, \ldots, t_n]$ is symmetric and let the s_i be the elementary symmetric polynomials in the t_j. Define the *rank* of a monomial $t_1^{a_1}t_2^{a_2} \ldots t_n^{a_n}$ to be $a_1 + 2a_2 + \cdots na_n$. Define the *rank* of p to be the maximum of the ranks of all monomials that occur in p, and let its part of highest rank be the sum of the terms whose ranks attain this maximum value. Find a polynomial q composed of terms of the form $ks_1^{b_1}s_2^{b_2} \ldots s_n^{b_n}$, where $k \in K$, such that the part of q of highest rank equals that of p. Observe that $p - q$ has smaller rank than p, and use induction on the rank to prove that p is a polynomial in the s_i.

8.6 Suppose that $f(t) = a_n t^n + \cdots + a_0 \in K[t]$, and suppose that in some subfield L of $\mathbb{C}$ such that $K \subset L$ we can factorise f as

$$f(t) = a_n(t - \alpha_1)\dots(t - \alpha_n)$$

Define

$$\lambda_j = \alpha_1^j + \cdots + \alpha_n^j$$

Prove *Newton's identities*

$$\begin{aligned} a_{n-1} + a_n\lambda_1 &= 0 \\ 2a_{n-2} + a_{n-1}\lambda_1 + a_n\lambda_2 &= 0 \\ &\dots \\ na_0 + a_1\lambda_1 + \cdots + a_{n-1}\lambda_{n-1} + a_n\lambda_n &= 0 \\ &\dots \\ a_0\lambda_k + a_1\lambda_{k+1} + \cdots + a_{n-1}\lambda_{k+n-1} + a_n\lambda_{k+n} &= 0 \quad (k \geq 1) \end{aligned}$$

Show how to use these identities inductively to obtain formulas for the λ_j.

8.7 Prove that the alternating group $\mathbb{A}_n$ is generated by 3-cycles.

8.8 Prove that every element of $\mathbb{A}_5$ is the product of two 5-cycles. Deduce that $\mathbb{A}_5$ is simple.

8.9 Solve the general quadratic by Ruffini radicals. (*Hint:* If the roots are α_1, α_2, show that $\alpha_1 - \alpha_2$ is a Ruffini radical.)

8.10 Solve the general cubic by Ruffini radicals. (*Hint:* If the roots are $\alpha_1, \alpha_2, \alpha_3$, show that $\alpha_1 + \omega\alpha_2 + \omega^2\alpha_3$ and $\alpha_1 + \omega^2\alpha_2 + \omega\alpha_3$ are Ruffini radicals.)

8.11 Suppose that $I \subseteq J$ are subfields of $\mathbb{C}(t_1, \dots, t_n)$ (that is, subsets closed under the operations $+, -, \times, \div$), and J is generated by $J_1, \dots, J_r$ where $I \subseteq J_j \subseteq J$ for each j and $J_j : I$ is radical. By induction on r, prove that $J : I$ is radical.

8.12 Mark the following true or false.

(a) The K-automorphisms of a field extension $L : K$ form a subfield of $\mathbb{C}$.

(b) The K-automorphisms of a field extension $L : K$ form a group.

(c) The fixed field of the Galois group of any finite extension $L : K$ contains K.

(d) The fixed field of the Galois group of any finite extension $L : K$ equals K.

(e) The alternating group $\mathbb{A}_5$ has a normal subgroup H with quotient isomorphic to $\mathbb{Z}_5$.

(f) The alternating group $\mathbb{A}_5$ has a normal subgroup H with quotient isomorphic to $\mathbb{Z}_3$.

(g) The alternating group $\mathbb{A}_5$ has a normal subgroup H with quotient isomorphic to $\mathbb{Z}_2$.

(h) The general quintic equation can be solved using radicals, but it cannot be solved using Ruffini radicals.

Chapter 9

Normality and Separability

In this chapter we define the important concepts of *normality* and *separability* for field extensions, and develop some of their key properties.

Suppose that K is a subfield of $\mathbb{C}$. Often a polynomial $p(t) \in K[t]$ has no zeros in K. But it must have zeros in $\mathbb{C}$, by the Fundamental Theorem of Algebra, Theorem 2.4. Therefore it may have at least some zeros in a given extension field L of K. For example $t^2+1 \in \mathbb{R}[t]$ has no zeros in $\mathbb{R}$, but it has zeros $\pm i \in \mathbb{C}$, in $\mathbb{Q}(i)$, and for that matter in any subfield containing $\mathbb{Q}(i)$. We shall study this phenomenon in detail, showing that every polynomial can be resolved into a product of linear factors (and hence has its full complement of zeros) if the ground field K is extended to a suitable 'splitting field' N, which has finite degree over K. An extension $N:K$ is normal if any irreducible polynomial over K with at least one zero in N splits into linear factors in N. We show that a finite extension is normal if and only if it is a splitting field.

Separability is a complementary property to normality. An irreducible polynomial is separable if its zeros in its splitting field are simple. It turns out that over $\mathbb{C}$, this property is automatic. We make it explicit because it is *not* automatic for more general fields, see Chapter 16.

9.1 Splitting Fields

The most tractable polynomials are products of linear ones, so we are led to single this property out:

Definition 9.1. If K is a subfield of $\mathbb{C}$ and f is a nonzero polynomial over K, then f *splits* over K if it can be expressed as a product of linear factors

$$f(t) = k(t-\alpha_1)\dots(t-\alpha_n)$$

where $k, \alpha_1, \dots, \alpha_n \in K$.

If this is the case, then the zeros of f in K are precisely $\alpha_1, \dots, \alpha_n$. The Fundamental Theorem of Algebra, Theorem 2.4, implies that f splits over K if and only if all of its zeros in $\mathbb{C}$ actually lie in K. Equivalently, K contains the subfield generated by all the zeros of f.

Examples 9.2. (1) The polynomial $f(t) = t^3 - 1 \in \mathbb{Q}[t]$ splits over $\mathbb{C}$, because it can be written as

$$f(t) = (t-1)(t-\omega)(t-\omega^2)$$

where $\omega = e^{2\pi i/3} \in \mathbb{C}$. Similarly, f splits over the subfield $\mathbb{Q}(i, \sqrt{3})$ since $\omega \in \mathbb{Q}(i, \sqrt{3})$, and indeed f splits over $\mathbb{Q}(\omega)$, the smallest subfield of $\mathbb{C}$ with that property.

(2) The polynomial $f(t) = t^4 - 4t^2 - 5$ splits over $\mathbb{Q}(i, \sqrt{5})$, because

$$f(t) = (t-i)(t+i)(t-\sqrt{5})(t+\sqrt{5})$$

However, over $\mathbb{Q}(i)$ the best we can do is factorise it as

$$(t-i)(t+i)(t^2-5)$$

with an irreducible factor $t^2 - 5$ of degree greater than 1. (It is easy to show that 5 is not a square in $\mathbb{Q}(i)$.)

So over $\mathbb{Q}(i)$, the polynomial f does not split. This shows that even if a polynomial $f(t)$ has some linear factors in an extension field L, it need not split over L.

If f is a polynomial over K and L is an extension field of K, then f is also a polynomial over L. It therefore makes sense to talk of f splitting over L, meaning that it is a product of linear factors with coefficients in L. We show that given K and f we can always construct an extension Σ of K such that f splits over Σ. It is convenient to require in addition that f does not split over any smaller field, so that Σ is as economical as possible.

Definition 9.3. A subfield Σ of $\mathbb{C}$ is a *splitting field* for the nonzero polynomial f over the subfield K of $\mathbb{C}$ if $K \subseteq \Sigma$ and

(1) f splits over Σ.

(2) If $K \subseteq \Sigma' \subseteq \Sigma$ and f splits over Σ' then $\Sigma' = \Sigma$.

The second condition is clearly equivalent to:

(2′) $\Sigma = K(\sigma_1, \ldots, \sigma_n)$ where $\sigma_1, \ldots, \sigma_n$ are the zeros of f in Σ.

Clearly every polynomial over a subfield K of $\mathbb{C}$ has a splitting field:

Theorem 9.4. *If K is any subfield of $\mathbb{C}$ and f is any nonzero polynomial over K, then there exists a unique splitting field Σ for f over K. Moreover, $[\Sigma : K]$ is finite.*

Proof. We can take $\Sigma = K(\sigma_1, \ldots, \sigma_n)$, where the σ_j are the zeros of f in $\mathbb{C}$. In fact, this is the only possibility, so Σ is unique. The degree $[\Sigma : K]$ is finite since $K(\sigma_1, \ldots, \sigma_n)$ is finitely generated and algebraic, so Lemma 6.11 applies. □

Isomorphic subfields of $\mathbb{C}$ have isomorphic splitting fields, in the following strong sense:

Lemma 9.5. *Suppose that $\iota : K \to K'$ is an isomorphism of subfields of $\mathbb{C}$. Let f be a nonzero polynomial over K and let $\Sigma \supseteq K$ be the splitting field for f. Let L be any extension field of K' such that $\iota(f)$ splits over L. Then there exists a monomorphism $j : \Sigma \to L$ such that $j|_K = \iota$.*

Proof. We have the following situation:

$$\begin{array}{ccc} K & \to & \Sigma \\ \iota \downarrow & & \downarrow j \\ K' & \to & L \end{array}$$

where j has yet to be found. We construct j using induction on ∂f. As a polynomial over Σ,

$$f(t) = k(t - \sigma_1) \dots (t - \sigma_n)$$

The minimal polynomial m of σ_1 over K is an irreducible factor of f. Now $\iota(m)$ divides $\iota(f)$ which splits over L, so that over L

$$\iota(m) = (t - \alpha_1) \dots (t - \alpha_r)$$

where $\alpha_1, \dots, \alpha_r \in L$. Since $\iota(m)$ is irreducible over K' it must be the minimal polynomial of α_1 over K'. So by Theorem 5.16 there is an isomorphism

$$j_1 : K(\sigma_1) \to K'(\alpha_1)$$

such that $j_1|_K = \iota$ and $j_1(\sigma_1) = \alpha_1$. Now Σ is a splitting field over $K(\sigma_1)$ of the polynomial $g = f/(t - \sigma_1)$. By induction there exists a monomorphism $j : \Sigma \to L$ such that $j|_{K(\sigma_1)} = j_1$. But then $j|_K = \iota$ and we are finished. □

This enables us to prove the uniqueness theorem.

Theorem 9.6. *Let $\iota : K \to K'$ be an isomorphism. Let Σ be the splitting field for f over K, and let Σ' be the splitting field for $\iota(f)$ over K'. Then there is an isomorphism $j : \Sigma \to \Sigma'$ such that $j|_K = \iota$. In other words, the extensions $\Sigma : K$ and $\Sigma' : K'$ are isomorphic.*

Proof. Consider the following diagram:

$$\begin{array}{ccc} K & \to & \Sigma \\ \iota \downarrow & & \downarrow j \\ K' & \to & \Sigma' \end{array}$$

We must find j to make the diagram commute, given the rest of the diagram. By Lemma 9.5 there is a monomorphism $j : \Sigma \to \Sigma'$ such that $j|_K = \iota$. But $j(\Sigma)$ is clearly the splitting field for $\iota(f)$ over K', and is contained in Σ'. Since Σ' is also the splitting field for $\iota(f)$ over K', we have $j(\Sigma) = \Sigma'$, so that j is onto. Hence j is an isomorphism, and the theorem follows. □

Examples 9.7. (1) Let $f(t) = (t^2 - 3)(t^3 + 1)$ over $\mathbb{Q}$. We can construct a splitting field for f as follows: over $\mathbb{C}$ the polynomial f splits into linear factors

$$f(t) = (t + \sqrt{3})(t - \sqrt{3})(t + 1)\left(t - \frac{1 + i\sqrt{3}}{2}\right)\left(t - \frac{1 - i\sqrt{3}}{2}\right)$$

so there exists a splitting field in $\mathbb{C}$, namely

$$\mathbb{Q}\left(\sqrt{3}, \frac{1 + i\sqrt{3}}{2}\right)$$

This is clearly the same as $\mathbb{Q}(\sqrt{3}, i)$.
(2) Let $f(t) = (t^2 - 2t - 2)(t^2 + 1)$ over $\mathbb{Q}$. The zeros of f in $\mathbb{C}$ are $1 \pm \sqrt{3}, \pm i$, so a splitting field is afforded by $\mathbb{Q}(1 + \sqrt{3}, i)$ which equals $\mathbb{Q}(\sqrt{3}, i)$. This is the same field as in the previous example, although the two polynomials involved are different.
(3) It is even possible to have two distinct irreducible polynomials with the same splitting field. For example $t^2 - 3$ and $t^2 - 2t - 2$ are both irreducible over $\mathbb{Q}$, and both have $\mathbb{Q}(\sqrt{3})$ as their splitting field over $\mathbb{Q}$.

9.2 Normality

The idea of a normal extension was explicitly recognised by Galois (but, as always, in terms of polynomials over $\mathbb{C}$). In the modern treatment it takes the following form:

Definition 9.8. An algebraic field extension $L : K$ is *normal* if every irreducible polynomial f over K that has at least one zero in L splits in L.

For example, $\mathbb{C} : \mathbb{R}$ is normal since every polynomial (irreducible or not) splits in $\mathbb{C}$. On the other hand, we can find extensions that are not normal. Let α be the real cube root of 2 and consider $\mathbb{Q}(\alpha) : \mathbb{Q}$. The irreducible polynomial $t^3 - 2$ has a zero, namely α, in $\mathbb{Q}(\alpha)$, but it does not split in $\mathbb{Q}(\alpha)$. If it did, then there would be three real cube roots of 2, not all equal. This is absurd.

Compare with the examples of Galois groups given in Chapter 8. The normal extension $\mathbb{C} : \mathbb{R}$ has a well-behaved Galois group, in the sense that the Galois correspondence is a bijection. The same goes for $\mathbb{Q}(\sqrt{2}, \sqrt{3}, \sqrt{5}) : \mathbb{Q}$. In contrast, the non-normal extension $\mathbb{Q}(\alpha) : \mathbb{Q}$ has a badly behaved Galois group. Although this is not the whole story, it illustrates the importance of normality.

There is a close connection between normal extensions and splitting fields which provides a wide range of normal extensions:

Theorem 9.9. *A field extension $L : K$ is normal and finite if and only if L is a splitting field for some polynomial over K.*

Proof. Suppose $L : K$ is normal and finite. By Lemma 6.11, $L = K(\alpha_1, \ldots, \alpha_s)$ for certain α_j algebraic over K. Let m_j be the minimal polynomial of α_j over K and let $f = m_1 \ldots m_s$. Each m_j is irreducible over K and has a zero $\alpha_j \in L$, so by normality each m_j splits over L. Hence f splits over L. Since L is generated by K and the zeros of f, it is the splitting field for f over K.

To prove the converse, suppose that L is the splitting field for some polynomial g over K. The extension $L : K$ is then obviously finite; we must show it is normal. To do this we must take an irreducible polynomial f over K with a zero in L and show that it splits in L. Let $M \supseteq L$ be a splitting field for fg over K. Suppose that θ_1 and θ_2 are zeros of f in M. By irreducibility, f is the minimal polynomial of θ_1 and θ_2 over K.

We claim that

$$[L(\theta_1) : L] = [L(\theta_2) : L]$$

This is proved by an interesting trick. We look at several subfields of M, namely $K, L, K(\theta_1), L(\theta_1), K(\theta_2), L(\theta_2)$. There are two towers

$$K \subseteq K(\theta_1) \subseteq L(\theta_1) \subseteq M$$
$$K \subseteq K(\theta_2) \subseteq L(\theta_2) \subseteq M$$

The claim will follow from a simple computation of degrees. For $j = 1$ or 2

$$[L(\theta_j) : L][L : K] = [L(\theta_j) : K] = [L(\theta_j) : K(\theta_j)][K(\theta_j) : K] \tag{9.1}$$

By Proposition 6.7, $[K(\theta_1) : K] = [K(\theta_2) : K]$. Clearly $L(\theta_j)$ is the splitting field for g over $K(\theta_j)$, and by Corollary 5.13 $K(\theta_1)$ is isomorphic to $K(\theta_2)$. Therefore by Theorem 9.6 the extensions $L(\theta_j) : K(\theta_j)$ are isomorphic for $j = 1, 2$, so they have the same degree. Substituting in (9.1) and cancelling,

$$[L(\theta_1) : L] = [L(\theta_2) : L]$$

as claimed. From this point on, the rest is easy. If $\theta_1 \in L$ then $[L(\theta_1) : L] = 1$, so $[L(\theta_2) : L] = 1$ and $\theta_2 \in L$ also. Hence $L : K$ is normal. □

9.3 Separability

Galois did not explicitly recognise the concept of separability, since he worked only with the complex field, where, as we shall see, separability is automatic. However, the concept is implicit in his work, and must be invoked when studying more general fields.

Definition 9.10. An irreducible polynomial f over a subfield K of $\mathbb{C}$ is *separable* over K if it has simple zeros in $\mathbb{C}$, or equivalently, simple zeros in its splitting field.

This means that over its splitting field, or over $\mathbb{C}$, f takes the form

$$f(t) = k(t-\sigma_1)\dots(t-\sigma_n)$$

where the σ_j are all different.

Example 9.11. The polynomial $t^4+t^3+t^2+t+1$ is separable over $\mathbb{Q}$, since its zeros in $\mathbb{C}$ are $e^{2\pi i/5}$, $e^{4\pi i/5}$, $e^{6\pi i/5}$, $e^{8\pi i/5}$, which are all different.

For polynomials over $\mathbb{R}$ there is a standard method for detecting multiple zeros by differentiation. To obtain maximum generality later, we redefine the derivative in a purely formal manner.

Definition 9.12. Suppose that K is a subfield of $\mathbb{C}$, and let

$$f(t) = a_0 + a_1 t + \cdots + a_n t^n \in K[t]$$

Then the *formal derivative* of f is the polynomial

$$Df = a_1 + 2a_2 t + \cdots + na_n t^{n-1} \in K[t]$$

For $K = \mathbb{R}$ (and indeed for $K = \mathbb{C}$) this is the usual derivative. Several useful properties of the derivative carry over to D. In particular, simple computations (Exercise 9.3) show that for all polynomials f and g over K,

$$\begin{aligned} D(f+g) &= Df + Dg \\ D(fg) &= (Df)g + f(Dg) \end{aligned}$$

Also, if $\lambda \in K$ then $D(\lambda) = 0$, so

$$D(\lambda f) = \lambda(Df)$$

These properties of D let us state a criterion for the existence of multiple zeros without knowing what the zeros are.

Lemma 9.13. *Let $f \neq 0$ be a polynomial over a subfield K of $\mathbb{C}$, and let Σ be its splitting field. Then f has a multiple zero (in $\mathbb{C}$ or Σ) if and only if f and Df have a common factor of degree ≥ 1 in $K[t]$.*

Proof. Suppose f has a repeated zero in Σ, so that over Σ

$$f(t) = (t-\alpha)^2 g(t)$$

where $\alpha \in \Sigma$. Then

$$Df = (t-\alpha)[(t-\alpha)Dg + 2g]$$

so f and Df have a common factor $(t-\alpha)$ in $\Sigma[t]$. Hence f and Df have a common factor in $K[t]$, namely the minimal polynomial of α over K.

Now suppose that f has no repeated zeros. Suppose that f and Df have a common factor, and let α be a zero of that factor. Then $f = (t-\alpha)g$ and $Df = (t-\alpha)h$. Differentiate the former to get $(t-\alpha)h = Df = g + (t-\alpha)Dg$, so $(t-\alpha)$ divides g, hence $(t-\alpha)^2$ divides f. □

We now prove that separability of an irreducible polynomial is automatic over subfields of $\mathbb{C}$.

Proposition 9.14. *If K is a subfield of $\mathbb{C}$ then every irreducible polynomial over K is separable.*

Proof. An irreducible polynomial f over K is inseparable if and only if f and Df have a common factor of degree ≥ 1. If so, then since f is irreducible the common factor must be f, but Df has smaller degree than f, and the only multiple of f having smaller degree is 0, so $Df = 0$. Thus if

$$f(t) = a_0 + \cdots + a_m t^m$$

then this is equivalent to $na_n = 0$ for all integers $n > 0$. For subfields of $\mathbb{C}$, this is equivalent to $a_n = 0$ for all $n > 0$. □

EXERCISES

9.1 Determine splitting fields over $\mathbb{Q}$ for the polynomials $t^3 - 1, t^4 + 5t^2 + 6, t^6 - 8$, in the form $\mathbb{Q}(\alpha_1, \ldots, \alpha_k)$ for explicit α_j.

9.2 Find the degrees of these fields as extensions of $\mathbb{Q}$.

9.3 Prove that the formal derivative D has the following properties:

(a) $D(f+g) = Df + Dg$
(b) $D(fg) = (Df)g + f(Dg)$
(c) If $f(t) = t^n$, then $Df(t) = nt^{n-1}$

9.4 Show that we can extend the definition of the formal derivative to $K(t)$ by defining

$$D(f/g) = (Df \cdot g - f \cdot Dg)/g^2$$

when $g \neq 0$. Verify the relevant properties of D.

9.5 Which of the following extensions are normal?

(a) $\mathbb{Q}(t) : \mathbb{Q}$
(b) $\mathbb{Q}(\sqrt{-5}) : \mathbb{Q}$
(c) $\mathbb{Q}(\alpha) : \mathbb{Q}$ where α is the real seventh root of 5
(d) $\mathbb{Q}(\sqrt{5}, \alpha) : \mathbb{Q}(\alpha)$, where α is as in (c)
(e) $\mathbb{R}(\sqrt{-7}) : \mathbb{R}$

9.6 Show that every extension in $\mathbb{C}$, of degree 2, is normal. Is this true if the degree is greater than 2?

9.7 If Σ is the splitting field for f over K and $K \subseteq L \subseteq \Sigma$, show that Σ is the splitting field for f over L.

9.8* Let f be a polynomial of degree n over K, and let Σ be the splitting field for f over K. Show that $[\Sigma : K]$ divides $n!$ (*Hint:* Use induction on n. Consider separately the cases when f is reducible or irreducible. Note that $a!b!$ divides $(a+b)!$ (why?).)

9.9 Mark the following true or false.

(a) Every polynomial over $\mathbb{Q}$ splits over some subfield of $\mathbb{C}$.

(b) Splitting fields in $\mathbb{C}$ are unique.

(c) Every finite extension is normal.

(d) $\mathbb{Q}(\sqrt{19}) : \mathbb{Q}$ is a normal extension.

(e) $\mathbb{Q}(\sqrt[4]{19}) : \mathbb{Q}$ is a normal extension.

(f) $\mathbb{Q}(\sqrt[4]{19}) : \mathbb{Q}(\sqrt{19})$ is a normal extension.

(g) A normal extension of a normal extension is a normal extension.

Chapter 10

Counting Principles

When proving the Fundamental Theorem of Galois theory in Chapter 12, we will need to show that if H is a subgroup of the Galois group of a finite normal extension $L:K$, then $H^{\dagger *}=H$. Here the maps $*$ and $\dagger$ are as defined in Section 8.6. Our method will be to show that H and $H^{\dagger *}$ are finite groups and have the same order. Since we already know that $H \subseteq H^{\dagger *}$, the two groups must be equal. This is an archetypal application of a *counting principle*: showing that two finite sets, one contained in the other, are identical, by counting how many elements they have, and showing that the two numbers are the same.

It is largely for this reason that we need to restrict attention to finite extensions and finite groups. If an infinite set is contained in another of the same cardinality, they need not be equal—for example, $\mathbb{Z} \subseteq \mathbb{Q}$ and both sets are countable, but $\mathbb{Z} \neq \mathbb{Q}$. So counting principles may fail for infinite sets.

The object of this chapter is to perform part of the calculation of the order of $H^{\dagger *}$. Namely, we find the degree $[H^{\dagger}:K]$ in terms of the order of H. In Chapter 11 we find the order of $H^{\dagger *}$ in terms of this degree; putting the pieces together will give the desired result.

10.1 Linear Independence of Monomorphisms

We begin with a theorem of Dedekind, who was the first to make a systematic study of field monomorphisms.

To motivate the theorem and its proof, we consider a special case. Suppose that K and L are subfields of $\mathbb{C}$, and let λ and μ be monomorphisms $K \to L$. We claim that λ cannot be a constant multiple of μ unless $\lambda = \mu$. By 'constant' here we mean an element of L. Suppose that there exists $a \in L$ such that

$$\mu(x) = a\lambda(x) \tag{10.1}$$

for all $x \in K$. Replace x by yx, where $y \in K$, to get

$$\mu(yx) = a\lambda(yx)$$

Since λ and μ are monomorphisms,

$$\mu(y)\mu(x) = a\lambda(y)\lambda(x)$$

Multiplying (10.1) by $\lambda(y)$, we also have

$$\lambda(y)\mu(x) = a\lambda(y)\lambda(x)$$

Comparing the two, $\lambda(y) = \mu(y)$ for all y, so $\lambda = \mu$.

In other words, if λ and μ are distinct monomorphisms $K \to L$, they must be *linearly independent* over L.

Next, suppose that $\lambda_1, \lambda_2, \lambda_3$ are three distinct monomorphisms $K \to L$, and assume that they are linearly dependent over L. That is,

$$a_1\lambda_1 + a_2\lambda_2 + a_3\lambda_3 = 0$$

for $a_j \in L$. In more detail,

$$a_1\lambda_1(x) + a_2\lambda_2(x) + a_3\lambda_3(x) = 0 \tag{10.2}$$

for all $x \in K$. If some $a_j = 0$ then we reduce to the previous case, so we may assume all $a_j \neq 0$.

Substitute yx for x in (10.2) to get

$$a_1\lambda_1(yx) + a_2\lambda_2(yx) + a_3\lambda_3(yx) = 0 \tag{10.3}$$

That is,

$$[a_1\lambda_1(y)]\lambda_1(x) + [a_2\lambda_2(y)]\lambda_2(x) + [a_3\lambda_3(y)]\lambda_3(x) = 0 \tag{10.4}$$

Relations (10.2) and (10.4) are independent—that is, they are not scalar multiples of each other—unless $\lambda_1(y) = \lambda_2(y) = \lambda_3(y)$, and we can choose y to prevent this. therefore we may eliminate one of the λ_j to deduce a linear relation between at most two of them, contrary to the previous case. Specifically, there exists $y \in K$ such that $\lambda_1(y) \neq \lambda_3(y)$. Multiply (10.2) by $\lambda_3(y)$ and subtract from (10.4) to get

$$[a_1\lambda_1(y) - a_1\lambda_3(y)]\lambda_1(x) + [a_2\lambda_2(y) - a_2\lambda_3(y)]\lambda_2(x) = 0$$

Then the coefficient of $\lambda_1(x)$ is $a_1(\lambda_1(y) - \lambda_3(y)) \neq 0$, a contradiction.

Dedekind realised that this approach can be used inductively to prove:

Lemma 10.1 (Dedekind). *If K and L are subfields of $\mathbb{C}$, then every set of distinct monomorphisms $K \to L$ is linearly independent over L.*

Proof. Let $\lambda_1, \dots, \lambda_n$ be distinct monomorphisms $K \to L$. To say these are linearly independent over L is to say that there do not exist elements $a_1, \dots, a_n \in L$ such that

$$a_1\lambda_1(x) + \cdots + a_n\lambda_n(x) = 0 \tag{10.5}$$

for all $x \in K$, unless all the a_j are 0.

Assume the contrary, so that (10.5) holds. At least one of the a_i is non-zero. Among all the valid equations of the form (10.5) with all $a_i \neq 0$, there must be at least one for which the number n of non-zero terms is least. Since all λ_j are non-zero, $n \neq 1$. We choose notation so that equation (10.5) is such as expression. Hence we may assume that *there does not exist an equation like* (10.5) *with fewer than n terms*. From this we deduce a contradiction.

Since $\lambda_1 \neq \lambda_n$, there exists $y \in K$ such that $\lambda_1(y) \neq \lambda_n(y)$. Therefore $y \neq 0$. Now (10.5) holds with yx in place of x, so

$$a_1\lambda_1(yx) + \cdots + a_n\lambda_n(yx) = 0$$

for all $x \in K$, whence

$$a_1\lambda_1(y)\lambda_1(x) + \cdots + a_n\lambda_n(y)\lambda_n(x) = 0 \tag{10.6}$$

for all $x \in K$. Multiply (10.5) by $\lambda_1(y)$ and subtract (10.6), so that the first terms cancel: we obtain

$$a_2[\lambda_2(x)\lambda_1(y) - \lambda_2(x)\lambda_2(y)] + \cdots + a_n[\lambda_n(x)\lambda_1(y) - \lambda_n(x)\lambda_n(y)] = 0$$

The coefficient of $\lambda_n(x)$ is $a_n[\lambda_1(y) - \lambda_n(y)] \neq 0$, so we have an equation of the form (10.5) with fewer terms. Deleting any zero terms does not alter this statement. This contradicts the italicised assumption above.

Consequently no equation of the form (10.5) exists, so and the monomorphisms are linearly independent. □

Example 10.2. Let $K = \mathbb{Q}(\alpha)$ where $\alpha = \sqrt[3]{2} \in \mathbb{R}$. There are three monomorphisms $K \to \mathbb{C}$, namely

$$\begin{aligned}
\lambda_1(p+q\alpha+r\alpha^2) &= p+q\alpha+r\alpha^2 \\
\lambda_2(p+q\alpha+r\alpha^2) &= p+q\omega\alpha+r\omega^2\alpha^2 \\
\lambda_3(p+q\alpha+r\alpha^2) &= p+q\omega^2\alpha+r\omega\alpha^2
\end{aligned}$$

where $p, q, r \in \mathbb{Q}$ and ω is a primitive cube root of unity. We prove by 'bare hands' methods that the λ_j are linearly independent. Suppose that $a_1\lambda_1(x) + a_2\lambda_2(x) + a_3\lambda_3(x) = 0$ for all $x \in K$. Set $x = 1, \alpha, \alpha^2$ respectively to get

$$\begin{aligned}
a_1 + a_2 + a_3 &= 0 \\
a_1 + \omega a_2 + \omega^2 a_3 &= 0 \\
a_1 + \omega^2 a_2 + \omega a_3 &= 0
\end{aligned}$$

The only solution of this system of linear equations is $a_1 = a_2 = a_3 = 0$.

For our next result we need two lemmas. The first is a standard theorem of linear algebra, which we quote without proof.

Lemma 10.3. *If $n > m$ then a system of m homogeneous linear equations*

$$a_{i1}x_1 + \cdots + a_{in}x_n = 0 \quad 1 \leq i \leq m$$

in n unknowns $x_1, \ldots, x_n$, with coefficients a_{ij} in a field K, has a solution in which the x_i are all in K and are not all zero.

This theorem is proved in most first-year undergraduate linear algebra courses, and can be found in any text of linear algebra, for example Anton (1987).

The second lemma states a useful general principle.

Lemma 10.4. *If G is a group whose distinct elements are $g_1,\dots,g_n$, and if $g \in G$, then as j varies from 1 to n the elements gg_j run through the whole of G, each element of G occurring precisely once.*

Proof. If $h \in G$ then $g^{-1}h = g_j$ for some j and $h = gg_j$. If $gg_i = gg_j$ then $g_i = g^{-1}gg_i = g^{-1}gg_j = g_j$. Thus the map $g_i \mapsto gg_i$ is a bijection $G \to G$, and the result follows. □

We also recall some standard notation. We denote the cardinality of a set S by $|S|$. Thus if G is a group, then $|G|$ is the *order* of G. For example, $|\mathbb{S}_n| = n!$ and $|\mathbb{A}_n| = n!/2$.

We now come to the main theorem of this chapter, whose proof is similar to that of Lemma 10.1, and which can be motivated in a similar manner.

Theorem 10.5. *Let G be a finite subgroup of the group of automorphisms of a field K, and let K_0 be the fixed field of G. Then $[K : K_0] = |G|$.*

Proof. Let $n = |G|$, and suppose that the elements of G are $g_1,\dots,g_n$, where $g_1 = 1$. We prove separately that $[K : K_0] < n$ and $[K : K_0] > n$ are impossible.

(1) Suppose that $[K : K_0] = m < n$. Let $\{x_1,\dots,x_m\}$ be a basis for K over K_0. By Lemma 10.3 there exist $y_1,\dots,y_n \in K$, not all zero, such that

$$y_1g_1(x_i) + \cdots + y_ng_n(x_i) = 0 \tag{10.7}$$

for $i = 1,\dots,m$. Let x be any element of K. Then

$$x = \alpha_1x_1 + \cdots + \alpha_mx_m$$

where $\alpha_1,\dots,\alpha_m \in K_0$. Hence

$$\begin{aligned} y_1g_1(x) + \cdots + y_ng_n(x) &= y_1g_1\left(\sum_l \alpha_lx_l\right) + \cdots + y_ng_n\left(\sum_l \alpha_lx_l\right) \\ &= \sum_l \alpha_l[y_1g_1(x_l) + \cdots + y_ng_n(x_l)] \\ &= 0 \end{aligned}$$

using (10.7). Hence the distinct monomorphisms $g_1,\dots,g_n$ are linearly dependent, contrary to Lemma 10.1. Therefore $m \geq n$.

(2) Next, suppose for a contradiction that $[K : K_0] > n$. Then there exists a set of $n+1$ elements of K that are linearly independent over K_0; let such a set be $\{x_1,\dots,x_{n+1}\}$. By Lemma 10.3 there exist $y_1,\dots,y_{n+1} \in K$, not all zero, such that for $j = 1,\dots,n$

$$y_1g_j(x_1) + \cdots + y_{n+1}g_j(x_{n+1}) = 0 \tag{10.8}$$

We subject this equation to a combinatorial attack, similar to that used in proving Lemma 10.1. Choose $y_1, \dots, y_{n+1}$ so that as few as possible are non-zero, and renumber so that

$$y_1, \dots, y_r \neq 0, \quad y_{r+1}, \dots, y_{n+1} = 0$$

Equation (10.8) now becomes

$$y_1 g_j(x_1) + \cdots + y_r g_j(x_r) = 0 \tag{10.9}$$

Let $g \in G$, and operate on (10.9) with g. This gives a system of equations

$$g(y_1) g g_j(x_1) + \cdots + g(y_r) g g_j(x_r) = 0$$

By Lemma 10.4, as j varies, this system of equations is equivalent to the system

$$g(y_1) g_j(x_1) + \cdots + g(y_r) g_j(x_r) = 0 \tag{10.10}$$

Multiply (10.9) by $g(y_1)$ and (10.10) by y_1 and subtract, to get

$$[y_2 g(y_1) - g(y_2) y_1] g_j(x_2) + \cdots + [y_r g(y_1) - g(y_r) y_1] g_j(x_r) = 0$$

This is a system of equations like (10.9) but with fewer terms, which gives a contradiction unless all the coefficients

$$y_i g(y_1) - y_1 g(y_i)$$

are zero. If this happens then

$$y_i y_1^{-1} = g(y_i y_1^{-1})$$

for all $g \in G$, so that $y_i y_1^{-1} \in K_0$. Thus there exist $z_1, \dots, z_r \in K_0$ and an element $k \in K$ such that $y_i = k z_i$ for all i. Then (10.9), with $j = 1$, becomes

$$x_1 k z_1 + \cdots + x_r k z_r = 0$$

and since $k \neq 0$ we may divide by k, which shows that the x_i are linearly dependent over K_0. This is a contradiction.

Therefore $[K : K_0]$ is not less than n and not greater than n, so $[K : K_0] = n = |G|$ as required. □

Corollary 10.6. *If G is the Galois group of the finite extension $L : K$, and H is a finite subgroup of G, then*

$$[H^\dagger : K] = [L : K]/|H|$$

Proof. By the Tower Law, $[L : K] = [L : H^\dagger][H^\dagger : K]$, so $[H^\dagger : K] = [L : K]/[L : H^\dagger]$. But this equals $[L : K]/|H|$ by Theorem 10.5. □

Examples 10.7. We illustrate Theorem 10.5 by two examples, one simple, the other more intricate.

(1) Let G be the group of automorphisms of $\mathbb{C}$ consisting of the identity and complex conjugation. The fixed field of G is $\mathbb{R}$, for if $x-iy=x+iy \ (x,y\in\mathbb{R})$ then $y=0$, and conversely. Hence $[\mathbb{C}:\mathbb{R}]=|G|=2$, a conclusion which is manifestly correct.

(2) Let $K=\mathbb{Q}(\zeta)$ where $\zeta=\exp(2\pi i/5)\in\mathbb{C}$. Now $\zeta^5=1$ and $\mathbb{Q}(\zeta)$ consists of all elements

$$p+q\zeta+r\zeta^2+s\zeta^3+t\zeta^4 \tag{10.11}$$

where $p,q,r,s,t\in\mathbb{Q}$. The Galois group of $\mathbb{Q}(\zeta):\mathbb{Q}$ is easy to find, for if α is a $\mathbb{Q}$-automorphism of $\mathbb{Q}(\zeta)$ then

$$(\alpha(\zeta))^5=\alpha(\zeta^5)=\alpha(1)=1,$$

so that $\alpha(\zeta)=\zeta,\zeta^2,\zeta^3$, or ζ^4. This gives four candidates for $\mathbb{Q}$-automorphisms:

$$\begin{aligned}
\alpha_1 &: p+q\zeta+r\zeta^2+s\zeta^3+t\zeta^4 \mapsto p+q\zeta+r\zeta^2+s\zeta^3+t\zeta^4\\
\alpha_2 &: \qquad\qquad\qquad\qquad\qquad\quad \mapsto p+s\zeta+q\zeta^2+t\zeta^3+r\zeta^4\\
\alpha_3 &: \qquad\qquad\qquad\qquad\qquad\quad \mapsto p+r\zeta+t\zeta^2+q\zeta^3+s\zeta^4\\
\alpha_4 &: \qquad\qquad\qquad\qquad\qquad\quad \mapsto p+t\zeta+s\zeta^2+r\zeta^3+q\zeta^4
\end{aligned}$$

It is easy to check that all of these are $\mathbb{Q}$-automorphisms. The only point to bear in mind is that $1,\zeta,\zeta^2,\zeta^3,\zeta^4$ are not linearly independent over $\mathbb{Q}$. However, their linear relations are generated by just one: $\zeta+\zeta^2+\zeta^3+\zeta^4=-1$, and this relation is preserved by all of the candidate $\mathbb{Q}$-automorphisms.

Alternatively, observe that $\zeta,\zeta^2,\zeta^3,\zeta^4$ all have the same minimal polynomial $t^4+t^3+t^2+t+1$ and use Corollary 5.13.

We deduce that the Galois group of $\mathbb{Q}(\zeta):\mathbb{Q}$ has order 4. It is easy to find the fixed field of this group: it turns out to be $\mathbb{Q}$. Therefore, by Theorem 10.5, $[\mathbb{Q}(\zeta):\mathbb{Q}]=4$. At first sight this might seem wrong, for equation (10.11) expresses each element in terms of five basic elements; the degree should be 5. In support of this contention, ζ is a zero of t^5-1. The astute reader will already have seen the source of this dilemma: t^5-1 is not the minimal polynomial of ζ over $\mathbb{Q}$, since it is reducible. The minimal polynomial is, as we have seen, $t^4+t^3+t^2+t+1$, which has degree 4. Equation (10.11) holds, but the elements of the supposed 'basis' are linearly dependent. Every element of $\mathbb{Q}(\zeta)$ can be expressed *uniquely* in the form

$$p+q\zeta+r\zeta^2+s\zeta^3$$

where $p,q,r,s\in\mathbb{Q}$. We did not use this expression because it lacks symmetry, making the computations formless and therefore harder.

EXERCISES

10.1 Check Theorem 10.5 for the extension $\mathbb{C}(t_1,\ldots,t_n):\mathbb{C}(s_1,\ldots,s_n)$ of Chapter 8 Section 8.7.

10.2 Find the fixed field of the subgroup $\{\alpha_1, \alpha_4\}$ for Example 10.7(2). Check that Theorem 10.5 holds.

10.3 Parallel the argument of Example 10.7(2) when $\zeta = e^{2\pi i/7}$.

10.4 Find all monomorphisms $\mathbb{Q} \to \mathbb{C}$.

10.5 Mark the following true or false.

(a) If $S \subseteq T$ is a finite set and $|S| = |T|$, then $S = T$.

(b) The same is true of infinite sets.

(c) There is only one monomorphism $\mathbb{Q} \to \mathbb{Q}$.

(d) If K and L are subfields of $\mathbb{C}$, then there exists at least one monomorphism $K \to L$.

(e) Distinct automorphisms of a field K are linearly independent over K.

(f) Linearly independent monomorphisms are distinct.

Chapter 11

Field Automorphisms

The theme of this chapter is the construction of automorphisms to given specifications. We begin with a generalisation of a K-automorphism, known as a K-monomorphism. For normal extensions we shall use K-monomorphisms to build up K-automorphisms. Using this technique, we can calculate the order of the Galois group of any finite normal extension, which combines with the result of Chapter 10 to give a crucial part of the fundamental theorem of Chapter 12.

We also introduce the concept of a normal closure of a finite extension. This useful device enables us to steer around some of the technical obstructions caused by non-normal extensions.

11.1 K-Monomorphisms

We begin by generalising the concept of a K-automorphism of a subfield L of $\mathbb{C}$, by relaxing the condition that the map should be onto. We continue to require it to be one-to-one.

Definition 11.1. *Suppose that K is a subfield of each of the subfields M and L of $\mathbb{C}$. Then a K-monomorphism of M into L is a field monomorphism $\phi : M \to L$ such that $\phi(k) = k$ for every $k \in K$.*

Example 11.2. Suppose that $K = \mathbb{Q}, M = \mathbb{Q}(\alpha)$ where α is a real cube root of 2, and $L = \mathbb{C}$. We can define a K-monomorphism $\phi : M \to L$ by insisting that $\phi(\alpha) = \omega\alpha$, where $\omega = e^{2\pi i/3}$. In more detail, every element of M is of the form $p + q\alpha + r\alpha^2$ where $p, q, r \in \mathbb{Q}$, and

$$\phi(p + q\alpha + r\alpha^2) = p + q\omega\alpha + r\omega^2\alpha^2$$

Since α and $\omega\alpha$ have the same minimal polynomial, namely $t^3 - 2$, Corollary 5.13 implies that ϕ is a K-monomorphism.

There are two other K-monomorphisms $M \to L$ in this case. One is the identity, and the other takes α to $\omega^2\alpha$ (see Figure 18).

In general if $K \subseteq M \subseteq L$ then any K-automorphism of L restricts to a K-monomorphism $M \to L$. We are particularly interested in when this process can be reversed.

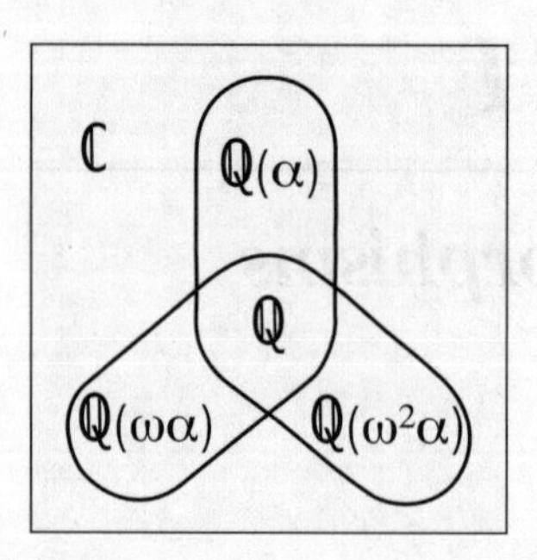

FIGURE 18: Images of $\mathbb{Q}$-monomorphisms of $\alpha = \mathbb{Q}(\sqrt[3]{2}) : \mathbb{Q}$.

Theorem 11.3. *Suppose that $L : K$ is a finite normal extension and $K \subseteq M \subseteq L$. Let τ be any K-monomorphism $M \to L$. Then there exists a K-automorphism σ of L such that $\sigma|_M = \tau$.*

Proof. By Theorem 9.9, L is the splitting field over K of some polynomial f over K. Hence it is simultaneously the splitting field over M for f and over $\tau(M)$ for $\tau(f)$. But $\tau|_K$ is the identity, so $\tau(f) = f$. We have the diagram

$$\begin{array}{ccc} M & \to & L \\ \tau\downarrow & & \downarrow\sigma \\ \tau(M) & \to & L \end{array}$$

with σ yet to be found. By Theorem 9.6, there is an isomorphism $\sigma : L \to L$ such that $\sigma|_M = \tau$. Therefore σ is an automorphism of L, and since $\sigma|_K = \tau|_K$ is the identity, σ is a K-automorphism of L. □

This result can be used to construct K-automorphisms:

Proposition 11.4. *Suppose that $L : K$ is a finite normal extension, and α, β are zeros in L of the irreducible polynomial p over K. Then there exists a K-automorphism σ of L such that $\sigma(\alpha) = \beta$.*

Proof. By Corollary 5.13 there is an isomorphism $\tau : K(\alpha) \to K(\beta)$ such that $\tau|_K$ is the identity and $\tau(\alpha) = \beta$. By Theorem 11.3, τ extends to a K-automorphism σ of L. □

11.2 Normal Closures

When extensions are not normal, we can try to recover normality by making the extensions larger.

Definition 11.5. Let L be a finite extension of K. A *normal closure* of $L : K$ is an extension N of L such that

(1) $N : K$ is normal;

(2) If $L \subseteq M \subseteq N$ and $M : K$ is normal, then $M = N$.

Thus N is the smallest extension of L that is normal over K.

The next theorem assures us of a sufficient supply of normal closures, and shows that (working inside $\mathbb{C}$) they are unique.

Theorem 11.6. *If $L : K$ is a finite extension in $\mathbb{C}$, then there exists a unique normal closure $N \subseteq \mathbb{C}$ of $L : K$, which is a finite extension of K.*

Proof. Let $x_1, \ldots, x_r$ be a basis for L over K, and let m_j be the minimal polynomial of x_j over K. Let N be the splitting field for $f = m_1 m_2 \ldots m_r$ over L. Then N is also the splitting field for f over K, so $N : K$ is normal and finite by Theorem 9.9. Suppose that $L \subseteq P \subseteq N$ where $P : K$ is normal. Each polynomial m_j has a zero $x_j \in P$, so by normality f splits in P. Since N is the splitting field for f, we have $P = N$. Therefore N is a normal closure.

Now suppose that M and N are both normal closures. The above polynomial f splits in M and in N, so each of M and N contain the splitting field for f over K. This splitting field contains L and is normal over K, so it must be equal to both M and N. □

Example 11.7. Consider $\mathbb{Q}(\alpha):\mathbb{Q}$ where α is the real cube root of 2. This extension is not normal, as we have seen. If we let K be the splitting field for $t^3 - 2$ over $\mathbb{Q}$, contained in $\mathbb{C}$, then $K = \mathbb{Q}(\alpha, \alpha\omega, \alpha\omega^2)$ where $\omega = (-1 + i\sqrt{3})/2$ is a complex cube root of unity. This is the same as $\mathbb{Q}(\alpha, \omega)$. Now K is the normal closure for $\mathbb{Q}(\alpha) : \mathbb{Q}$. So here we obtain the normal closure by adjoining *all* the 'missing' zeros.

Normal closures let us place restrictions on the image of a monomorphism.

Lemma 11.8. *Suppose that $K \subseteq L \subseteq N \subseteq M$ where $L : K$ is finite and N is the normal closure of $L : K$. Let τ be any K-monomorphism $L \to M$. Then $\tau(L) \subseteq N$.*

Proof. Let $\alpha \in L$. Let m be the minimal polynomial of α over K. Then $m(\alpha) = 0$ so $\tau(m(\alpha)) = 0$. But $\tau(m(\alpha)) = m(\tau(\alpha))$ since τ is a K-monomorphism, so $m(\tau(\alpha)) = 0$ and $\tau(\alpha)$ is a zero of m. Therefore $\tau(\alpha)$ lies in N since $N : K$ is normal. Therefore $\tau(L) \subseteq N$.

This result often lets us restrict attention to the normal closure of a given extension when discussing monomorphisms. The next theorem provides a sort of converse. □

Theorem 11.9. *For a finite extension $L : K$ the following are equivalent:*

(1) *$L : K$ is normal.*

(2) *There exists a finite normal extension N of K containing L such that every K-monomorphism $\tau : L \to N$ is a K-automorphism of L.*

(3) *For every finite extension M of K containing L, every K-monomorphism $\tau : L \to M$ is a K-automorphism of L.*

Proof. We show that (1) $\Rightarrow$ (3) $\Rightarrow$ (2) $\Rightarrow$ (1).

(1) $\Rightarrow$ (3). If $L : K$ is normal then L is the normal closure of $L : K$, so by Lemma 11.8, $\tau(L) \subseteq L$. But τ is a K-linear map defined on the finite-dimensional vector space L over K, and is a monomorphism. Therefore $\tau(L)$ has the same dimension as L, whence $\tau(L) = L$ and τ is a K-automorphism of L.

(3) $\Rightarrow$ (2). Let N be the normal closure for $L : K$. Then N exists by Theorem 11.6, and has the requisite properties by (3).

(2) $\Rightarrow$ (1). Suppose that f is any irreducible polynomial over K with a zero $\alpha \in L$. Then f splits over N by normality, and if β is any zero of f in N, then by Proposition 11.4 there exists an automorphism σ of N such that $\sigma(\alpha) = \beta$. By hypothesis, σ is a K-automorphism of L, so $\beta = \sigma(\alpha) \in \sigma(L) = L$. Therefore f splits over L and $L : K$ is normal. □

Our next result is of a more computational nature.

Theorem 11.10. *Suppose that $L : K$ is a finite extension of degree n. Then there are precisely n distinct K-monomorphisms of L into the normal closure N of $L : K$, and hence into any given normal extension M of K containing L.*

Proof. Use induction on $[L : K]$. If $[L : K] = 1$, then the result is clear. Suppose that $[L : K] = k > 1$. Let $\alpha \in L \backslash K$ with minimal polynomial m over K. Then

$$\partial m = [K(\alpha) : K] = r > 1$$

Now m is an irreducible polynomial over a subfield of $\mathbb{C}$ with one zero in the normal extension N, so m splits in N and its zeros $\alpha_1, \ldots, \alpha_r$ are distinct. By induction there are precisely s distinct $K(\alpha)$-monomorphisms $\rho_1, \ldots, \rho_s : L \to N$, where $s = [L : K(\alpha)] = k/r$. By Proposition 11.4, there are r distinct K-automorphisms $\tau_1, \ldots, \tau_r$ of N such that $\tau_i(\alpha) = \alpha_i$. The maps

$$\phi_{ij} = \tau_i \rho_j \quad (1 \leq i \leq r, 1 \leq j \leq s)$$

are K-monomorphisms $L \to N$.

We claim they are distinct. Suppose $\phi_{ij} = \phi_{kl}$. Then $\tau_k^{-1}\tau_i = \rho_l \rho_j^{-1}$. The ρ_j fix $K(\alpha)$, so they map α to itself. But ρ_j is defined by its action on α, so $\rho_l \rho_j^{-1}$ is the identity. That is, $\rho_l = \rho_j$. So $\tau_k^{-1}\tau_i$ is the identity, and $\tau_k = \tau_i$. Therefore $i = k, j = l$, so the ϕ_{ij} are distinct. They therefore provide $rs = k$ distinct K-monomorphisms $L \to N$.

Finally, we show that these are all of the K-monomorphisms $L \to N$. Let $\tau : L \to N$ be a K-monomorphism. Then $\tau(\alpha)$ is a zero of m in N, so $\tau(\alpha) = \alpha_i$ for some i. The map $\phi = \tau_i^{-1}\tau$ is a $K(\alpha)$-monomorphism $L \to N$, so by induction $\phi = \rho_j$ for some j. Hence $\tau = \tau_i \rho_j = \phi_{ij}$ and the theorem is proved. □

We can now calculate the order of the Galois group of a finite normal extension, a result of fundamental importance.

Corollary 11.11. *If $L : K$ is a finite normal extension inside $\mathbb{C}$, then there are precisely $[L : K]$ distinct K-automorphisms of L. That is,*

$$|\Gamma(L : K)| = [L : K]$$

Proof. Use Theorem 11.10. □

From this we easily deduce the important:

Theorem 11.12. *Let $L : K$ be a finite extension with Galois group G. If $L : K$ is normal, then K is the fixed field of G.*

Proof. Let K_0 be the fixed field of G, and let $[L:K] = n$. Corollary 11.11 implies that $|G| = n$. By Theorem 10.5, $[L:K_0] = n$. Since $K \subseteq K_0$ we must have $K = K_0$. □

An alternative and in some ways simpler approach to Corollary 11.11 and Theorem 11.12 can be found in Geck (2014).

There is a converse to Theorem 11.12, which shows why we must consider normal extensions in order to make the Galois correspondence a bijection. Before we can prove the converse, we need a theorem whose statement and proof closely resemble those of Theorem 11.10.

Theorem 11.13. *Suppose that $K \subseteq L \subseteq M$ and $M : K$ is finite. Then the number of distinct K-monomorphisms $L \to M$ is at most $[L : K]$.*

Proof. Let N be a normal closure of $M : K$. Then the set of K-monomorphisms $L \to M$ is contained in the set of K-monomorphisms $L \to N$, and by Theorem 11.10 there are precisely $[L : K]$ of those. □

Theorem 11.14. *If L is any field, G any finite group of automorphisms of L, and K is its fixed field, then $L : K$ is finite and normal, with Galois group G.*

Proof. By Theorem 10.5, $[L : K] = |G| = n$, say. There are exactly n distinct K-monomorphisms $L \to L$, namely, the elements of the Galois group.

We prove normality using Theorem 11.9. Thus let N be an extension of K containing L, and let τ be a K-monomorphism $L \to N$. Since every element of the Galois group of $L : K$ defines a K-monomorphism $L \to N$, the Galois group provides n distinct K-monomorphisms $L \to N$, and these are automorphisms of L. But by Theorem 11.13 there are at most n distinct K-monomorphisms $L \to N$, so τ must be one of these monomorphisms. Hence τ is an automorphism of L. Finally, $L : K$ is normal by Theorem 11.9. □

If the Galois correspondence is a bijection, then K must be the fixed field of the Galois group of $L : K$, so by the above $L : K$ must be normal. That these hypotheses are also *sufficient* to make the Galois correspondence bijective (for subfields of $\mathbb{C}$) will be proved in Chapter 12. For general fields we need the additional concept of 'separability', see Chapter 17.

EXERCISES

11.1 Suppose that $L : K$ is finite. Show that every K-monomorphism $L \to L$ is an automorphism. Does this result hold if the extension is not finite?

11.2 Construct the normal closure N for the following extensions:

(a) $\mathbb{Q}(\alpha):\mathbb{Q}$ where α is the real fifth root of 3

(b) $\mathbb{Q}(\beta):\mathbb{Q}$ where β is the real seventh root of 2

(c) $\mathbb{Q}(\sqrt{2},\sqrt{3}):\mathbb{Q}$

(d) $\mathbb{Q}(\alpha,\sqrt{2}):\mathbb{Q}$ where α is the real cube root of 2

(e) $\mathbb{Q}(\gamma):\mathbb{Q}$ where γ is a zero of t^3-3t^2+3

11.3 Find the Galois groups of the extensions (a), (b), (c), (d) in Exercise 11.2.

11.4 Find the Galois groups of the extensions $N:\mathbb{Q}$ for their normal closures N.

11.5 Show that Lemma 11.8 fails if we do not assume that $N:K$ is normal, but is true for any extension N of L such that $N:K$ is normal, rather than just for a normal closure.

11.6 Use Corollary 11.11 to find the order of the Galois group of the extension $\mathbb{Q}(\sqrt{3},\sqrt{5},\sqrt{7}):\mathbb{Q}$. (*Hint:* Argue as in Example 6.8.)

11.7 Mark the following true or false.

(a) Every K-monomorphism is a K-automorphism.

(b) Every finite extension has a normal closure.

(c) If $K \subseteq L \subseteq M$ and σ is a K-automorphism of M, then the restriction $\sigma|_L$ is a K-automorphism of L.

(d) An extension having Galois group of order 1 is normal.

(e) A finite normal extension has finite Galois group.

(f) Every Galois group is abelian (commutative).

(g) The Galois correspondence fails to be bijective for non-normal extensions.

(h) A finite normal extension inside $\mathbb{C}$, of degree n, has Galois group of order n.

(i) The Galois group of a normal extension is cyclic.

Chapter 12

The Galois Correspondence

We are at last in a position to establish the fundamental properties of the Galois correspondence between a field extension and its Galois group. Most of the work has already been done, and all that remains is to put the pieces together.

12.1 The Fundamental Theorem of Galois Theory

Let us recall a few points of notation from Chapter 8. Let $L:K$ be a field extension in $\mathbb{C}$ with Galois group G, which consists of all K-automorphisms of L. Let $\mathscr{F}$ be the set of intermediate fields, that is, subfields M such that $K \subseteq M \subseteq L$, and let $\mathscr{G}$ be the set of all subgroups H of G. We have defined two maps

$$^* : \mathscr{F} \to \mathscr{G}$$
$$^\dagger : \mathscr{G} \to \mathscr{F}$$

as follows: if $M \in \mathscr{F}$, then M^* is the group of all M-automorphisms of L. If $H \in \mathscr{G}$, then $H^\dagger$ is the fixed field of H. We have observed in (8.4) that the maps * and † reverse inclusions.

Before proceeding to the main theorem, we need a lemma:

Lemma 12.1. *Suppose that $L:K$ is a field extension, M is an intermediate field, and τ is a K-automorphism of L. Then $\tau(M)^* = \tau M^* \tau^{-1}$.*

Proof. Let $M' = \tau(M)$, and take $\gamma \in M^*, x_1 \in M'$. Then $x_1 = \tau(x)$ for some $x \in M$. Compute:

$$(\tau\gamma\tau^{-1})(x_1) = \tau\gamma(x) = \tau(x) = x_1$$

so $\tau M^* \tau^{-1} \subseteq M'^*$. Similarly $\tau^{-1} M'^* \tau \subseteq M^*$, so $\tau M^* \tau^{-1} \supseteq M'^*$, and the lemma is proved. □

We are now ready to prove the main result:

Theorem 12.2 (Fundamental Theorem of Galois Theory). *If $L:K$ is a finite normal field extension inside $\mathbb{C}$, with Galois group G, and if $\mathscr{F}, \mathscr{G}, ^*, ^\dagger$ are defined as above, then:*

(1) *The Galois group G has order $[L:K]$.*

(2) *The maps* * *and* † *are mutual inverses, and set up an order-reversing one-to-one correspondence between* $\mathscr{F}$ *and* $\mathscr{G}$.

(3) *If* M *is an intermediate field, then*

$$[L:M] = |M^*| \qquad [M:K] = |G|/|M^*|$$

(4) *An intermediate field* M *is a normal extension of* K *if and only if* M^* *is a normal subgroup of* G.

(5) *If an intermediate field* M *is a normal extension of* K*, then the Galois group of* $M:K$ *is isomorphic to the quotient group* G/M^*.

Proof. Part (1) is a restatement of Corollary 11.11.

For part (2), suppose that M is an intermediate field, and let $[L:M] = d$. Then $|M^*| = d$ by Theorem 10.5. On the other hand, if H is a subgroup of G of order d, then $[L:H^\dagger] = d$ by Corollary 11.11. Hence the composite operators $*\dagger$ and $\dagger*$ preserve $[L:M]$ and $|H|$ respectively.

From their definitions, $M^{*\dagger} \supseteq M$ and $H^{\dagger *} \supseteq H$. Therefore these inclusions are equalities.

For part (3), again note that $L:M$ is normal. Corollary 11.11 states that $[L:M] = |M^*|$, and the other equality follows immediately.

We now prove part (4). If $M:K$ is normal, let $\tau \in G$. Then $\tau|_M$ is a K-monomorphism $M \to L$, so is a K-automorphism of M by Theorem 11.9. Hence $\tau(M) = M$. By Lemma 12.1, $\tau M^* \tau^{-1} = M^*$, so M^* is a normal subgroup of G.

Conversely, suppose that M^* is a normal subgroup of G. Let σ be any K-monomorphism $M \to L$. By Theorem 11.3, there is a K-automorphism τ of L such that $\tau|_M = \sigma$. Now $\tau M^* \tau^{-1} = M^*$ since M^* is a normal subgroup of G, so by Lemma 12.1, $\tau(M)^* = M^*$. By part 2 of Theorem 12.2, $\tau(M) = M$. Hence $\sigma(M) = M$ and σ is a K-automorphism of M. By Theorem 11.9, $M:K$ is normal.

Finally we prove part (5). Let G' be the Galois group of $M:K$. We can define a map $\phi: G \to G'$ by

$$\phi(\tau) = \tau|_M \qquad \tau \in G$$

This is clearly a group homomorphism $G \to G'$, for by Theorem 11.9 $\tau|_M$ is a K-automorphism of M. By Theorem 11.3, ϕ is onto. The kernel of ϕ is obviously M^*, so by standard group theory

$$G' = \mathrm{im}(\phi) \cong G/\ker(\phi) = G/M^*$$

where im is the image and ker the kernel. □

Note how Theorem 10.5 is used in the proof of part (2) of Theorem 12.2: its use is crucial. Many of the most beautiful results in mathematics hang by equally slender threads.

Parts (4) and (5) of Theorem 12.2 can be generalized: see Exercise 12.2. Note that the proof of part (5) provides an explicit isomorphism between $\Gamma(M:K)$ and G/M^*, namely, restriction to M.

The importance of the Fundamental Theorem of Galois Theory derives from its potential as a tool rather than its intrinsic merit. It enables us to apply group theory to otherwise intractable problems about polynomials over $\mathbb{C}$ and associated subfields of $\mathbb{C}$, and we shall spend most of the remaining chapters exploiting such applications.

EXERCISES

12.1 Work out the details of the Galois correspondence for the extension

$$\mathbb{Q}(i, \sqrt{5}) : \mathbb{Q}$$

whose Galois group is $G = \{I, R, S, T\}$ as in Chapter 8.

12.2 Let $L : K$ be a finite normal extension in $\mathbb{C}$ with Galois group G. Suppose that M, N are intermediate fields with $M \subseteq N$. Prove that $N : M$ is normal if and only if N^* is a normal subgroup of M^*. In this case prove that the Galois group of $N : M$ is isomorphic to M^*/N^*.

12.3* Let $\gamma = \sqrt{2+\sqrt{2}}$. Show that $\mathbb{Q}(\gamma)$: $\mathbb{Q}$ is normal, with cyclic Galois group. Show that $\mathbb{Q}(\gamma, i)$=$\mathbb{Q}(\mu)$ where $\mu^4 = i$.

12.4* Find the Galois group of $t^6 - 7$ over $\mathbb{Q}$.

12.5* Find the Galois group of $t^6 - 2t^3 - 1$ over $\mathbb{Q}$.

12.6 Let $\zeta = e^{\pi i/6}$ be a primitive 12th root of unity. Find the Galois group $\Gamma(\mathbb{Q}(\zeta) : \mathbb{Q})$ as follows.

(a) Prove that ζ is a zero of the polynomial $t^4 - t^2 + 1$, and that the other zeros are $\zeta^5, \zeta^7, \zeta^{11}$.

(b) Prove that $t^4 - t^2 + 1$ is irreducible over $\mathbb{Q}$, and is the minimal polynomial of ζ over $\mathbb{Q}$.

(c) Prove that $\Gamma(\mathbb{Q}(\zeta) : \mathbb{Q})$ consists of four $\mathbb{Q}$-automorphisms ϕ_j, defined by

$$\phi_j(\zeta) = \zeta^j \qquad j = 1, 5, 7, 11$$

(d) Prove that $\Gamma(\mathbb{Q}(\zeta) : \mathbb{Q}) \cong \mathbb{Z}_2 \times \mathbb{Z}_2$.

12.7 Using the subgroup structure of $\mathbb{Z}_2 \times \mathbb{Z}_2$ as in Exercise 12.6, find all intermediate fields between $\mathbb{Q}$ and $\mathbb{Q}(\zeta)$. [*Hint*: Calculate the fixed fields of the subgroups.]

12.8 Mark the following true or false.

(a) If $L : K$ is a finite normal extension inside $\mathbb{C}$, then the order of the Galois group of $L : K$ is equal to the dimension of L considered as a vector space over K.

(b) If M is any intermediate field of a finite normal extension inside $\mathbb{C}$, then $M^{\dagger *} = M$.

(c) If M is any intermediate field of a finite normal extension inside $\mathbb{C}$, then $M^{*\dagger} = M$.

(d) If M is any intermediate field of a finite normal extension $L : K$ inside $\mathbb{C}$, then the Galois group of $M : K$ is a subgroup of the Galois group of $L : K$.

(e) If M is any intermediate field of a finite normal extension $L : K$ inside $\mathbb{C}$, then the Galois group of $L : M$ is a quotient of the Galois group of $L : K$.

Chapter 13

A Worked Example

The Fundamental Theorem of Galois theory is quite a lot to take in at one go, so it is worth spending some time thinking it through. We therefore analyse how the Galois correspondence works out on an extended example.

The extension that we discuss is a favourite with writers on Galois theory, because of its archetypal quality. A simpler example would be too small to illustrate the theory adequately, and anything more complicated would be unwieldy. The example is the Galois group of the splitting field of $t^4 - 2$ over $\mathbb{Q}$.

The discussion will be cut into small pieces to make it more easily digestible.

(1) Let $f(t) = t^4 - 2$ over $\mathbb{Q}$, and let K be a splitting field for f such that $K \subseteq \mathbb{C}$. We can factorise f as follows:

$$f(t) = (t - \xi)(t + \xi)(t - i\xi)(t + i\xi)$$

where $\xi = \sqrt[4]{2}$ is real and positive. Therefore $K = \mathbb{Q}(\xi, i)$. Since K is a splitting field, $K : \mathbb{Q}$ is finite and normal. We are working in $\mathbb{C}$, so separability is automatic.

(2) We find the degree of $K : \mathbb{Q}$. By the Tower Law,

$$[K : \mathbb{Q}] = [\mathbb{Q}(\xi, i) : \mathbb{Q}(\xi)][\mathbb{Q}(\xi) : \mathbb{Q}]$$

The minimal polynomial of i over $\mathbb{Q}(\xi)$ is $t^2 + 1$, since $i^2 + 1 = 0$ but $i \notin \mathbb{R} \supseteq \mathbb{Q}(\xi)$. So $[\mathbb{Q}(\xi, i) : \mathbb{Q}(\xi)] = 2$.

Now ξ is a zero of f over $\mathbb{Q}$, and f is irreducible by Eisenstein's Criterion, Theorem 3.19. Hence f is the minimal polynomial of ξ over $\mathbb{Q}$, and $[\mathbb{Q}(\xi) : \mathbb{Q}] = 4$. Therefore

$$[K : \mathbb{Q}] = 2.4 = 8$$

(3) We find the elements of the Galois group of $K : \mathbb{Q}$. By a direct check, or by Corollary 5.13, there are $\mathbb{Q}$-automorphisms σ, τ of K such that

$$\begin{aligned} \sigma(i) &= i & \sigma(\xi) &= i\xi \\ \tau(i) &= -i & \tau(\xi) &= \xi \end{aligned}$$

Products of these yield eight distinct $\mathbb{Q}$-automorphisms of K:

Automorphism	*Effect on* ξ	*Effect on i*
1	ξ	i
σ	$i\xi$	i
σ^2	$-\xi$	i
σ^3	$-i\xi$	i
τ	ξ	$-i$
$\sigma\tau$	$i\xi$	$-i$
$\sigma^2\tau$	$-\xi$	$-i$
$\sigma^3\tau$	$-i\xi$	$-i$

Other products do not give new automorphisms, since $\sigma^4 = 1, \tau^2 = 1,\ \tau\sigma = \sigma^3\tau,\ \tau\sigma^2 = \sigma^2\tau,\ \tau\sigma^3 = \sigma\tau$. (The last two relations follows from the first three.) Any $\mathbb{Q}$-automorphism of K sends i to some zero of t^2+1, so $i \mapsto \pm i$; similarly ξ is mapped to $\xi, i\xi, -\xi$, or $-i\xi$. All possible combinations of these (eight in number) appear in the above list, so these are precisely the $\mathbb{Q}$-automorphisms of K.
(4) The abstract structure of the Galois group G can be found. The generator-relation presentation

$$G = \langle \sigma, \tau : \sigma^4 = \tau^2 = 1,\ \tau\sigma = \sigma^3\tau \rangle$$

shows that G is the dihedral group of order 8, which we write as $\mathbb{D}_4$. (In some books the notation $\mathbb{D}_8$ is used instead. It depends on what you think is important: the order is 8 or there is a normal subgroup $\mathbb{Z}_4$.)

The group $\mathbb{D}_4$ has a geometric interpretation as the symmetry group of a square. In fact we can label the four vertices of a square with the zeros of t^4-2, in such a way that the geometric symmetries are precisely the permutations of the zeros that occur in the Galois group (Figure 19).

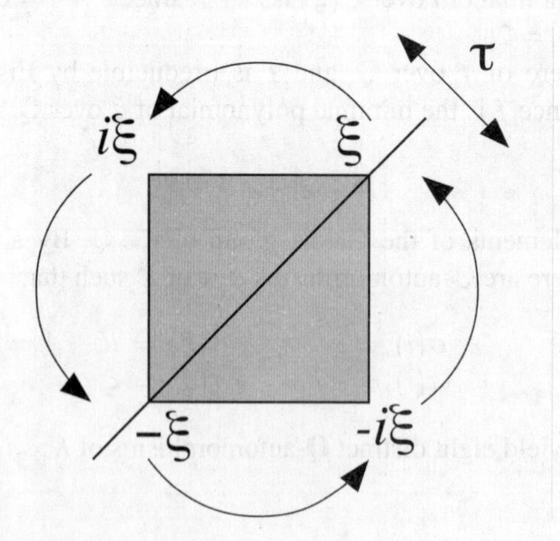

FIGURE 19: The Galois group $\mathbb{D}_4$ interpreted as the symmetry group of a square.

(5) It is an easy exercise to find the subgroups of G. If as usual we let $\mathbb{Z}_n$ denote the cyclic group of order n, and $\times$ the direct product, then the subgroups are as follows:

Order 8:	G	$G \cong \mathbb{D}_4$
Order 4:	$\{1, \sigma, \sigma^2, \sigma^3\}$	$S \cong \mathbb{Z}_4$
	$\{1, \sigma^2, \tau, \sigma^2\tau\}$	$T \cong \mathbb{Z}_2 \times \mathbb{Z}_2$
	$\{1, \sigma^2, \sigma\tau, \sigma^3\tau\}$	$U \cong \mathbb{Z}_2 \times \mathbb{Z}_2$
Order 2:	$\{1, \sigma^2\}$	$A \cong \mathbb{Z}_2$
	$\{1, \tau\}$	$B \cong \mathbb{Z}_2$
	$\{1, \sigma\tau\}$	$C \cong \mathbb{Z}_2$
	$\{1, \sigma^2\tau\}$	$D \cong \mathbb{Z}_2$
	$\{1, \sigma^3\tau\}$	$E \cong \mathbb{Z}_2$
Order 1:	$\{1\}$	$I \cong 1$

(6) The inclusion relations between the subgroups of G can be summed up by the *lattice diagram* of Figure 20. In such diagrams, $X \subseteq Y$ if there is a sequence of upward-sloping lines from X to Y.

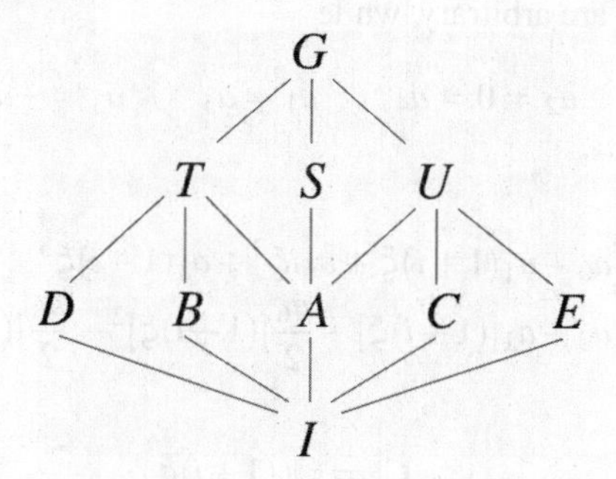

FIGURE 20: Lattice of subgroups.

(7) Under the Galois correspondence we obtain the intermediate fields. Since the correspondence reverses inclusions, we obtain the lattice diagram in Figure 21.

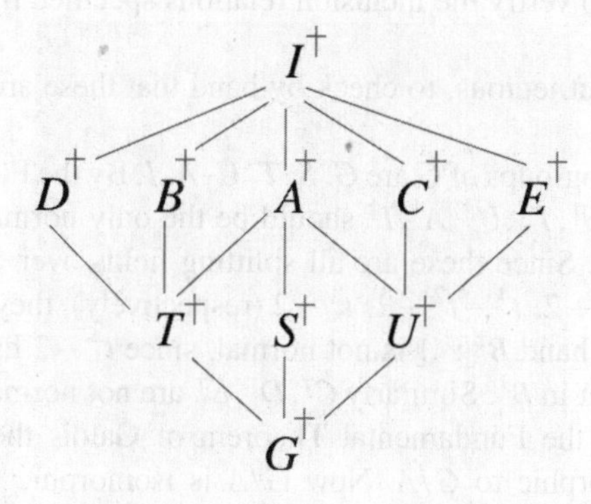

FIGURE 21: Lattice of subfields.

(8) We now describe the elements of these intermediate fields. There are three obvious subfields of K of degree 2 over $\mathbb{Q}$, namely $\mathbb{Q}(i)$, $\mathbb{Q}(\sqrt{2})$, $\mathbb{Q}(i\sqrt{2})$. These are clearly the fixed fields $S^\dagger, T^\dagger$, and $U^\dagger$, respectively. The other fixed fields are less obvious. To illustrate a possible approach we shall find $C^\dagger$. Any element of K can be expressed uniquely in the form

$$x = a_0 + a_1\xi + a_2\xi^2 + a_3\xi^3 + a_4 i + a_5 i\xi + a_6 i\xi^2 + a_7 i\xi^3$$

where $a_0, \ldots, a_7 \in \mathbb{Q}$. Then

$$\begin{aligned}\sigma\tau(x) &= a_0 + a_1 i\xi - a_2\xi^2 - a_3 i\xi^3 - a_4 i + a_5(-i)i\xi - a_6 i(i\xi)^2 - a_7 i(i\xi)^3 \\ &= a_0 + a_5\xi - a_2\xi^2 - a_7\xi^3 - a_4 i + a_1 i\xi + a_6 i\xi^2 - a_3 i\xi^3\end{aligned}$$

The element x is fixed by $\sigma\tau$ (and hence by C) if and only if

$$\begin{array}{llll} a_0 = a_0 & a_1 = a_5 & a_2 = -a_2 & a_3 = -a_7 \\ a_4 = -a_4 & a_5 = a_1 & a_6 = a_6 & a_7 = -a_3 \end{array}$$

Therefore a_0 and a_6 are arbitrary, while

$$a_2 = 0 = a_4 \qquad a_1 = a_5 \qquad a_3 = -a_7$$

It follows that

$$\begin{aligned} x &= a_0 + a_1(1+i)\xi + a_6 i\xi^2 + a_3(1-i)\xi^3 \\ &= a_0 + a_1[(1+i)\xi] + \frac{a_6}{2}[(1+i)\xi]^2 - \frac{a_3}{2}[(1+i)\xi]^3 \end{aligned}$$

which shows that

$$C^\dagger = \mathbb{Q}((1+i)\xi)$$

Similarly,

$$A^\dagger = \mathbb{Q}(i,\sqrt{2}) \qquad B^\dagger = \mathbb{Q}(\xi) \qquad D^\dagger = \mathbb{Q}(i\xi) \qquad E^\dagger = \mathbb{Q}((1-i)\xi)$$

It is now easy to verify the inclusion relations specified by the lattice diagram in Figure 21.

(9) It is possible, but tedious, to check by hand that these are the only intermediate fields.

(10) The normal subgroups of G are G, S, T, U, A, I. By the Fundamental Theorem of Galois theory, $G^\dagger, S^\dagger, T^\dagger, U^\dagger, A^\dagger, I^\dagger$ should be the only normal extensions of $\mathbb{Q}$ that are contained in K. Since these are all splitting fields over $\mathbb{Q}$, for the polynomials t, t^2+1, t^2-2, t^2+2, t^4-t^2-2, t^4-2 (respectively), they are normal extensions of $\mathbb{Q}$. On the other hand $B^\dagger : \mathbb{Q}$ is not normal, since t^4-2 has a zero, namely ξ, in $B^\dagger$ but does not split in $B^\dagger$. Similarly $C^\dagger, D^\dagger, E^\dagger$ are not normal extensions of $\mathbb{Q}$.

(11) According to the Fundamental Theorem of Galois theory, the Galois group of $A^\dagger : \mathbb{Q}$ is isomorphic to G/A. Now G/A is isomorphic to $\mathbb{Z}_2 \times \mathbb{Z}_2$. We calculate directly the Galois group of $A^\dagger : \mathbb{Q}$. Since $A^\dagger = \mathbb{Q}(i,\sqrt{2})$ there are four $\mathbb{Q}$-automorphisms:

Automorphism	*Effect on i*	*Effect on* $\sqrt{2}$
1	i	$\sqrt{2}$
α	i	$-\sqrt{2}$
β	$-i$	$\sqrt{2}$
$\alpha\beta$	$-i$	$-\sqrt{2}$

and since $\alpha^2 = \beta^2 = 1$ and $\alpha\beta = \beta\alpha$, this group is $\mathbb{Z}_2 \times \mathbb{Z}_2$ as expected.
(12) The lattice diagrams for $\mathscr{F}$ and $\mathscr{G}$ do *not* look the same unless one of them is turned upside-down. Hence there does not exist a correspondence like the Galois correspondence but preserving inclusion relations. It may seem a little odd at first that the Galois correspondence reverses inclusions, but in fact it is entirely natural, and quite as useful a property as preservation of inclusions.

It is in general a difficult problem to compute the Galois group of a given field extension, particularly when there is no explicit representation for the elements of the large field. See Chapter 22.

EXERCISES

13.1 Find the Galois groups of the following extensions:

(a) $\mathbb{Q}(\sqrt{2}, \sqrt{5}) : \mathbb{Q}$

(b) $\mathbb{Q}(\alpha) : \mathbb{Q}$ where $\alpha = e^{2\pi i/3}$.

(c) $K : \mathbb{Q}$ where K is the splitting field over $\mathbb{Q}$ for $t^4 - 3t^2 + 4$.

13.2 Find all subgroups of these Galois groups.

13.3 Find the corresponding fixed fields.

13.4 Find all normal subgroups of the above Galois groups.

13.5 Check that the corresponding extensions are normal.

13.6 Verify that the Galois groups of these normal extensions are the relevant quotient groups.

13.7* Consider the Galois group of $t^6 - 7$ over $\mathbb{Q}$, found in Exercise 12.4. Use the Galois correspondence to find all intermediate fields.

13.8* Consider the Galois group of $t^6 - 2t^3 - 1$ over $\mathbb{Q}$, found in Exercise 12.5. Use the Galois correspondence to find all intermediate fields.

13.9 Find the Galois group of $t^8 - i$ over $\mathbb{Q}(i)$.

13.10 Find the Galois group of $t^8 + t^4 + 1$ over $\mathbb{Q}(i)$.

3.11 Use the Galois group $\mathbb{Z}_2 \times \mathbb{Z}_2 \times \mathbb{Z}_2$ of $\mathbb{Q}(\sqrt{2}, \sqrt{3}, \sqrt{5}) : \mathbb{Q}$ to find all intermediate fields. Which of these are normal over $\mathbb{Q}$?

13.12 Mark the following true or false.

(a) A 3×3 square has exactly 9 distinct symmetries.

(b) The symmetry group of a square is isomorphic to $\mathbb{Z}_8$.

(c) The symmetry group of a square is isomorphic to $\mathbb{S}_8$.

(d) The symmetry group of a square is isomorphic to a subgroup of $\mathbb{S}_8$.

(e) The group $\mathbb{D}_4$ has 10 distinct subgroups.

(f) The Galois correspondence preserves inclusion relations.

(g) The Galois correspondence reverses inclusion relations.

Chapter 14

Solubility and Simplicity

In order to apply the Galois correspondence, in particular to solving equations by radicals, we need to have at our fingertips a number of group-theoretic concepts and theorems. We have already assumed familiarity with elementary group theory: subgroups, normal subgroups, quotient groups, conjugates, permutations (up to cycle decomposition): to these we now add the standard isomorphism theorems. The relevant theory, along with most of the material in this chapter, can be found in any basic textbook on group theory, for example Fraleigh (1989), Humphreys (1996), or Neumann, Stoy, and Thompson (1994).

We start by defining soluble groups and proving some basic properties. These groups are of cardinal importance for the theory of the solution of equations by radicals. Next, we discuss simple groups, the main target being a proof of the simplicity of the alternating group of degree 5 or more. We end by proving Cauchy's Theorem: if a prime p divides the order of a finite group, then the group has an element of order p.

14.1 Soluble Groups

Soluble groups were first defined and studied (though not in the current abstract way) by Galois in his work on the solution of equations by radicals. They have since proved extremely important in many branches of mathematics.

In the following definition, and thereafter, the notation $H \lhd G$ will mean that H is a normal subgroup of the group G. Recall that an *abelian* (or *commutative*) group is one in which $gh = hg$ for all elements g, h.

Definition 14.1. A group G is *soluble* (in the US: *solvable*) if it has a finite series of subgroups

$$1 = G_0 \subseteq G_1 \subseteq \cdots \subseteq G_n = G \tag{14.1}$$

such that

(1) $G_i \lhd G_{i+1}$ for $i = 0, \ldots, n-1$.

(2) G_{i+1}/G_i is abelian for $i = 0, \ldots, n-1$.

Condition (14.1) does not imply that $G_i \lhd G$, since $G_i \lhd G_{i+1} \lhd G_{i+2}$ does not imply $G_i \lhd G_{i+2}$. See Exercise 14.10.

Examples 14.2. (1) Every abelian group G is soluble, with series $1 \lhd G$.
(2) The symmetric group $\mathbb{S}_3$ of degree 3 is soluble, since it has a cyclic normal subgroup of order 3 generated by the cycle (123) whose quotient is cyclic of order 2. All cyclic groups are abelian.
(3) The dihedral group $\mathbb{D}_8$ of order 8 is soluble. In the notation of Chapter 13, it has a normal subgroup S of order 4 whose quotient has order 2, and S is abelian.
(4) The symmetric group $\mathbb{S}_4$ of degree 4 is soluble, having a series

$$1 \lhd \mathbb{V} \lhd \mathbb{A}_4 \lhd \mathbb{S}_4$$

where $\mathbb{A}_4$ is the alternating group of order 12, and $\mathbb{V}$ is the Klein four-group, which we recall consists of the permutations 1, (12)(34),(13)(24), (14)(23) and hence is a direct product of two cyclic groups of order 2. The quotient groups are

$$\begin{array}{rl} \mathbb{V}/1 \cong \mathbb{V} & \text{abelian of order 4} \\ \mathbb{A}_4/\mathbb{V} \cong \mathbb{Z}_3 & \text{abelian of order 3} \\ \mathbb{S}_4/\mathbb{A}_4 \cong \mathbb{Z}_2 & \text{abelian of order 2.} \end{array}$$

(5) The symmetric group $\mathbb{S}_5$ of degree 5 is not soluble. This follows from Lemma 8.11 with a bit of extra work. See Corollary 14.8.

We recall the following isomorphism theorems:

Lemma 14.3. *Let G, H, and A be groups.*

(1) *If $H \lhd G$ and $A \subseteq G$ then $H \cap A \lhd A$ and*

$$\frac{A}{H \cap A} \cong \frac{HA}{H}$$

(2) *If $H \lhd G$, and $H \subseteq A \lhd G$ then $H \lhd A$, $A/H \lhd G/H$ and*

$$\frac{G/H}{A/H} \cong \frac{G}{A}$$

(3) *If $H \lhd G$ and $A/H \lhd G/H$ then $A \lhd G$.*

Parts (1) and (2) are respectively the *First* and *Second Isomorphism Theorems.* They are the translation into normal subgroup language of two straightforward facts: restricting a homomorphism to a subgroup yields a homomorphism, and composing two homomorphisms yields a homomorphism. See Exercise 14.11. Part (3) is a converse to part (2) and is easy to prove.

Judicious use of these isomorphism theorems lets us prove that soluble groups persist in being soluble even when subjected to quite drastic treatment.

Theorem 14.4. *Let G be a group, H a subgroup of G, and N a normal subgroup of G.*

(1) *If G is soluble, then H is soluble.*

(2) *If G is soluble, then G/N is soluble.*

(3) *If N and G/N are soluble, then G is soluble.*

Proof. (1) Let

$$1 = G_0 \lhd G_1 \lhd \ldots \lhd G_r = G$$

be a series for G with abelian quotients G_{i+1}/G_i. Let $H_i = G_i \cap H$. Then H has a series

$$1 = H_0 \lhd \ldots \lhd H_r = H$$

We show the quotients are abelian. Now

$$\frac{H_{i+1}}{H_i} = \frac{G_{i+1} \cap H}{G_i \cap H} = \frac{G_{i+1} \cap H}{G_i \cap (G_{i+1} \cap H)} \cong \frac{G_i(G_{i+1} \cap H)}{G_i}$$

by the first isomorphism theorem. But this latter group is a subgroup of G_{i+1}/G_i which is abelian. Hence H_{i+1}/H_i is abelian for all i, and H is soluble.
(2) Take G_i as before. Then G/N has a series

$$N/N = G_0N/N \lhd G_1N/N \lhd \ldots \lhd G_rN/N = G/N$$

A typical quotient is

$$\frac{G_{i+1}N/N}{G_iN/N}$$

which by the second isomorphism theorem is isomorphic to

$$\frac{G_{i+1}N}{G_iN} = \frac{G_{i+1}(G_iN)}{G_iN} \cong \frac{G_{i+1}}{G_{i+1} \cap (G_iN)} \cong \frac{G_{i+1}/G_i}{(G_{i+1} \cap (G_iN))/G_i}$$

which is a quotient of the abelian group G_{i+1}/G_i, so is abelian. Therefore G/N is soluble.
(3) There exist two series

$$1 = N_0 \lhd N_1 \lhd \ldots \lhd N_r = N$$
$$N/N = G_0/N \lhd G_1/N \lhd \ldots \lhd G_s/N = G/N$$

with abelian quotients. Consider the series of G given by combining them:

$$1 = N_0 \lhd N_1 \lhd \ldots \lhd N_r = N = G_0 \lhd G_1 \lhd \ldots \lhd G_s = G$$

The quotients are either N_{i+1}/N_i (which is abelian) or G_{i+1}/G_i, which is isomorphic to

$$\frac{G_{i+1}/N}{G_i/N}$$

and again is abelian. Therefore G is soluble. □

A group G is an *extension* of a group A by a group B if G has a normal subgroup N isomorphic to A such that G/N is isomorphic to B. We may sum up the three properties of the above theorem as: the class of soluble groups is closed under taking subgroups, quotients, and extensions. The class of abelian groups is closed under taking subgroups and quotients, but not extensions. It is largely for this reason that Galois was led to define soluble groups.

14.2 Simple Groups

We turn to groups that are, in a sense, the opposite of soluble.

Definition 14.5. A group G is *simple* if it is nontrivial and its only normal subgroups are 1 and G.

Every cyclic group $\mathbb{Z}_p$ of prime order is simple, since it has no subgroups other than 1 and $\mathbb{Z}_p$, hence in particular no other normal subgroups. These groups are also abelian, hence soluble. They are in fact the only soluble simple groups:

Theorem 14.6. *A soluble group is simple if and only if it is cyclic of prime order.*

Proof. Since G is soluble group, it has a series

$$1 = G_0 \lhd G_1 \lhd \ldots \lhd G_n = G$$

where by deleting repeats we may assume $G_{i+1} \neq G_i$. Then G_{n-1} is a proper normal subgroup of G. However, G is simple, so $G_{n-1} = 1$ and $G = G_n/G_{n-1}$, which is abelian. Since every subgroup of an abelian group is normal, and every element of G generates a cyclic subgroup, G must be cyclic with no non-trivial proper subgroups. Hence G has prime order.

The converse is trivial. □

Simple groups play an important role in finite group theory. They are in a sense the fundamental units from which all finite groups are made. Indeed the Jordan–Hölder theorem, which we do not prove, states that every finite group has a series of subgroups like (14.1) whose quotients are simple, and these simple groups depend only on the group and not on the series chosen.

We do not need to know much about simple groups, intriguing as they are. We require just one result:

Theorem 14.7. *If $n \geq 5$, then the alternating group $\mathbb{A}_n$ of degree n is simple.*

Proof. We use much the same strategy as in Lemma 8.11, but we are proving a rather stronger property, so we have to work a bit harder.

Suppose that $1 \neq N \lhd \mathbb{A}_n$. Our strategy will be as follows: first, observe that if N contains a 3-cycle then it contains all 3-cycles, and since the 3-cycles generate $\mathbb{A}_n$, we must have $N = \mathbb{A}_n$. Second, prove that N must contain a 3-cycle. It is here that we need $n \geq 5$.

Suppose then, that N contains a 3-cycle; without loss of generality N contains (123). Now for any $k > 3$ the cycle $(32k)$ is an even permutation, so lies in $\mathbb{A}_n$, and therefore

$$(32k)(123)(32k)^{-1} = (1k2)$$

lies in N. Hence N contains $(1k2)^2 = (12k)$ for all $k \geq 3$. We claim that $\mathbb{A}_n$ is generated by all 3-cycles of the form $(12k)$. If $n = 3$ then we are done. If $n > 3$ then for all $a, b > 2$ the permutation $(1a)(2b)$ is even, so lies in $\mathbb{A}_n$, and then $\mathbb{A}_n$ contains

$$(1a)(2b)(12k)((1a)(2b))^{-1} = (abk)$$

if $k \neq a, b$. Since $\mathbb{A}_n$ is generated by all 3-cycles (Exercise 8.7), it follows that $N = \mathbb{A}_n$.

It remains to show that N must contain at least one 3-cycle. We do this by an analysis into cases.

(1) Suppose that N contains an element $x = abc\ldots$, where $a, b, c, \ldots$ are disjoint cycles and

$$a = (a_1 \ldots a_m) \quad (m \geq 4)$$

Let $t = (a_1 a_2 a_3)$. Then N contains $t^{-1}xt$. Since t commutes with $b, c, \ldots$ (disjointness of cycles) it follows that

$$t^{-1}xt = (t^{-1}at)bc\ldots = z \quad \text{(say)}$$

so that N contains

$$zx^{-1} = (a_1 a_3 a_m)$$

which is a 3-cycle.

(2) Now suppose N contains an element involving at least two 3-cycles. Without loss of generality N contains

$$x = (123)(456)y$$

where y is a permutation fixing 1, 2, 3, 4, 5, 6. Let $t = (234)$. Then N contains

$$(t^{-1}xt)x^{-1} = (12436)$$

Then by case (1) N contains a 3-cycle.

(3) Now suppose that N contains an element x of the form $(ijk)p$, where p is a product of 2-cycles disjoint from each other and from (ijk). Then N contains $x^2 = (ikj)$, which is a 3-cycle.

(4) There remains the case when every element of N is a product of disjoint 2-cycles. (This actually occurs when $n = 4$, giving the four-group $\mathbb{V}$.) But as $n \geq 5$, we can assume that N contains

$$x = (12)(34)p$$

where p fixes 1, 2, 3, 4. If we let $t = (234)$ then N contains

$$(t^{-1}xt)x^{-1} = (14)(23)$$

and if $u = (145)N$ contains

$$u^{-1}(t^{-1}xtx^{-1})u = (45)(23)$$

so that N contains

$$(45)(23)(14)(23) = (145)$$

contradicting the assumption that every element of N is a product of disjoint 2-cycles.

Hence $\mathbb{A}_n$ is simple if $n \geq 5$. □

In fact $\mathbb{A}_5$ is the smallest non-abelian simple group. This result is often attributed to Galois, but Neumann (2011), in his translation of Galois's mathematical writings, points out on pages 384–385 that alternating groups are not mentioned in any significant work by Galois, and that the methods available to him were inadequate to eliminate various orders for a potential simple group, such as 56. Although it seems plausible that Galois knew that $\mathbb{A}_n$ is simple for $n \geq 5$, there is no clear evidence that he did. Indeed, his proof that the quintic cannot be solved by radicals uses other special features of the Galois group of an equation of prime degree: see Neumann (2011) chapter IV. We discuss this point further in Chapter 25.

From this theorem we deduce:

Corollary 14.8. *The symmetric group $\mathbb{S}_n$ of degree n is not soluble if $n \geq 5$.*

Proof. If $\mathbb{S}_n$ were soluble then $\mathbb{A}_n$ would be soluble by Theorem 14.4, and simple by Theorem 14.7, hence of prime order by Theorem 14.6. But $|\mathbb{A}_n| = \frac{1}{2}(n!)$ is not prime if $n \geq 5$. □

14.3 Cauchy's Theorem

We next prove Cauchy's Theorem: if a prime p divides the order of a finite group, then the group has an element of order p. We begin by recalling several ideas from group theory.

Definition 14.9. Elements a and b of a group G are *conjugate* in G if there exists $g \in G$ such that $a = g^{-1}bg$.

Conjugacy is an equivalence relation; the equivalence classes are the *conjugacy classes* of G.

If the conjugacy classes of G are $C_1, \ldots, C_r$, then one of them, say C_1, contains only the identity element of G. Therefore $|C_1| = 1$. Since the conjugacy classes form a partition of G we have

$$|G| = 1 + |C_2| + \cdots + |C_r| \tag{14.2}$$

which is the *class equation* for G.

Definition 14.10. If G is a group and $x \in G$, then the *centraliser* $C_G(x)$ of x in G is the set of all $g \in G$ for which $xg = gx$. It is always a subgroup of G.

There is a useful connection between centralisers and conjugacy classes.

Lemma 14.11. *If G is a group and $x \in G$, then the number of elements in the conjugacy class of x is the index of $C_G(x)$ in G.*

Proof. The equation $g^{-1}xg = h^{-1}xh$ holds if and only if $hg^{-1}x = xhg^{-1}$, which means that $hg^{-1} \in C_G(x)$, that *is*, h and g lie in the same coset of $C_G(x)$ in G. The number of these cosets is the index of $C_G(x)$ in G, so the lemma is proved. □

Corollary 14.12. *The number of elements in any conjugacy class of a finite group G divides the order of G.*

Definition 14.13. The *centre* $Z(G)$ of a group G is the set of all elements $x \in G$ such that $xg = gx$ for all $g \in G$.

The centre of G is a normal subgroup of G. Many groups have trivial centre, for example $Z(\mathbb{S}_3) = 1$. Abelian groups go to the other extreme and have $Z(G) = G$.

Lemma 14.14. *If A is a finite abelian group whose order is divisible by a prime p, then A has an element of order p.*

Proof. Use induction on $|A|$. If $|A|$ is prime the result follows. Otherwise take a proper subgroup M of A whose order m is maximal. If p divides m we are home by induction, so we may assume that p does not divide m. Let b be in A but not in M, and let B be the cyclic subgroup generated by b. Then MB is a subgroup of A, larger than M, so by maximality $A = MB$. From the First Isomorphism Theorem, Lemma 14.3(1),

$$|MB| = |M||B|/|M \cap B|$$

so p divides the order r of B. Since B is cyclic, the element $b^{r/p}$ has order p. □

From this result we can derive a more general theorem of Cauchy in which the group need not be abelian:

Theorem 14.15 (Cauchy's Theorem). *If a prime p divides the order of a finite group G, then G has an element of order p.*

Proof. We prove the theorem by induction on the order $|G|$. The first few cases $|G| = 1, 2, 3$ are obvious. For the induction step, start with the class equation

$$|G| = 1 + |C_2| + \cdots + |C_r|$$

Since $p \mid |G|$, we must have $p \nmid |C_j|$ for some $j \geq 2$. If $x \in C_j$ it follows that $p \mid |C_G(x)|$, since $|C_j| = |G|/|C_G(x)|$.

If $C_G(x) \neq G$ then by induction $C_G(x)$ contains an element of order p, and this element also belongs to G.

Otherwise $C_G(x) = G$, which implies that $x \in Z(G)$, and by choice $x \neq 1$, so $Z(G) \neq 1$.

Either $p||Z(G)|$ or $p\nmid|Z(G)|$. In the first case the proof reduces to the abelian case, Lemma 14.14. In the second case, by induction there exists $x \in G$ such that the image $\bar{x} \in G/Z(G)$ has order p. That is, $x^p \in Z(G)$ but $x \notin Z(G)$. Let X be the cyclic group generated by x. Now $XZ(G)$ is abelian and has order divisible by p, so by Lemma 14.14 it has an element of order p, and again this element also belongs to G.

This completes the induction step, and with it the proof. □

Cauchy's Theorem does not work for composite divisors of $|G|$. See Exercise 14.6.

EXERCISES

14.1 Show that the general dihedral group

$$\mathbb{D}_n = \langle a, b : a^n = b^2 = 1, b^{-1}ab = a^{-1} \rangle$$

is a soluble group. Here a, b are generators and the equalities are relations between them.

14.2 Prove that $\mathbb{S}_n$ is not soluble for $n \geq 5$, using only the simplicity of $\mathbb{A}_5$.

14.3 Prove that a normal subgroup of a group is a union of conjugacy classes. Find the conjugacy classes of $\mathbb{A}_5$, using the cycle type of the permutations, and hence show that $\mathbb{A}_5$ is simple.

14.4 Prove that $\mathbb{S}_n$ is generated by the 2-cycles $(12), \ldots, (1n)$.

14.5 If the point $\alpha \in \mathbb{C}$ is constructible by ruler and compasses, show that the Galois group of $\mathbb{Q}(\alpha) : \mathbb{Q}$ is soluble.

14.6 Show that $\mathbb{A}_5$ has no subgroup of order 15, even though 15 divides its order.

14.7 Show that $\mathbb{S}_n$ has trivial centre if $n \geq 3$.

14.8 Find the conjugacy classes of the dihedral group $\mathbb{D}_n$ defined in Exercise 14.1. Work out the centralisers of selected elements, one from each conjugacy class, and check Lemma 13.7.

14.9 If G is a group and $x, g \in G$, show that $C_G(g^{-1}xg) = g^{-1}C_G(x)g$.

14.10 Show that the relation 'normal subgroup of' is not transitive. (*Hint:* Consider the subgroup $G \subseteq \mathbb{V} \subseteq \mathbb{S}_4$ generated by the element (12)(34).)

14.11 There are (at least) two distinct ways to think about a group homomorphism. One is the definition as a structure-preserving mapping, the other is in terms of a quotient group by a normal subgroup. The relation between these is as follows. If $\phi : G \to H$ is a homomorphism then

$$\ker(\phi) \lhd G \quad \text{and} \quad G/\ker(\phi) \cong \mathrm{im}(\phi)$$

If $N \lhd G$ then there is a natural surjective homomorphism

$$\phi : G \to G/N \quad \text{with } \ker(\phi) = N$$

Show that the first and second isomorphism theorems are the translations into 'quotient group' language of two facts that are trivial in 'structure-preserving mapping' language:

(1) The restriction of a homomorphism to a subgroup is a homomorphism.

(2) The composition of two homomorphisms is a homomorphism.

14.12* By counting the sizes of conjugacy classes, prove that the group of rotational symmetries of a regular icosahedron is simple. Show that it is isomorphic to $\mathbb{A}_5$.

14.13 Mark the following true or false.

(a) The direct product of two soluble groups is soluble.

(b) Every simple soluble group is cyclic.

(c) Every cyclic group is simple.

(d) The symmetric group $\mathbb{S}_n$ is simple if $n \geq 5$.

(e) Every conjugacy class of a group G is a subgroup of G.

Chapter 15

Solution by Radicals

The historical aspects of the problem of solving polynomial equations by radicals have been discussed in the introduction. Early in his career, Galois briefly thought that he had solved the quintic equation by radicals, Figure 22. However, he found a mistake when it was suggested that he should try some numerical examples. This motivated his work on solubility by radicals.

The object of this chapter is to use the Galois correspondence to derive a condition that must be satisfied by any polynomial equation that is soluble by radicals, namely: the associated Galois group must be a soluble group. We then construct a quintic polynomial equation whose Galois group is not soluble, namely the disarmingly straightforward-looking $t^5 - 6t + 3 = 0$, which shows that the quintic equation cannot be solved by radicals.

Solubility of the Galois group is also a sufficient condition for an equation to be soluble by radicals, but we defer this result to Chapter 18.

15.1 Radical Extensions

Some care is needed in formalising the idea of 'solubility by radicals'. We begin from the point of view of field extensions.

Informally, a radical extension is obtained by a sequence of adjunctions of nth roots, for various n. For example, the following expression is radical:

$$\sqrt[3]{11}\sqrt[5]{\frac{7+\sqrt{3}}{2}} + \sqrt[4]{1+\sqrt[3]{4}} \tag{15.1}$$

To find an extension of $\mathbb{Q}$ that contains this element we may adjoin in turn elements

$$\alpha = \sqrt[3]{11} \qquad \beta = \sqrt{3} \qquad \gamma = \sqrt[5]{(7+\beta)/2} \qquad \delta = \sqrt[3]{4} \qquad \varepsilon = \sqrt[4]{1+\delta}$$

Recall Definition 8.12, which formalises the idea of a radical extension: $L : K$ is radical if $L = K(\alpha_1, \ldots, \alpha_m)$ where for each $j = 1, \ldots, m$ there exists n_j such that

$$\alpha_j^{n_j} \in K(\alpha_1, \ldots, \alpha_{j-1}) \qquad (j \geq 1)$$

The elements α_j form a radical sequence for $L : K$, and the *radical degree* of α_j is n_j.

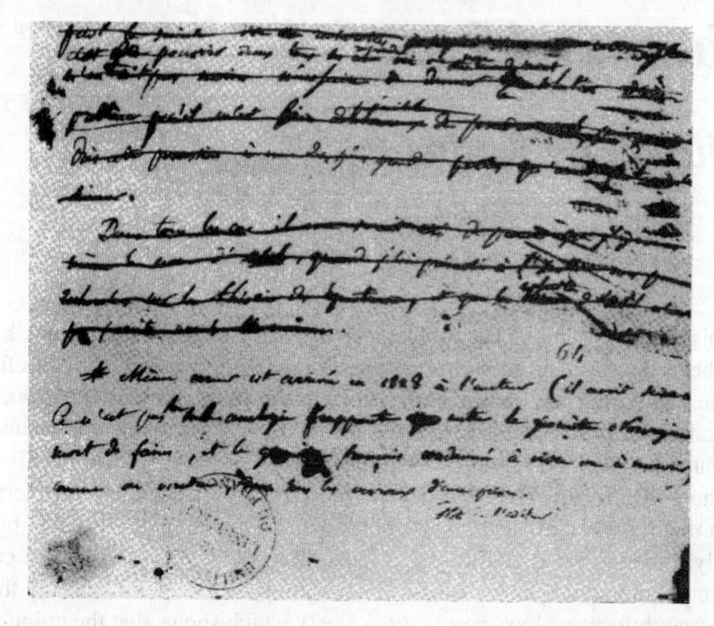

FIGURE 22: Galois thought he had solved the quintic... but changed his mind.

For example, the expression (15.1) is contained in a radical extension of the form $\mathbb{Q}(\alpha,\beta,\gamma,\delta,\varepsilon)$ of $\mathbb{Q}$, where $\alpha^3 = 11$, $\beta^2 = 3$, $\gamma^5 = (7+\beta)/2$, $\delta^3 = 4$, $\varepsilon^4 = 1+\delta$.

It is clear that any radical expression, in the sense of the introduction, is contained in some radical extension.

A polynomial should be considered soluble by radicals provided *all* of its zeros are radical expressions over the ground field.

Definition 15.1. *Let f be a polynomial over a subfield K of $\mathbb{C}$, and let Σ be the splitting field for f over K. We say that f is soluble by radicals if there exists a field M containing Σ such that $M : K$ is a radical extension.*

We emphasise that in the definition, we do not require the splitting field extension $\Sigma : K$ to be radical. There is a good reason for this. We want everything in the splitting field Σ to be expressible by radicals, but it is pointless to expect everything expressible by the same radicals to be inside the splitting field. Indeed,if $M : K$ is radical and L is an intermediate field, then $L : K$ need not be radical: see Exercise 15.6.

Note also that we require *all* zeros of f to be expressible by radicals. It is possible for some zeros to be expressible by radicals, while others are not—simply take a product of two polynomials, one soluble by radicals and one not. However, if an *irreducible* polynomial f has one zero expressible by radicals, then all the zeros must be so expressible, by a simple argument based on Corollary 5.13.

The main theorem of this chapter is:

Theorem 15.2. *If K is a subfield of $\mathbb{C}$ and $K \subseteq L \subseteq M \subseteq \mathbb{C}$ where $M : K$ is a radical extension, then the Galois group of $L : K$ is soluble.*

The otherwise curious word 'soluble' for groups arises in this context: a soluble (by radicals) polynomial has a soluble Galois group (of its splitting field over the base field).

The proof of this result is not entirely straightforward, and we must spend some time on preliminaries.

Lemma 15.3. *If $L : K$ is a radical extension in $\mathbb{C}$ and M is the normal closure of $L : K$, then $M : K$ is radical.*

Proof. Let $L = K(\alpha_1, \ldots, \alpha_r)$ with $\alpha_i^{n_i} \in K(\alpha_1, \ldots, \alpha_{i-1})$. Let f_i be the minimal polynomial of α_i over K. Then $M \supseteq L$ is clearly the splitting field of $\prod_{i=1}^{r} f_i$. For every zero β_{ij} of f_i in M there exists an isomorphism $\sigma : K(\alpha_i) \to K(\beta_{ij})$ by Corollary 5.13. By Proposition 11.4, σ extends to a K-automorphism $\tau : M \to M$. Since α_i is a member of a radical sequence for a subfield of M, so is β_{ij}. By combining the sequences, we get a radical sequence for M. □

The next two lemmas show that certain Galois groups are abelian.

Lemma 15.4. *Let K be a subfield of $\mathbb{C}$, and let L be the splitting field for $t^p - 1$ over K, where p is prime. Then the Galois group of $L : K$ is abelian.*

Proof. The derivative of $t^p - 1$ is pt^{p-1}, which is prime to $t^p - 1$, so by Lemma 9.13 the polynomial has no multiple zeros in L. Clearly its zeros form a group under multiplication; this group has prime order p since the zeros are distinct, so is cyclic. Let ε be a generator of this group. Then $L = K(\varepsilon)$ so that any K-automorphism of L is determined by its effect on ε. Further, K-automorphisms permute the zeros of $t^p - 1$. Hence any K-automorphism of L is of the form

$$\alpha_j : \varepsilon \mapsto \varepsilon^j$$

and is uniquely determined by this condition.

But then $\alpha_i \alpha_j$ and $\alpha_j \alpha_i$ both map ε to ε^{ij}, so the Galois group is abelian. □

It is possible to determine the precise structure of the above Galois group, and to remove the condition that p be prime. However, this needs extra work and is not needed at this stage. See Theorem 21.9.

Lemma 15.5. *Let K be a subfield of $\mathbb{C}$ in which $t^n - 1$ splits. Let $a \in K$, and let L be a splitting field for $t^n - a$ over K. Then the Galois group of $L : K$ is abelian.*

Proof. Let α be any zero of $t^n - a$. Since $t^n - 1$ splits in K, the general zero of $t^n - a$ is $\varepsilon\alpha$ where ε is a zero of $t^n - 1$ in K. Since $L = K(\alpha)$, any K-automorphism of L is determined by its effect on α. Given two K-automorphisms

$$\phi : \alpha \mapsto \varepsilon\alpha \qquad \psi : \alpha \mapsto \eta\alpha$$

where ε and $\eta \in K$ are zeros of $t^n - 1$, then

$$\phi\psi(\alpha) = \varepsilon\eta\alpha = \eta\varepsilon\alpha = \psi\phi(\alpha)$$

As before, the Galois group is abelian. □

The main work in proving Theorem 15.2 is done in the next lemma.

Lemma 15.6. *If K is a subfield of $\mathbb{C}$ and $L : K$ is normal and radical, then $\Gamma(L : K)$ is soluble.*

Proof. Suppose that $L = K(\alpha_1, \ldots, \alpha_n)$ with $\alpha_j^{n_j} \in K(\alpha_1, \ldots, \alpha_{j-1})$. By Proposition 8.9 we may assume that n_j *is prime for all* j. In particular there is a prime p such that $\alpha_1^p \in K$.

We prove the result by induction on n, using the additional hypothesis that all n_j are prime. The case $n = 0$ is trivial, which gets the induction started.

If $\alpha_1 \in K$, then $L = K(\alpha_2, \ldots, \alpha_n)$ and $\Gamma(L : K)$ is soluble by induction.

We may therefore assume that $\alpha_1 \notin K$. Let f be the minimal polynomial of α_1 over K. Since $L : K$ is normal, f splits in L; since $K \subseteq \mathbb{C}$, f has no repeated zeros. Since $\alpha_1 \notin K$, the degree of f is at least 2. Let β be a zero of f different from α_1, and put $\varepsilon = \alpha_1/\beta$. Then $\varepsilon^p = 1$ and $\varepsilon \neq 1$. Thus ε has order p in the multiplicative group of L, so the elements $1, \varepsilon, \varepsilon^2, \ldots, \varepsilon^{p-1}$ are distinct pth roots of unity in L. Therefore $t^p - 1$ splits in L.

Let $M \subseteq L$ be the splitting field for $t^p - 1$ over K, that is, let $M = K(\varepsilon)$. Consider the chain of subfields $K \subseteq M \subseteq M(\alpha_1) \subseteq L$. The strategy of the remainder of the proof is illustrated in the following diagram:

$$\begin{array}{l}
L \\
\big| \longleftarrow \Gamma(L : M(\alpha_1)) \text{ soluble by induction} \\
M(\alpha_1) \\
\big| \longleftarrow \Gamma(M(\alpha_1) : M) \text{ abelian by Lemma 15.5} \\
M \\
\big| \longleftarrow \Gamma(M : K) \text{ abelian by Lemma 15.4} \\
K
\end{array}$$

Observe that $L : K$ is finite and normal, hence so is $L : M$, therefore Theorem 12.2 applies to $L : K$ and to $L : M$.

Since $t^p - 1$ splits in M and $\alpha_1^p \in M$, the proof of Lemma 15.5 implies that $M(\alpha_1)$ is a splitting field for $t^p - \alpha_1^p$ over M. Thus $M(\alpha_1) : M$ is normal, and by Lemma 15.5 $\Gamma(M(\alpha_1) : M)$ is abelian. Apply Theorem 12.2 to $L : M$ to deduce that

$$\Gamma(M(\alpha_1) : M) \cong \Gamma(L : M)/\Gamma(L : M(\alpha_1))$$

Now

$$L = M(\alpha_1)(\alpha_2, \ldots, \alpha_n)$$

so that $L : M(\alpha_1)$ is a normal radical extension. By induction $\Gamma(L : M(\alpha_1))$ is soluble. Hence by Theorem 14.4(3), $\Gamma(L : M)$ is soluble.

Since M is the splitting field for $t^p - 1$ over K, the extension $M : K$ is normal. By Lemma 15.4, $\Gamma(M : K)$ is abelian. Theorem 12.2 applied to $L : K$ yields

$$\Gamma(M : K) \cong \Gamma(L : K)/\Gamma(L : M)$$

Now Theorem 14.4(3) shows that $\Gamma(L : K)$ is soluble, completing the induction step. □

We can now complete the proof of the main result:

Proof of Theorem 15.2. Let K_0 be the fixed field of $\Gamma(L : K)$, and let $N : M$ be the normal closure of $M : K_0$. Then

$$K \subseteq K_0 \subseteq L \subseteq M \subseteq N$$

Since $M : K_0$ is radical, Lemma 15.3 implies that $N : K_0$ is a normal radical extension. By Lemma 15.6, $\Gamma(N : K_0)$ is soluble.

By Theorem 11.14, the extension $L : K_0$ is normal. By Theorem 12.2

$$\Gamma(L : K_0) \cong \Gamma(N : K_0)/\Gamma(N : L)$$

Theorem 14.4(2) implies that $\Gamma(L : K_0)$ is soluble. But $\Gamma(L : K) = \Gamma(L : K_0)$, so $\Gamma(L : K)$ is soluble. □

The idea of this proof is simple: a radical extension is a series of extensions by nth roots; such extensions have abelian Galois groups; so the Galois group of a radical extension is made up by fitting together a sequence of abelian groups. Unfortunately there are technical problems in carrying out the proof; we need to throw in roots of unity, and we have to make various extensions normal before the Galois correspondence can be used. These obstacles are similar to those encountered by Abel and overcome by his Theorem on Natural Irrationalities in Section 8.8.

Now we translate back from fields to polynomials, and in doing so revert to Galois's original viewpoint.

Definition 15.7. Let f be a polynomial over a subfield K of $\mathbb{C}$, with splitting field Σ over K. The *Galois group* of f over K is the Galois group $\Gamma(\Sigma : K)$.

Let G be the Galois group of a polynomial f over K and let $\partial f = n$. If $\alpha \in \Sigma$ is a zero of f, then $f(\alpha) = 0$, so for any $g \in G$

$$f(g(\alpha)) = g(f(\alpha)) = 0$$

Hence each element $g \in G$ induces a permutation g' of the set of zeros of f in Σ. Distinct elements of G induce distinct permutations, since Σ is generated by the zeros of f. It follows easily that the map $g \mapsto g'$ is a group monomorphism of G into the group $\mathbb{S}_n$ of all permutations of the zeros of f. In other words, we can think of G as a group of permutations on the zeros of f. This, in effect, was how Galois thought of the Galois group, and for many years afterwards the only groups considered by mathematicians were permutation groups and groups of transformations of variables. Arthur Cayley was the first to propose a definition for an abstract group, although it seems that the earliest satisfactory axiom system for groups was given by Leopold Kronecker in 1870 (Huntingdon 1905).

We may restate Theorem 15.2 as:

Theorem 15.8. *Let f be a polynomial over a subfield K of $\mathbb{C}$. If f is soluble by radicals, then the Galois group of f over K is soluble.*

The converse also holds: see Theorem 18.21.

Thus to find a polynomial not soluble by radicals it suffices to find one whose Galois group is not soluble. There are two main ways of doing this. One is to look at the general polynomial of degree n, which we introduced in Chapter 8 Section 8.7, but this approach has the disadvantage that it does not show that there are specific polynomials with rational coefficients that are insoluble by radicals. The alternative approach, which we now pursue, is to exhibit a specific polynomial with rational coefficients whose Galois group is not soluble. Since Galois groups are hard to calculate, a little low cunning is necessary, together with some knowledge of the symmetric group.

15.2 An Insoluble Quintic

Watch carefully; there is nothing up my sleeve...

Lemma 15.9. *Let p be a prime, and let f be an irreducible polynomial of degree p over $\mathbb{Q}$. Suppose that f has precisely two non-real zeros in $\mathbb{C}$. Then the Galois group of f over $\mathbb{Q}$ is isomorphic to the symmetric group $\mathbb{S}_p$.*

Proof. By the Fundamental Theorem of Algebra, Theorem 2.4, $\mathbb{C}$ contains the splitting field Σ of f. Let G be the Galois group of f over $\mathbb{Q}$, considered as a permutation group on the zeros of f. These are distinct by Proposition 9.14, so G is (isomorphic to) a subgroup of $\mathbb{S}_p$. When we construct the splitting field of f we first adjoin an element of degree p, so $[\Sigma : \mathbb{Q}]$ is divisible by p. By Theorem 12.2(1), p divides the order of G. By Cauchy's Theorem 14.15, G has an element of order p. But the only elements of $\mathbb{S}_p$ having order p are the p-cycles. Therefore G contains a p-cycle.

Complex conjugation is a $\mathbb{Q}$-automorphism of $\mathbb{C}$, and therefore induces a $\mathbb{Q}$-automorphism of Σ. This leaves the $p-2$ real zeros of f fixed, while transposing the two non-real zeros. Therefore G contains a 2-cycle.

By choice of notation for the zeros, and if necessary taking a power of the p-cycle, we may assume that G contains the 2-cycle (12) and the p-cycle $(12\dots p)$. We

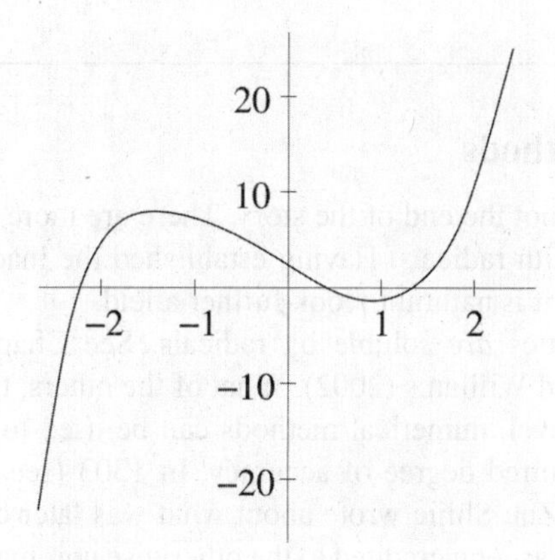

FIGURE 23: A quintic with three real zeros.

claim that these generate the whole of $\mathbb{S}_p$, which will complete the proof. To prove the claim, let $c = (12\ldots p), t = (12)$, and let G be the group generated by c and t. Then G contains $c^{-1}tc = (23)$, hence $c^{-1}(23)c = (34), \ldots$ and hence all transpositions $(m, m+1)$. Then G contains

$$(12)(23)(12) = (13) \qquad (13)(34)(13) = (14)$$

and so on, and therefore contains all transpositions $(1m)$. Finally, G contains all products $(1m)(1r)(1m) = (mr)$ with $1 < m < r$. But every element of $\mathbb{S}_n$ is a product of transpositions, so $G = \mathbb{S}_p$. □

We can now exhibit a specific quintic polynomial over $\mathbb{Q}$ that is not soluble by radicals.

Theorem 15.10. *The polynomial $t^5 - 6t + 3$ over $\mathbb{Q}$ is not soluble by radicals.*

Proof. Let $f(t) = t^5 - 6t + 3$. By Eisenstein's Criterion, f is irreducible over $\mathbb{Q}$. We shall show that f has precisely three real zeros, each with multiplicity 1, and hence has two non-real zeros. Since 5 is prime, by Lemma 15.9 the Galois group of f over $\mathbb{Q}$ is $\mathbb{S}_5$. By Corollary 14.8, $\mathbb{S}_5$ is not soluble. By Theorem 15.8, $f(t) = 0$ is not soluble by radicals.

It remains to show that f has exactly three real zeros, each of multiplicity 1. Now $f(-2) = -17$, $f(-1) = 8$, $f(0) = 3$, $f(1) = -2$, and $f(2) = 23$. A rough sketch of the graph of $y = f(x)$ looks like Figure 23. This certainly appears to give only three real zeros, but we must be rigorous. By Rolle's theorem, the zeros of f are separated by zeros of Df. Moreover, $Df = 5t^4 - 6$, which has two zeros at $\pm\sqrt[4]{6/5}$. Clearly f and Df are coprime, so f has no repeated zeros (this also follows by irreducibility) so f has at most three real zeros. But certainly f has at least three real zeros, since a continuous function defined on the real line cannot change sign except by passing through 0. Therefore f has precisely three real zeros, and the result follows. □

15.3 Other Methods

Of course this is not the end of the story. There are more ways of killing a quintic than choking it with radicals. Having established the inadequacy of radicals for solving the problem, it is natural to look further afield.

First, some quintics *are* soluble by radicals. See Chapter 1 Section 1.4 and Berndt, Spearman and Williams (2002). What of the others, though?

On a mundane level, numerical methods can be used to find the zeros (real or complex) to any required degree of accuracy. In 1303 (see Joseph 2000) the Chinese mathematician Zhu Shijie wrote about what was later called Horner's method in the West; there it was long credited to the otherwise unremarkable William George Horner, who discovered it in 1819. For hand calculations it is a useful practical method, but there are many others. The mathematical theory of such numerical methods can be far from mundane—but from the algebraic point of view it is unilluminating.

Another way of solving the problem is to say, in effect, 'What's so special about radicals?' Suppose for any real number a we define the *ultraradical* of a to be the real zero of t^5+t-a. It was shown by G.B. Jerrard (see Kollros 1949, p. 19) that the quintic equation can be solved by the use of radicals and ultraradicals. See King (1996).

Instead of inventing new tools we can refashion existing ones. Charles Hermite made the remarkable discovery that the quintic equation can be solved in terms of 'elliptic modular functions', special functions of classical mathematics which arose in a quite different context, the integration of algebraic functions. The method is analogous to the trigonometric solution of the cubic equation, Exercise 1.8. In a triumph of mathematical unification, Klein (1913) succeeded in connecting together the quintic equation, elliptic functions, and the rotation group of the regular icosahedron. The latter is isomorphic to the alternating group $\mathbb{A}_5$, which we have seen plays a key part in the theory of the quintic. Klein's work helped to explain the unexpected appearance of elliptic functions in the theory of polynomial equations; these ideas were subsequently generalised by Henri Poincaré to cover polynomials of arbitrary degree.

EXERCISES

15.1 Find radical extensions of $\mathbb{Q}$ containing the following elements of $\mathbb{C}$, by exhibiting suitable radical sequences (See Definition 8.12):

(a) $(\sqrt{11}-\sqrt[7]{23})/\sqrt[4]{5}$

(b) $(\sqrt{6}+2\sqrt[3]{5})^4$

(c) $(2\sqrt[5]{5}-4)/\sqrt{1+\sqrt{99}}$

15.2 What is the Galois group of t^p-1 over $\mathbb{Q}$ for prime p?

15.3 Show that the polynomials t^5-4t+2, t^5-4t^2+2, t^5-6t^2+3, and $t^7-10t^5+15t+5$ over $\mathbb{Q}$ are not soluble by radicals.

15.4 Solve the sextic equation

$$t^6-t^5+t^4-t^3+t^2-t+1=0$$

satisfied by a primitive 14th root of unity, in terms of radicals (*Hint:* Put $u=t+1/t$.)

15.5 Solve the sextic equation

$$t^6+2t^5-5t^4+9t^3-5t^2+2t+1=0$$

by radicals (*Hint:* Put $u=t+1/t$.)

15.6* If $L:K$ is a radical extension in $\mathbb{C}$ and M is an intermediate field, show that $M:K$ need not be radical.

15.7 If p is an irreducible polynomial over $K\subseteq\mathbb{C}$ and at least one zero of p is expressible by radicals, prove that every zero of p is expressible by radicals.

15.8* If $K\subseteq\mathbb{C}$ and $\alpha^2=a\in K$, $\beta^2=b\in K$, and none of a, b, ab are squares in K, prove that $K(\alpha,\beta)$: K has Galois group $\mathbb{Z}_2\times\mathbb{Z}_2$.

15.9* Show that if N is an integer such that $|N|>1$, and p is prime, then the quintic equation

$$x^5-Npx+p=0$$

cannot be solved by radicals.

15.10* Suppose that a quintic equation $f(t)=0$ over $\mathbb{Q}$ is irreducible, and has one real root and two complex conjugate pairs. Does an argument similar to that of Lemma 15.9 prove that the Galois group contains $\mathbb{A}_5$? If so, why? If not, why not?

15.11 Prove the Theorem on Natural Irrationalities using the Galois correspondence.

15.12 Mark the following true or false.

(a) Every quartic equation over a subfield of $\mathbb{C}$ can be solved by radicals.

(b) Every radical extension is finite.

(c) Every finite extension is radical.

(d) The order of the Galois group of a polynomial of degree n divides $n!$

(e) Any reducible quintic polynomial can be solved by radicals.

(f) There exist quartics with Galois group $\mathbb{S}_4$.

(g) An irreducible polynomial of degree 11 with exactly two non-real zeros has Galois group $\mathbb{S}_{11}$.

(h) The normal closure of a radical extension is radical.

(i) $\mathbb{A}_5$ has 50 elements.

Chapter 16

Abstract Rings and Fields

Having seen how Galois Theory works in the context assumed by its inventor, we can generalise everything to a much broader context. Instead of subfields of $\mathbb{C}$, we can consider arbitrary fields. This step goes back to Weber in 1895, but first achieved prominence in the work of Emil Artin in lectures of 1926, later published as Artin (1948). With the increased generality, new phenomena arise, and these must be dealt with.

One such phenomenon relates to the Fundamental Theorem of Algebra, which does not hold in an arbitrary field. We *could* get round this by constructing an analogue, the 'algebraic closure' of a field, in which *every* polynomial splits into linear factors. However, the machinery needed to prove the existence of an algebraic closure is powerful enough to make the concept of an algebraic closure irrelevant anyway. So we concentrate on developing that machinery, which centres on the abstract properties of field extensions, especially finite ones.

A more significant problem is that a general field K need not contain $\mathbb{Q}$ as a subfield. The reason is that sums $1+1+\cdots+1$ can behave in novel ways. In particular, such a sum may be zero. If it is, then the smallest number of 1s involved must be a prime p, and K contains a subfield isomorphic to $\mathbb{Z}_p$, the integers modulo p. Such fields are said to have 'characteristic' p, and they introduce significant complications into the theory. The most important complication is that irreducible polynomials need not be separable; that is, they may have multiple zeros. Separability is automatic for subfields of $\mathbb{C}$, so it has not been seen to play a major role up to this point. However, behind the scenes it has been one of the two significant constraints that make Galois theory work, the other being normality. From now on, separability has to be taken a lot more seriously, and it has a substantial effect.

Rethinking the old results in the new context provides good revision and reinforcement, and it explains where the general concepts come from. Nonetheless, if you seriously work through the material and do not just accept that everything works, you will come to appreciate that Bourbaki had a point.

16.1 Rings and Fields

Today's concepts of 'ring' and 'field' are the brainchildren of Dedekind, who introduced them as a way of systematising algebraic number theory; their influence then spread as was reinforced by the growth of abstract algebra under the influence of Weber, Hilbert, Emmy Noether, and Bartel Leenert van der Waerden. These concepts are motivated by the observation that the classical number systems $\mathbb{Z}$, $\mathbb{Q}$, $\mathbb{R}$, and $\mathbb{C}$ enjoy a long list of useful algebraic properties. Specifically, $\mathbb{Z}$ is a 'ring' and the others are 'fields'.

The formal definition of a ring is:

Definition 16.1. a *ring* R is a set, equipped with two operations of addition (denoted $a+b$) and multiplication (denoted ab), satisfying the following axioms:

(A1) $a+b=b+a$ for all $a,b \in R$.

(A2) $(a+b)+c=a+(b+c)$ for all $a,b,c \in R$.

(A3) There exists $0 \in R$ such that $0+a=a$ for all $a \in R$.

(A4) Given $a \in R$, there exists $-a \in R$ such that $a+(-a)=0$.

(M1) $ab=ba$ for all $a,b \in R$.

(M2) $(ab)c=a(bc)$ for all $a,b,c \in R$.

(M3) There exists $1 \in R$ such that $1a=a$ for all $a \in R$.

(D) $a(b+c)=ab+ac$ for all $a,b,c \in R$.

(The standard definition of a ring omits (M3): with that condition, the standard term is 'ring-with-1' or 'unital ring' or various similar phrases. Since nearly all rings that we need have a 1, it seems simpler to require (M3). Occasionally, we dispense with it.)

When we say that addition and multiplication are 'operations' on R, we automatically imply that if $a,b \in R$ then $a+b, ab \in R$, so R is 'closed' under each of these operations. Some axiom systems for rings include these conditions as explicit axioms.

Axioms (A1) and (M1) are the *commutative laws* for addition and multiplication, respectively. Axioms (A2) and (M2) are the *associative laws* for addition and multiplication, respectively. Axiom (D) is the *distributive law*. The element 0 is called the *additive identity* or *zero element*; the element 1 is called the *multiplicative identity* or *unity element*. The element $-a$ is the *additive inverse* or *negative* of a. The word 'the' is justified here because 0 is unique, and for any given $a \in F$ the inverse $-a$ is unique. The condition $1 \neq 0$ in (M3) excludes the trivial ring with one element.

The modern convention is that axioms (M1) and (M3) are optional for rings. Any ring that satisfies (M1) is said to be *commutative*, and any ring that satisfies (M3) is a *ring with* 1. However, in this book the phrase 'commutative ring with 1' is shortened to 'ring', because we do not require greater generality.

Examples 16.2. (1) The classical number systems $\mathbb{Z}$, $\mathbb{Q}$, $\mathbb{R}$, $\mathbb{C}$ are all rings.
(2) The set of natural numbers $\mathbb{N}$ is not a ring, because axiom (A4) fails.
(3) The set $\mathbb{Z}[i]$ of all complex numbers of the form $a+bi$, with $a,b \in \mathbb{Z}$, is a ring.

(4) The set of polynomials $\mathbb{Z}[t]$ over $\mathbb{Z}$ is a ring, as the usual name 'ring of polynomials' indicates.
(5) The set of polynomials $\mathbb{Z}[t_1,\ldots,t_n]$ in n indeterminates over $\mathbb{Z}$ is a ring.
(6) If n is any integer, the set $\mathbb{Z}_n$ of integers modulo n is a ring.

If R is a ring, then we can define subtraction by

$$a-b=a+(-b) \qquad a,b\in R$$

The axioms ensure that all of the usual algebraic rules of manipulation, except those for division, hold in any ring.

Two extra axioms are required for a field:

Definition 16.3. *A* field *is a ring F satisfying the extra axiom*

(M4) *Given $a\in F$, with $a\neq 0$, there exists $a^{-1}\in F$ such that $aa^{-1}=1$.*

(M4) $1\neq 0$.

Without condition (M5) the set $\{0\}$ would be a field with one element: this causes problems and is usually avoided.

We call a^{-1} the *multiplicative inverse* of $a\neq 0$. This inverse also unique. If F is a field, then we can define division by

$$a/b=ab^{-1} \qquad a,b\in F, b\neq 0$$

The axioms ensure that all the usual algebraic rules of manipulation, including those for division, hold in any field.

Examples 16.4. (1) The classical number systems $\mathbb{Q}$, $\mathbb{R}$, $\mathbb{C}$ are all fields.
(2) The set of integers $\mathbb{Z}$ is not a field, because axiom (M4) fails.
(3) The set $\mathbb{Q}[i]$ of all complex numbers of the form $a+bi$, with $a,b\in\mathbb{Q}$, is a field.
(4) The set of polynomials $\mathbb{Q}[t]$ over $\mathbb{Q}$ is not a field, because axiom (M4) fails.
(5) The set of rational functions $\mathbb{Q}(t)$ over $\mathbb{Q}$ is a field.
(6) The set of rational functions $\mathbb{Q}(t_1,\ldots,t_n)$ in n indeterminates over $\mathbb{Q}$ is a field.
(7) The set $\mathbb{Z}_2$ of integers modulo 2 is a field. The multiplicative inverses of the only nonzero element 1 is $1^{-1}=1$. In this field, $1+1=0$. So $1+1\neq 0$ does not count as one of the 'usual laws of algebra'. Note that it involves an inequality; the statement $1+1=2$ is true in $\mathbb{Z}_2$. What is not true is that $2\neq 0$.
(8) The set $\mathbb{Z}_6$ of integers modulo 6 is not a field, because axiom (M4) fails. In fact, the elements $2,3,4$ do not have multiplicative inverses. Indeed, $2.3=0$ but $2,3\neq 0$, a phenomenon that cannot occur in a field: if F is a field, and $a,b\neq 0$ in F but $ab=0$, then $a=abb^{-1}=0b^{-1}=0$, a contradiction.
(9) The set $\mathbb{Z}_5$ of integers modulo 5 is a field. The multiplicative inverses of the nonzero elements are $1^{-1}=1, 2^{-1}=3, 3^{-1}=2, 4^{-1}=4$. In this field, $1+1+1+1+1=0$.
(10) The set $\mathbb{Z}_1$ of integers modulo 1 is *not* a field. It consists of the single element 0, and so violates (M3) which states that $1\neq 0$. This is a sensible convention since 1 is *not* prime.

The fields $\mathbb{Z}_2$ and $\mathbb{Z}_5$, or more generally $\mathbb{Z}_p$ where p is prime (see Theorem 16.7 below), are prototypes for an entirely new kind of field, with unusual properties. For example, the formula for solving quadratic equations fails spectacularly over $\mathbb{Z}_2$. Suppose that we want to solve

$$t^2 + at + b = 0$$

where $a, b \in \mathbb{Z}_2$. Completing the square involves rewriting the equation in terms of $(t + a/2)$. But $a/2 = a/0$, which makes no sense. The standard quadratic formula involves division by 2 and also makes no sense. Nevertheless, many choices of a, b here lead to soluble equations:

$$\begin{aligned} t^2 &= 0 \quad \text{has solution } t = 0 \\ t^2 + 1 &= 0 \quad \text{has solution } t = 1 \\ t^2 + t &= 0 \quad \text{has solutions } t = 0, 1 \\ t^2 + t + 1 &= 0 \quad \text{has no solution} \end{aligned}$$

16.2 General Properties of Rings and Fields

We briefly develop some of the basic properties of rings and fields, with emphasis on structural features that will allow us to construct examples of fields. Among these features are the presence or absence of 'divisors of zero' (like $2, 3 \in \mathbb{Z}_6$), leading to the concept of an integral domain, and the notion of an ideal in a ring, leading to quotient rings and a general construction for interesting fields. Most readers will have encountered these ideas before; if not, it may be a good idea to find an introductory textbook and work through the first two or three chapters. For example, Fraleigh (1989) and Sharpe (1987) cover the relevant material.

Definition 16.5. (1) A *subring* of a ring R is a non-empty subset S of R such that if a, $b \in S$ then $a + b \in S$, $a - b \in S$, and $ab \in S$.

Note that by this definition a subring need not satisfy (M3). This is one of the disadvantages of simplifying 'ring-with-1' to 'ring'. Perhaps we ought to define 'ring-without-a-1'.

(2) A *subfield* of a field F is a subset S of F containing the elements 0 and 1, such that if a, $b \in S$ then $a + b$, $a - b$, $ab \in S$, and further if $a \neq 0$ then $a^{-1} \in S$.

(3) An *ideal* of a ring R is a subring I such that if $i \in I$ and $r \in R$ then ir and ri lie in I.

Thus $\mathbb{Z}$ is a subring of $\mathbb{Q}$, and $\mathbb{R}$ is a subfield of $\mathbb{C}$, while the set $2\mathbb{Z}$ of even integers is an ideal of $\mathbb{Z}$.

If R, S are rings, then a *ring homomorphism* $\phi : R \to S$ is a map that satisfies three conditions:

$$\phi(1) = 1 \quad \phi(r_1 + r_2) = \phi(r_1) + \phi(r_2) \quad \phi(r_1 r_2) = \phi(r_1)\phi(r_2) \quad \text{for all } r_1, r_2 \in R$$

The *kernel* $\ker\phi$ of ϕ is $\{r : \phi(r) = 0\}$. It is an ideal of R. An *isomorphism* is a homomorphism that is one-to-one and onto; a *monomorphism* is a homomorphism that is one-to-one. A homomorphism is a monomorphism if and only if its kernel is zero.

The most important property of an ideal is the possibility of working modulo that ideal, or, more abstractly, constructing the 'quotient ring' by that ideal. Specifically, if I is an ideal of the ring R, then the *quotient ring* R/I consists of the cosets $I+s$ of I in R (considering R as a group under addition) The operations in the quotient ring are:

$$\begin{aligned}(I+r)+(I+s) &= I+(r+s)\\ (I+r)(I+s) &= I+(rs)\end{aligned}$$

where $r, s \in R$ and $I+r$ is the coset $\{i+r : i \in I\}$.

Examples 16.6. (1) Let $n\mathbb{Z}$ be the set of integers divisible by a fixed integer n. This is an ideal of $\mathbb{Z}$, and the quotient ring $\mathbb{Z}_n = \mathbb{Z}/n\mathbb{Z}$ is the ring of integers modulo n, that is, $\mathbb{Z}_n$.
(2) Let $R = K[t]$ where K is a subfield of $\mathbb{C}$, and let $m(t)$ be an irreducible polynomial over K. Define $I = \langle m(t)\rangle$ to be the set of all multiples of $m(t)$. Then I is an ideal, and R/I is what we previously denoted by $K[t]/\langle m\rangle$ in Chapter 5. This quotient is a field.
(3) We can perform the same construction as in Example 2, without taking m to be irreducible. We still get a quotient ring, but if m is reducible the quotient is no longer a field.

When I is an ideal of R, there is a natural ring homomorphism $\phi : R \to R/I$, defined by $\phi(r) = I+r$. Its kernel is I.

We shall need the following property of $\mathbb{Z}_n$, which explains the differences we found among $\mathbb{Z}_2$, $\mathbb{Z}_5$, and $\mathbb{Z}_6$.

Theorem 16.7. *The ring $\mathbb{Z}_n$ is a field if and only if n is a prime number.*

Proof. First suppose that n is not prime. If $n = 1$, then $\mathbb{Z}_n = \mathbb{Z}/\mathbb{Z}$, which has only one element and so cannot be a field. If $n > 1$ then $n = rs$ where r and s are integers less than n. Putting $I = n\mathbb{Z}$,

$$(I+r)(I+s) = I+rs = I$$

But I is the zero element of $\mathbb{Z}/I$, while $I+r$ and $I+s$ are non-zero. Since in a field the product of two non-zero elements is non-zero, $\mathbb{Z}/I$ cannot be a field.

Now suppose that n is prime. Let $I+r$ be a non-zero element of $\mathbb{Z}/I$. Then r and n are coprime, so by standard properties of $\mathbb{Z}$ there exist integers a and b such that $ar+bn = 1$. Therefore

$$(I+a)(I+r) = (I+1) - (I+n)(I+b) = I+1$$

and similarly

$$(I+r)(I+a) = I+1$$

Since $I+1$ is the identity element of $\mathbb{Z}/I$, we have found a multiplicative inverse for the given element $I+r$. Thus every non-zero element of $\mathbb{Z}/I$ has an inverse, so that $\mathbb{Z}_n = \mathbb{Z}/I$ is a field. □

From now on, when dealing with $\mathbb{Z}_n$, we revert to the usual convention and write the elements as $0, 1, 2, \ldots, n-1$ rather than $I, I+1, I+2, \ldots, I+n-1$.

16.3 Polynomials Over General Rings

We now introduce polynomials with coefficients in a given ring. The main point to bear in mind is that identifying polynomials with functions, as we cheerfully did in Chapter 2 for coefficients in $\mathbb{C}$, is no longer a good idea, because Proposition 2.3, which states that polynomials defining the same function are equal, need not be true when the coefficients belong to a general ring.

Indeed, consider the ring $\mathbb{Z}_2$. Suppose that $f(t) = t^2+1, g(t) = t^4+1$. There are numerous reasons to want these to be different polynomials, the most obvious being that they have different coefficients. But if we interpret them as functions from $\mathbb{Z}_2$ to itself, we find that $f(0) = 1 = g(0)$ and $f(1) = 0 = g(1)$. As functions, f and g are equal.

It turns out that a problem arises here because the ring is finite. Since finite rings (especially finite fields) are important, we need a definition of 'polynomials' that does not rely on interpreting them as functions. We did this in Section 2.1 for polynomials over $\mathbb{C}$, and the same idea works for any ring.

To be specific, let R be a ring. We define a *polynomial over R in the indeterminate t* to be an expression

$$r_0 + r_1 t + \cdots + r_n t^n$$

where $r_0, \ldots, r_n \in R$, $0 \leq n \in \mathbb{Z}$, and t is undefined. Again, for set-theoretic purity we can replace such an expression by the sequence $(r_0, \ldots, r_n)$, as in Exercise 2.2. The elements $r_0, \ldots, r_n$ are the *coefficients* of the polynomial.

Two polynomials are defined to be equal if and only if the corresponding coefficients are equal (with the understanding that powers of t not occurring in the polynomial may be taken to have zero coefficient). The sum and the product of two polynomials are defined using the same formulas (2.3, 2.4) as in Section 2.1, but now the r_i belong to a general ring. It is straightforward to check that the set of all polynomials over R in the indeterminate t is a ring—the *ring of polynomials over R in the indeterminate t*. As before, we denote this by the symbol $R[t]$. We can also define polynomials in several indeterminates $t_1, t_2, \ldots$ and obtain the polynomial ring $R[t_1, t_2, \ldots]$. Again, each polynomial $f \in R[t]$ defines a function from R to R. We use the same symbols f, to denote this function. If $f(t) = \sum r_i t^i$ then $f(\alpha) = \sum r_i \alpha^i$, for $\alpha \in R$. We reiterate that two distinct polynomials over R may give rise to the same function on R.

Proposition 2.3 is still true when $R = \mathbb{R}, \mathbb{Q}$, or $\mathbb{Z}$, with the same proof. And the definition of 'degree' applies without change, as does the proof of Proposition 2.2.

16.4 The Characteristic of a Field

In Proposition 4.4 we observed that every subfield of $\mathbb{C}$ must contain $\mathbb{Q}$. The main step in the proof was that the subfield contains all elements $1+1+\cdots+1$, that is, it contains $\mathbb{N}$, hence $\mathbb{Z}$, hence $\mathbb{Q}$.

The same idea *nearly* works for any field. However, a finite field such as $\mathbb{Z}_5$ cannot contain $\mathbb{Q}$, or even anything isomorphic to $\mathbb{Q}$, because $\mathbb{Q}$ is infinite. How does the proof fail? As we have already seen, in $\mathbb{Z}_5$ the equation $1+1+1+1+1 = 0$ holds. So we can build up a unique smallest subfield just as before—but now it need not be isomorphic to $\mathbb{Q}$.

Pursuing this line of thought leads to:

Definition 16.8. The *prime subfield* of a field K is the intersection of all subfields of K.

It is easy to see that the intersection of any collection of subfields of K is a subfield (the intersection is not empty since every subfield contains 0 and 1), and therefore the prime subfield of K is the *unique* smallest subfield of K. The fields $\mathbb{Q}$ and $\mathbb{Z}_p$ (p prime) have no proper subfields, so are equal to their prime subfields. The next theorem shows that these are the only fields that can occur as prime subfields.

Theorem 16.9. *For every field K, the prime subfield of K is isomorphic either to the field $\mathbb{Q}$ of rationals or the field $\mathbb{Z}_p$ of integers modulo a prime number p.*

Proof. Let K be a field, P its prime subfield. Then P contains 0 and 1, and therefore contains the elements n^* $(n \in \mathbb{Z})$ defined by

$$n^* = \begin{cases} 1+1+\ldots+1 \ (n \text{ times}) & \text{if } n > 0 \\ 0 & \text{if } n = 0 \\ -(-n)^* & \text{if } n < 0 \end{cases}$$

A short calculation using the distributive law (D) and induction shows that the map $^* : \mathbb{Z} \to P$ so defined is a ring homomorphism. Two distinct cases arise.

(1) $n^* = 0$ for some $n \neq 0$. Since also $(-n)^* = 0$, there exists a smallest positive integer p such that $p^* = 0$. If p is composite, say $p = rs$ where r and s are smaller positive integers, then $r^*s^* = p^* = 0$, so either $r^* = 0$ or $s^* = 0$, contrary to the definition of p. Therefore p is prime. The elements n^* form a ring isomorphic to $\mathbb{Z}_p$, which is a field by Theorem 16.7. This must be the whole of P, since P is the smallest subfield of K.

(2) $n^* \neq 0$ if $n \neq 0$. Then P must contain all the elements m^*/n^* where m, n are integers and $n \neq 0$. These form a subfield isomorphic to $\mathbb{Q}$ (by the map which sends m^*/n^* to m/n) which is necessarily the whole of P. □

The distinction among possible prime subfields is summed up by:

Definition 16.10. The *characteristic* of a field K is 0 if the prime subfield of K is isomorphic to $\mathbb{Q}$, and p if the prime subfield of K is isomorphic to $\mathbb{Z}_p$.

For example, the fields $\mathbb{Q}, \mathbb{R}, \mathbb{C}$ all have characteristic zero, since in each case the prime subfield is $\mathbb{Q}$. The field $\mathbb{Z}_p$ (p prime) has characteristic p. We shall see later that there are other fields of characteristic p: for an example, see Exercise 16.6.

The elements n^* defined in the proof of Theorem 16.9 are of considerable importance in what follows. It is conventional to omit the asterisk and write n instead of n^*. This abuse of notation will cause no confusion as long as it is understood that n may be zero in the field without being zero as an integer. Thus in $\mathbb{Z}_5$ we have $10 = 0$ and $2 = 7 = -3$. This difficulty does not arise in fields of characteristic zero.

With this convention, a product nk ($n \in \mathbb{Z}$, $k \in K$) makes sense, and

$$nk = \pm(k + \cdots + k)$$

Lemma 16.11. *If K is a subfield of L, then K and L have the same characteristic.*

Proof. In fact, K and L have the same prime subfield. □

Lemma 16.12. *If k is a non-zero element of the field K, and if n is an integer such that $nk = 0$, then n is a multiple of the characteristic of K.*

Proof. We must have $n = 0$ in K, that is, in old notation, $n^* = 0$. If the characteristic is 0, then this implies that $n = 0$ as an integer. If the characteristic is $p > 0$, then it implies that n is a multiple of p. □

16.5 Integral Domains

The ring $\mathbb{Z}$ has an important property, which is shared by many of the other rings that we shall be studying: if $mn = 0$ where m, n are integers, then $m = 0$ or $n = 0$. We abstract this property as:

Definition 16.13. A ring R is an *integral domain* if $rs = 0$, for $r, s \in R$, implies that $r = 0$ or $s = 0$.

We often express this condition as 'D has no zero-divisors', where a *zero-divisor* is a non-zero element $a \in D$ for which there exists a non-zero element $b \in D$ such that $ab = 0$.

Examples 16.14. (1) The integers $\mathbb{Z}$ form an integral domain.

(2) Any field is an integral domain. For suppose K is a field and $rs = 0$. Then either $s = 0$, or $r = rss^{-1} = 0s^{-1} = 0$.

(3) The ring $\mathbb{Z}_6$ is not an integral domain. As observed earlier, in this ring $2.3 = 0$ but $2, 3 \neq 0$.

(4) The polynomial ring $\mathbb{Z}[t]$ is an integral domain. If $f(t)g(t) = 0$ as polynomials, but $f(t), g(t) \neq 0$, then we can find an element $x \in \mathbb{Z}$ such that $f(x) \neq 0, g(x) \neq 0$. (Just choose x different from the finite set of zeros of f together with zeros of g.) But then $f(x)g(x) \neq 0$, a contradiction.

It turns out that a ring is an integral domain if and only if it is (isomorphic to) a subring of some field. To understand how this comes about, we analyse when it is possible to *embed* a ring R in a field—that is, find a field containing a subring isomorphic to R. Thus $\mathbb{Z}$ can be embedded in $\mathbb{Q}$. This particular example has the property that every element of $\mathbb{Q}$ is a fraction whose numerator and denominator lie in $\mathbb{Z}$. We wish to generalise this situation.

Definition 16.15. A *field of fractions* of the ring R is a field K containing a subring R' isomorphic to R, such that every element of K can be expressed in the form r/s for $r, s \in R'$, where $s \neq 0$.

To see how to construct a field of fractions for R, we analyse how $\mathbb{Z}$ is embedded in $\mathbb{Q}$. We can think of a rational number, written as a fraction r/s, as an ordered pair (r, s) of integers. However, the same rational number corresponds to many distinct fractions: for instance $\frac{2}{3} = \frac{4}{6} = \frac{10}{15}$ and so on. Therefore the pairs (2, 3), (4, 6), and (10, 15) must be treated as if they are 'the same'. The way to achieve this is to define an equivalence relation that makes them equivalent to each other. In general (r, s) represents the same rational as (t, u) if and only if $r/s = t/u$, that is, $ru = st$. In this form the condition involves only the arithmetic of $\mathbb{Z}$. By generalising these ideas we obtain:

Theorem 16.16. *Every integral domain possesses a field of fractions.*

Proof. Let R be an integral domain, and let S be the set of all ordered pairs (r, s) where r and s lie in R and $s \neq 0$. Define a relation $\sim$ on S by

$$(r,s) \sim (t,u) \iff ru = st$$

It is easy to verify that $\sim$ is an equivalence relation; we denote the equivalence class of (r, s) by $[r, s]$. The set F of equivalence classes will provide the required field of fractions. First we define the operations on F by

$$\begin{aligned} [r,s]+[t,u] &= [ru+ts,su] \\ [r,s][t,u] &= [rt,su] \end{aligned}$$

Then we perform a long series of computations to show that F has all the required properties. Since these computations are routine we shall not perform them here, but if you've never seen them, you should check them for yourself, see Exercise 16.7. What you have to prove is:

(1) The operations are well defined. That is to say, if $(r, s) \sim (r', s')$ and $(t, u) \sim (t', u')$, then

$$\begin{aligned} [r,s]+[t,u] &= [r',s']+[t',u'] \\ [r,s][t,u] &= [r',s'][t',u'] \end{aligned}$$

(2) They are operations on F (this is where we need to know that R is an integral domain).

(3) F is a field.

(4) The map $R \to F$ which sends $r \to [r, 1]$ is a monomorphism.

(5) $[r, s] = [r, 1]/[s, 1]$.

□

It can be shown (Exercise 16.8) that for a given integral domain R, all fields of fractions are isomorphic. We can therefore refer to the field constructed above as *the* field of fractions of R. It is customary to identify an element $r \in R$ with its image $[r, 1]$ in F, whereupon $[r, s] = r/s$.

A short calculation reveals a useful property:

Lemma 16.17. *If R is an integral domain and t is an indeterminate, then $R[t]$ is an integral domain.*

Proof. Suppose that

$$f = f_0 + f_1 t + \cdots + f_n t^n \qquad g = g_0 + g_1 t + \cdots + g_m t^m$$

where $f_n \neq 0 \neq g_m$ and all the coefficients lie in R. The coefficient of t^{m+n} in fg is $f_n g_m$, which is non-zero since R is an integral domain. Thus if f, g are non-zero then fg is non-zero. This implies that $R[t]$ is an integral domain, as claimed. □

Corollary 16.18. *If F is a field, then the polynomial ring $F[t_1, \dots, t_n]$ in n indeterminates is an integral domain for any n.*

Proof. Write $F[t_1, \dots, t_n] = F[t_1][t_2, \dots, t_n]$ and use induction. □

Proposition 2.2 applies to polynomials over any integral domain.

Theorem 16.16 implies that when R is an integral domain, $R[t]$ has a field of fractions. We call this the *field of rational expressions in t over R* and denote by $R(t)$. Its elements are of the form $p(t)/q(t)$ where p and q are polynomials and q is not the zero polynomial. Similarly $R[t_1, \dots, t_n]$ has a field of fractions $R(t_1, \dots, t_n)$. Rational expressions can be considered as fractions $p(t)/q(t)$, where $p, q \in R[t]$ and q is not the zero polynomial. If we add two such fractions together, or multiply them, the result is another such fraction. In fact, by the usual rules of algebra,

$$\frac{p(t)}{q(t)} \frac{r(t)}{s(t)} = \frac{p(t)r(t)}{q(t)s(t)}$$

$$\frac{p(t)}{q(t)} + \frac{r(t)}{s(t)} = \frac{p(t)s(t) + q(t)r(t)}{q(t)s(t)}$$

We can also divide and subtract such expressions:

$$\frac{p(t)}{q(t)} / \frac{r(t)}{s(t)} = \frac{p(t)s(t)}{q(t)r(t)}$$

$$\frac{p(t)}{q(t)} - \frac{r(t)}{s(t)} = \frac{p(t)s(t) - q(t)r(t)}{q(t)s(t)}$$

where in the first equation we assume $r(t)$ is not the zero polynomial.

The Division Algorithm and the Euclidean Algorithm work for polynomials over any field, without change. Therefore the entire theory of factorisation of polynomials, including irreducibles, works for polynomials in $K[t]$ whose coefficients lie in any field K.

EXERCISES

16.1 Show that $15\mathbb{Z}$ is an ideal of $5\mathbb{Z}$, and that $5\mathbb{Z}/15\mathbb{Z}$ is isomorphic to $\mathbb{Z}_3$.

16.2 Are the rings $\mathbb{Z}$ and $2\mathbb{Z}$ isomorphic?

16.3 Write out addition and multiplication tables for $\mathbb{Z}_6$, $\mathbb{Z}_7$, and $\mathbb{Z}_8$. Which of these rings are integral domains? Which are fields?

16.4 Define a *prime field* to be a field with no proper subfields. Show that the prime fields (up to isomorphism) are precisely $\mathbb{Q}$ and $\mathbb{Z}_p$ (p prime).

16.5 Find the prime subfield of $\mathbb{Q}$, $\mathbb{R}$, $\mathbb{C}$, $\mathbb{Q}(t)$, $\mathbb{R}(t)$, $\mathbb{C}(t)$, $\mathbb{Z}_5(t)$, $\mathbb{Z}_{17}(t_1,t_2)$.

16.6 Show that the following tables define a field.

$+$	0	1	α	β
0	0	1	α	β
1	1	0	β	α
α	α	β	0	1
β	β	α	1	0

$\cdot$	0	1	α	β
0	0	0	0	0
1	0	1	α	β
α	0	α	β	1
β	0	β	1	α

Find its prime subfield P.

16.7 Prove properties (1–5) listed in the construction of the field of fractions of an integral domain in Theorem 16.16.

16.8 Let D be an integral domain with a field of fractions F. Let K be any field. Prove that any monomorphism $\phi : D \to K$ has a unique extension to a monomorphism $\psi : F \to K$ defined by

$$\psi(a/b) = \phi(a)/\phi(b)$$

for $a, b \in D$. By considering the case where K is another field of fractions for D and ϕ is the inclusion map show that fields of fractions are unique up to isomorphism.

16.9 Let $K = \mathbb{Z}_2$. Describe the subfields of $K(t)$ of the form:

(a) $K(t^2)$

(b) $K(t+1)$

(c) $K(t^5)$

(d) $K(t^2+1)$

16.10 Does the condition $\partial(f+g) \leq \max(\partial f, \partial g)$ hold for polynomials f, g over a general ring?

By considering the polynomials $3t$ and $2t$ over $\mathbb{Z}_6$ show that the equality $\partial(fg) = \partial f + \partial g$ fails for polynomials over a general ring R. What if R is an integral domain?

16.11 Mark the following true or false:

(a) Every integral domain is a field.

(b) Every field is an integral domain.

(c) If F is a field, then $F[t]$ is a field.

(d) If F is a field, then $F(t)$ is a field.

(e) $\mathbb{Z}(t)$ is a field.

Chapter 17

Abstract Field Extensions

Having defined rings and fields, and equipped ourselves with several methods for constructing them, we are now in a position to attack the general structure of an abstract field extension. Our previous work with subfields of $\mathbb{C}$ paves the way, and most of the effort goes into making minor changes to terminology and checking carefully that the underlying ideas generalise in the obvious manner.

We begin by extending the classification of simple extensions to general fields. Having done that, we assure ourselves that the theory of normal extensions, including their relation to splitting fields, carries over to the general case. A new issue, separability, comes into play when the characteristic of the field is not zero. The main result is that the Galois correspondence can be set up for any finite separable normal extension, and it then has exactly the same properties that we have already proved over $\mathbb{C}$.

Convention on Generalisations. Much of this chapter consists of routine verification that theorems previously stated and proved for subfields or subrings of $\mathbb{C}$ remain valid for general rings and fields—and have essentially the same proofs. As a standing convention, we refer to 'Lemma X.Y (generalised)' to mean the generalisation to an arbitrary ring or field of Lemma X.Y; usually we do not restate Lemma X.Y in its new form. In cases where the proof requires a new method, or extra hypotheses, we will be more specific. Moreover, some of the most important theorems will be restated explicitly.

17.1 Minimal Polynomials

Definition 17.1. A *field extension* is a monomorphism $\iota : K \to L$, where K, L are fields.

Usually we identify K with its image $\iota(K)$, and in this case K becomes a subfield of L.

We write $L : K$ for an extension where K is a subfield of L. In this case, ι is the inclusion map.

We define the *degree* $[L : K]$ of an extension $L : K$ exactly as in Chapter 6. Namely, consider L as a vector space over K and take its dimension. The Tower Law remains valid and has exactly the same proof.

In Chapter 16 we observed that all of the usual properties of factorisation of polynomials over $\mathbb{C}$ carry over, without change, to general polynomials. (Even Gauss's Lemma and Eisenstein's Criterion can be generalised to polynomials over suitable rings, but we do not discuss such generalisations here.) Specifically, the definitions of reducible and irreducible polynomials, uniqueness of factorisation into irreducibles, and the concept of a highest common factor, or hcf, carry over to the general case. Moreover, if K is a field and $h \in K[t]$ is an hcf of $f, g \in K[t]$, then there exist $a, b \in K[t]$ such that $h = af + bg$. As before, a polynomial is *monic* if its term of highest degree has coefficient 1.

If $L : K$ is a field extension and $\alpha \in L$, the same dichotomy arises: either α is a zero of some polynomial $f \in K[t]$, or it is not. In the first case α is *algebraic* over K; in the second case α is *transcendental* over K.

An element $\alpha \in L$ that is algebraic over K has a well-defined minimal polynomial $m(t) \in K[t]$; this is the unique monic polynomial over K of smallest degree such that $m(\alpha) = 0$.

17.2 Simple Algebraic Extensions

As before, we can define the subfield of L *generated by* a subset $X \subseteq L$, together with some subfield K, and we employ the same notation $K(X)$ for this field. We say that it is obtained by *adjoining* X to K. The terms *finitely generated extension* and *simple extension* generalise without change.

We mimic the classification of simple extensions in $\mathbb{C}$ of Chapter 5. Simple transcendental extensions are easy to analyse, and we obtain the same result: every simple transcendental extension $K(\alpha)$ of K is isomorphic to $K(t) : K$, the field of rational expressions in one indeterminate t. Moreover, there is an isomorphism that carries t to α.

The algebraic case is slightly trickier: again the key is irreducible polynomials. The result that opens up the whole area is:

Theorem 17.2. *Let K be a field and suppose that $m \in K[t]$ is irreducible and monic. Let I be the ideal of $K[t]$ consisting of all multiples of m. Then $K[t]/I$ is a field, and there is a natural monomorphism $\iota : K \to K[t]/I$ such that $\iota(k) = I + k$. Morover, $I + k$ is a zero of m, which is its minimal polynomial.*

Proof. First, observe that I really is an ideal (Exercise 17.1). We know on general nonsense grounds that $K[t]/I$ is a ring. So suppose that $I + f \in K[t]/I$ is not the zero element, which in this case means that $f \notin I$. Then f is not a multiple of m, and since m is irreducible, the hcf of f and m is 1. Therefore there exist $a, b \in K[t]$ such that $af + bm = 1$. We claim that the multiplicative inverse of $I + f$ is $I + a$. To prove this, compute:

$$(I + f)(I + a) = I + fa = I + (1 - bm) = I + 1$$

since $bm \in I$ by definition. But $I+1$ is the multiplicative identity of $K[t]/I$. Therefore $K[t]/I$ is a field.

Define $\iota : K \to K[t]/I$ by $\iota(k) = I+k$. It is easy to check that ι is a homomorphism. We show that it is one-to-one. If $a \neq b \in K$ then clearly $a-b \notin \langle m \rangle$, so $\iota(a) \neq \iota(b)$. Therefore ι is a monomorphism. □

It is easy to see that the minimal polynomial of $I+t \in K[t]/I$ over K is $m(t)$. Indeed, $m(I+t) = I+m(t) = I+0$. (This is the only place we use the fact that m is monic. But if m is irreducible and not monic, then some multiple km, with $k \in K$, is irreducible and monic; moreover, m and km determine the same ideal I.)

This proof can be made more elegant and more general: see Exercise 17.2. We can (and do) identify K with its image $\iota(K)$, so we can assume without loss of generality that $K \subseteq K[t]/I$. We now prove a classification theorem for simple algebraic extensions:

Theorem 17.3. *Let $K(\alpha) : K$ be a simple algebraic extension, where α has minimal polynomial m over K. Then $K(\alpha) : K$ is isomorphic to $K[t]/I : K$, where I is the ideal of $K[t]$ consisting of all multiples of m. Moreover, there is a natural isomorphism in which $\alpha \mapsto$ the coset $I+t$.*

Proof. Define a map $\phi : K[t] \to K(\alpha)$ by $\phi(f(t)) = f(\alpha)$. This is clearly a ring homomorphism. Its image is the whole of $K(\alpha)$, and its kernel consists of all multiples of $m(t)$ by Lemma 5.6 (generalised). Now $K(\alpha) = \mathrm{im}(\phi) \cong K[t]/\ker(\phi) = K[t]/I$, as required. □

We can now prove a preliminary version of the result that K and m between them determine the extension $K(\alpha)$.

Theorem 17.4. *Suppose $K(\alpha) : K$ and $K(\beta) : K$ are simple algebraic extensions, such that α and β have the same minimal polynomial m over K. Then the two extensions are isomorphic, and the isomorphism of the large fields can be taken to map α to β.*

Proof. This is an immediate corollary of Theorem 17.3. □

17.3 Splitting Fields

In Chapter 9 we defined the term 'splitting field': a polynomial $f \in K[t]$ splits in L if it can be expressed as a product of linear factors over L, and the splitting field Σ of f is the smallest such L. There, we appealed to the Fundamental Theorem of Algebra to construct the splitting field for any given complex polynomial. In the general case, the Fundamental Theorem of Algebra is not available to us. (There is a version of it, Exercise 17.3, but in order to prove that version, we must be able to construct splitting fields *without* appealing to that version of the Fundamental Theorem of

Algebra.) And there is no longer a unique splitting field—though splitting fields are unique up to isomorphism.

We start by generalising Definitions 9.1 and 9.3.

Definition 17.5. If K is a field and f is a nonzero polynomial over K, then f *splits* over K if it can be expressed as a product of linear factors

$$f(t) = k(t-\alpha_1)\dots(t-\alpha_n)$$

where $k, \alpha_1, \dots, \alpha_n \in K$.

Definition 17.6. Let K be a field and let Σ be an extension of K. Then Σ is a *splitting field* for the polynomial f over K if

(1) f splits over Σ.

(2) If $K \subseteq \Sigma' \subseteq \Sigma$ and f splits over Σ' then $\Sigma' = \Sigma$.

Our aim is to show that for any field K, any polynomial over K has a splitting field Σ, and this splitting field is unique up to isomorphism of extensions.

The work that we have already done allows us to construct, in the abstract, any *simple* extension of a field K. Specifically, any simple transcendental extension $K(\alpha)$ of K is isomorphic to the field $K(t)$ of rational expressions in t over K. And if $m \in K[t]$ is irreducible and monic, and I is the ideal of $K[t]$ consisting of all multiples of m, then $K[t]/I$ is a simple algebraic extension $K(\alpha)$ of K where $\alpha = I + t$ has minimal polynomial m over K. Moreover, all simple algebraic extensions of K arise (up to isomorphism) by this construction.

Definition 17.7. We refer to these constructions as *adjoining* α to K.

When we were working with subfields K of $\mathbb{C}$, we could assume that the element(s) being adjoined were in $\mathbb{C}$, so all we had to do was take the field they generate, together with K. Now we do not have a big field in which to work, so we have to create the fields along with the elements we need.

We construct a splitting field by adjoining to K elements that are to be thought of as the zeros of f. We already know how to do this for irreducible polynomials, see Theorem 17.2, so we split f into irreducible factors and work on these separately.

Theorem 17.8. *If K is any field and f is any nonzero polynomial over K, then there exists a splitting field for f over K.*

Proof. Use induction on the degree ∂f. If $\partial f = 1$ there is nothing to prove, for f splits over K. If f does not split over K then it has an irreducible factor f_1 of degree > 1. Using Theorem 5.7 (generalised) we adjoin σ_1 to K, where $f_1(\sigma_1) = 0$. Then in $K(\sigma_1)[t]$ we have $f = (t-\sigma_1)g$ where $\partial g = \partial f - 1$. By induction, there is a splitting field Σ for g over $K(\sigma_1)$. But then Σ is clearly a splitting field for f over K. □

It would appear at first sight that we might construct different splitting fields for f by varying the choice of irreducible factors. In fact splitting fields (for given f and K) are unique up to isomorphism. The statements and proofs are exactly as in Lemma 9.5 and Theorem 9.6, and we do not repeat them here.

17.4 Normality

As before, the key properties that drive the Galois correspondence are normality and separability. We discuss normality in this section, and separability in the next.

Because we suppressed explicit use of 'over $\mathbb{C}$' from our earlier definition, it remains seemingly unchanged:

Definition 17.9. A field extension $L : K$ is *normal* if every irreducible polynomial f over K that has at least one zero in L splits in L.

So does the proof of the main result about normality and splitting fields:

Theorem 17.10. *A field extension $L : K$ is normal and finite if and only if L is a splitting field for some polynomial over K.*

Proof. The same as for Theorem 9.9, except that 'the splitting field' becomes 'a splitting field'. □

Finally we need to discuss the concept of a normal closure in the abstract context. For subfields of $\mathbb{C}$ the normal closure of an extension $L : K$ is an extension N of L such that $N : K$ is normal, and N is as small as possible subject to this condition. We proved existence by taking a suitable splitting field, yielding a normal extension of K containing L, and then finding the unique smallest subfield with those two properties.

For abstract fields, we have to proceed in a similar but technically different manner. The proof of Theorem 11.6 still constructs a normal closure, because this is defined there using a splitting field, which we construct using Theorem 17.8. The only difference is that the normal closure is now unique *up to isomorphism*. That is, if $N_1 : K$ and $N_2 : K$ are normal closures of $L : K$, then the extensions $N_1 : L$ and $N_2 : L$ are isomorphic. This follows because splitting fields are unique up to isomorphism, as remarked immediately after Theorem 17.8.

17.5 Separability

We generalise Definition 9.10:

Definition 17.11. An irreducible polynomial f over a field K is *separable* over K if it has no multiple zeros in a splitting field.

Since the splitting field is unique up to isomorphism, it is irrelevant which splitting field we use to check this property.

Example 17.12. Consider $f(t) = t^2 + t + 1$ over $\mathbb{Z}_2$. This time we cannot use $\mathbb{C}$, so we must go back to the basic construction for a splitting field. The field $\mathbb{Z}_2$ has two elements, 0 and 1. We note that f is irreducible, so we may adjoin an element ζ such that ζ has minimal polynomial f over $\mathbb{Z}_2$. Then $\zeta^2+\zeta+1=0$ so that $\zeta^2=1+\zeta$ (remember, the characteristic is 2) and the elements $0,1,\zeta,1+\zeta$ form a field. This follows from Theorem 5.10 (generalised). It can also be verified directly by working out addition and multiplication tables:

$+$	0	1	ζ	$1+\zeta$
0	0	1	ζ	$1+\zeta$
1	1	0	$1+\zeta$	ζ
ζ	ζ	$1+\zeta$	0	1
$1+\zeta$	$1+\zeta$	ζ	1	0

$\cdot$	0	1	ζ	$1+\zeta$
0	0	0	0	0
1	0	1	ζ	$1+\zeta$
ζ	0	ζ	$1+\zeta$	1
$1+\zeta$	0	$1+\zeta$	1	ζ

A typical calculation for the second table runs like this:

$$\zeta(1+\zeta)=\zeta+\zeta^2=\zeta+\zeta+1=1$$

Therefore $\mathbb{Z}_2(\zeta)$ is a field with four elements. Now f splits over $\mathbb{Z}_2(\zeta)$:

$$t^2+t+1=(t-\zeta)(t-1-\zeta)$$

but over no smaller field. Hence $\mathbb{Z}_2(\zeta)$ is a splitting field for f over $\mathbb{Z}_2$.

We have now reached the point at which the theory of fields of prime characteristic p starts to differ markedly from that for characteristic zero. A major difference is that separability (see Definition 9.10) can, and often does, fail. To investigate this phenomenon, we introduce a new term:

Definition 17.13. An irreducible polynomial over a field K is *inseparable* over K if it is not separable over K.

We are now ready to prove the existence of a very useful map.

Lemma 17.14. *Let K be a field of characteristic $p>0$. Then the map $\phi: K\to K$ defined by $\phi(k)=k^p$ $(k\in K)$ is a field monomorphism. If K is finite, ϕ is an automorphism.*

Proof. Let $x, y\in K$. Then

$$\phi(xy)=(xy)^p=x^py^p=\phi(x)\phi(y)$$

By the binomial theorem,

$$\phi(x+y)=(x+y)^p=x^p+px^{p-1}y+\binom{p}{2}x^{p-2}y^2+\cdots+pxy^{p-1}+y^p \qquad (17.1)$$

Since the characteristic is p, Lemma 3.21 implies that the sum in (17.1) reduces to its first and last terms, and

$$\phi(x+y) = x^p + y^p = \phi(x) + \phi(y)$$

We have now proved that ϕ is a homomorphism.

To show that ϕ is one-to-one, suppose that $\phi(x) = \phi(y)$. Then $\phi(x-y) = 0$. So $(x-y)^p = 0$, so $x = y$. Therefore ϕ is a monomorphism.

If K is finite, then any monomorphism $K \to K$ is automatically onto by counting elements, so ϕ is an automorphism in this case. □

Definition 17.15. If K is a field of characteristic $p > 0$, the map $\phi : K \to K$ defined by $\phi(k) = k^p$ $(k \in K)$ is the *Frobenius monomorphism* or *Frobenius map* of K. When K is finite, ϕ is called the *Frobenius automorphism* of K.

If you try this on the field $\mathbb{Z}_5$, it turns out that ϕ is the identity map, which is not very inspiring. The same goes for $\mathbb{Z}_p$ for any prime p. But for the field of Example 17.12 we have $\phi(0) = 0$, $\phi(1) = 1$, $\phi(\zeta) = 1+\zeta$, $\phi(1+\zeta) = \zeta$, so that ϕ is not always the identity.

Example 17.16. We use the Frobenius map to give an example of an inseparable polynomial. Let $K_0 = \mathbb{Z}_p$ for prime p. Let $K = K_0(u)$ where u is transcendental over K_0, and let

$$f(t) = t^p - u \in K[t]$$

Let Σ be a splitting field for f over K, and let τ be a zero of f in Σ. Then $\tau^p = u$. Now use the Frobenius map:

$$(t-\tau)^p = t^p - \tau^p = t^p - u = f(t)$$

Thus if $\sigma^p - u = 0$ then $(\sigma-\tau)^p = 0$ so that $\sigma = \tau$; all the zeros of f in Σ are *equal*.

It remains to show that f is irreducible over K. Suppose that $f = gh$ where $g, h \in K[t]$, and g and h are monic and have lower degree than f. We must have $g(t) = (t-\tau)^s$ where $0 < s < p$ by uniqueness of factorisation. Hence the constant coefficient $(-\tau)^s$ of g lies in K. This implies that $\tau \in K$, for there exist integers a and b such that $as + bp = 1$, and since $\tau^{as+bp} \in K$ it follows that $\tau \in K$. Then $\tau = v(u)/w(u)$ where $v, w \in K_0[u]$, so

$$v(u)^p - u(w(u))^p = 0$$

But the terms of highest degree cannot cancel. Hence f is irreducible.

The formal derivative Df of a polynomial f can be defined for any underlying field K:

Definition 17.17. Suppose that K is a field, and let

$$f(t) = a_0 + a_1 t + \cdots + a_n t^n \in K[t]$$

Then the *formal derivative* of f is the polynomial

$$Df = a_1 + 2a_2 t + \cdots + n a_n t^{n-1}$$

Note that here the elements $2,\ldots,n$ belong to K, not $\mathbb{Z}$. In fact they are what we briefly wrote as $2^*,\ldots,n^*$ in the proof of Theorem 16.9.

Lemma 9.13 states that a polynomial $f \neq 0$ has a multiple zero in a splitting field if and only if f and Df have a common factor of degree ≥ 1. This lemma remains valid over any field, and has the same proof. Using the formal derivative, we can characterise inseparable irreducible polynomials:

Proposition 17.18. *If K is a field of characteristic 0, then every irreducible polynomial over K is separable over K.*

If K has characteristic $p > 0$, then an irreducible polynomial f over K is inseparable if and only if

$$f(t) = k_0 + k_1 t^p + \cdots + k_r t^{rp}$$

where $k_0,\ldots,k_r \in K$.

Proof. By Lemma 9.13 (generalised), an irreducible polynomial f over K is inseparable if and only if f and Df have a common factor of degree ≥ 1. If so, then since f is irreducible and Df has smaller degree than f, we must have $Df = 0$. Thus if

$$f(t) = a_0 + \cdots + a_m t^m$$

then $na_n = 0$ for all integers $n > 0$. For characteristic 0 this is equivalent to $a_n = 0$ for all n. For characteristic $p > 0$ it is equivalent to $a_n = 0$ if p does not divide n. Let $k_i = a_{ip}$, and the result follows. □

The condition on f for inseparability over fields of characteristic p can be expressed by saying that only powers of t that are multiples of p occur. That is $f(t) = g(t^p)$ for some polynomial g over K.

We now define two more uses of the word 'separable'.

Definition 17.19. If $L : K$ is an extension then an algebraic element $\alpha \in L$ is *separable* over K if its minimal polynomial over K is separable over K.

An algebraic extension $L : K$ is a *separable extension* if every $\alpha \in L$ is separable over K.

For algebraic extensions, separability carries over to intermediate fields.

Lemma 17.20. *Let $L : K$ be a separable algebraic extension and let M be an intermediate field. Then $M : K$ and $L : M$ are separable.*

Proof. Clearly $M : K$ is separable. Let $\alpha \in L$, and let m_K and m_M be its minimal polynomials over K, M respectively. Now $m_M | m_K$ in $M[t]$. But α is separable over K so m_K is separable over K, hence m_M is separable over M. Therefore $L : M$ is a separable extension. □

We end this section by proving that an extension generated by the zeros of a separable polynomial is separable. To prove this, we first prove:

Lemma 17.21. *Let $L : K$ be a field extension where the fields have characteristic p, and let $\alpha \in L$ be algebraic over K. Then α is separable over K if and only if $K(\alpha^p) = K(\alpha)$.*

Proof. Since α is a zero of $t^p - \alpha^p \in K(\alpha^p)[t]$, which equals $(t-\alpha)^p$ by the Frobenius map, the minimal polynomial of α over $K(\alpha^p)$ must divide $(t-\alpha)^p$ and hence be $(t-\alpha)^s$ for some $s \leq p$.

If α is separable over K then it is separable over $K(\alpha^p)$. Therefore $(t-\alpha)^s$ has simple zeros, so $s = 1$. Therefore $\alpha \in K(\alpha^p)$, so $K(\alpha^p) = K(\alpha)$.

For the converse, suppose that α is inseparable over K. Then its minimal polynomial over K has the form $g(t^p)$ for some $g \in K[t]$. Thus α has degree $p\partial g$ over K. In contrast, α^p is a zero of g, which has smaller degree ∂g. Thus $K(\alpha^p)$ and $K(\alpha)$ have different degrees over K, so cannot be equal. □

Theorem 17.22. *If $L : K$ is a field extension such that L is generated over K by a set of separable algebraic elements, then $L : K$ is separable.*

Proof. We may assume that K has characteristic p. It is sufficient to prove that the set of elements of L that are separable over K is closed under addition, subtraction, multiplication, and division. (Indeed, subtraction and division are enough.) We give the proof for addition: the other cases are similar.

Suppose that $\alpha, \beta \in L$ are separable over K. Observe that

$$K(\alpha+\beta,\beta) = K(\alpha,\beta) = K(\alpha^p,\beta^p) = K(\alpha^p+\beta^p,\beta^p) \tag{17.2}$$

using Lemma 17.21 for the middle equality. Now consider the towers

$$K \subseteq K(\alpha+\beta) \subseteq K(\alpha+\beta,\beta)$$
$$K \subseteq K(\alpha^p+\beta^p) \subseteq K(\alpha^p+\beta^p,\beta^p)$$

and consider the corresponding degrees. Apply the Frobenius map to minimal polynomials to see that

$$[K(\alpha^p+\beta^p,\beta^p) : K(\alpha^p+\beta^p)] \leq [K(\alpha+\beta,\beta) : K(\alpha+\beta)]$$

and

$$[K(\alpha^p+\beta^p) : K] \leq [K(\alpha+\beta) : K]$$

However,

$$[K(\alpha^p+\beta^p,\beta^p) : K] = [K(\alpha+\beta,\beta) : K]$$

by (17.2). Now the Tower Law implies that the above inequalities of degrees must actually be equalities. The result follows. □

17.6 Galois Theory for Abstract Fields

Finally, we can set up the Galois correspondence as in Chapter 12. Everything works, provided that we work with a normal separable field extension rather than just a normal one. As we remarked in that context, separability is automatic for subfields of $\mathbb{C}$. So there should be no difficulty in reworking the theory in the more general context.

Note in particular that Theorem 11.14 (generalised) requires separability for fields of prime characteristic.

Because of its importance, we restate the Fundamental Theorem of Galois Theory:

Theorem 17.23 (Fundamental Theorem of Galois Theory, General Case). *If $L:K$ is a finite separable normal field extension, with Galois group G, and if $\mathscr{F}, \mathscr{G}, ^*, ^\dagger$ are defined as before, then:*

(1) *The Galois group G has order $[L:K]$.*

(2) *The maps * and † are mutual inverses, and set up an order-reversing one-to-one correspondence between $\mathscr{F}$ and $\mathscr{G}$.*

(3) *If M is an intermediate field, then*

$$[L:M] = |M^*| \qquad [M:K] = |G|/|M^*|$$

(4) *An intermediate field M is a normal extension of K if and only if M^* is a normal subgroup of G.*

(5) *If an intermediate field M is a normal extension of K, then the Galois group of $M:K$ is isomorphic to the quotient group G/M^*.*

Proof. Mimic the proof of Theorem 12.2 and look out for steps that require separability. □

Another thing to look out for is the uniqueness of the splitting field of a polynomial: now it is unique only up to isomorphism. For example, we defined the Galois group of a polynomial f over K to be the Galois group of $\Sigma : K$, where Σ is the splitting field of f. When K is a subfield of $\mathbb{C}$, the subfield Σ is unique. In general it is unique up to isomorphism, so the Galois group of f is unique up to isomorphism. That suits us fine.

What about radical extensions? In characteristic p, inseparability raises its ugly head, and its effect is serious. For example, $t^p - 1 = (t-1)^p$, by the Frobenius map, so the only pth root of unity is 1. The definition of 'radical extension' has to be changed in characteristic p, and we shall not go into the details. However, everything carries through unchanged to fields with characteristic 0.

We have now reworked the entire theory established in previous chapters, generalising from subfields of $\mathbb{C}$ to arbitrary fields. Now we can pick up the thread again, but from now on, the abstract formalism is there if we need it.

EXERCISES

17.1 Let K be a field, and let $f(t) \in K[t]$. Prove that the set of all multiples of f is an ideal of $K[t]$.

17.2 Let $\phi : K \to R$ be a ring homomorphism, where K is a field and R is a ring. Prove that ϕ is one-to-one. (Note that in this book rings have identity elements 1 and homomorphisms preserve such elements.)

17.3* Prove by transfinite induction that every field can be embedded in an algebraically closed field, its *algebraic closure*. (*Hint:* Keep adjoining zeros of irreducible polynomials until there are none left.)

17.4* Prove that algebraic closures are unique up to isomorphism. More strongly, if K is any field, and A, B are algebraic closures of K, show that the extensions $A : K$ and $B : K$ are isomorphic.

17.5 Let $\mathbb{A}$ denote the set of all complex numbers that are algebraic over $\mathbb{Q}$. The elements of $\mathbb{A}$ are called *algebraic numbers*. Show that $\mathbb{A}$ is a field, as follows.

(a) Prove that a complex number $\alpha \in \mathbb{A}$ if and only if $[\mathbb{Q}(\alpha) : \mathbb{Q}] < \infty$.

(b) Let $\alpha, \beta \in \mathbb{A}$. Use the Tower Law to show that $\mathbb{Q}(\alpha, \beta) : \mathbb{Q}] < \infty$.

(c) Use the Tower Law to show that $[\mathbb{Q}(\alpha + \beta) : \mathbb{Q}] < \infty$, $[\mathbb{Q}(-\alpha) : \mathbb{Q}] < \infty$, $[\mathbb{Q}(\alpha\beta) : \mathbb{Q}] < \infty$, and if $\alpha \neq 0$ then $[\mathbb{Q}(\alpha^{-1}) : \mathbb{Q}] < \infty$.

(d) Therefore $\mathbb{A}$ is a field.

17.6 Prove that $\mathbb{R}[t]/\langle t^2 + 1 \rangle$ is isomorphic to $\mathbb{C}$.

17.7 Find the minimal polynomials over the small field of the following elements in the following extensions:

(a) α in $K : P$ where K is the field of Exercise 16.2 and P is its prime subfield.

(b) α in $\mathbb{Z}_3(t)(\alpha) : \mathbb{Z}_3(t)$ where t is indeterminate and $\alpha^2 = t + 1$.

17.8 For which of the following values of $m(t)$ do there exist extensions $K(\alpha)$ of K for which α has minimal polynomial $m(t)$?

(a) $m(t) = t^2 + 1, K = \mathbb{Z}_3$

(b) $m(t) = t^2 + 1, K = \mathbb{Z}_5$

(c) $m(t) = t^7 - 3t^6 + 4t^3 - t - 1, K = \mathbb{R}$

17.9 Show that for fields for characteristic 2 there may exist quadratic equations that cannot be solved by adjoining square roots of elements in the field. (*Hint:* Try $\mathbb{Z}_2$.)

17.10 Show that we can solve quadratic equations over a field of characteristic 2 if as well as square roots we adjoin elements $\sqrt[*]{k}$ defined to be solutions of the equation

$$(\sqrt[*]{k})^2 + \sqrt[*]{k} = k.$$

17.11 Show that the two zeros of $t^2 + t - k = 0$ in the previous question are $\sqrt[*]{k}$ and $1 + \sqrt[*]{k}$.

17.12 Let $K = \mathbb{Z}_3$. Find all irreducible quadratics over K, and construct all possible extensions of K by an element with quadratic minimal polynomial. Into how many isomorphism classes do these extensions fall? How many elements do they have?

17.13 Mark the following true or false.

(a) The minimal polynomial over a field K of any element of an algebraic extension of K is irreducible over K.

(b) Every monic irreducible polynomial over a field K can be the minimum polynomial of some element α in a simple algebraic extension of K.

(c) A transcendental element does not have a mimimum polynomial.

(d) Any field has infinitely many non-isomorphic simple transcendental extensions.

(e) Splitting fields for a given polynomial are unique.

(f) Splitting fields for a given polynomial are unique up to isomorphism.

(g) The polynomial $t^6 - t^3 + 1$ is separable over $\mathbb{Z}_3$.

Chapter 18

The General Polynomial Equation

As we saw in Chapter 8, the so-called 'general' polynomial is in fact very special. It is a polynomial whose coefficients do not satisfy any algebraic relations. This property makes it in some respects simpler to work with than, say, a polynomial over $\mathbb{Q}$, and in particular it is easier to calculate its Galois group. As a result, we can show that the general quintic polynomial is not soluble by radicals without assuming as much group theory as we did in Chapter 15, and without having to prove the Theorem on Natural Irrationalities, Theorem 8.15.

Chapter 15 makes it clear that the Galois group of the general polynomial of degree n should be the whole symmetric group $\mathbb{S}_n$, and we will show that this contention is correct. This immediately leads to the insolubility of the general quintic. Moreover, our knowledge of the structure of $\mathbb{S}_2$, $\mathbb{S}_3$, and $\mathbb{S}_4$ can be used to find a unified method to solve the general quadratic, cubic, and quartic equations. Further work, not described here, leads to a method for solving any quintic that *is* soluble by radicals, and finding out whether this is the case: see Berndt, Spearman and Williams (2002).

18.1 Transcendence Degree

Previously, we have avoided transcendental extensions. Indeed the assumption that extensions are finite has been central to the theory. We now need to consider a wider class of extensions, which still have a flavour of finiteness.

Definition 18.1. An extension $L:K$ is *finitely generated* if $L = K(\alpha_1, \ldots, \alpha_n)$ where n is finite.

Here the α_j may be either algebraic or transcendental over K.

Definition 18.2. If $\alpha_1, \ldots, \alpha_n$ are transcendental elements over a field K, all lying inside some extension L of K, then they are *independent* if there is no non-trivial polynomial $p \in K[t_1, \ldots, t_n]$ such that $p(\alpha_1, \ldots, \alpha_n) = 0$ in L.

Thus, for example, if t is transcendental over K and u is transcendental over $K(t)$, then $K(t,u)$ is a finitely generated extension of K, and t,u are independent. On the other hand, t and $u = t^2+1$ are both transcendental over K, but are connected by the polynomial equation $t^2+1-u = 0$, so are not independent.

We now prove a condition for a set to consist of independent transcendental elements.

Lemma 18.3. *Let $K \subseteq M$ be fields, $\alpha_1, \ldots, \alpha_r \in M$, and suppose that $\alpha_1, \ldots, \alpha_{r-1}$ are independent transcendental elements over K. Then the following conditions are equivalent:*

(1) *α_r is transcendental over $\alpha_1, \ldots, \alpha_{r-1}$*

(2) *$\alpha_1, \ldots, \alpha_r$ are independent transcendental elements over K.*

Proof. We show that (1) is false if and only if (2) is false, which is equivalent to the above statement.

Suppose (2) is false. Let $p(t_1, \ldots, t_r) \in K[t_1, \ldots, t_r]$ be a nonzero polynomial such that $p(\alpha_1, \ldots, \alpha_r) = 0$. Write $p = \sum_{j=1}^{n} p_j t_r^j$ where each $p_j \in K[t_1, \ldots, t_{r-1}]$. That is, think of p as a polynomial in t_r with coefficients not evolving t_r. Since p is nonzero, some p_j must be nonzero. Because $\alpha_1, \ldots, \alpha_{r-1}$ are independent transcendental elements over K, the polynomial p_j remains nonzero when we substitute α_i for t_1, with $1 \leq i \leq r-1$. This substitution turns p into a nonzero polynomial over $K(\alpha_1, \ldots, \alpha_{r-1})$ satisifed by α_r, so (1) fails.

The converse uses essentially the same idea. If (1) fails, then α_r satisfies a polynomial in t_r with coefficients in $K(\alpha_1, \ldots, \alpha_{r-1})$. Multiplying by the denominators of the coefficients we may assume the coefficients lie in $K[\alpha_1, \ldots, \alpha_{r-1}]$. But now we have constructed a nonzero polynomial in $K[t_1, \ldots, t_r]$ satisfied by the α_j, so (2) fails. □

The next result describes the structure of a finitely generated extension. The main point is that we can adjoin a number of independent transcendental elements first, with algebraic ones coming afterwards.

Lemma 18.4. *If $L : K$ is finitely generated, then there exists an intermediate field M such that*

(1) *$M = K(\alpha_1, \ldots, \alpha_r)$ where the α_i are independent transcendental elements over K.*

(2) *$L : M$ is a finite extension.*

Proof. We know that $L = K(\beta_1, \ldots, \beta_n)$. If all the β_j are algebraic over K, then $L : K$ is finite by Lemma 6.11 (generalised) and we may take $M = K$. Otherwise some β_i is transcendental over K. Call this α_1. If $L : K(\alpha_1)$ is not finite, there exists some β_k transcendental over $K(\alpha_1)$. Call this α_2. We may continue this process until $M = K(\alpha_1, \ldots, \alpha_r)$ is such that $L : M$ is finite. By Lemma 18.3, the α_j are independent transcendental elements over K. □

A result due to Ernst Steinitz says that the integer r that gives the number of independent transcendental elements does not depend on the choice of M.

Lemma 18.5 (Steinitz Exchange Lemma). *With the notation of Lemma* 18.4*, if there is another intermediate field* $N = K(\beta_1, \dots, \beta_s)$ *such that* $\beta_1, \dots, \beta_s$ *are independent transcendental elements over* K *and* $L : N$ *is finite, then* $r = s$.

Proof. The idea of the proof is that if there is a nontrivial polynomial relation involving α_i and β_j, then we can swap them, leaving the field concerned the same except for some finite extension. Inductively, we replace successive α_i by β_j until all β_j have been used, proving that $s \leq r$. By symmetry, $r \leq s$ and we are finished.

The details require some care. We claim inductively on m, that:
If $0 \leq m \leq s$, then renumbering the α_j if necessary,
(1) $L : K(\beta_1, \dots, \beta_m, \alpha_{m+1}, \dots, \alpha_r)$ is finite.
(2) $\beta_1, \dots, \beta_m, \alpha_{m+1}, \dots, \alpha_r$ are independent transcendental elements over K.
The renumbering simplifies the notation, and is also carried out inductively. No α_j is renumbered more than once.

Claims (1, 2) are true when $m = 0$; in this case, no β_i occurs, and the conditions are the same as those in Lemma 18.4.

Assuming (1, 2), we must prove the corresponding claims for $m+1$. To be explicit, these are:
(1′) $L : K(\beta_1, \dots, \beta_{m+1}, \alpha_{m+2}, \dots, \alpha_r)$ is finite.
(2′) $\beta_1, \dots, \beta_{m+1}, \alpha_{m+2}, \dots, \alpha_r$ are independent transcendental elements over K.
We have $m+1 \leq s$, so β_{m+1} exists. It is algebraic over $K(\beta_1, \dots, \beta_m, \alpha_{m+1}, \dots, \alpha_r)$ by (1). Therefore there is some polynomial equation

$$p(\beta_1 \dots, \beta_{m+1}, \alpha_{m+1}, \dots, \alpha_r) = 0 \tag{18.1}$$

in which both β_{m+1} and some α_j actually occur. (That is, each appears in some term with a nonzero coefficient.) Renumbering if necessary, we can assume that this α_j is α_{m+1}. Define four fields:

$$\begin{aligned} K_0 &= K(\beta_1 \dots, \beta_{m+1}, \alpha_{m+1}, \dots, \alpha_r) \\ K_1 &= K(\beta_1 \dots, \beta_m, \alpha_{m+1}, \dots, \alpha_r) \\ K_2 &= K(\beta_1 \dots, \beta_{m+1}, \alpha_{m+2}, \dots, \alpha_r) \\ K_3 &= K(\beta_1 \dots, \beta_m, \alpha_{m+2}, \dots, \alpha_r) \end{aligned}$$

Then $K_3 \subseteq K_1$, $K_3 \subseteq K_2$, $K_1 \subseteq K_0$, $K_2 \subseteq K_0$.

To prove (1′), observe that $K_0 \supseteq K$, and $L : K_1$ is finite by (2), so $L : K_0$ is finite. But $K_0 : K_2$ is finite by (18.1). By the Tower Law, $L : K_2$ is finite. This is (2′).

To prove (2′), suppose it is false. Then there is a polynomial equation

$$p(\beta_1 \dots, \beta_{m+1}, \alpha_{m+2}, \dots, \alpha_r) = 0$$

The element β_{m+1} actually occurs in some nonzero term, otherwise (2) is false. Therefore β_{m+1} is algebraic over K_3, so $K_2 : K_3$ is finite, so $L : K_3$ is finite by (1′) which we have already proved. Therefore $K_1 : K_3$ is finite, but this contradicts (1).

This completes the induction. Continuing up to $m = s$ we deduce that $s \leq r$. Similarly $r \leq s$, so $r = s$. □

Definition 18.6. The integer r defined in Lemma 15.1 is the *transcendence degree* of $L:K$. By Lemma 18.5, the value of r is well-defined.

For example consider $K(t,\alpha,u):K$, where t is transcendental over K, $\alpha^2=t$, and u is transcendental over $K(t,\alpha)$. Then $M=K(t,u)$ where t and u are independent transcendental elements over K, and

$$K(t,\alpha,u):M=M(\alpha):M$$

is finite. The transcendence degree is 2.

The degree $[L:M]$ of the algebraic part is not an invariant, see Exercise 18.3.

It is straightforward to show that an extension $K(\alpha_1,\ldots,\alpha_r):K$ by independent transcendental elements α_i is isomorphic to $K(t_1,\ldots,t_r):K$ where $K(t_1,\ldots,t_r)$ is the field of rational expressions in the indeterminates t_i. In consequence:

Proposition 18.7. *A finitely generated extension $L:K$ has transcendence degree r if and only if there is an intermediate field M such that L is a finite extension of M and $M:K$ is isomorphic to $K(t_1,\ldots,t_r):K$.*

Corollary 18.8. *if $L:K$ is a finitely generated extension, and E is a finite extension of L, then the transcendence degrees of E and L over K are equal.*

18.2 Elementary Symmetric Polynomials

Usually we are given a polynomial and wish to find its zeros. But it is also possible to work in the opposite direction: given the zeros and their multiplicities, reconstruct the polynomial. This is a far easier problem which has a complete general solution, as we saw in Section 8.7 for complex polynomials. We recap the main ideas.

Consider a monic polynomial of degree n having its full quota of n zeros (counting multiplicities). It is therefore a product of n linear factors

$$f(t)=(t-\alpha_1)\ldots(t-\alpha_n)$$

where the α_j are the zeros in K (not necessarily distinct). Suppose that

$$f(t)=a_0+a_1t+\cdots+a_{n-1}t^{n-1}+t^n$$

If we expand the first product and equate coefficients with the second expression, we get the expected result:

$$\begin{aligned}a_{n-1}&=-(\alpha_1+\cdots+\alpha_n)\\a_{n-2}&=(\alpha_1\alpha_2+\alpha_1\alpha_3+\cdots+\alpha_{n-1}\alpha_n)\\&\cdots\\a_0&=(-1)^n\alpha_1\alpha_2\ldots\alpha_n\end{aligned}$$

The expressions in $\alpha_1,\ldots,\alpha_n$ on the right are the elementary symmetric polynomials of Chapter 8, but now they are more generally interpreted as elements of $K[t_1,\ldots,t_n]$ and evaluated at $t_j=\alpha_j$, for $1\le j\le n$.

The elementary symmetric polynomials are symmetric in the sense that they are unchanged by permuting the indeterminates t_j. This property suggests:

Definition 18.9. A polynomial $q\in K[t_1,\ldots,t_n]$ is *symmetric* if

$$q(t_{\sigma(1)},\ldots,t_{\sigma(n)})=q(t_1,\ldots,t_n)$$

for all permutations $\sigma\in\mathbb{S}_n$.

There are other symmetric polynomials apart from the elementary ones, for example $t_1^2+\cdots+t_n^2$, but they can all be expressed in terms of elementary symmetric polynomials:

Theorem 18.10. *Over a field K, any symmetric polynomial in $t_1,\ldots,t_n$ can be expressed as a polynomial of smaller or equal degree in the elementary symmetric polynomials $s_r(t_1,\ldots,t_n)(r=0,\ldots,n)$.*

Proof. See Exercise 8.4 (generalised to any field). □

A slightly weaker version of this result is proved in Corollary 18.12. We need Theorem 18.10 to prove that π is transcendental (Chapter 24). The quickest proof of Theorem 18.10 is by induction, and full details can be found in any of the older algebra texts (such as Salmon 1885 page 57, Van der Waerden 1953 page 81).

18.3 The General Polynomial

Let K be any field, and let $t_1,\ldots,t_n$ be independent transcendental elements over K. The symmetric group $\mathbb{S}_n$ can be made to act as a group of K-automorphisms of $K(t_1,\ldots,t_n)$ by defining

$$\sigma(t_i)=t_{\sigma(i)}$$

for all $\sigma\in\mathbb{S}_n$, and extending any rational expressions ϕ by defining

$$\sigma(\phi(t_1,\cdots,t_n))=\phi(t_{\sigma(1)},\cdots,t_{\sigma(n)})$$

It is easy to prove that σ, extended in this way, is a K-automorphism.

For example, if $n=4$ and σ is the permutation

$$\begin{pmatrix}1234\\2431\end{pmatrix}$$

then $\sigma(t_1)=t_2$, $\sigma(t_2)=t_4$, $\sigma(t_3)=t_3$, and $\sigma(t_4)=t_1$. Moreover, as a typical case,

$$\sigma\left(\frac{t_1^5t_4}{t_2^4-7t_3}\right)=\frac{t_2^5t_1}{t_4^4-7t_3}$$

Clearly distinct elements of $\mathbb{S}_n$ give rise to distinct K-automorphisms.

The fixed field F of $\mathbb{S}_n$ obviously contains all the symmetric polynomials in the t_i, and in particular the elementary symmetric polynomials $s_r = s_r(t_1,\ldots,t_n)$. We show that these generate F.

Lemma 18.11. *With the above notation, $F = K(s_1,\ldots,s_n)$. Moreover,*

$$[K(t_1,\ldots,t_n) : K(s_1,\ldots,s_n)] = n! \tag{18.2}$$

Proof. Clearly $L = K(t_1,\ldots,t_n)$ is a splitting field of $f(t)$ over both $K(s_1,\ldots,s_n)$ and the possibly larger field F. Since $\mathbb{S}_n$ fixes both of these fields, the Galois group of each extension contains $\mathbb{S}_n$, so must equal $\mathbb{S}_n$. Therefore the fields F and $K(s_1,\ldots,s_n)$ are equal. Equation (18.2) follows by the Galois correspondence. □

Corollary 18.12. *Every symmetric polynomial in $t_1,\ldots,t_n$ over K can be written as a rational expression in $s_1,\ldots,s_n$.*

Proof. By definition, symmetric polynomials are precisely those that lie inside the fixed field F of $\mathbb{S}_n$. By Lemma 18.11, $F = K(s_1,\ldots,s_n)$. □

Compare this result with Theorem 18.10.

Lemma 18.13. *With the above notation, $s_1,\ldots,s_n$ are independent transcendental elements over K.*

Proof. By 18.2, $K(t_1,\ldots,t_n)$ is a finite extension of $K(s_1,\ldots,s_n)$. By Corollary 18.8 they both have the same transcendence degree over K, namely n. Therefore the s_j are independent, for otherwise the transcendence degree of $K(s_1,\ldots,s_n) : K$ would be smaller than n. □

Definition 18.14. Let K be a field and let $s_1,\ldots,s_n$ be independent transcendental elements over K. The *general polynomial of degree n* 'over' K is the polynomial

$$t^n - s_1 t^{n-1} + s_2 t^{n-2} - \cdots + (-1)^n s_n$$

over the field $K(s_1,\ldots,s_n)$.

The quotation marks are used because technically the polynomial is over the field $K(s_1,\ldots,s_n)$, not over K.

Theorem 18.15. *For any field K let g be the general polynomial of degree n 'over' K, and let Σ be a splitting field for g over $K(s_1,\ldots,s_n)$. Then the zeros $t_1,\ldots,t_n$ of g in Σ are independent transcendental elements over K, and the Galois group of $\Sigma : K(s_1,\ldots,s_n)$ is the symmetric group $\mathbb{S}_n$.*

Proof. The extension $\Sigma : K(s_1,\ldots,s_n)$ is finite by Theorem 9.9, so the transcendence degree of $\Sigma : K$ is equal to that of $K(s_1,\ldots,s_n) : K$, namely n. Since $\Sigma = K(t_1,\ldots,t_n)$, the t_j are independent transcendental elements over K, since any algebraic relation between them would lower the transcendence degree. The s_j are now the elementary symmetric polynomials in $t_1,\ldots,t_n$ by Theorem 18.10. As above, $\mathbb{S}_n$ acts as a group of automorphisms of $\Sigma = K(t_1,\dots,t_n)$, and by Lemma 15.3 the fixed field is $K(s_1,\dots,s_n)$. By Theorem 11.14, $\Sigma : K(s_1,\dots,s_n)$ is separable and normal (normality also follows from the definition of Σ as a splitting field), and by Theorem 10.5 its degree is $|\mathbb{S}_n| = n!$. Then by Theorem 17.23(1) the Galois group has order $n!$, and contains $\mathbb{S}_n$, so it equals $\mathbb{S}_n$. □

Theorem 15.8 and Corollary 14.8 imply:

Theorem 18.16. *If K is a field of characteristic zero and $n \geq 5$, the general polynomial of degree n 'over' K is not soluble by radicals.*

18.4 Cyclic Extensions

Theorem 18.16 does not imply that any particular polynomial over K of degree $n \geq 5$ is not soluble by radicals, because the general polynomial 'over' K is actually a polynomial over the extension field $K(s_1,\dots,s_n)$, with n independent transcendental elements s_j. For example, the theorem does not rule out the possibility that every quintic over 'K might be soluble by radicals, but that the formula involved varies so much from case to case that no 'general' formula holds.

However, when the general polynomial of degree n 'over' K can be solved by radicals, it is easy to deduce a solution by radicals of *any* polynomial of degree n over K, by substituting elements of K for $s_1,\dots,s_n$ in that solution. This is the source of the 'generality' of the general polynomial. From Theorem 18.16, the best that we can hope for using radicals is a solution of polynomials of degree ≤ 4. We fulfil this hope by analysing the structure of $\mathbb{S}_n$ for $n \leq 4$, and appealing to a converse to Theorem 15.8. This converse is proved by showing that 'cyclic extensions'—extensions with cyclic Galois group—are closely linked to radicals.

Definition 18.17. Let $L : K$ be a finite normal extension with Galois group G. The *norm* of an element $a \in L$ is

$$N(a) = \tau_1(a)\tau_2(a)\dots\tau_n(a)$$

where $\tau_1,\dots,\tau_n$ are the elements of G.

Clearly $N(a)$ lies in the fixed field of G (use Lemma 10.4) so if the extension is also separable, then $N(a) \in K$.

The next result is traditionally referred to as Hilbert's Theorem 90 from its appearance in his 1893 report on algebraic numbers.

Theorem 18.18 (Hilbert's Theorem 90). *Let $L : K$ be a finite normal extension with cyclic Galois group G generated by an element τ. Then $a \in L$ has norm $N(a) = 1$ if and only if*

$$a = b/\tau(b)$$

for some $b \in L$, where $b \neq 0$.

Proof. Let $|G| = n$. If $a = b/\tau(b)$ and $b \neq 0$ then

$$\begin{aligned} N(a) &= a\tau(a)\tau^2(a)\dots\tau^{n-1}(a) \\ &= \frac{b}{\tau(b)}\frac{\tau(b)}{\tau^2(b)}\frac{\tau^2(b)}{\tau^3(b)}\cdots\frac{\tau^{n-1}(b)}{\tau^n(b)} \\ &= 1 \end{aligned}$$

since $\tau^n = 1$.

Conversely, suppose that $N(a) = 1$. Let $c \in L$, and define

$$\begin{aligned} d_0 &= ac \\ d_1 &= (a\tau(a))\tau(c) \\ &\dots \\ d_j &= [a\tau(a)\dots\tau^i(a)]\tau^i(c) \end{aligned}$$

for $0 \leq j \leq n-1$. Then

$$d_{n-1} = N(a)\tau^{n-1}(c) = \tau^{n-1}(c)$$

Further,

$$d_{j+1} = a\tau(d_j) \qquad (0 \leq j \leq n-2)$$

Define

$$b = d_0 + d_1 + \dots + d_{n-1}$$

We choose c to make $b \neq 0$. Suppose on the contrary that $b = 0$ for all choices of c. Then for any $c \in L$

$$\lambda_0\tau^0(c) + \lambda_1\tau(c) + \dots + \lambda_{n-1}\tau^{n-1}(c) = 0$$

where

$$\lambda_j = a\tau(a)\dots\tau^j(a)$$

belongs to L. Hence the distinct automorphisms τ^j are linearly dependent over L, contrary to Lemma 10.1.

Therefore we can choose c so that $b \neq 0$. But now

$$\begin{aligned} \tau(b) &= \tau(d_0) + \dots + \tau(d_{n-1}) \\ &= (1/a)(d_1 + \dots + d_{n-1}) + \tau^n(c) \\ &= (1/a)(d_0 + \dots + d_{n-1}) \\ &= b/a \end{aligned}$$

Thus $a = b/\tau(b)$ as claimed. □

Theorem 18.19. *Suppose that $L : K$ is a finite separable normal extension whose Galois group G is cyclic of prime order p, generated by τ. Assume that the characteristic of K is 0 or is prime to p, and that $t^p - 1$ splits in K. Then $L = K(\alpha)$, where α is a zero of an irreducible polynomial $t^p - a$ over K for some $a \in K$.*

Proof. The p zeros of $t^p - 1$ from a group of order p, which must therefore be cyclic, since any group of prime order is cyclic. Because a cyclic group consists of powers of a single element, the zeros of $t^p - 1$ are the powers of some $\varepsilon \in K$ where $\varepsilon^p = 1$. But then

$$N(\varepsilon) = \varepsilon \dots \varepsilon = 1$$

since $\varepsilon \in K$, so $\tau^i(\varepsilon) = \varepsilon$ for all i. By Theorem 18.18, $\varepsilon = \alpha/\tau(\alpha)$ for some $\alpha \in L$. Therefore

$$\tau(\alpha) = \varepsilon^{-1}\alpha \qquad \tau^2(\alpha) = \varepsilon^{-2}\alpha \qquad \dots \qquad \tau^j(\alpha) = \varepsilon^{-j}\alpha$$

and $a = \alpha^p$ is fixed by G, so lies in K. Now $K(\alpha)$ is a splitting field for $t^p - a$ over K. The K-automorphisms $1, \tau, \dots, \tau^{p-1}$ map α to distinct elements, so they give p distinct K-automorphisms of $K(\alpha)$. By Theorem 17.23(1) the degree $[K(\alpha) : K] \geq p$. But $[L : K] = |G| = p$, so $L = K(\alpha)$. Hence $t^p - a$ is the minimal polynomial of α over K, otherwise we would have $[K(\alpha) : K] < p$. Being a minimal polynomial, $t^p - a$ is irreducible over K. □

We can now prove the promised converse to Theorem 15.8. Compare with Lemma 8.17(2).

Theorem 18.20. *Let K be a field of characteristic* 0 *and let $L : K$ be a finite normal extension with soluble Galois group G. Then there exists an extension R of L such that $R : K$ is radical.*

Proof. All extensions are separable since the characteristic is 0. Use induction on $|G|$. The result is clear when $|G| = 1$. If $|G| \neq 1$, consider a maximal proper normal subgroup H of G, which exists since G is a finite group. Then G/H is simple, since H is maximal, and is also soluble by Theorem 14.4(2). By Theorem 14.6, G/H is cyclic of prime order p. Let N be a splitting field over L of $t^p - 1$. Then $N : K$ is normal, for by Theorem 9.9 L is a splitting field over K of some polynomial f, so N is a splitting field over L of $(t^p - 1)f$, which implies that $N : K$ is normal by Theorem 9.9.

The Galois group of $N : L$ is abelian by Lemma 15.6, and by Theorem 17.23(5) $\Gamma(L : K)$ is isomorphic to $\Gamma(N : K)/\Gamma(N : L)$. By Theorem 14.4(3), $\Gamma(N : K)$ is soluble. Let M be the subfield of N generated by K and the zeros of $t^p - 1$. Then $N : M$ is normal. Now $M : K$ is clearly radical, and since $L \subseteq N$ the desired result will follow provided we can find an extension R of N such that $R : M$ is radical.

We claim that the Galois group of $N : M$ is isomorphic to a subgroup of G. Let us map any M-automorphism τ of N into its restriction $\tau|_L$. Since $L : K$ is normal, $\tau|_L$ is a K-automorphism of L, and there is a group homomorphism

$$\phi : \Gamma(N : M) \to \Gamma(L : K).$$

If $\tau \in \ker(\phi)$ then τ fixes all elements of M and L, which generate N. Therefore $\tau = 1$, so ϕ is a monomorphism, which implies that $\Gamma(N : M)$ is isomorphic to a subgroup J of $\Gamma(L : K)$.

If $J = \phi(\Gamma(N : M))$ is a proper subgroup of G, then by induction there is an extension R of N such that $R : M$ is radical.

The remaining possibility is that $J = G$. Then we can find a subgroup $H \lhd \Gamma(N : M)$ of index p, namely $H = \phi^{-1}(H)$. Let P be the fixed field $H^\dagger$. Then $[P : M] = p$ by Theorem 17.23(3), $P : M$ is normal by Theorem 17.23(4), and $t^p - 1$ splits in M. By Theorem 18.19 (generalised), $P = M(\alpha)$ where $\alpha^p = a \in M$. But $N : P$ is a normal extension with soluble Galois group of order smaller than $|G|$, so by induction there exists an extension R of N such that $R : P$ is radical. But then $R : M$ is radical, and the theorem is proved. □

To extend this result to fields of characteristic $p > 0$, radical extensions must be defined differently. As well as adjoining elements α such that α^n lies in the given field, we must also allow adjunction of elements α such that $\alpha^p - \alpha$ lies in the given field (where p is the same as the characteristic). It is then true that a polynomial is soluble by radicals if and only if its Galois group is soluble. The proof is different because we have to consider extensions of degree p over fields of characteristic p. Then Theorem 18.19 (generalised) breaks down, and extensions of the second type above come in. If we do not modify the definition of solubility by radicals then although every soluble polynomial has soluble group, the converse need not hold—indeed some quadratic polynomials with abelian Galois group are not soluble by radicals, see Exercises 18.13 and 18.14.

Since a splitting field is always a normal extension, we have:

Theorem 18.21. *Over a field of characteristic zero, a polynomial is soluble by radicals if and only if it has a soluble Galois group.*

Proof. Use Theorems 15.8 and 18.20. □

18.5 Solving Equations of Degree Four or Less

The general polynomial of degree n has Galois group $\mathbb{S}_n$, and we know that for $n \leq 4$ this is soluble (Chapter 14). Theorem 18.21 therefore implies that for a field K of characteristic zero, the general polynomial of degree ≤ 4 can be solved by radicals. We already know this from the classical tricks in Chapter 1, but now we can use the structure of the symmetric group to explain, in a unified way, why those tricks work.

Linear Equations

The general linear polynomial is

$$t - s_1$$

Trivially $t_1 = s_1$ is a zero.

The Galois group here is trivial, and adds little to the discussion except to confirm that the zero must lie in K.

Quadratic Equations

The general quadratic polynomial is

$$t^2 - s_1 t + s_2$$

Let the zeros be t_1 and t_2. The Galois group $\mathbb{S}_2$ consists of the identity and a map interchanging t_1 and t_2. By Hilbert's Theorem 90, Theorem 18.18, there must exist an element which, when acted on by the nontrivial element of $\mathbb{S}_2$, is multiplied by a primitive square root of 1; that is, by -1. Obviously $t_1 - t2$ has this property. Therefore

$$(t_1 - t_2)^2$$

is fixed by $\mathbb{S}_2$, so lies in $K(s_1, s_2)$. By explicit calculation

$$(t_1 - t_2)^2 = s_1^2 - 4s_2$$

Hence

$$\begin{aligned} t_1 - t_2 &= \pm\sqrt{s_1^2 - 4s_2} \\ t_1 + t_2 &= s_1 \end{aligned}$$

and we have the familiar formula

$$t_1, t_2 = \frac{s_1 \pm \sqrt{s_1^2 - 4s_2}}{2}$$

Cubic Equations

The general cubic polynomial is

$$t^3 - s_1 t^2 + s_2 t - s_3$$

Let the zeros be t_1, t_2, t_3. The Galois group $\mathbb{S}_3$ has a series

$$1 \lhd \mathbb{A}_3 \lhd \mathbb{S}_3$$

with abelian quotients.

Motivated once more by Hilbert's Theorem 90, Theorem 18.18, we adjoin an element $\omega \neq 1$ such that $\omega^3 = 1$. Consider

$$y = t_1 + \omega t_2 + \omega^2 t_3$$

The elements of $\mathbb{A}_3$ permute t_1, t_2, and t_3 cyclically, and therefore multiply y by a power of ω. Hence y^3 is fixed by $\mathbb{A}_3$. Similarly if

$$z = t_1 + \omega^2 t_2 + \omega t_3$$

then z^3 is fixed by $\mathbb{A}_3$. Now any odd permutation in $\mathbb{S}_3$ interchanges y^3 and z^3, so that y^3+z^3 and y^3z^3 are fixed by the whole of $\mathbb{S}_3$, hence lie in $K(s_1,s_2,s_3)$. (Explicit formulas are given in the final section of this chapter.) Hence y^3 and z^3 are zeros of a quadratic over $K(s_1,s_2,s_3)$ which can be solved as in part (b). Taking cube roots we know y and z. But since

$$s_1 = t_1+t_2+t_3$$

it follows that

$$\begin{aligned} t_1 &= \tfrac{1}{3}(s_1+y+z) \\ t_2 &= \tfrac{1}{3}(s_1+\omega^2 y+\omega z) \\ t_3 &= \tfrac{1}{3}(s_1+\omega y+\omega^2 z) \end{aligned}$$

Quartic Equations

The general quartic polynomial is

$$t^4 - s_1t^3 + s_2t^2 - s_3t + s_4$$

Let the zeros be t_1,t_2,t_3,t_4. The Galois group $\mathbb{S}_4$ has a series

$$1 \lhd \mathbb{V} \lhd \mathbb{A}_4 \lhd \mathbb{S}_4$$

with abelian quotients, where

$$\mathbb{V} = \{1,(12)(34),(13)(24),(14)(23)\}$$

is the Klein four-group. It is therefore natural to consider the three expressions

$$\begin{aligned} y_1 &= (t_1+t_2)(t_3+t_4) \\ y_2 &= (t_1+t_3)(t_2+t_4) \\ y_3 &= (t_1+t_4)(t_2+t_3) \end{aligned}$$

These are permuted among themselves by any permutation in $\mathbb{S}_4$, so that all the elementary symmetric polynomials in y_1,y_2,y_3 lie in $K(s_1,s_2,s_3,s_4)$. (Explicit formulas are indicated below). Then y_1,y_2,y_3 are the zeros of a certain cubic polynomial over $K(s_1,s_2,s_3,s_4)$ called the *resolvent cubic*. Since

$$t_1+t_2+t_3+t_4 = s_1$$

we can find three quadratic polynomials whose zeros are t_1+t_2 and t_3+t_4, t_1+t_3 and t_2+t_4, t_1+t_4 and t_2+t_3. From these it is easy to find t_1,t_2,t_3,t_4.

Explicit Formulas

For completeness, we now state, for degrees 3 and 4, the explicit formulas whose existence is alluded to above. Figure 24 shows a picture of Cardano, who first published them. For details of the calculations, see Van der Waerden (1953, pages 177-182). Compare with Chapter 1 Section 1.4.

Cubic. The Tschirnhaus transformation

$$u = t - \tfrac{1}{3}s_1$$

converts the general cubic polynomial to

$$u^3 + pu + q$$

If we can find the zeros of this it is an easy matter to find them for the general cubic. The above procedure for this polynomial leads to

$$\begin{aligned} y^3 + z^3 &= -27q \\ y^3z^3 &= -27p^3 \end{aligned}$$

implying that y^3 and z^3 are the zeros of the quadratic polynomial

$$t^2 + 27qt - 27p^3$$

This yields Cardano's formula (1.8).

Quartic. The Tschirnhaus transformation

$$u = t - \tfrac{1}{4}s_1$$

reduces the quartic to the form

$$t^4 + pt^2 + qt + r$$

In the above procedure,

$$\begin{aligned} y_1 + y_2 + y_3 &= 2p \\ y_1y_2 + y_1y_3 + y_2y_3 &= p^2 - 4r \\ y_1y_2y_3 &= -q^2 \end{aligned}$$

The resolvent cubic is

$$t^3 - 2pt^2 + (p^2 - 4r)t + q^2$$

(a thinly disguised form of (1.12) with $t = -2u$). Its zeros are y_1, y_2, y_3, and

$$\begin{aligned} t_1 &= \tfrac{1}{2}(\sqrt{-y_1} + \sqrt{-y_2} + \sqrt{-y_3}) \\ t_2 &= \tfrac{1}{2}(\sqrt{-y_1} - \sqrt{-y_2} - \sqrt{-y_3}) \\ t_3 &= \tfrac{1}{2}(-\sqrt{-y_1} + \sqrt{-y_2} - \sqrt{-y_3}) \\ t_4 &= \tfrac{1}{2}(-\sqrt{-y_1} - \sqrt{-y_2} + \sqrt{-y_3}) \end{aligned}$$

Here the signs of the square roots must be chosen so that

$$\sqrt{-y_1}\sqrt{-y_2}\sqrt{-y_3} = -q$$

FIGURE 24: Cardano, the first person to publish solutions of cubic and quartic equations.

EXERCISES

18.1 If K is a countable field and $L:K$ is finitely generated, show that L is countable. Hence show that $\mathbb{R}:\mathbb{Q}$ and $\mathbb{C}:\mathbb{Q}$ are not finitely generated.

18.2 Calculate the transcendence degrees of the following extensions:

(a) $\mathbb{Q}(t,u,v,w):\mathbb{Q}$ where t,u,v,w are independent transcendental elements over $\mathbb{Q}$.

(b) $\mathbb{Q}(t,u,v,w):\mathbb{Q}$ where $t^2=2$, u is transcendental over $\mathbb{Q}(t)$, $v^3=t+5$, and w is transcendental over $\mathbb{Q}(t,u,v)$.

(c) $\mathbb{Q}(t,u,v):\mathbb{Q}$ where $t^2=u^3=v^4=7$.

18.3 Show that in Lemma 18.4 the degree $[L:M]$ is not independent of the choice of M. (*Hint:* Consider $K(t^2)$ as a subfield of $K(t)$.)

18.4 Suppose that $K\subseteq L\subseteq M$, and each of $M:K$, $L:K$ is finitely generated. Show that $M:K$ and $L:K$ have the same transcendence degree if and only if $M:L$ is finite.

18.5* For any field K show that t^3-tx+1 is either irreducible or splits in K. (*Hint:* Show that any zero is a rational expression in any other zero.)

18.6 Suppose that $L : K$ is finite, normal, and separable with Galois group G. For any $a \in L$ define the *trace*

$$T(a) = \tau_1(a) + \cdots + \tau_n(a)$$

where $\tau_1, \ldots, \tau_n$ are the distinct elements of G. Show that $T(a) \in K$ and that T is a surjective map $L \to K$.

18.7 If in the previous exercise G is cyclic with generator τ, show that $T(a) = 0$ if and only if $a = b - \tau(b)$ for some $b \in L$.

18.8 Solve by radicals the following polynomial equations over $\mathbb{Q}$:

(a) $t^3 - 7t + 5 = 0$

(b) $t^3 - 7t + 6 = 0$

(c) $t^4 + 5t^3 - 2t - 1 = 0$

(d) $t^4 + 4t + 2 = 0$

18.9 Show that a finitely generated algebraic extension is finite, and hence find an algebraic extension that is not finitely generated.

18.10* Let θ have minimal polynomial

$$t^3 + at^2 + bt + c$$

over $\mathbb{Q}$. Find necessary and sufficient conditions in terms of a, b, c such that $\theta = \phi^2$ where $\phi \in \mathbb{Q}(\theta)$. (*Hint:* Consider the minimal polynomial of ϕ.) Hence or otherwise express $\sqrt[3]{28} - 3$ as a square in $\mathbb{Q}(\sqrt[3]{28})$, and $\sqrt[3]{5} - \sqrt[3]{4}$ as a square in $\mathbb{Q}(\sqrt[3]{5}, \sqrt[3]{2})$. (See Ramanujan 1962 page 329.)

18.11 Let Γ be a finite group of automorphisms of K with fixed field K_0. Let t be transcendental over K. For each $\sigma \in \Gamma$ show there is a unique automorphism σ' of $K(t)$ such that

$$\sigma'(k) = \sigma(k) \quad (k \in K)$$
$$\sigma'(t) = t$$

Show that the σ' form a group Γ' isomorphic to Γ, with fixed field $K_0(t)$.

18.12 Let K be a field of characteristic p. Suppose that $f(t) = t^p - t - \alpha \in K[t]$. If β is a zero of f, show that the zeros of f are $\beta + k$ where $k = 0, 1, \ldots, p-1$. Deduce that either f is irreducible over K or f splits in K.

18.13* If f in Exercise 18.13 is irreducible over K, show that the Galois group of f is cyclic. State and prove a characterisation of finite normal separable extensions with soluble Galois group in characteristic p.

18.14 Mark the following true or false.

(a) Every finite extension is finitely generated.

(b) Every finitely generated extension is algebraic.

(c) The transcendence degree of a finitely generated extension is invariant under isomorphism.

(d) If $t_1, \ldots, t_n$ are independent transcendental elements, then their elementary symmetric polynomials are also independent transcendental elements.

(e) The Galois group of the general polynomial of degree n is soluble for all n.

(f) The general quintic polynomial is soluble by radicals.

(g) The only proper subgroups of $\mathbb{S}_3$ are 1 and $\mathbb{A}_3$.

(h) The transcendence degree of $\mathbb{Q}(t) : \mathbb{Q}$ is 1.

(i) The transcendence degree of $\mathbb{Q}(t^2) : \mathbb{Q}$ is 2.

Chapter 19

Finite Fields

Fields that have finitely many elements are important in many branches of mathematics, including number theory, group theory, and projective geometry. They also have practical applications, especially to the coding of digital communications, see Lidl and Niederreiter (1986), and, especially for the history, Thompson (1983).

The most familiar examples of such fields are the fields $\mathbb{Z}_p$ for prime p, but these are not all. In this chapter we give a complete classification of all finite fields. It turns out that a finite field is uniquely determined up to isomorphism by the number of elements that it contains, that this number must be a power of a prime, and that for every prime p and integer $n > 0$ there exists a field with p^n elements. All these facts were discovered by Galois, though not in this terminology.

19.1 Structure of Finite Fields

We begin by proving the second of these three statements.

Theorem 19.1. *If F is a finite field, then F has characteristic $p > 0$, and the number of elements of F is p^n where n is the degree of F over its prime subfield.*

Proof. Let P be the prime subfield of F. By Theorem 16.9, P is isomorphic either to $\mathbb{Q}$ or to $\mathbb{Z}_p$ for prime p. Since $\mathbb{Q}$ is infinite, $P \cong \mathbb{Z}_p$. Therefore F has characteristic p. By Theorem 6.1, F is a vector space over P. This vector space has finitely many elements, so $[F : P] = n$ is finite. Let $x_1, \ldots, x_n$ be a basis for F over P. Every element of F is uniquely expressible in the form

$$\lambda_1 x_1 + \cdots + \lambda_n x_n$$

where $\lambda_1, \ldots, \lambda_n \in P$. Each λ_j may be chosen in p ways since $|P| = p$, hence there are p^n such expressions. Therefore $|F| = p^n$. □

Thus there do not exist fields with $6, 10, 12, 14, 18, 20, \ldots$ elements. Notice the contrast with group theory, where there exist groups of any given order. However, there exist non-isomorphic groups with equal orders. To show that this cannot happen for finite fields, we recall the Frobenius map, Definition 17.15, which maps x to

x^p, and is an automorphism when the field is finite by Lemma 17.14. We use the Frobenius automorphism to establish a basic uniqueness theorem for finite fields:

Theorem 19.2. *Let p be any prime number and let $q = p^n$ where n is any integer > 0. A field F has q elements if and only if it is a splitting field for $f(t) = t^q - t$ over the prime subfield $P \cong \mathbb{Z}_p$ of F.*

Proof. Suppose that $|F| = q$. The set $F \setminus \{0\}$ forms a group under multiplication, of order $q-1$, so if $0 \neq x \in F$ then $x^{q-1} = 1$. Hence $x^q - x = 0$. Since $0^q - 0 = 0$, every element of F is a zero of $t^q - t$, so $f(t)$ splits in F. Since the zeros of f exhaust F, they certainly generate it, so F is a splitting field for f over P.

Conversely, let K be a splitting field for f over $\mathbb{Z}_p$. Since $Df = -1$, which is prime to f, all the zeros of f in K are distinct, so f has exactly q zeros. The set of zeros is precisely the set of elements fixed by ϕ^n, that is, its fixed field. So the zeros form a field, which must therefore be the whole splitting field K. Therefore $|K| = q$.

$\square$

Since splitting fields exist and are unique up to isomorphism, we deduce a complete classification of finite fields:

Theorem 19.3. *A finite field has $q = p^n$ elements where p is a prime number and n is a positive integer. For each such q there exists, up to isomorphism, precisely one field with q elements, which can be constructed as a splitting field for $t^q - t$ over $\mathbb{Z}_p$.*

Definition 19.4. The *Galois Field* $\mathbb{GF}(q)$ is the unique field with q elements.

19.2 The Multiplicative Group

The above classification of finite fields, although a useful result in itself, does not give any detailed information on their deeper structure. There are many questions we might ask—what are the subfields? How many are there? What are the Galois groups? We content ourselves with proving one important theorem, which gives the structure of the multiplicative group $F \setminus \{0\}$ of any finite field F. First we need to know a little more about abelian groups.

Definition 19.5. The *exponent* $e(G)$ of a finite group G is the least common multiple of the orders of the elements of G.

The order of any element of G divides the order $|G|$, so $e(G)$ divides $|G|$. In general, G need not possess an element of order $e(G)$. For example if $G = \mathbb{S}_3$ then $e(G) = 6$, but G has no element of order 6. Abelian groups are better behaved in this respect:

Lemma 19.6. *Any finite abelian group G contains an element of order $e(G)$.*

Proof. Let $e = e(G) = p_1^{\alpha_1} \dots p_n^{\alpha_n}$ where the p_j are distinct primes and $\alpha_j \geq 1$. The definition of $e(G)$ implies that for each j, the group G must possess an element g_j whose order is divisible by $p_j^{\alpha_j}$. Then a suitable power a_j of g_j has order $p_j^{\alpha_j}$. Define

$$g = a_1 a_2 \dots a_n \tag{19.1}$$

Suppose that $g^m = 1$ where $m \geq 1$. Then

$$a_j^m = a_1^{-m} \dots a_{j-1}^{-m} a_{j+1}^{-m} \dots a_n^{-m}$$

So if

$$q = p_1^{\alpha_1} \dots p_{j-1}^{\alpha_{j-1}} p_{j+1}^{\alpha_{j+1}} \dots p_n^{\alpha_n}$$

then $a_j^{mq} = 1$. But q is prime to the order of a_j, so $p_j^{\alpha_j}$ divides m. Hence e divides m. But clearly $g^e = 1$. Hence g has order e, which is what we want. □

Corollary 19.7. *If G is a finite abelian group such that $e(G) = |G|$, then G is cyclic.*

Proof. The element g in (19.1) generates G. □

We can apply this corollary immediately.

Theorem 19.8. *If G is a finite subgroup of the multiplicative group $K \backslash \{0\}$ of a field K, then G is cyclic.*

Proof. Since multiplication in K is commutative, G is an abelian group. Let $e = e(G)$. For any $x \in G$ we have $x^e = 1$, so that x is a zero of the polynomial $t^e - 1$ over K. By Theorem 3.28 (generalised) there are at most e zeros of this polynomial, so $|G| \leq e$. But $e \leq |G|$, hence $e = |G|$; by Corollary 19.7, G is cyclic. □

Corollary 19.9. *The multiplicative group of a finite field is cyclic.*

Therefore for any finite field F there is at least one element x such that every non-zero element of F is a power of x. We give two examples.

Examples 19.10. (1) The field $\mathbb{GF}(11)$. The powers of 2, in order, are

$$1, 2, 4, 8, 5, 10, 9, 7, 3, 6, 1$$

so 2 generates the multiplicative group. On the other hand, the powers of 4 are

$$1, 4, 5, 9, 3, 1$$

so 4 does not generate the group.
(2) The field $\mathbb{GF}(25)$. This can be constructed as a splitting field for $t^2 - 2$ over $\mathbb{Z}_5$, since $t^2 - 2$ is irreducible and of degree 2. We can therefore represent the elements of $\mathbb{GF}(25)$ in the form $a + b\alpha$ where $\alpha^2 = 2$. There is no harm in writing $\alpha = \sqrt{2}$.

By trial and error we are led to consider the element $2+\sqrt{2}$. Successive powers of this are

$$\begin{array}{cccccc} 1 & 2+\sqrt{2} & 1+4\sqrt{2} & 4\sqrt{2} & 3+3\sqrt{2} & 2+4\sqrt{2} \quad 2 \\ 4+2\sqrt{2} & 2+3\sqrt{2} & 3\sqrt{2} & 1+\sqrt{2} & 4+3\sqrt{2} & 4 \\ 3+4\sqrt{2} & 4+\sqrt{2} & \sqrt{2} & 2+2\sqrt{2} & 3+\sqrt{2} & 3 \\ 1+3\sqrt{2} & 3+2\sqrt{2} & 2\sqrt{2} & 4+4\sqrt{2} & 1+2\sqrt{2} & 1 \end{array}$$

Hence $2+\sqrt{2}$ generates the multiplicative group.

There is no known procedure for finding a generator other than enlightened trial and error. Fortunately the existence of a generator is usually sufficient information.

19.3 Application to Solitaire

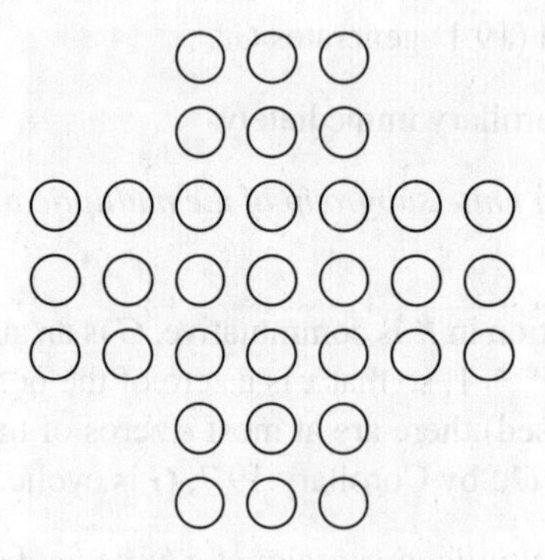

FIGURE 25: The solitaire board

Finite fields have an unexpected application to the recreational pastime of solitaire (de Bruijn 1972). Solitaire is played on a board with holes arranged like Figure 25. A peg is placed in each hole, except the centre one, and play proceeds by jumping any peg horizontally or vertically over an adjacent peg into an empty hole; the peg that is jumped over is removed. The player's objective is to remove all pegs except one, which—traditionally—is the peg that occupies the central hole. Can it be another hole? Experiment shows that it can, but suggests that the final peg cannot occupy *any* hole. Which holes are possible?

De Bruijn's idea is to use the field $\mathbb{GF}(4)$, whose addition and multiplication tables are given in Exercise 16.6, in terms of elements $0, 1, \alpha, \beta$. Consider the holes as a subset of the integer lattice $\mathbb{Z}^2$, with the origin $(0,0)$ at the centre and the axes horizontal and vertical as usual. If X is a set of pegs, define

$$A(X) = \sum_{(x,y)\in X} \alpha^{x+y} \qquad B(X) = \sum_{(x,y)\in X} \alpha^{x-y}$$

Observe that if a legal move changes X to Y, then $A(Y) = A(X), B(Y) = B(X)$. This follows easily from the equation $\alpha^2 + \alpha + 1 = 0$, which in turn follows from the tables. Thus the pair $(A(X), B(X))$ is invariant under any sequence of legal moves.

The starting position X has $A(X) = B(X) = 1$. Therefore any position Y that arises during the game must satisfy $A(Y) = B(Y) = 1$. If the game ends with a single peg on (x, y) then $\alpha^{x+y} = \alpha^{x-y} = 1$. Now $\alpha^3 = 1$, so α has order 3 in the multiplicative group of nonzero elements of $\mathbb{GF}(4)$. Therefore $x+y, x-y$ are multiples of 3, so x, y are multiples of 3. Thus the only possible end positions are $(-3,0), (0,-3), (0,0), (0,3), (3,0)$. Experiment (by symmetry, only $(0,0)$, the traditional finish, and $(3,0)$ need be attempted; moreover, the same penultimate move must lead to both, depending on which peg is moved) shows that all five of these positions can be obtained by a series of legal moves.

EXERCISES

19.1 For which of the following values of n does there exist a field with n elements?

$$1, 2, 3, 4, 5, 6, 17, 24, 312, 65536,$$
$$65537, 83521, 103823, 2^{13466917} - 1$$

(*Hint:* See 'Mersenne primes' under 'Internet' in the References.)

19.2 Construct fields having 8, 9, and 16 elements.

19.3 Let ϕ be the Frobenius automorphism of $\mathbb{GF}(p^n)$. Find the smallest value of $m > 0$ such that ϕ^m is the identity map.

19.4 Show that the subfields of $\mathbb{GF}(p^n)$ are isomorphic to $\mathbb{GF}(p^r)$ where r divides n, and there exists a unique subfield for each such r.

19.5 Show that the Galois group of $\mathbb{GF}(p^n) : \mathbb{GF}(p)$ is cyclic of order n, generated by the Frobenius automorphism ϕ. Show that for finite fields the Galois correspondence is a bijection, and find the Galois groups of

$$\mathbb{GF}(p^n) : \mathbb{GF}(p^m)$$

whenever m divides n.

19.6 Are there any composite numbers r that divide all the binomial coefficients $\binom{r}{s}$ for $1 \leq s \leq r-1$?

19.7 Find generators for the multiplicative groups of $\mathbb{GF}(p^n)$ when $p^n = 8, 9, 13, 17, 19, 23, 29, 31, 37, 41$, and 49.

19.8 Show that the additive group of $\mathbb{GF}(p^n)$ is a direct product of n cyclic groups of order p.

19.9 By considering the field $\mathbb{Z}_2(t)$, show that the Frobenius monomorphism is not always an automorphism.

19.10* For which values of n does $\mathbb{S}_n$ contain an element of order $e(\mathbb{S}_n)$?
(*Hint:* Use the cycle decomposition to estimate the maximum order of an element of $\mathbb{S}_n$, and compare this with an estimate of $e(\mathbb{S}_n)$. You may need estimates on the size of the nth prime: for example, 'Bertrand's Postulate', which states that the interval $[n, 2n]$ contains a prime for any integer $n \geq 1$.)

19.11* Prove that in a finite field every element is a sum of two squares.

19.12 Mark the following true or false.

(a) There is a finite field with 124 elements.

(b) There is a finite field with 125 elements.

(c) There is a finite field with 126 elements.

(d) There is a finite field with 127 elements.

(e) There is a finite field with 128 elements.

(f) The multiplicative group of $\mathbb{GF}(19)$ contains an element of order 3.

(g) $\mathbb{GF}(2401)$ has a subfield isomorphic to $\mathbb{GF}(49)$.

(h) Any monomorphism from a finite field to itself is an automorphism.

(i) The additive group of a finite field is cyclic.

Chapter 20

Regular Polygons

We return with more sophisticated weapons to the time-honoured problem of ruler-and-compass construction, introduced in Chapter 7. We consider the following question: for which values of n can the regular n-sided polygon be constructed by ruler and compass?

The ancient Greeks knew of constructions for 3-, 5-, and 15-gons; they also knew how to construct a $2n$-gon given an n-gon, by the obvious method of bisecting the angles. We describe these constructions in Section 20.1. For about two thousand years little progress was made beyond the Greeks. If you answered Exercises 7.16 or 7.17 you got further than they did. It seemed 'obvious' that the Greeks had found all the constructible regular polygons ... Then, on 30 March 1796, Gauss made the remarkable discovery that the regular 17-gon can be constructed (Figure 26). He was nineteen years old at the time. So pleased was he with this discovery that he resolved to dedicate the rest of his life to mathematics, having until then been unable to decide between that and the study of languages. In his *Disquisitiones Arithmeticae*, reprinted as Gauss (1966), he stated necessary and sufficient conditions for constructibility of the regular n-gon, and proved their sufficiency; he claimed to have a proof of necessity although he never published it. Doubtless he did: Gauss knew a proof when he saw one.

20.1 What Euclid Knew

Euclid's *Elements* gets down to business straight away. The first regular polygon constructed there is the equilateral triangle, in Book 1 Proposition 1. Figure 27 (left) makes the construction fairly clear.

The square also makes its appearance in Book 1:

Proposition 46 (Euclid) *On a given straight line to describe a square.*

In the proof, which we give in detail to illustrate Euclid's style, notation such as [1,31] refers to Proposition 31 of Book 1 of the *Elements*. The proof is taken from Heath (1956), the classic edition of Euclid's *Elements*. Refer to Figure 27 (right) for the lettering.

Proof. Let AB be the given straight line; thus it is required to describe a square on the straight line AB.

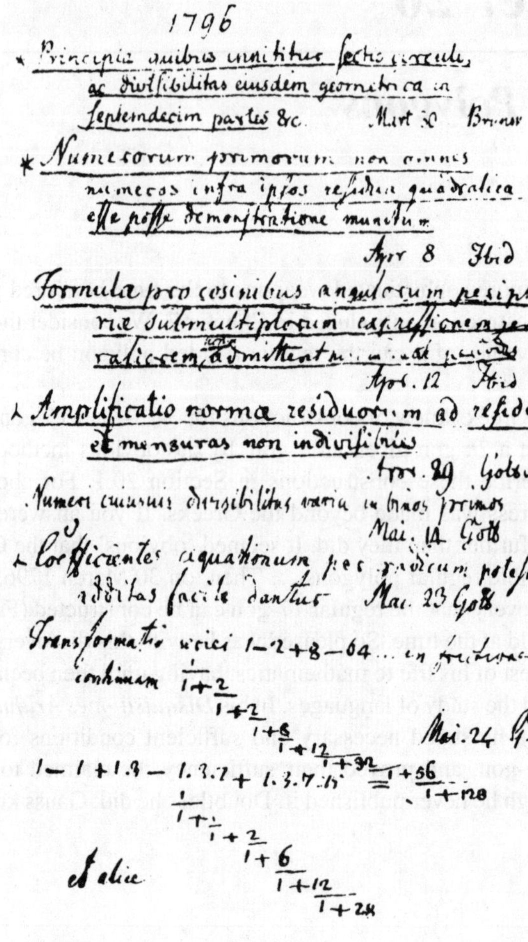

FIGURE 26: The first entry in Gauss's notebook records his discovery that the regular 17-gon can be constructed.

Let AC be drawn at right angles to the straight line AB from the point A on it [1, 11], and let AD be made equal to AB;
through the point D let DE be drawn parallel to AB,
and through the point B let BE be drawn parallel to AD. [1,31]

Therefore ADEB is a parallelogram;

therefore AB is equal to DE, and AD to BE. [1, 34]

But AB is equal to AD;

therefore the four straight lines BA, AD, DE, EB are equal to one another;
therefore the parallelogram ADEB is equilateral.

I say next that it is also right-angled.

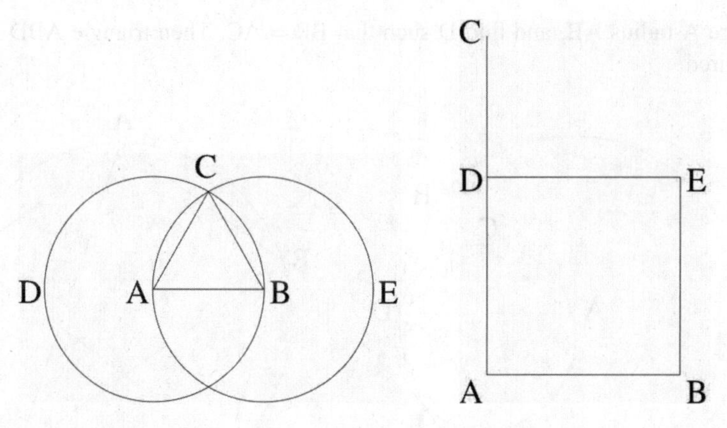

FIGURE 27: *Left*: Euclid's construction of an equilateral triangle. *Right*: Euclid's construction of a square.

For, since the straight line AD falls upon the parallels AB, DE,
the angles BAD, ADE are equal to two right angles. [1, 29]
But the angle BAD is also right;
therefore the angle ADE is also right.
And in parallelogrammic areas the opposite sides and angles are equal to one another; [1, 34]
therefore each of the opposite angles ABE, BED is also right.
Therefore ADEB is right-angled.
And it was also proved equilateral.
Therefore it is a square; and it is described on the straight line AB.
Q.E.F.

□

Here Q.E.F. (quod erat faciendum—that which was to be done) replaces the familiar Q.E.D. (quod erat demonstrandum—that which was to be proved) because this is not a theorem but a construction. In any case, the Latin phrase occurs in later translations: Euclid wrote in Greek. Now imagine you are a Victorian schoolboy—it always *was* a schoolboy in those days—trying to learn Euclid's proof by heart, including the exact choice of letters in the diagrams...

The construction of the regular pentagon has to wait until Book 4 Proposition 11, because it depends on some quite sophisticated ideas, notably Proposition 10 of Book 4: *To construct an isosceles triangle having each of the angles at the base double of the remaining one.* In modern terms: construct a triangle with angles $2\pi/5, 2\pi/5, \pi/5$. Euclid's method for doing this is shown in Figure 28. Given AB, find C so that $AB \times BC = CA^2$. To do that, see Book 2 Proposition 11, which is itself quite complicated—the construction here is essentially the famous 'golden section', a name that seems to have been introduced in 1835 by Martin Ohm (Herz-Fischler 1998, Livio 2002). Euclid's method is given in Exercise 19.10. Next, draw the circle centre A radius AB, and find D such that BD = AC. Then triangle ABD is the one required.

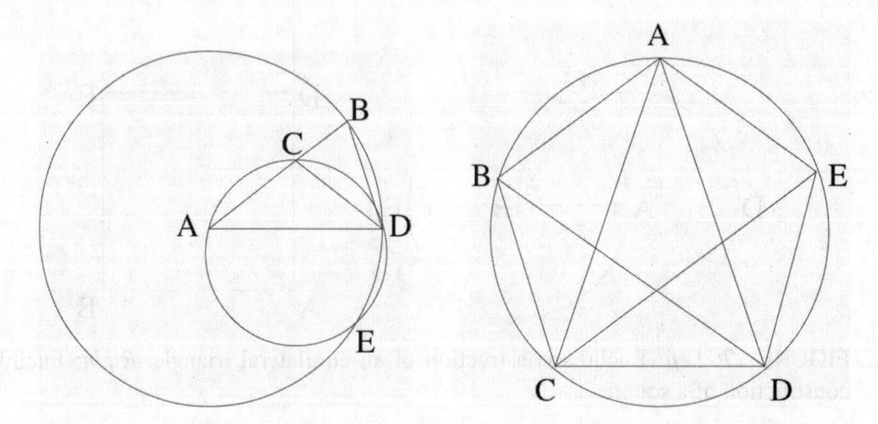

FIGURE 28: *Left*: Euclid's construction of an isosceles triangle with base angles $4\pi/5$. *Right*: Euclid's construction of a regular pentagon. Make ACD similar to triangle ABD in the left-hand Figure and proceed from there.

With this shape of triangle under his belt, Euclid then constructs the regular pentagon: Figure 28 (right) shows his method.

The hexagon occurs in Book 4 Proposition 15, the 15-gon in Book 4 Proposition 16. Bisection of any angle, Book 1 Proposition 9, effectively completes the Euclidean catalogue of constructible regular polygons.

20.2 Which Constructions are Possible?

That, however, was not the end of the story.

We derived necessary and sufficient conditions for the existence of a ruler-and-compass construction in Theorem 7.11. We restate it here for convenience as:

Theorem 20.1. *Suppose that K is a subfield of $\mathbb{C}$, generated by points in a subset $P \subseteq \mathbb{C}$. Let α lie in an extension L of K such that there exists a finite series of subfields*

$$K = K_0 \subseteq K_1 \subseteq \cdots \subseteq K_r = L$$

such that $[K_{j+1} : K_j] = 2$ for $j = 0, \ldots, r-1$. Then the point $\alpha \in \mathbb{C}$ is constructible from P. The converse is also valid.

There is a more useful, but weaker, version of Theorem 20.1. To prove it, we first need:

Lemma 20.2. *If G is a finite group and $|G| = 2^r$ for $r \geq 1$, then its centre $Z(G)$ contains an element of order 2.*

Proof. Use the class equation (14.2). We have

$$1 + C_2 + \cdots + C_k = 2^r$$

so some C_j is odd. By Corollary 14.12 this C_j also divides 2^r, so we must have $|C_j| = 1$. Hence $Z(G) \neq 1$. Now apply Lemma 14.14. □

Corollary 20.3. *If G is a finite group and* $|G| = 2^r$ *then there exists a series*

$$1 = G_0 \subseteq G_1 \subseteq \cdots \subseteq G_r = G$$

of normal subgroups of G, such that $|G_j| = 2^j$ *for* $0 \leq j \leq r$.

Proof. Use Lemma 20.2 and induction. □

Now we can state and prove the promised modification of Theorem 20.1.

Proposition 20.4. *If K is a subfield of* $\mathbb{C}$, *generated by points in a subset* $P \subseteq \mathbb{C}$, *and if* α *lies in a normal extension L of K such that* $[L : K] = 2^r$ *for some integer r, then* α *is constructible from P.*

Proof. $L : K$ is separable since the characteristic is zero. Let G be the Galois group of $L : K$. By Theorem 12.2(1) $|G| = 2^r$. By Corollary 20.3, G has a series of normal subgroups

$$1 = G_0 \subseteq G_1 \subseteq \cdots \subseteq G_r = G$$

such that $|G_j| = 2^j$. Let K_j be the fixed field $G^{\dagger}_{r-j}$. By Theorem 12.2(3) $[K_{j+1} : K_j] = 2$ for all j. By Theorem 20.1, α is constructible from P. □

20.3 Regular Polygons

We shall use a mixture of algebraic and geometric ideas to find those values of n for which the regular n-gon is constructible. To save breath, let us make the following (non-standard):

Definition 20.5. The positive integer n is *constructive* if the regular n-gon is constructible by ruler and compasses.

The first step is to reduce the problem to prime-power values of n.

Lemma 20.6. *If n is constructive and m divides n, then m is constructive. If m and n are coprime and constructive, then mn is constructive.*

Proof. If m divides n, then we can construct a regular m-gon by joining every dth vertex of a regular n-gon, where $d = n/m$.

If m and n are coprime, then there exist integers a, b such that $am + bn = 1$. Therefore

$$\frac{1}{mn} = a\frac{1}{n} + b\frac{1}{m}$$

Hence from angles $2\pi/m$ and $2\pi/n$ we can construct $2\pi/mn$, and from this we obtain a regular mn-gon. □

Corollary 20.7. *Suppose that $n = p_1^{m_1} \dots p_r^{m_r}$ where $p_1, \dots, p_r$ are distinct primes. Then n is constructive if and only if each $p_j^{m_j}$ is constructive.*

Another obvious result:

Lemma 20.8. *For any positive integer m, the number 2^m is constructive.*

Proof. Any angle can be bisected by ruler and compasses, and the result follows by induction on m. □

This reduces the problem of constructing regular polygons to the case when the number of sides is an odd prime power. Now we bring in the algebra. In the complex plane, the set of nth roots of unity forms the vertices of a regular n-gon. Further, as we have seen repeatedly, these roots of unity are the zeros in $\mathbb{C}$ of the polynomial

$$t^n - 1 = (t-1)(t^{n-1} + t^{n-2} + \cdots + t + 1)$$

We concentrate on the second factor on the right-hand side: $f(t) = t^{n-1} + t^{n-2} + \cdots + t + 1$. Its zeros are the powers ζ^k for $1 \leq k \leq n-1$ of a primitive nth root of unity

$$\zeta = e^{2\pi i/n}$$

Lemma 20.9. *Let p be a prime such that p^n is constructive. Let ζ be a primitive p^nth root of unity in $\mathbb{C}$. Then the degree of the minimal polynomial of ζ over $\mathbb{Q}$ is a power of 2.*

Proof. Take $\zeta = e^{2\pi i/p^n}$. The number p^n is constructive if and only if we can construct ζ from $\mathbb{Q}$. Hence by Theorem 7.12 $[\mathbb{Q}(\zeta) : \mathbb{Q}]$ is a power of 2. Hence the degree of the minimal polynomial of ζ over $\mathbb{Q}$ is a power of 2. □

The next step is to calculate the relevant minimal polynomials to find their degrees. It turns out to be sufficient to consider p and p^2 only.

Lemma 20.10. *If p is a prime and ζ is a primitive pth root of unity in $\mathbb{C}$, then the minimal polynomial of ζ over $\mathbb{Q}$ is*

$$f(t) = 1 + t + \cdots + t^{p-1}$$

Proof. This polynomial is irreducible over $\mathbb{Q}$ by Lemma 3.22. Clearly ζ is a zero. Therefore it is the minimal polynomial of ζ. □

To prove the case p^2, we apply the method of Lemma 3.22.

Lemma 20.11. *If p is a prime and ζ is a primitive p^2th root of unity in $\mathbb{C}$, then the minimal polynomial of ζ over $\mathbb{Q}$ is*

$$g(t) = 1 + t^p + \cdots + t^{p(p-1)}$$

Proof. Note that $g(t) = (t^{p^2} - 1)/(t^p - 1)$. Now $\zeta^{p^2} - 1 = 0$ but $\zeta^p - 1 \neq 0$ so $g(\zeta) = 0$. It suffices to show that $g(t)$ is irreducible over $\mathbb{Q}$. As before we make the substitution $t = 1 + u$. Then

$$g(1+u) = \frac{(1+u)^{p^2} - 1}{(1+u)^p - 1}$$

and modulo p this is

$$\frac{(1+u^{p^2}) - 1}{(1+u^p) - 1} = u^{p(p-1)}$$

Therefore $g(1+u) = u^{p(p-1)} + pk(u)$ where k is a polynomial in u over $\mathbb{Z}$. From the alternative expression

$$g(1+u) = 1 + (1+u)^p + \cdots + (1+u)^{p(p-1)}$$

it follows that k has constant term 1. By Eisenstein's Criterion, $g(1+u)$ is irreducible over $\mathbb{Q}$. □

We can now obtain a more specific result than Lemma 15.4 for pth roots of unity over $\mathbb{Q}$:

Theorem 20.12. *Let p be prime and let ζ be a primitive pth root of unity in $\mathbb{C}$. Then the Galois group of $\mathbb{Q}(\zeta) : \mathbb{Q}$ is cyclic of order $p - 1$.*

Proof. This follows the same lines as the proof of Lemma 15.4, but now we can say a little more.

The zeros in $\mathbb{C}$ of $t^p - 1$ are ζ^j, where $0 \leq j \leq p-1$, and these are distinct. These zeros form a group under multiplication, and this group is cyclic, generated by ζ. Therefore any $\mathbb{Q}$-automorphism of $\mathbb{Q}(\zeta)$ is determined by its effect on ζ. Further, $\mathbb{Q}$-automorphisms permute the zeros of $t^p - 1$. Hence any $\mathbb{Q}$-automorphism of $\mathbb{Q}(\zeta)$ has the form

$$\alpha_j : \zeta \mapsto \zeta^j$$

and is uniquely determined by this condition.

We claim that every α_j is, in fact, a $\mathbb{Q}$-automorphism of $\mathbb{Q}(\zeta)$. The ζ^j with $j > 0$ are the zeros of $1 + t + \cdots + t^{p-1}$. This polynomial is irreducible over $\mathbb{Q}$ by Lemma 3.22. Therefore it is the minimal polynomial of any of its zeros, namely ζ^j where $1 \leq j \leq p-1$. By Proposition 11.4, every α_j is a $\mathbb{Q}$-automorphism of $\mathbb{Q}(\zeta)$, as claimed.

Clearly $\alpha_i\alpha_j = \alpha_{ij}$, where the product ij is taken modulo p. Therefore the Galois group of $\mathbb{Q}(\zeta) : \mathbb{Q}$ is isomorphic to the multiplicative group $\mathbb{Z}_p^*$. This is cyclic by Corollary 19.9. □

We now come to the main result of this chapter.

Theorem 20.13 (Gauss). *The regular n-gon is constructible by ruler and compasses if and only if*

$$n = 2^r p_1 \dots p_s$$

where r and s are integers ≥ 0, *and* $p_1, \dots, p_s$ *are distinct odd primes of the form*

$$p_j = 2^{2^{r_j}} + 1$$

for positive integers r_j.

Proof. Let n be constructive. Then $n = 2^r p_1^{m_1} \dots p_s^{m_s}$ where $p_1, \dots, p_s$ are distinct odd primes. By Corollary 20.7, each $p_j^{m_j}$ is constructive. If $m_j \geq 2$ then p_j^2 is constructive by Theorem 20.1. Hence the degree of the minimal polynomial of a primitive p_j^2th root of unity over $\mathbb{Q}$ is a power of 2 by Lemma 20.9. By Lemma 20.11, $p_j(p_j - 1)$ is a power of 2, which cannot happen since p_j is odd. Therefore $m_j = 1$ for all j. Therefore p_j is constructive. By Lemma 3.22

$$p_j - 1 = 2^{s_j}$$

for suitable s_j. Suppose that s_j has an odd divisor $a > 1$, so that $s_j = ab$. Then

$$p_j = (2^b)^a + 1$$

which is divisible by $2^b + 1$ since

$$t^a + 1 = (t+1)(t^{a-1} - t^{a-2} + \dots + 1)$$

when a is odd. So p_j cannot be prime. Hence s_j has no odd factors, so

$$s_j = 2^{r_j}$$

for some $r_j > 0$.

This establishes the necessity of the given form of n. Now we prove sufficiency. By Corollary 20.7 we need consider only prime-power factors of n. By Lemma 20.8, 2^r is constructive. We must show that each p_j is constructive. Let ζ be a primitive p_jth root of unity. By Theorem 20.12

$$[\mathbb{Q}(\zeta) : \mathbb{Q}] = p_j - 1 = 2^{s_j}$$

Now $\mathbb{Q}(\zeta)$ is a splitting field for $f(t) = 1 + \dots + t^{p-1}$ over $\mathbb{Q}$, so that $\mathbb{Q}(\zeta) : \mathbb{Q}$ is normal. It is also separable since the characteristic is zero. By Lemma 15.5, the Galois group $\Gamma(\mathbb{Q}(\zeta) : \mathbb{Q})$ is abelian, and by Theorem 20.12 or an appeal to the Galois correspondence it has order 2^{s_j}. By Proposition 20.4, $\zeta \in \mathbb{C}$ is constructible. □

20.4 Fermat Numbers

The problem of finding all constructible regular polygons now reduces to number theory, and there the question has a longer history. In 1640 Pierre de Fermat wondered when $2^k + 1$ is prime, and proved that a necessary condition is for k to be a power of 2. Thus we are led to:

Definition 20.14. The nth *Fermat number* is $F_n = 2^{2^n} + 1$.

The question becomes: when is F_n prime?

Fermat noticed that $F_0 = 3, F_1 = 5, F_2 = 17, F_3 = 257$, and $F_4 = 65537$ are all prime. He conjectured that F_n is prime for all n, but this was disproved by Euler in 1732, who proved that F_5 is divisible by 641 (Exercise 20.5). Knowledge of factors of Fermat numbers is changing almost daily, thanks to the prevalence of fast computers and special algorithms for primality testing of Fermat numbers: see References under 'Internet'. At the time of writing, the largest known composite Fermat number was $F_{3329780}$, with a factor $193.2^{3329782} + 1$. This was proved by Raymond Ottusch in July 2014 as a contribution to PrimeGrid's Proth Prime Search. At that time, 277 Fermat numbers were known to be composite.

No new Fermat primes have been found, so the only known Fermat primes are still those found by Fermat himself: 2, 3, 5, 17, 257, and 65537. We sum up the current state of knowledge as:

Proposition 20.15. *If p is a prime, then the regular p-gon is constructible for $p = 2, 3, 5, 17, 257, 65537$.*

20.5 How to Draw a Regular 17-gon

Many constructions for the regular 17-gon have been devised, the earliest published being that of Huguenin (see Klein 1913) in 1803. For several of these constructions there are proofs of their correctness which use only synthetic geometry (ordinary Euclidean geometry without coordinates). A series of papers giving a construction for the regular 257-gon was published by F.J. Richelot (1832), under one of the longest titles I have ever seen. Bell (1965) tells of an over-zealous research student being sent away to find a construction for the 65537-gon, and reappearing with one twenty years later. This story, though apocryphal, is not far from the truth; Professor Hermes of Lingen spent ten years on the problem, and his manuscripts are still preserved at Göttingen.

One way to construct a regular 17-gon is to follow faithfully the above theory, which in fact provides a perfectly definite construction after a little extra calculation. With ingenuity it is possible to shorten the work. The construction that we now describe is taken from Hardy and Wright (1962).

Our immediate object is to find radical expressions for the zeros of the polynomial

$$\frac{t^{17}-1}{t-1} = t^{16}+\cdots+t+1 \tag{20.1}$$

over $\mathbb{C}$. We know the zeros are ζ^k, where $\zeta = e^{2\pi i/17}$ and $1 \le k \le 16$. To simplify notation, let

$$\theta = 2\pi/17$$

so that $\zeta^k = \cos k\theta + i\sin k\theta$.

Theorem 20.12 for $n = 17$ implies that the Galois group $\Gamma(\mathbb{Q}(\zeta):\mathbb{Q})$ consists of the $\mathbb{Q}$-automorphisms γ_j defined by

$$\gamma_j(\zeta) = \zeta^j \quad 1 \le j \le 16$$

and this is isomorphic to the multiplicative group $\mathbb{Z}_{17}^*$. By Theorem 19.8 $\mathbb{Z}_{17}^*$ is cyclic of order 16.

Galois theory now implies that ζ is constructible. In fact, there must exist a generator α for $\mathbb{Z}_{17}^*$. Then $\alpha^{16} = 1$ and no smaller power of α is 1. Consider the series of subgroups

$$1 = \langle \alpha^{16} \rangle \lhd \langle \alpha^8 \rangle \lhd \langle \alpha^4 \rangle \lhd \langle \alpha^2 \rangle \lhd \langle \alpha \rangle = \mathbb{Z}_{17}^* \tag{20.2}$$

The Galois correspondence leads to a tower of subfields from $\mathbb{Q}$ to $\mathbb{Q}(\zeta)$ in which each step is an extension of degree 2. By Theorem 7.11, ζ is constructible, so the regular 17-gon is constructible.

To convert this to an explicit construction we must find a generator for $\mathbb{Z}_{17}^*$. Experimenting with small values, $\alpha = 2$ is not a generator (it has order 8), but $\alpha = 3$ is a generator. In fact, the powers of 3 modulo 17 are:

m	0	1	2	3	4	5	6	7	8	9	10	11	12	13	14	15
3^m	1	3	9	10	13	5	15	11	16	14	8	7	4	12	2	6

Motivated by (20.2), define

$$\begin{aligned}
x_1 &= \zeta + \zeta^9 + \zeta^{13} + \zeta^{15} + \zeta^{16} + \zeta^8 + \zeta^4 + \zeta^2 \\
x_2 &= \zeta^3 + \zeta^{10} + \zeta^5 + \zeta^{11} + \zeta^{14} + \zeta^7 + \zeta^{12} + \zeta^6 \\
y_1 &= \zeta + \zeta^{13} + \zeta^{16} + \zeta^4 \\
y_2 &= \zeta^9 + \zeta^{15} + \zeta^8 + \zeta^2 \\
y_3 &= \zeta^3 + \zeta^5 + \zeta^{14} + \zeta^{12} \\
y_4 &= \zeta^{10} + \zeta^{11} + \zeta^7 + \zeta^6
\end{aligned}$$

By definition, these lie in various fixed fields in the aforementioned tower. Now

$$\zeta^k + \zeta^{17-k} = 2\cos k\theta \tag{20.3}$$

for $k = 1, \ldots, 16$, so

$$\begin{aligned} x_1 &= 2(\cos\theta + \cos 8\theta + \cos 4\theta + \cos 2\theta) \\ x_2 &= 2(\cos 3\theta + \cos 7\theta + \cos 5\theta + \cos 6\theta) \\ y_1 &= 2(\cos\theta + \cos 4\theta) \\ y_2 &= 2(\cos 8\theta + \cos 2\theta) \\ y_3 &= 2(\cos 3\theta + \cos 5\theta) \\ y_4 &= 2(\cos 7\theta + \cos 6\theta) \end{aligned} \tag{20.4}$$

Equation (20.1) implies that

$$x_1 + x_2 = -1$$

Now (20.4) and the identity

$$2\cos m\theta \, \cos n\theta = \cos(m+n)\theta + \cos(m-n)\theta$$

imply that

$$x_1 x_2 = 4(x_1 + x_2) = -4$$

using (20.3). Hence x_1 and x_2 are zeros of the quadratic polynomial

$$t^2 + t - 4 \tag{20.5}$$

Further, $x_1 > 0$ so that $x_1 > x_2$. By further trigonometric expansions,

$$y_1 + y_2 = x_1 \qquad y_1 y_2 = -1$$

and y_1, y_2 are the zeros of

$$t^2 - x_1 t - 1 \tag{20.6}$$

Further, $y_1 > y_2$. Similarly, y_3 and y_4 are the zeros of

$$t^2 - x_2 t - 1 \tag{20.7}$$

and $y_3 > y_4$. Now

$$\begin{aligned} 2\cos\theta + 2\cos 4\theta &= y_1 \\ 4\cos\theta \, \cos 4\theta = 2\cos 5\theta + 2\cos 3\theta &= y_3 \end{aligned}$$

so

$$z_1 = 2\cos\theta \qquad z_2 = 2\cos 4\theta$$

are the zeros of

$$t^2 - y_1 t + y_3 \tag{20.8}$$

and $z_1 > z_2$.

Solving the series of quadratics (20.5–20.8) and using the inequalities to decide which zero is which, we obtain

$$\cos\theta = \frac{1}{16}\Bigg(-1+\sqrt{17}+\sqrt{34-2\sqrt{17}} \qquad (20.9)$$

$$+\sqrt{68+12\sqrt{17}-16\sqrt{34+2\sqrt{17}}-2(1-\sqrt{17})\sqrt{34-2\sqrt{17}}}\Bigg)$$

where the square roots are the positive ones.

From this we can deduce a geometric construction for the 17-gon by constructing the relevant square roots. This procedure is animated in an iPad app, Stewart (2014), and can also be found on the web. By using greater ingenuity it is possible to obtain an aesthetically more satisfying construction. The following method (Figure 29) is due to Richmond (1893).

Let ϕ be the smallest positive acute angle such that $\tan 4\phi = 4$. Then $\phi, 2\phi$, and 4ϕ are all acute. Expression (20.5) can be written

$$t^2 + 4t\cot 4\phi - 4$$

whose zeros are

$$2\tan 2\phi \qquad -2\cot 2\phi$$

Hence

$$x_1 = 2\tan 2\phi \qquad x_2 = -2\cot 2\phi$$

This implies that

$$y_1 = \tan\left(\phi+\frac{\pi}{4}\right) \qquad y_2 = \tan\left(\phi-\frac{\pi}{4}\right) \qquad y_3 = \tan\phi \qquad y_4 = -\cot\phi$$

so that

$$\begin{aligned} 2(\cos 3\theta + \cos 5\theta) &= \tan\phi \\ 4\cos 3\theta\cos 5\theta &= \tan\left(\phi-\frac{\pi}{4}\right) \end{aligned}$$

In Figure 29, let OA, OB be two perpendicular radii of a circle. Make $\mathrm{OI} = \frac{1}{4}\mathrm{OB}$ and $\angle\mathrm{OIE} = \frac{1}{4}\angle\mathrm{OIA}$. Find F on AO produced to make $\angle\mathrm{EIF} = \frac{\pi}{4}$. Let the circle on AF as diameter cut OB in K, and let the circle centre E through K cut OA in N_3 and N_5 as shown. Draw $\mathrm{N}_3\mathrm{P}_3$ and $\mathrm{N}_5\mathrm{P}_5$ perpendicular to OA. Then $\angle\mathrm{OIA} = 4\phi$ and $\angle\mathrm{OIE} = \phi$. Also

$$\begin{aligned} 2(\cos\angle\mathrm{AOP}_3 + \cos\angle\mathrm{AOP}_5) &= 2\frac{\mathrm{ON}_3 - \mathrm{ON}_5}{\mathrm{OA}} \\ &= 4\frac{\mathrm{OE}}{\mathrm{OA}} + \frac{\mathrm{OE}}{\mathrm{OI}} = \tan\phi \end{aligned}$$

and

$$
\begin{aligned}
4\cos\angle \mathrm{AOP}_3 \cos\angle \mathrm{AOP}_5 &= -4\frac{\mathrm{ON}_3 \times \mathrm{ON}_5}{\mathrm{OA}\times\mathrm{OA}} \\
&= -4\frac{\mathrm{OK}^2}{\mathrm{OA}^2} \\
&= -4\frac{\mathrm{OF}}{\mathrm{OA}} \\
&= -\frac{\mathrm{OF}}{\mathrm{OI}} = \tan\left(\phi - \frac{\pi}{4}\right)
\end{aligned}
$$

Comparing these with equation (17.8) we see that

$$\angle \mathrm{AOP}_3 = 3\theta \qquad \angle \mathrm{AOP}_5 = 5\theta$$

Hence A, P_3, P_5 are the zeroth, third, and fifth vertices of a regular 17-gon inscribed in the given circle. The other vertices are now easily found.

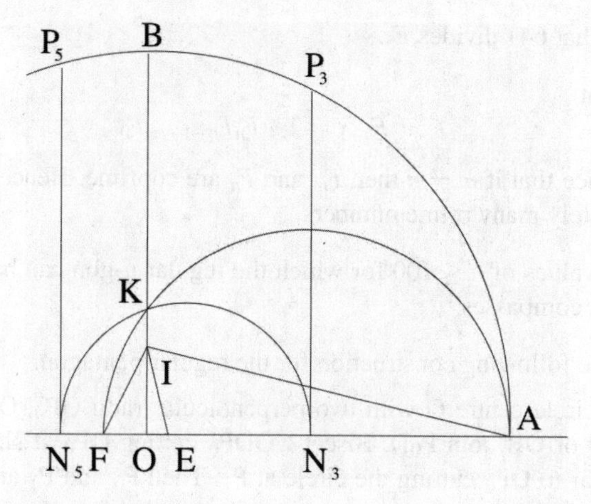

FIGURE 29: Richmond's construction for a regular 17-gon.

In Chapter 21 we return to topics associated with regular polygons, especially so-called cyclotomic polynomials. We end that chapter by investigating the construction of regular polygons when an angle-trisector is permitted, as well as the traditional ruler and compass.

EXERCISES

20.1 Using only the operations 'ruler' and 'compass', show how to draw a parallel to a given line through a given point.

20.2 Verify the following approximate constructions for regular n-gons found by Oldroyd (1955):

(a) 7-*gon*. Construct $\cos^{-1}\frac{4+\sqrt{5}}{10}$ giving an angle of approximately $2\pi/7$.

(b) 9-*gon*. Construct $\cos^{-1}\frac{5\sqrt{3}-1}{10}$.

(c) 11-*gon*. Construct $\cos^{-1}\frac{8}{9}$ and $\cos^{-1}\frac{1}{2}$ and take their difference.

(d) 13-*gon*. Construct $\tan^{-1}1$ and $\tan^{-1}\frac{4+\sqrt{5}}{20}$ and take their difference.

20.3 Show that for n odd the only known constructible n-gons are precisely those for which n is a divisor of $2^{32}-1=4294967295$.

20.4 Work out the approximate size of F_{382449}, which is known to be composite. Explain why it is no easy task to find factors of Fermat numbers.

20.5 Use the equations

$$641=5^4+2^4=5.2^7+1$$

to show that 641 divides F_5.

20.6 Show that

$$F_{n+1}=2+F_nF_{n-1}\ldots F_0$$

and deduce that if $m\neq n$ then F_m and F_n are coprime. Hence show that there are infinitely many prime numbers.

20.7 List the values of $n\leq 100$ for which the regular n-gon can be constructed by ruler and compasses.

20.8 Verify the following construction for the regular pentagon.

Draw a circle centre O with two perpendicular radii OP_0, OB. Let D be the midpoint of OB, join P_0D. Bisect $\angle ODP_0$ cutting OP_0 at N. Draw NP_1 perpendicular to OP_0 cutting the circle at P_1. Then P_0 and P_1 are the zeroth and first vertices of a regular pentagon inscribed in the circle.

20.9 Euclid's construction for an isosceles triangle with angles $4\pi/5, 4\pi/5, 2\pi/5$ depends on constructing the so-called golden section: that is, *To construct a given straight line so that the rectangle contained by the whole and one of the segments is equal to the square on the other segment.* The Greek term was 'extreme and mean ratio'. In Book 2 Proposition 11 of the *Elements* Euclid solves this problem as in Figure 30.

Let AB be the given line. Make ABDC a square. Bisect AC at E, and make EF = BE. Now find H such that AH = AF. Then the square on AH has the same area as the rectangle with sides AB and BH, as required.

Prove that Euclid was right.

20.10 Mark the following true or false.

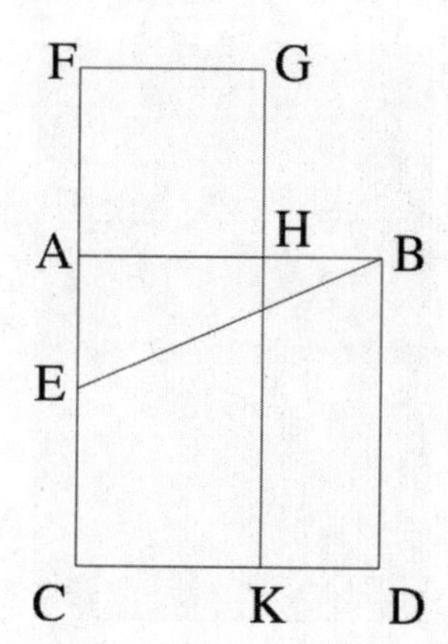

FIGURE 30: Cutting a line in extreme and mean ratio.

(a) $2^n + 1$ cannot be prime unless n is a power of 2.

(b) If n is a power of 2 then $2^n + 1$ is always prime.

(c) The regular 771-gon is constructible using ruler and compasses.

(d) The regular 768-gon is constructible using ruler and compasses.

(e) The regular 51-gon is constructible using ruler and compasses.

(f) The regular 25-gon is constructible using ruler and compasses.

(g) For an odd prime p, the regular p^2-gon is never constructible using ruler and compasses.

(h) If n is an integer > 0 then a line of length $\sqrt{n}$ can always be constructed from $\mathbb{Q}$ using ruler and compass.

(i) If n is an integer > 0 then a line of length $\sqrt[4]{n}$ can always be constructed from $\mathbb{Q}$ using ruler and compass.

(j) A point whose coordinates lie in a normal extension of $\mathbb{Q}$ whose degree is a power of 2 is constructible using ruler and compasses.

(k) If p is a prime, then $t^{p^2} - 1$ is irreducible over $\mathbb{Q}$.

Chapter 21

Circle Division

To halt the story of regular polygons at the stage of ruler-and-compass constructions would leave a small but significant gap in our understanding of the solution of polynomial equations by radicals. Our definition of 'radical extension' involves a slight cheat, which becomes evident if we ask what the expression of a root of unity looks like. Specifically, what does the radical expression of the primitive 11th root of unity

$$\zeta_{11} = \cos\frac{2\pi}{11} + i\sin\frac{2\pi}{11}$$

look like?

As the theory stands, the best we can offer is

$$\sqrt[11]{1} \tag{21.1}$$

which is not terribly satisfactory, because the obvious interpretation of $\sqrt[11]{1}$ is 1, not ζ_{11}. Gauss's theory of the 17-gon hints that there might be a more impressive answer. In place of $\sqrt[17]{1}$ Gauss has a marvellously complicated system of nested square roots, which we repeat from equation (20.9):

$$\begin{aligned}\cos\frac{2\pi}{17} = \frac{1}{16}\Bigg(& -1+\sqrt{17}+\sqrt{34-2\sqrt{17}} \\ & + \sqrt{68+12\sqrt{17}-16\sqrt{34+2\sqrt{17}}-2(1-\sqrt{17})\sqrt{34-2\sqrt{17}}}\Bigg)\end{aligned}$$

with a similar expression for $\sin\frac{2\pi}{17}$, and hence an even more impressive formula for $\zeta_{17} = \cos\frac{2\pi}{17} + i\sin\frac{2\pi}{17}$.

Can something similar be done for the 11th root of unity? For *all* roots of unity? The answer to both questions is 'yes', and we are getting the history back to front, because Gauss gave that answer as part of his work on the 17-gon. Indeed, Vandermonde came very close to the same answer 25 years earlier, in 1771, and in particular he managed to find an expression by radicals for ζ_{11} that is less disappointing than (21.1). He, in turn, built on the epic investigations of Lagrange.

The technical term for this area is 'cyclotomy', from the Greek for 'circle cutting'. In particular, pursuing Gauss's and Vandermonde's line of enquiry will lead us to some fascinating properties of the 'cyclotomic polynomial' $\Phi_d(t)$, which is the minimal polynomial over $\mathbb{Q}$ of a primitive dth root of unity in $\mathbb{C}$.

21.1 Genuine Radicals

Of course, we can 'solve' the entire problem at a stroke if we *define* $\sqrt[n]{1}$ to be the primitive nth root of unity

$$\cos\frac{2\pi}{n}+i\sin\frac{2\pi}{n}$$

instead of defining it to be 1. In a sense, this is what Definition 15.1 does. However, there is a better solution, as we shall see. What makes the above interpretation of $\sqrt[n]{1}$ unsatisfactory? Consider the typical case of $\zeta_{17}=\sqrt[17]{1}$. The minimal polynomial of ζ_{17} is not $t^{17}-1$, as the notation $\sqrt[17]{1}$ suggests; instead, it has degree 16, being equal to

$$t^{16}+t^{15}+\cdots t+1$$

It would be reasonable to seek to determine the zeros of this 16th degree equation using radicals of degree 16 or less, but a 17th root seems rather out of place. Especially since we know from Gauss that in this case (nested) square roots are enough.

However, that is a rather special example. What about other nth roots of unity? Can they also be expressed by what we might informally call 'genuine' radicals, those not employing the $\sqrt[n]{1}$ trick? (We pin down this concept formally in Definition 21.1.) Classically, the answer was found to be 'yes' for $2\le n\le 10$, as we now indicate.

When $n=2$, the primitive square root of unity is -1. This lies in $\mathbb{Q}$, so no radicals are needed.

When $n=3$, the primitive cube roots of unity are solutions of the *quadratic* equation

$$t^2+t+1=0$$

and so are of the form ω,ω^2 where

$$\omega=-\frac{1}{2}+i\frac{\sqrt{3}}{2}$$

involving only a square root.

When $n=4$, a primitive 4th root of unity is i, which again can be represented using only a square root, since $i=\sqrt{-1}$.

When $n=5$, we have to solve

$$t^4+t^3+t^2+t+1=0 \tag{21.2}$$

We know from Chapter 18 that any quartic can be solved by radicals; indeed only square and cube roots are required (in part because $\sqrt[4]{x}=\sqrt{\sqrt{x}}$). But we can do better. There is a standard trick that applies to equations of even degree that are *palindromic*—the list of coefficients is symmetric about the central term. We encountered this trick in Exercises 15.4 and 15.5: express the equations in terms of a new variable

$$u=t+\frac{1}{t} \tag{21.3}$$

Then

$$u^2 = t^2 + 2 + \frac{1}{t^2}$$
$$u^3 = t^3 + 3t + \frac{3}{t} + \frac{1}{t^3}$$

and so on. Rewrite (21.2) by dividing by t^2:

$$t^2 + t + 1 + \frac{1}{t} + \frac{1}{t^2} = 0$$

which in terms of u becomes

$$u^2 + u - 1 = 0$$

which is quadratic in u. Solving for u:

$$u = \frac{-1 \pm \sqrt{5}}{2}$$

Now we recover t from u by solving a second quadratic equation. From (21.3)

$$t^2 - ut + 1 = 0$$

so

$$t = \frac{u \pm \sqrt{u^2 - 4}}{2}$$

Explicitly, we get four zeros:

$$t = \frac{-1 \pm \sqrt{5} \pm \sqrt{-10 \pm 2\sqrt{5}}}{4} \tag{21.4}$$

with independent choices of the first two $\pm$ signs, and the third equalling the first. So we can express primitive 5th roots of unity using nothing worse than square roots.

Continuing in this way, we can find a radical expression for a primitive 6th root of unity (it is $-\omega$); a primitive 7th root of unity (use the $t + 1/t$ trick to reduce to a cubic); a primitive 8th root of unity ($\sqrt{i}$ is one possibility, $\frac{1+i}{\sqrt{2}}$ is perhaps better); a primitive 9th root of unity ($\sqrt[3]{\omega}$); and a primitive 10th root of unity ($-\zeta_5$). The first case that baffled mathematicians prior to 1771 was therefore the primitive 11th root of unity, which leads to a *quintic* if we try the $t + 1/t$ trick. But in that year, Vandermonde obtained the explicit radical expression

$$\begin{aligned}
\zeta_{11} = \frac{1}{5}\Bigg[& \sqrt[5]{\frac{11}{4}\left(89 + 25\sqrt{5} - 5\sqrt{-5 + 2\sqrt{5}} + 45\sqrt{-5 - 2\sqrt{5}}\right)} \\
& + \sqrt[5]{\frac{11}{4}\left(89 + 25\sqrt{5} + 5\sqrt{-5 + 2\sqrt{5}} - 45\sqrt{-5 - 2\sqrt{5}}\right)} \\
& + \sqrt[5]{\frac{11}{4}\left(89 - 25\sqrt{5} - 5\sqrt{-5 + 2\sqrt{5}} - 45\sqrt{-5 - 2\sqrt{5}}\right)} \\
& + \sqrt[5]{\frac{11}{4}\left(89 - 25\sqrt{5} + 5\sqrt{-5 + 2\sqrt{5}} + 45\sqrt{-5 - 2\sqrt{5}}\right)}\Bigg]
\end{aligned}$$

He stated that his method would work for any primitive nth root of unity, but he did not give a proof. That was supplied by Gauss in 1796, with a gap in the proof, see below, and it was published in 1801 in his *Disquisitiones Arithmeticae*. It is not known whether Gauss was aware of Vandermonde's pioneering work.

21.2 Fifth Roots Revisited

Before proving a version of Gauss's theorem on the representability of roots of unity by genuine radicals, it helps to have an example. We can explain Vandermonde's approach in the simpler case $n = 5$, where explicit calculations are not too lengthy.

As before, we want to solve

$$t^4 + t^3 + t^2 + t + 1 = 0$$

by radicals. We know that the zeros are

$$\zeta \qquad \zeta^2 \qquad \zeta^3 \qquad \zeta^4$$

where $\zeta = \cos\frac{2\pi}{5} + i\sin\frac{2\pi}{5}$. The exponents $1,2,3,4$ can be considered as elements of the multiplicative group of the field $\mathbb{Z}_5$. By Theorem 20.12 the Galois group of $\mathbb{Q}(\zeta):\mathbb{Q}$ consists of the $\mathbb{Q}$-automorphisms

$$\phi_j : \zeta \mapsto \zeta^j \quad 1 \leq j \leq 4$$

The Galois group is therefore isomorphic to is $\mathbb{Z}_5^*$, which is cyclic of order 4 by Theorem 19.8. Experiment quickly shows that it is generated by the element 2 (mod 5). Indeed, modulo 5 the powers of 2 are

$$2^0 = 1 \qquad 2^1 = 2 \qquad 2^2 = 4 \qquad 2^3 = 3 \tag{21.5}$$

Hilbert's Theorem 90, Theorem 18.18, leads us to consider the expression

$$\alpha_1 = \zeta + i\zeta^2 - \zeta^4 - i\zeta^3$$

and compute its fourth power. We find (suppressing some details) that

$$\alpha_1^2 = -(1+2i)(\zeta - \zeta^2 + \zeta^4 - \zeta^3)$$

so, squaring again,

$$\alpha_1^4 = -15 + 20i$$

Therefore we can express α_1 by radicals:

$$\alpha_1 = \sqrt[4]{-15+20i}$$

We can play a similar game with

$$\alpha_3 = \zeta - i\zeta^2 - \zeta^4 + i\zeta^3$$

to get

$$\alpha_3 = \sqrt[4]{15 - 20i}$$

The calculation of α_1^4 also draws attention to

$$\alpha_2 = \zeta - \zeta^2 + \zeta^4 - \zeta^3$$

and shows that $\alpha_2^2 = 5$, so

$$\alpha_2 = \sqrt{5}$$

Summarising:

$$\begin{aligned}
\alpha_0 &= \zeta + \zeta^2 + \zeta^4 + \zeta^3 &&= -1 \\
\alpha_1 &= \zeta + i\zeta^2 - \zeta^4 - i\zeta^3 &&= \sqrt[4]{-15 + 20i} \\
\alpha_2 &= \zeta - \zeta^2 + \zeta^4 - \zeta^3 &&= \sqrt{5} \\
\alpha_3 &= \zeta - i\zeta^2 - \zeta^4 + i\zeta^3 &&= \sqrt[4]{-15 - 20i}
\end{aligned}$$

Thus we find four *linear* equations in $\zeta, \zeta^2, \zeta^3, \zeta^4$. These equations are independent, and we can solve them. In particular,

$$\alpha_0 + \alpha_1 + \alpha_2 + \alpha_3$$

is equal to

$$\zeta(1+1+1+1) + \zeta^2(1+i-1-i) + \zeta^4(1-1+1-1) + \zeta^3(1-i-1+i) = 4\zeta$$

Therefore

$$\zeta = \frac{1}{4}\left(-1 - \sqrt{5} + \sqrt{\sqrt{-15+20i}} + \sqrt{\sqrt{-15-20i}}\right)$$

This expression is superficially different from (21.4), but in fact the two are equivalent. Both use nothing worse than square roots.

This calculation is too remarkable to be mere coincidence. It must work out nicely because of some hidden structure. What lies behind it?

The general idea behind Vandermonde's calculation, as isolated by Gauss, is the following. Recall Definition 21.7, which introduces the group of units $\mathbb{Z}_n^*$ of the ring $\mathbb{Z}_n$. This consists of all elements that have a multiplicative inverse (mod n), and it is a group under multiplication. When n is prime, this consists of all nonzero elements. In general, it consists of those elements that are prime to n.

The multiplicative group $\mathbb{Z}_5^*$ is cyclic of order 4, and the number 2 (modulo 5) is a generator. It has order 4 in $\mathbb{Z}_5^*$. The complex number i is a primitive 4th root of unity, so i has order 4 in the multiplicative group of 4th roots of unity, namely $1, i, -1, -i$. These two facts conspire to make the algebra work.

To see how, we apply a little Galois theory—a classic case of being wise after the event. By Theorem 21.9, the Galois group Γ of $\mathbb{Q}(\zeta):\mathbb{Q}$ has order 4 and comprises the $\mathbb{Q}$-automorphisms generated by the maps

$$\rho_k : \zeta \mapsto \zeta^k$$

for $k = 1,2,3,4$. The group Γ is isomorphic to $\mathbb{Z}_5^*$ by the map $\rho_k \mapsto k \pmod 5$. Therefore ρ_2 has order 4 in Γ, hence generates Γ, and Γ is cyclic of order 4.

The extension is normal, since it is a splitting field for an irreducible polynomial, and we are working over $\mathbb{C}$ so the extension is separable. By the Galois correspondence, any rational function of ζ that is fixed by ρ_2 is in fact a rational number.

Consider as a typical case the expression α_1 above. Write this as

$$\alpha_1 = \zeta + i\rho_2(\zeta) + i^2\rho_2^2(\zeta) + i^3\rho_2^3(\zeta)$$

Then

$$\rho_2(\alpha_1) = \rho_2(\zeta) + i\rho_2^2(\zeta) + i^2\rho_2^3(\zeta) + i^3\zeta$$

since $\rho_4^4(\zeta) = \zeta$. Therefore

$$\rho_2(\alpha_1) = i^{-1}\alpha_1$$

so

$$\rho_2(\alpha_1^4) = (i^{-1}\alpha_1)^4 = \alpha_1^4$$

Thus α_1^4 lies in the fixed field of ρ_2, that is, the fixed field of Γ, which is $\mathbb{Q}$...

Hold it.

The *idea* is right, but the argument has a flaw. The explicit calculation shows that $\alpha_1^4 = -15 + 20i$, which lies in $\mathbb{Q}(i)$, not $\mathbb{Q}$. What was the mistake? The problem is that α_1 is not an element of $\mathbb{Q}(\zeta)$. It belongs to the larger field $\mathbb{Q}(\zeta)(i)$, which equals $\mathbb{Q}(i,\zeta)$. So we have to do the Galois theory for $\mathbb{Q}(i,\zeta):\mathbb{Q}$, not $\mathbb{Q}(\zeta):\mathbb{Q}$.

It is fairly straightforward to do this. Since 4 and 5 are coprime, the product $\xi = i\zeta$ is a primitive 20th root of unity. Moreover, $\xi^5 = i$ and $\xi^{16} = \zeta$. Therefore $\mathbb{Q}(i,\zeta) = \mathbb{Q}(\xi)$. Since 20 is not prime, we do not know that this group is cyclic, so we have to work out its structure. In fact, it is the group of units $\mathbb{Z}_{20}^*$ of the ring $\mathbb{Z}_{20}$, which is isomorphic to $\mathbb{Z}_2 \times \mathbb{Z}_4$, not $\mathbb{Z}_8$. By considering the tower of fields

$$\mathbb{Q} \subseteq \mathbb{Q}(i) \subseteq \mathbb{Q}(\xi)$$

and using the structure of $\mathbb{Z}_{20}^*$, it can be shown that the Galois group of $\mathbb{Q}(\xi):\mathbb{Q}(i)$ is the subgroup of $\mathbb{Z}_{20}^*$ isomorphic to $\mathbb{Z}_4$, generated by the $\mathbb{Q}(i)$-automorphism $\tilde{\rho}_2$ that sends ζ to ζ^2 and *fixes* $\mathbb{Q}(i)$. We prove a more general result in Theorem 21.3 below.

Having made the switch to $\mathbb{Q}(\xi)$, the above calculation shows that α_1^4 lies in the fixed field of the Galois group $\Gamma(\mathbb{Q}(\xi):\mathbb{Q}(i))$. This field is $\mathbb{Q}(i)$, because the extension is normal and separable. So without doing the explicit calculations, we can see in advance that α_1^4 must lie in $\mathbb{Q}(i)$. The same goes for α_2^4, α_3^4, and (trivially) α_0^4.

21.3 Vandermonde Revisited

Vandermonde was very competent, but a bit of a plodder; he did not follow up his idea in full generality, and thereby missed a major discovery. He could well have anticipated Gauss, possibly even Galois, if he had found a proof that his method was a completely general way to express roots of unity by genuine radicals, instead of just asserting that it was.

As preparation, we now establish Vandermonde's main point about the primitive 11th roots of unity. Any unproved assertions about Galois groups will be dealt with in the general case, see Section 21.4. Let $\zeta = \zeta_{11}$. Vandermonde started with the identity

$$\zeta^{10} + \zeta^9 + \cdots + \zeta + 1 = 0$$

and played the $u = \zeta + 1/\zeta$ trick to reduce the problem to a quintic, but with hindsight this step is not necessary and if anything makes the idea more obscure. Introduce a primitive 10th root of unity θ, so that $\theta\zeta$ is a primitive 110th root of unity. Consider the field extension $\mathbb{Q}(\theta\zeta) : \mathbb{Q}(\theta)$, which turns out to be of degree 10, with a cyclic Galois group of order 10 that is isomorphic to $\mathbb{Z}_{11}^*$. A generator for $\mathbb{Z}_{11}^*$ is readily found, and turns out to be the number 2, whose successive powers are

$$1, 2, 4, 8, 5, 10, 9, 7, 3, 6$$

Therefore $\Gamma = \Gamma(\mathbb{Q}(\theta\zeta) : \mathbb{Q}(\theta))$ consists of the $\mathbb{Q}(\theta)$-automorphisms ρ_k, for $k = 1, \ldots, 10$, that map

$$\zeta \mapsto \zeta^k \qquad \theta \mapsto \theta$$

Let l be any integer, $0 \leq l \leq 9$, and define

$$\begin{aligned} \alpha_l &= \zeta + \theta^l \zeta^2 + \theta^{2l} \zeta^4 + \cdots + \theta^{9l} \zeta^6 \\ &= \textstyle\sum_{j=0}^{9} \theta^{jl} \zeta^{2^j} \end{aligned} \tag{21.6}$$

Consider the effect of ρ_2, which sends $\zeta \mapsto \zeta^2$ and fixes θ. We have

$$\rho_2(\alpha_l) = \sum_{j=0}^{9} \theta^{jl} \zeta^{2^{j+1}} = \theta^{-l} \alpha_l$$

so

$$\rho_2(\alpha_l^{10}) = \theta^{-10l} \alpha_l^{10} = \alpha_l^{10}$$

and α_l^{10} lies in the fixed field of Γ, which is $\mathbb{Q}(\theta)$). Thus there is some polynomial $f_l(\theta)$, of degree ≤ 9 over $\mathbb{Q}$, with

$$\alpha_l^{10} = f_l(\theta)$$

With effort, we can compute $f_l(\theta)$ explicitly. Short cuts help. At any rate,

$$\alpha_l = \sqrt[10]{f_l(\theta)} \tag{21.7}$$

We already know how to express θ by genuine radicals since it is a primitive 10th root of unity, so we have expressed α_l by radicals—in fact, only square roots and fifth roots are needed, since $\sqrt[10]{\ } = \sqrt[5]{\sqrt{\ }}$ and fifth roots of unity require only square roots.

Finally, the ten equations (21.6) for the α_l can be interpreted as a system of 10 linear equations for the powers $\zeta, \zeta^2, \ldots, \zeta^{10}$ over $\mathbb{C}$. These equations are independent, so the system can be solved. Indeed, using elementary properties of 10th roots of unity, it can be shown that

$$\zeta^{2^j} = \frac{1}{10}\left(\sum_{l=0}^{9} \theta^{-jl}\alpha_l\right)$$

In particular,

$$\zeta = \frac{1}{10}\left(\sum_{l=0}^{9} \alpha_l\right) = \frac{1}{10}\left(\sum_{l=0}^{9} \sqrt[10]{f_l(\theta)}\right)$$

Thus we have expressed ζ_{11} in terms of radicals, using only square roots and fifth roots.

Vandermonde's answer also uses only square roots and fifth roots, and can be deduced from the above formula. Because he used a variant of the above strategy, his answer does not immediately look the same as ours, but it is equivalent. To go beyond Vandermonde, we must prove that his method works for *all* primitive nth roots of unity. This we now establish.

21.4 The General Case

The time has come to define what we mean by a 'genuine' radical expression. Recall from Definition 8.12 that the *radical degree* of the radical $\sqrt[n]{\ }$ is n, and define the radical degree of a radical expression to be the maximum radical degree of the radicals that appear in it.

Definition 21.1. A number $\alpha \in \mathbb{C}$ has a *genuine radical expression* if α belongs to a radical extension of $\mathbb{Q}$ formed by successive adjunction of kth roots of elements β, where at every step the polynomial $t^k - \beta$ is irreducible over the field to which the root is adjoined.

This definition rules out $\sqrt[11]{1}$ as a genuine radical expression for ζ_{11}, but it permits $\sqrt{-1}$ as a genuine radical expression for i, and $\sqrt[3]{2}$ as a genuine radical expression for—well, $\sqrt[3]{2}$.

Our aim is to prove a theorem that was effectively stated by Vandermonde, and proved in full rigour (and greater generality, but we have to stop *somewhere*) by Gauss. The name 'Vandermonde-Gauss Theorem' is not standard, but it ought to be, so we shall use it.

Theorem 21.2 (Vandermonde-Gauss Theorem). *For any $n \geq 1$, any nth root of unity has a genuine radical expression.*

The aim of this section is to prove the Vandermonde-Gauss Theorem. In fact we prove something distinctly stronger: see Exercise 21.3. We prove the theorem by induction on n. It is easy to see that the induction step reduces to the case where n is prime and the nth root of unity concerned is therefore primitive, because if n is composite we can write it as $n = pq$ where p is prime, and $\sqrt[n]{\;} = \sqrt[q]{\sqrt[p]{\;}}$.

Let $n = p$ be prime and focus attention on a primitive pth root of unity ζ_p, which for simplicity we denote by ζ. In trigonometric terms,

$$\zeta = \cos\frac{2\pi}{p} + i\sin\frac{2\pi}{p}$$

but we do not actually use this formula.

We already know the minimal polynomial of ζ over $\mathbb{Q}$, from Lemma 3.22. It is

$$m(t) = t^{p-1} + t^{p-2} + \cdots + t + 1 = \frac{t^p - 1}{t - 1}$$

Let

$$\theta = \cos\frac{2\pi}{p-1} + i\sin\frac{2\pi}{p-1}$$

be a primitive $(p-1)$th root of unity. Since $p-1$ is composite (except when $p = 2, 3$) the minimal polynomial of θ over $\mathbb{Q}$ is *not* equal to

$$c(t) = t^{p-2} + t^{p-3} + \cdots + t + 1 = \frac{t^{p-1} - 1}{t-1}$$

but instead it is some irreducible divisor of $c(t)$.

We work not with $\mathbb{Q}(\zeta) : \mathbb{Q}$, but with $\mathbb{Q}(\theta, \zeta) : \mathbb{Q}$. Since $p, p-1$ are coprime, this extension is the same as

$$\mathbb{Q}(\theta\zeta) : \mathbb{Q}$$

where $\theta\zeta$ is a primitive $p(p-1)$th root of unity. A general element of $\mathbb{Q}(\theta\zeta)$ can be written as a linear combination over $\mathbb{Q}(\theta)$ of the powers $1, \zeta, \zeta^2, \ldots, \zeta^{p-2}$. It is convenient to throw in ζ^{p-1} as well, but now we must always bear in mind the relation $1 + \zeta + \zeta^2 + \cdots + \zeta^{p-1} = 0$.

We base the deduction on the following result, which we prove in Section 21.7 to avoid technical distractions.

Theorem 21.3. *The Galois group of $\mathbb{Q}(\theta\zeta) : \mathbb{Q}(\theta)$ is cyclic of order $p-1$. It comprises the $\mathbb{Q}(\theta)$-automorphisms of the form ρ_j, $\quad (j = 1, 2, \ldots p-1)$, where*

$$\rho_j : \zeta \mapsto \zeta^j$$
$$\theta \mapsto \theta$$

The main technical issue in proving this theorem is that although we know that $\zeta, \zeta^2, \dots, \zeta^{p-2}$ are linearly independent over $\mathbb{Q}$, we do not (yet) know that they are linearly independent over $\mathbb{Q}(\theta)$. Even Gauss omitted the proof of this fact from his discussion in the *Disquisitiones Arithmeticae*, but that may have been because to him it was obvious. He never published a proof of this particular fact, though he must have known one. So in a sense the first complete proof should probably be credited to Galois.

Assuming Theorem 21.3, we can follow Vandermonde's method in complete generality, using a few simple facts about roots of unity.

Proof of the Vandermonde-Gauss Theorem. We prove the theorem by induction on n. The cases $n = 1, 2$ are trivial since the roots of unity concerned are $1, -1$. As explained above, the induction step reduces to the case where n is prime and the nth root of unity concerned is therefore primitive. Throughout the proof it helps to bear in mind the above examples when $n = 5, 11$.

We write $n = p$ to remind us that n is prime. Let ζ be a primitive pth root of unity and let θ be a primitive $(p-1)$th root of unity as above. Then $\theta\zeta$ is a primitive $p(p-1)$th root of unity.

By Theorem 21.3, the Galois group of $\mathbb{Q}(\theta\zeta) : \mathbb{Q}$ is isomorphic to $\mathbb{Z}_p^*$, and is thus cyclic of order $p-1$ by Corollary 19.9. It comprises the automorphisms ρ_j for $j = 1, \dots, p-1$. Since $\mathbb{Z}_p^*$ is cyclic, there exists a generator a. That is, every $j \in \mathbb{Z}_p^*$ can be expressed as a power $j = a^l$ of a. Then $\rho_j = \rho_a^l$, so ρ_a generates $\Gamma = \Gamma(\mathbb{Q}(\theta\zeta) : \mathbb{Q}(\theta))$.

By Theorem 21.3 and Proposition 17.18, $\mathbb{Q}(\theta\zeta) : \mathbb{Q}(\theta)$ is normal and separable, so in particular the fixed field of Γ is $\mathbb{Q}(\theta)$ by Theorem 12.2(2). Since ρ_a generates Γ, any element of $\mathbb{Q}(\theta\zeta)$ that is fixed by ρ_a must lie in $\mathbb{Q}(\theta)$.

We construct elements fixed by ρ_a as follows. Define

$$\begin{aligned} \alpha_l &= \zeta + \theta^l \zeta^a + \theta^{2l} \zeta^{a^2} + \cdots + \theta^{(p-2)l} \zeta^{a^{p-2}} \\ &= \sum_{j=0}^{p-2} \theta^{jl} \zeta^{a^j} \end{aligned} \tag{21.8}$$

for $0 \leq l \leq p-2$. Then

$$\rho_a(\alpha_l) = \sum_{j=0}^{p-2} \theta^{jl} \zeta^{a^{j+1}} = \theta^{-l} \alpha_l$$

Therefore

$$\rho_a(\alpha_l^{p-1}) = (\theta^{-l}\alpha_l)^{p-1} = (\theta^{p-1})^{-l} \alpha_l^{p-1} = 1 \cdot \alpha_l^{p-1} = \alpha_l^{p-1}$$

so α_l^{p-1} is fixed by ρ_a, hence lies in $\mathbb{Q}(\theta)$. Say

$$\alpha_l^{p-1} = \beta_l \in \mathbb{Q}(\theta)$$

Therefore

$$\alpha_l = \sqrt[p-1]{\beta_l} \qquad (0 \leq l \leq p-2)$$

Recall (Exercise 21.5) the following property of roots of unity:

$$1+\theta^j+\theta^{2j}+\cdots+\theta^{(p-2)j}=\begin{cases} p-1 & \text{if } j=0 \\ 0 & \text{if } 1\leq j\leq p-2 \end{cases}$$

Therefore, from (21.8),

$$\begin{aligned} \zeta &= \tfrac{1}{p-1}[\alpha_0+\alpha_1+\cdots+\alpha_{p-2}] \\ &= \tfrac{1}{p-1}[\sqrt[p-1]{\beta_0}+\sqrt[p-1]{\beta_1}+\cdots+\sqrt[p-1]{\beta_{p-2}}] \end{aligned} \tag{21.9}$$

which expresses ζ by radicals over $\mathbb{Q}(\theta)$.

Now, θ is a primitive $(p-1)$th root of unity, so by induction θ is a radical expression over $\mathbb{Q}$ of maximum radical degree $\leq p-2$. Each β_l is also a radical expression over $\mathbb{Q}$ of maximum radical degree $\leq p-2$, since β_l is a polynomial in θ with rational coefficients. (Actually we can say more: if $p>2$ then $p-1$ is even, so the maximum radical degree is $\max(2,(p-1)/2)$. Note that when $p=3$ we require a square root, but $(p-1)/2=1$. See Exercise 21.3.)

Substituting the rational expressions in (21.9) we see that ζ is a radical expression over $\mathbb{Q}$ of maximum radical degree $\leq p-1$. (Again, this can be improved to $\max(2,(p-1)/2)$ for $p>2$, see Exercise 21.3.)

Therefore, in particular, (21.9) yields a genuine radical expression for ζ according to the definition, and the Vandermonde-Gauss Theorem is proved. □

21.5 Cyclotomic Polynomials

In order to fill in the technical gap we first need:

Theorem 21.4. *Any two primitive nth roots of unity in $\mathbb{C}$ have the same minimal polynomial over $\mathbb{Q}$.*

We proved this in Lemma 20.10 when n is prime, but the composite case is more difficult. Before starting on the proof, some motivation will be useful.

Consider the case $n=12$. Let $\zeta=e^{\pi i/6}$ be a primitive 12th root of unity. We can classify its powers ζ^j according to their minimal power d such that $(\zeta^j)^d=1$. That is, we consider when they are *primitive* dth roots of unity. It is easy to see that in this case the primitive dth roots of unity are:

$$\begin{array}{ll} d=1 & 1 \\ d=2 & \zeta^6(=-1) \\ d=3 & \zeta^4,\zeta^8(=\omega,\omega^2) \\ d=4 & \zeta^3,\zeta^9(=i,-i) \\ d=6 & \zeta^2,\zeta^{10}(=-\omega,-\omega^2) \\ d=12 & \zeta,\zeta^5,\zeta^7,\zeta^{11} \end{array}$$

We can factorise $t^{12}-1$ by grouping corresponding zeros:

$$\begin{aligned} t^{12}-1 = & (t-1)\times \\ & (t-\zeta^6)\times \\ & (t-\zeta^4)(t-\zeta^8)\times \\ & (t-\zeta^3)(t-\zeta^9)\times \\ & (t-\zeta^2)(t-\zeta^{10})\times \\ & (t-\zeta)(t-\zeta^5)(t-\zeta^7)(t-\zeta^{11}) \end{aligned}$$

which simplifies to

$$t^{12}-1=(t-1)(t+1)(t^2+t+1)(t^2+1)(t^2-t+1)F(t)$$

where

$$F(t)=(t-\zeta)(t-\zeta^5)(t-\zeta^7)(t-\zeta^{11})$$

whose explicit form is not immediately obvious. One way to work out $F(t)$ is to use trigonometry (Exercise 21.4). The other is to divide $t^{12}-1$ by all the other factors, which leads rapidly to

$$F(t)=t^4-t^2+1$$

If we let $\Phi_d(t)$ be the factor corresponding to primitive dth roots of unity, we have proved that

$$t^{12}-1=\Phi_1\Phi_2\Phi_3\Phi_4\Phi_6\Phi_{12}$$

Our computations show that every factor Φ_j lies in $\mathbb{Z}[t]$. In fact, it turns out that the factors are all *irreducible* over $\mathbb{Z}$. This is obvious for all factors except t^4-t^2+1, where it can be proved by considering the factorisation $(t-\zeta)(t-\zeta^5)(t-\zeta^7)(t-\zeta^{11})$ (Exercise 21.5).

This calculation generalises, as the following proof (eventually) shows.

Proof of Theorem 21.4. Factorise t^n-1 into monic irreducible factors in $\mathbb{Q}[t]$. By Corollary 3.18 these actually lie in $\mathbb{Z}[t]$. By the derivative test, t^n-1 has no multiple zeros. So each zero is a zero of exactly one of these factors, and that factor is its minimal polynomial. Hence two zeros of t^n-1 have the same minimal polynomial if and only if they are zeros of the same irreducible factor. Denote the factor of which an nth root of unity ε is a zero by $m_{[\varepsilon]}(t)$, where the square brackets remind us that different ε can be zeros of the same polynomial.

We claim that if p is any prime that does not divide n, then ε and ε^p have the same minimal polynomial. This step, which is not at all obvious, is the heart of the proof.

We prove the claim by contradiction. If it is false, then $m_{[\varepsilon^p]}(t)\neq m_{[\varepsilon]}(t)$. Define

$$k(t)=m_{[\varepsilon^p]}(t^p)\in\mathbb{Z}[t]$$

so

$$k(\varepsilon)=m_{[\varepsilon^p]}(\varepsilon^p)=0$$

Therefore $m_{[\varepsilon]}(t)$ divides $k(t)$ in $\mathbb{Z}[t]$, so there exists $q(t) \in \mathbb{Z}[t]$ such that

$$m_{[\varepsilon]}(t)q(t) = k(t)$$

Reduce coefficients modulo p as in Section 3.5. Using bars to denote images modulo p,

$$\bar{m}_{[\varepsilon]}(t)\bar{q}(t) = \bar{k}(t) = \bar{m}_{[\varepsilon^p]}(t^p) = (\bar{m}_{[\varepsilon^p]}(t))^p$$

since the Frobenius map is a monomorphism in characteristic p by Lemma 17.14. Therefore $\bar{m}_{[\varepsilon^p]}(t)$ and $\bar{m}_{[\varepsilon]}(t)$ have a common zero in some extension field of $\mathbb{Z}_p$, so that

$$\overline{t^n - 1} = \prod_{[\varepsilon]} \bar{m}_{[\varepsilon]}(t)$$

has a repeated zero in some extension field of $\mathbb{Z}_p$. By Lemma 9.13 (generalised), $\overline{t^n - 1}$ and its formal derivative have a common zero. However, the formal derivative of $\overline{t^n - 1}$ is $\bar{n}t^{n-1}$ and $\bar{n} \neq 0$ since $p \nmid n$. Now

$$\frac{t}{\bar{n}}(\bar{n}t^{n-1}) - \overline{t^n - 1} = \bar{1}$$

so no such common zero exists (that is, $\bar{n}t^{n-1}$ and $\overline{t^n - 1}$ are coprime). This contradiction shows that ε and ε^p have the same minimal polynomial.

It follows that ε and ε^u have the same minimal polynomial for every $u = p_1 \dots p_l$, where the p_j are primes not dividing n. These u are precisely the natural numbers that are prime to n, so modulo n they form the group of units $\mathbb{Z}_n^*$. However, the primitive nth roots of unity are precisely the elements ε^u for such u. □

Definition 21.5. The polynomial $\Phi_d(t)$ defined by

$$\Phi_n(t) = \prod_{a \in \mathbb{Z}_n, (a,n)=1} (t - \zeta^a) \tag{21.10}$$

is the nth *cyclotomic polynomial* over $\mathbb{C}$.

Corollary 21.6. *For all $n \in \mathbb{N}$, the polynomial $\Phi_n(t)$ lies in $\mathbb{Z}[t]$ and is monic and irreducible.*

21.6 Galois Group of $\mathbb{Q}(\zeta):\mathbb{Q}$

In Theorem 20.12 we described the Galois group of $\mathbb{Q}(\zeta):\mathbb{Q}$ when ζ is a primitive pth root of unity, p prime. We now generalise this result to the composite case.

Let $f(t) = t^n - 1 \in \mathbb{Q}[t]$. The zeros in $\mathbb{C}$ are $1, \zeta, \zeta^1, \dots, \zeta^{n-1}$ where $\zeta = e^{2\pi i/n}$ is a primitive nth root of unity. The splitting field of f is clearly $\mathbb{Q}(\zeta)$. Theorem 9.9 implies that the extension $\mathbb{Q}(\zeta):\mathbb{Q}$ is normal. By Proposition 9.14 it is separable.

We will need:

Definition 21.7. The *group of units* $\mathbb{Z}_n^*$ of $\mathbb{Z}_n$ consists of the elements $a \in \mathbb{Z}_n$ such that $1 \leq a \leq n$ and a is prime to n, under the operation of multiplication.

The order of this group is given by an important number-theoretic function:

Definition 21.8. The *Euler function* $\phi(n)$ is the number of integers a, with $1 \leq a \leq n-1$, such that a is prime to n.

Definition 21.8 implies immediately that the order of $\mathbb{Z}_n^*$ is equal to $\phi(n)$.
The Euler function $\phi(n)$ has numerous interesting properties. In particular

$$\phi(p^k) = (p-1)p^{k-1}$$

if p is prime, and

$$\phi(r)\phi(s) = \phi(rs)$$

when r,s are coprime. See Exercise 12.4.
We can now prove:

Theorem 21.9. (1) *The Galois group* $\Gamma(\mathbb{Q}(\zeta):\mathbb{Q})$ *consists of the* $\mathbb{Q}$*-automorphisms* ψ_j *defined by*

$$\psi_j(\zeta) = \zeta^j$$

where $0 \leq j \leq n-1$ *and* j *is prime to* n.

(2) $\Gamma(\mathbb{Q}(\zeta):\mathbb{Q})$ *is isomorphic to* $\mathbb{Z}_n^*$ *and in particular is an abelian group.*

(3) *Its order is* $\phi(n)$.

(4) *If* n *is prime,* $\mathbb{Z}_n^*$ *is cyclic.*

Proof. (1) Let $\gamma \in \Gamma(\mathbb{Q}(\zeta):\mathbb{Q})$. Since $\gamma(\zeta)$ is a zero of $t^n - 1$, $\gamma = \psi_j$ for some j.
If j and n have a common factor $d > 1$ then ψ_j is not onto and hence not a $\mathbb{Q}$-automorphism.
If j and n are coprime, there exist integers a,b such that $aj + bn = 1$. Then

$$\zeta = \zeta^{aj+bn} = \zeta^{aj}\zeta^{bn} = (\zeta^j)^a$$

so ζ lies in the image of ψ_j. It follows that ψ_j is a $\mathbb{Q}$-automorphism.
(2) Clearly $\psi_j\psi_k = \psi_{jk}$, so the map $\psi_j \mapsto j$ is an isomorphism from $\Gamma(\mathbb{Q}(\zeta):\mathbb{Q})$ to $\mathbb{Z}_n^*$.
(3) $|\Gamma(\mathbb{Q}(\zeta):\mathbb{Q})| = |\mathbb{Z}_n^*| = \phi(n)$.
(4) This follows from Corollary 19.9. □

21.7 The Technical Lemma

We can now fill in the technical gap in the proof of the Vandermonde-Gauss Theorem in Section 21.4.

Theorem 21.10. *Let K be the splitting field of $\Phi_n(t)$ over $\mathbb{Q}$. Then the Galois group of the extension $K : \mathbb{Q}$ is isomorphic to the group of units $\mathbb{Z}_n^*$ of the ring $\mathbb{Z}_n$.*

Proof. The zeros of $\Phi_n(t)$ in $\mathbb{C}$ are powers ζ^a of a primitive nth root of unity ζ, where a ranges through the integers modulo n that are prime to n. The result is then a direct consequence of Theorem 21.9. □

We can now give the

Proof of Theorem 21.3. Since $\mathbb{Q}(\zeta) : \mathbb{Q}$ is normal, every automorphism of $\mathbb{Q}(\theta\zeta)$ over $\mathbb{Q}(\theta)$ carries $\mathbb{Q}(\zeta)$ to itself. Therefore restriction of automorphisms gives a homomorphism

$$\psi : \Gamma(\mathbb{Q}(\theta\zeta) : \mathbb{Q}(\theta)) \to \Gamma(\mathbb{Q}(\zeta) : \mathbb{Q})$$

Now $\Gamma(\mathbb{Q}(\zeta) : \mathbb{Q})$ is cyclic of order $p-1$, so it suffices to prove that ψ is an isomorphism. Since $\mathbb{Q}(\theta\zeta) = \mathbb{Q}(\theta)(\zeta)$, every automorphism of this field over $\mathbb{Q}(\theta)$ is determined by its effect on ζ. Therefore distinct automorphisms induce distinct automorphisms of $\mathbb{Q}(\zeta)$, showing that ψ is one-to-one.

To show it is onto, it suffices to prove that $\Gamma(\mathbb{Q}(\theta\zeta) : \mathbb{Q}(\theta))$ and $\Gamma(\mathbb{Q}(\zeta) : \mathbb{Q})$ have the same order.

Denote a primitive nth root of unity by ζ_n. By Theorem 21.10, for every n the order of $\Gamma(\mathbb{Q}(\zeta_n) : \mathbb{Q}) = |\mathbb{Z}_n^*| = \phi(n)$. The tower law implies that if $0 < r, s \in \mathbb{N}$ then

$$|\Gamma(\mathbb{Q}(\zeta_{rs}) : \mathbb{Q}(\zeta_s)| = \phi(rs)/\phi(s)$$

But when r, s are coprime, $\phi(rs) = \phi(r)\phi(s)$, so $\phi(rs)/\phi(s) = \phi(r) = |\Gamma(\mathbb{Q}(\zeta_r) : \mathbb{Q})|$. Set $r = p, s = p-1$ to get what we require. □

21.8 More on Cyclotomic Polynomials

It seems a shame to stop without saying a little more about the cyclotomic polynomials, because they are fascinating.

Theorem 21.10 shows that the cyclotomic polynomial $\Phi_n(t)$ is intimately associated with the ring $\mathbb{Z}_n$ and its group of units $\mathbb{Z}_n^*$, which we discussed briefly in Chapter 3. In particular, the order of this group is

$$|\mathbb{Z}_n^*| = \phi(n)$$

where ϕ is the Euler function, Definition 21.8, so $\phi(n)$ is the number of integers a, with $1 \leq a \leq n-1$, such that a is prime to n.

The most basic property of the cyclotomic polynomials is the identity

$$t^n - 1 = \prod_{d|n} \Phi_d(t) \tag{21.11}$$

which is a direct consequence of their definition. We can use this identity recursively to compute $\Phi_n(t)$. Thus

$$\Phi_1(t) = t - 1$$

so

$$t^2 - 1 = \Phi_2(t)\Phi_1(t)$$

which implies that

$$\Phi_2(t) = \frac{t^2-1}{\Phi_1(t)} = \frac{t^2-1}{t-1} = t+1$$

Similarly

$$\Phi_3(t) = \frac{t^3-1}{t-1} = t^2+t+1$$

and

$$\Phi_4(t) = \frac{t^4-1}{(t-1)(t+1)} = t^2+1$$

and so on. Table 21.8 shows the first 15 cyclotomic polynomials, computed in this manner. A curiosity of the table is that the coefficients of Φ_n always seem to be $0, 1$, or -1. Is this always true? See Exercise 21.11.

n	$\Phi_n(t)$
1	$t-1$
2	$t+1$
3	t^2+t+1
4	t^2+1
5	$t^4+t^3+t^2+t+1$
6	t^2-t+1
7	$t^6+t^5+t^4+t^3+t^2+t+1$
8	t^4+1
9	t^6+t^3+1
10	$t^4-t^3+t^2-t+1$
11	$t^{10}+t^9+t^8+t^7+t^6+t^5+t^4+t^3+t^2+t+1$
12	t^4-t^2+1
13	$t^{12}+t^{11}+t^{10}+t^9+t^8+t^7+t^6+t^5+t^4+t^3+t^2+t+1$
14	$t^6-t^5+t^4-t^3+t^2-t+1$
15	$t^8-t^7+t^5-t^4+t^3-t+1$

21.9 Constructions Using a Trisector

For a final flourish, we apply our results to the construction of regular polygons when an angle-trisector is permitted, as well as the traditional ruler and compass. The results are instructive, amusing, and slightly surprising. For example, the regular 7-gon can now be constructed. It is not immediately clear why the angle $\frac{2\pi}{7}$ arises from trisections. Other regular polygons, such as the 13-gon and 19-gon, also become constructible. On the other hand, the regular 11-gon still cannot be constructed.

The main point is the link between trisection and irreducible cubic equations. The trigonometric solution of cubics, Exercise 1.8, shows that an angle-trisector can be used to solve some cubic equations: those in the 'irreducible case', with three distinct real roots. Specifically, we use the trigonometric identity $\cos 3\theta = 4\cos^3\theta - 3\cos\theta$ to solve the cubic equation $t^3 + pt + q = 0$ when $27q^2 + 4p^3 < 0$. This is the condition for three distinct real roots. The method is as follows.

The inequality $27pq^2 + 4p^3 < 0$ implies that $p < 0$, so we can find a, b such that $p = -3a^2, q = -a^2b$. The cubic becomes

$$t^3 - 3a^2t = a^2b$$

and the inequality becomes $a > |b|/2$.

Substitute $t = 2a\cos\theta$, and observe that

$$t^3 - 3a^2t = 8a^3\cos^3\theta - 6a^3\cos\theta = 2a^3\cos 3\theta$$

The cubic thus reduces to

$$\cos 3\theta = \frac{b}{2a}$$

which we can solve using $\cos^{-1}$ because $|\frac{b}{2a}| \leq 1$, getting

$$\theta = \frac{1}{3}\cos^{-1}\frac{b}{2a}$$

There are three possible values of θ, the other two being obtained by adding $\frac{2\pi}{3}$ or $\frac{4\pi}{3}$. Finally, eliminate θ to get

$$t = 2a\cos\left(\frac{1}{3}\cos^{-1}\frac{b}{2a}\right)$$

where $a = \sqrt{\frac{-p}{3}}, b = \frac{3q}{p}$.

Conversely, solving cubics with real coefficients and three distinct real roots lets us trisect angles. So when a trisector is made available, the constructible numbers now lie in a series of extensions, starting with $\mathbb{Q}$, such that each successive extension has degree 2 or 3.

The use of a trisector motivates a generalisation of Fermat primes, named after the mathematician James Pierpont.

Definition 21.11. A *Pierpont prime* is a prime p of the form

$$p = 2^a 3^b + 1$$

where $a \geq 1, b \geq 0$.

(Here we exclude $a = 0$ because in this case $2^a 3^b + 1 = 3^b + 1$ is even.)

The Pierpont primes up to 100 are 3, 5, 7, 13, 17, 19, 37, 73, and 97. So they appear to be more common than Fermat primes, a point to which we return later.

Andrew Gleason (1988) proved the following theorem characterising those regular n-gons that can be constructed when the traditional instruments of Euclid are supplemented by an angle-trisector. He also gave explicit constructions of that kind for the regular 7-gon and 13-gon.

Theorem 21.12. *The regular n-gon can be constructed using ruler, compass, and trisector, if and only if n is of the form $2^r 3^s p_1 \cdots p_k$ where $r, s \geq 0$ and the p_j are distinct Pierpont primes > 3.*

Proof. First, suppose that the regular n-gon can be constructed using ruler, compass, and trisector. As remarked above, this implies that the primitive nth root of unity $\zeta = e^{2\pi i/n}$ lies in the largest field in some series of extensions, which starts with $\mathbb{Q}$, such that each successive extension has degree 2 or 3. Therefore

$$[\mathbb{Q}(\zeta) : \mathbb{Q}] = 2^c 3^d$$

for $c, d \in \mathbb{N}$.

The degree $[\mathbb{Q}(\zeta) : \mathbb{Q}]$ equals $\phi(n)$, where ϕ is the Euler function. This is the degree of the cyclotomic polynomial $\Phi_n(t)$, which is irreducible over $\mathbb{Q}$. Therefore a necessary condition for constructibility with ruler, compass, and trisector is $\phi(n) = 2^a 3^b$ for $a, b \in \mathbb{N}$. What does this imply about n?

Write n as a product of distinct prime powers $p_j^{m_j}$. Then $\phi(p_j^{m_j})$ must be of the form $2^{a_j} 3^{b_j}$. Since $\phi(p^m) = (p-1)p^{m-1}$ when p is prime, we require $(p_{j-1}p_j^{m_j-1}$ to be of the form $2^{a_j} 3^{b_j}$.

Either $m_j = 1$ or $p_j = 2, 3$. If $p_j = 2$ then $\phi(p_j^{m_j}) = 2^{m_j - 1}$ and any m_j can occur. If $p_j = 3$ then $\phi(p_j^{m_j}) = 2 \cdot 3^{m_j - 1}$ and again any power of 3 can occur. Otherwise $m_j = 1$ so $\phi(p_j^{m_j}) = \phi(p_j) = p_j - 1$, and $p_j = 2^{a_j} 3^{b_j} + 1$. Thus p_j is a Pierpont prime.

We have now proved the theorem in one direction: in order for the regular n-gon to be constructible by ruler, compass, and trisector, n must be a product of powers of 2, powers of 3, and distinct Pierpont primes > 3.

We claim that the converse is also true.

The proof is a simple application of Galois theory. Let $p = 2^a 3^b + 1$ be an odd prime. Let $\zeta = e^{2\pi i/p}$. Then $[\mathbb{Q}(\zeta) : \mathbb{Q}] = p - 1 = 2^a 3^b$. The extension $\mathbb{Q}(\zeta) : \mathbb{Q}$ is normal and separable, so the Galois correspondence is a bijection, and the Galois group $\Gamma = \Gamma(\mathbb{Q}(\zeta) : \mathbb{Q})$ has order $m = 2^a 3^b$. By Theorem 21.10 it is abelian, isomorphic to $\mathbb{Z}_m^*$. Therefore it has a series of normal subgroups

$$1 = \Gamma_0 \lhd \Gamma_1 \lhd \cdots \lhd \Gamma_r = \Gamma$$

where each factor Γ_{j+1}/Γ_j is isomorphic either to $\mathbb{Z}_2$ or $\mathbb{Z}_3$. In fact, $r = a+b$.

Let

$$\theta = \zeta + \zeta^1 = \zeta + \zeta^{p-1} = \zeta + \overline{\zeta} = 2\cos 2\pi/p$$

where the bar indicates complex conjugate. Then $\theta \in \mathbb{R}$. Consider the tower of subfields

$$\mathbb{Q} \subseteq \mathbb{Q}(\theta) \subseteq \mathbb{Q}(\zeta)$$

Clearly $\mathbb{Q}(\theta) \subseteq \mathbb{R}$. We have $\zeta + \zeta^{-1} = \theta, \zeta \cdot \zeta^{-1} = 1$, so ζ and ζ^{-1} are the zeros of $t^2 - \theta t + 1$ over $\mathbb{Q}(\theta)$. Therefore $[\mathbb{Q}(\zeta) : \mathbb{Q}(\theta)] \leq 2$, but $\zeta \notin \mathbb{R} \supseteq \mathbb{Q}(\theta)$ so $[\mathbb{Q}(\zeta) : \mathbb{Q}(\theta)] = 2$.

The group Λ of order 2 generated by complex conjugation is a subgroup of Γ, and it is a normal subgroup since Γ is abelian. We claim that the fixed field $\Lambda^\dagger = \mathbb{Q}(\theta) = \mathbb{Q}(\zeta) \cap \mathbb{R}$. We have $\mathbb{Q}(\zeta) \subseteq \mathbb{R}$ so $\mathbb{Q}(\zeta) \subseteq \Lambda^\dagger$. Since $[\mathbb{Q}(\zeta) : \mathbb{Q}(\theta)] = 2$ the only subfield properly containing $\mathbb{Q}(\zeta)$ is $\mathbb{Q}(\theta)$, and this is not fixed by Λ. Therefore $\mathbb{Q}(\zeta) = \Lambda^\dagger$. (It is easy to see that in fact, $\mathbb{Q}(\theta) = \mathbb{Q}(\zeta) \cap \mathbb{R}$.)

Therefore the Galois group of $\mathbb{Q}(\theta) : \mathbb{Q}$ is isomorphic to the quotient group $\Delta = \Gamma/\Lambda$, so it is cyclic of order $m/2 = 2^{a-1}3^b$. It has a series of normal subgroups

$$1 = \Delta_0 \lhd \Delta_1 \lhd \cdots \lhd \Delta_{r-1} = \Delta$$

where each factor Δ_{j+1}/Δ_j is isomorphic either to $\mathbb{Z}_2$ or $\mathbb{Z}_3$.

The corresponding fixed subfields $K_j = \Delta_j^\dagger$ form a tower

$$\mathbb{Q}(\theta) = K_0 \supseteq K_1 \supseteq \cdots \supseteq K_{r-1} = \mathbb{Q}$$

and each degree $[K_j : K_{j+1}]$ is either 2 or 3. So K_j can be obtained from K_{j+1} by adjoining either:

a root of a quadratic over K_{j+1}, or

a root of an irreducible cubic over K_{j+1} with all three roots real

(the latter because $\mathbb{Q}(\theta) \subseteq \mathbb{R}$).

In the quadratic case, any $z \in K_j$ can be constructed from K_{j+1} by ruler and compass. In the cubic case, any $z \in K_j$ can be constructed from K_{j+1} by trisector (plus ruler and compass for field operations). By backwards induction from $K_{r-1} = \mathbb{Q}$, we see that any element of K_0 can be constructed from $\mathbb{Q}$ by ruler, compass, and trisector. Finally, any element of $\mathbb{Q}(\zeta)$ can be constructed from $\mathbb{Q}$ by ruler, compass, and trisector. In particular, ζ can be so constructed, which gives a construction for a regular p-gon. □

This is a remarkable result, since at first sight there is no obvious link between regular polygons with (say) 7, 13, or 19 sides and angle-trisection. They appear to need division of an angle by 7, 13, or 19. So we give further detail for the first two cases, the 7-gon and the 13-gon.

$p = 7$: Let $\zeta = e^{2\pi i/7}$. Recall the basic relation

$$1 + \zeta + \zeta^2 + \zeta^3 + \zeta^4 + \zeta^5 + \zeta^6 = 0 \qquad (21.12)$$

Define

$$r_1 = \zeta + \zeta^6 = 2\cos\frac{2\pi}{7} \in \mathbb{R}$$
$$r_2 = \zeta^2 + \zeta^5 = 2\cos\frac{4\pi}{7} \in \mathbb{R}$$
$$r_3 = \zeta^3 + \zeta^4 = 2\cos\frac{6\pi}{7} \in \mathbb{R}$$

Compute the elementary symmetric functions of the r_j. By (21.12)

$$r_1 + r_2 + r_3 = -1$$

Next,

$$\begin{aligned} r_1r_2r_3 &= (\zeta+\zeta^6)(\zeta^2+\zeta^5)(\zeta^3+\zeta^4) \\ &= \zeta^6+\zeta^0+\zeta^2+\zeta^3+\zeta^4+\zeta^5+\zeta^0+\zeta \\ &= 1+1-1=1 \end{aligned}$$

Finally,

$$\begin{aligned} r_1r_2+r_1r_3+r_2r_3 &= (\zeta+\zeta^6)(\zeta^2+\zeta^5)+(\zeta+\zeta^6)(\zeta^3+\zeta^4)+(\zeta^2+\zeta^5)(\zeta^3+\zeta^4) \\ &= \zeta^3+\zeta^6+\zeta+\zeta^4+\zeta^4+\zeta^5+\zeta^2+\zeta^3+\zeta^5+\zeta^6+\zeta+\zeta^2 \\ &= -2 \end{aligned}$$

Therefore the r_j are roots of the cubic $t^3+2t^2+t-1=0$. This is irreducible (exercise) and the roots r_j are real. So they can be constructed using a trisector (plus ruler and compass for field operations). We omit details; an explicit construction can be found in Gleason (1988) and Conway and Guy (1996) page 200.

$p = 13$: Let $\zeta = e^{2\pi i/13}$. Recall the basic relation

$$1+\zeta+\zeta^2+\cdots+\zeta^{12} = 0 \tag{21.13}$$

Define $r_j = \zeta^j + \zeta^{-j} = 2\cos\frac{2\pi j}{13}$ for $1 \leq j \leq 6$.

It turns out that 2 is primitive root modulo 13. That is, the powers of 2 (mod 13) are, in order,

1 2 4 8 3 6 12 11 9 5 10 7

and then repeat: these are all the nonzero elements of $\mathbb{Z}_{13}$.

Add powers of ζ corresponding to every third number in this sequence:

$$\begin{aligned} s_1 &= \zeta+\zeta^8+\zeta^{12}+\zeta^5 = r_1+r_5 \\ s_2 &= \zeta^2+\zeta^3+\zeta^{11}+\zeta^{10} = r_2+r_3 \\ s_3 &= \zeta^4+\zeta^6+\zeta^9+\zeta^7 = r_4+r_6 \end{aligned}$$

Tedious but routine calculations show that the s_j are the three roots of the cubic

$$t^3+t^2-4t+1=0$$

which is irreducible (exercise) and has all roots real. Therefore the s_j can be constructed using trisector, ruler, and compass.

Then, for example,

$$\begin{aligned} r_1 + r_5 &= s_1 \\ r_1 r_5 &= (\zeta + \zeta^{12})(\zeta^5 + \zeta^8) \\ &= \zeta^6 + \zeta^9 + \zeta^4 + \zeta^7 = s_3 \end{aligned}$$

so r_1, r_5 are roots of a quadratic over $\mathbb{Q}(s_1, s_2, s_3)$. The same goes for the other pairs of r_j. Therefore we can construct the r_j by ruler and compass from the s_j. Finally, we can construct ζ from the r_j by solving a quadratic, hence by ruler and compass.

An explicit construction can again be found in Gleason (1988) and Conway and Guy (1996) page 200.

Earlier, I said that the Pierpont primes $p = 2^a 3^b + 1$ form a much richer set than the Fermat primes. It is worth expanding on that statement. It is generally believed that the only Fermat primes are the known ones, 2, 3, 5, 17, 257, and 65537, though this has not been proved. In contrast, Gleason (1988) conjectured that Pierpont primes are so common that there should be *infinitely many*; he suggested that there should be about $9k$ of them less than 10^k. More formally, the number of Pierpont primes less than N should be asymptotic to a constant times $\log N$. This conjecture remains open, but with modern computer algebra it is easy to explore larger values. For example, a quick, unsystematic search turned up the Pierpont prime

$$\begin{aligned} 2^{148} 3^{95} + 1 = {} & 756\,760\,676\,272\,923\,020\,551\,154\,471\,073 \\ & 240\,459\,834\,492\,063\,891\,235\,892\,290\,277 \\ & 703\,256\,956\,240\,171\,581\,788\,957\,704\,193 \end{aligned}$$

with 90 digits. There are 789 Pierpont primes up to 10^{100}. Currently, the largest known Pierpont prime is $3 \times 2^{7033641} + 1$, proved prime by Michael Herder in 2011.

EXERCISES

21.1 Prove that, in the notation of Section 21.4,

$$\zeta^j = \frac{1}{p-1}\left(\sum_{l=0}^{p-2} \theta^{-jl} \alpha_l\right)$$

21.2 Prove that $\Phi_{24}(t) = t^8 - t^4 + 1$.

21.3 Show that the zeros of the dth cyclotomic polynomial can be expressed by radicals of degree at most $\max(2, (d-1)/2)$. (The 2 occurs because of the case $d = 3$.)

21.4 Use trigonometric identities to prove directly from the definition that $\Phi_{12}(t) = t^4 - t^2 + 1$.

21.5 Prove that $\Phi_{12}(t)$ is irreducible over $\mathbb{Q}$.

21.6 Prove that if θ is a primitive $(p-1)$th root of unity, then

$$1+\theta^j+\theta^{2j}+\cdots+\theta^{(p-2)j} = \begin{cases} p-1 & \text{if } j=0 \\ 0 & \text{if } j \leq l \leq p-2 \end{cases}$$

21.7 Prove that the coefficients of $\Phi_p(t)$ are all contained in $\{-1,0,1\}$ when p is prime.

21.8 Prove that the coefficients of $\Phi_{p^k}(t)$ are all contained in $\{-1,0,1\}$ when p is prime and $k > 1$.

21.9 If m is odd, prove that $\Phi_{2m}(t) = \Phi_m(-t)$, and deduce that the coefficients of $\Phi_{2p^k}(t)$ are contained in $\{-1,0,1\}$ when p is an odd prime and $k > 1$.

21.10 If p,q are distinct odd primes, find a formula for $\Phi_{pq}(t)$ and deduce that the coefficients of $\Phi_{pq}(t)$ are all contained in $\{-1,0,1\}$.

21.11 Relate $\Phi_{pa}(t)$ and $\Phi_{p^k a}(t)$ when a,p are odd, p is prime, p and a are coprime, and $k > 1$. Deduce that if the coefficients of $\Phi_{pa}(t)$ are all contained in $\{-1,0,1\}$, so are those of $\Phi_{p^k a}(t)$.

21.12 Show that the smallest n such that the coefficients of $\Phi_m(t)$ might *not* all be contained in $\{-1,0,1\}$ is $n = 105$. If you have access to symbolic algebra software, or have an evening to spare, lots of paper, and are willing to be very careful checking your arithmetic, compute $\Phi_{105}(t)$ and see if some coefficient is not contained in $\{-1,0,1\}$.

21.13 Let $\phi(n)$ be the Euler function. Prove that

$$\phi(p^k) = (p-1)p^{k-1}$$

if p is prime, and

$$\phi(r)\phi(s) = \phi(rs)$$

when r,s are coprime. Deduce a formula for $\phi(n)$ in terms of the prime factorisation of n.

12.14 Prove that

$$\phi(n) = n \prod_{p \text{ prime},\ p|n} \left(1 - \frac{1}{p}\right)$$

12.15 If a is prime to n, where both are integers, prove that $a^{\phi(n)} \equiv 0 \pmod{n}$.

12.16 Prove that for any $m \in \mathbb{N}$ the equation $\phi(n) = m$ has only finitely many solutions n. Find examples to show that there may be more than one solution.

12.17 Experiment, make an educated guess, and prove a formula for $\sum_{d|n} \phi(d)$.

12.18 If n is odd, prove that $\phi(4n) = 2\phi(n)$.

12.19 Check that

$$\begin{aligned}
1+2 &= \frac{3}{2}\phi(3) \\
1+3 &= \frac{4}{2}\phi(4) \\
1+2+3+4 &= \frac{5}{2}\phi(5) \\
1+5 &= \frac{6}{2}\phi(6) \\
1+2+3+4+5+6 &= \frac{7}{2}\phi(7)
\end{aligned}$$

What is the theorem? Prove it.

12.20* Prove that if $g \in \mathbb{Z}_{24}^*$ then $g^2 = 1$, so g has order 2 or is the identity. Show that 24 is the largest value of n for which every non-identity element of $\mathbb{Z}_n^*$ has order 2. Which are the others?

21.21 Outline how to construct a regular 19-gon using ruler, compass, and trisector, along the lines discussed for the 7-gon and 13-gon.

21.22 Extend the list of Pierpont primes up to 1000.

21.23 If you have access to a computer algebra package, use it to extend the list of Pierpont primes up to 1,000,000.

21.24 (1) Prove that $2^a 3^b + 1$ is composite if a and b have an odd common factor greater than 1.

(2) Prove that $2^a 3^b + 1$ is divisible by 5 if and only if $a - b \equiv 2 \pmod 4$.

(3) Prove that $2^a 3^b + 1$ is divisible by 7 if and only if $a + 2b \equiv 0 \pmod 3$.

(4) Find similar necessary and sufficient conditions for $2^a 3^b + 1$ to be divisible by 11, 13, 17, 19.

(5) Prove that $2^a 3^b + 1$ is never divisible by 23.

[*Hint*: For (2, 3, 4, 5) prove that if p is prime then $2^a 3^b + 1 \equiv 0 \pmod p$ if and only if $2^a + 3^{-b} \equiv 0 \pmod p$, and look at powers of 2 and 3 modulo p.]

21.25 Mark the following true or false.

(a) Every root of unity in $\mathbb{C}$ has a expression by genuine radicals.

(b) A primitive 11th root of unity in $\mathbb{C}$ can be expressed in terms of rational numbers using only square roots and fifth roots.

(c) Any two primitive roots of unity in $\mathbb{C}$ have the same minimal polynomial over $\mathbb{Q}$.

(d) The Galois group of $\Phi_n(t)$ over $\mathbb{Q}$ is cyclic for all n.

(e) The Galois group of $\Phi_n(t)$ over $\mathbb{Q}$ is abelian for all n.

(f) The coefficients of any cyclotomic polynomial are all equal to $0, \pm 1$.

(g) The regular 483729409-gon can be constructed using ruler, compass, and trisector. (*Hint*: This number is prime, and you may assume this without further calculation.)

Chapter 22

Calculating Galois Groups

In order to apply Galois theory to specific polynomials, it is necessary to compute the corresponding Galois group. This was the weak point in the memoir that Galois submitted to the French Academy of Sciences, as Poisson and Lacroix pointed out in their referees' report.

However, the computation is possible—at least in principle. It becomes practical only with modern computers. It is neither simple nor straightforward, and until now we have emulated Galois and strenuously avoided it. Instead we have either studied special equations whose Galois group is relatively easy to find (I did say 'relatively'), resorted to special tricks, or obtained results that require only partial knowledge of the Galois group. The time has now come to face up squarely to the problem. This chapter contains relatively complete discussions for cubic and quartic polynomials. It also provides a general algorithm for equations of any degree, which is of theoretical importance but is too cumbersome to use in practice. More practical methods do exist, but they go beyond the scope of this book, see Soicher and McKay (1985) and the two references for Hulpke (Internet). The packages Maple and GAP can compute Galois groups for relatively small degrees.

22.1 Transitive Subgroups

We know that the Galois group $\Gamma(f)$ of a polynomial f with no multiple zeros of degree n is (isomorphic to) a subgroup of the symmetric group $\mathbb{S}_n$. In classical terminology, $\Gamma(f)$ permutes the roots of the equation $f(t) = 0$. Renumbering the roots changes $\Gamma(f)$ to some conjugate subgroup of $\mathbb{S}_n$, so we need consider only the conjugacy classes of subgroups. However, $\mathbb{S}_n$ has rather a lot of conjugacy classes of subgroups, even for moderate n (say $n \geq 6$). So the list of cases rapidly becomes unmanageable.

However, if f is irreducible (which we may always assume when solving $f(t) = 0$) we can place a fairly stringent restriction on the subgroups that can occur. To state it we need:

Definition 22.1. Let G be a permutation group; that is, a subgroup of the group of all permutations on a set S. We say that G is *transitive* (or *transitive on S*) if for all $s,t \in S$ there exists $\gamma \in G$ such that $\gamma(s) = t$.

To prove G transitive it is enough to show that for some fixed $s_0 \in S$, and any $s \in S$, there exists $\gamma \in G$ such that $\gamma(s_0) = s$. For if this holds, then given $t \in S$ there also exists $\delta \in G$ such that $\delta(s_0) = t$, so $(\delta\gamma^{-1})(s) = t$.

Examples 22.2. (1) The Klein four-group $\mathbb{V}$ is transitive on $\{1, 2, 3, 4\}$. The element 1 is mapped to:

1 by the identity

2 by (12)(34)

3 by (13)(24)

4 by (14)(23)

(2) The cyclic group generated by $\alpha = (1234)$ is transitive on $\{1, 2, 3, 4\}$. In fact, α^i maps 1 to i for $i = 1, 2, 3, 4$.
(3) The cyclic group generated by $\beta = (123)$ is not transitive on $\{1, 2, 3, 4\}$. There is no power of β that maps 1 to 4.

Proposition 22.3. *The Galois group of an irreducible polynomial f is transitive on the set of zeros of f.*

Proof. If α and β are two zeros of f then they have the same minimal polynomial, namely f. By Theorem 17.4 and Proposition 11.4 there exists γ in the Galois group such that $\gamma(\alpha) = \beta$. □

Listing the (conjugacy classes of) transitive subgroups of $\mathbb{S}_n$ is not as formidable as listing all (conjugacy classes of) subgroups. The transitive subgroups, up to conjugacy, have been classified for low values of n by Conway, Hulpke, and MacKay (1998). The GAP data library

http://www.gap-system.org/Datalib/trans.html

contains all transitive subgroups of $\mathbb{S}_n$ for $n \leq 30$. The methods used can be found in Hulpke (1996). There is only one such subgroup when $n = 2$, two when $n = 3$, and five when $n = 4, 5$. The magnitude of the task becomes apparent when $n = 6$: in this case there are 16 transitive subgroups up to conjugacy. The number drops to seven when $n = 7$; in general prime n lead to fewer conjugacy classes of transitive subgroups than composite n of similar size.

22.2 Bare Hands on the Cubic

As motivation, we begin with a cubic equation over $\mathbb{Q}$, where the answer can be obtained by direct 'bare hands' methods. Consider a cubic polynomial

$$f(t) = t^3 - s_1t^2 + s_2t - s_3 \in \mathbb{Q}[t]$$

The coefficient s_j are the elementary symmetric polynomials in the zeros $\alpha_1, \alpha_2, \alpha_3$, as in Section 18.2. If f is reducible then the calculation of its Galois group is easy: it is the trivial group, which we denote by 1, if all zeros are rational, and $\mathbb{S}_2$ otherwise. Thus we may assume that f is irreducible over $\mathbb{Q}$.

Let Σ be the splitting field of f,

$$\Sigma = \mathbb{Q}(\alpha_1,\ \alpha_2,\ \alpha_3)$$

By Proposition 22.3 the Galois group of f is a transitive subgroup of $\mathbb{S}_3$, hence is either $\mathbb{S}_3$ or $\mathbb{A}_3$. Suppose for argument's sake that it is $\mathbb{A}_3$. What does this imply about the zeros $\alpha_1,\ \alpha_2,\ \alpha_3$? By the Galois correspondence, the fixed field $\mathbb{A}_3^\dagger$ of $\mathbb{A}_3$ is $\mathbb{Q}$. Now $\mathbb{A}_3$ consists of the identity, and the two cyclic permutations (123) and (132). Any expression in $\alpha_1,\ \alpha_2,\ \alpha_3$ that is invariant under cyclic permutations must therefore lie in $\mathbb{Q}$. Two obvious expressions of this type are

$$\phi = \alpha_1^2\alpha_2 + \alpha_2^2\alpha_3 + \alpha_3^2\alpha_1$$

and

$$\psi = \alpha_1^2\alpha_3 + \alpha_2^2\alpha_1 + \alpha_3^2\alpha_2$$

Indeed it can, with a little effort, be shown that

$$\mathbb{A}_3^\dagger = \mathbb{Q}(\phi, \psi)$$

(see Exercise 22.3). In other words, the Galois group of f is $\mathbb{A}_3$ if and only if ϕ and ψ are rational.

This is useful only if we can calculate ϕ and ψ, which we now do. Because $\mathbb{S}_3$ is generated by $\mathbb{A}_3$ together with the transposition (12), which interchanges ϕ and ψ, it follows that both $\phi + \psi$ and $\phi\psi$ are symmetric polynomials in $\alpha_1,\ \alpha_2,\ \alpha_3$. By Theorem 18.10 they are therefore polynomials in s_1, s_2, and s_3. We can compute these polynomials explicitly, as follows. We have

$$\phi + \psi = \sum_{i \neq j} \alpha_i^2 \alpha_j$$

Compare this with

$$s_1 s_2 = (\alpha_1 + \alpha_2 + \alpha_3)(\alpha_1\alpha_2 + \alpha_2\alpha_3 + \alpha_3\alpha_1) = \sum_{i \neq j} \alpha_i^2\alpha_j + 3\alpha_1\alpha_2\alpha_3$$

Since $\alpha_1\alpha_2\alpha_3 = s_3$ we deduce that

$$\phi + \psi = s_1 s_2 - 3s_3$$

Similarly

$$\begin{aligned}\phi\psi &= \alpha_1^4\alpha_2\alpha_3 + \alpha_2^4\alpha_3\alpha_1 + \alpha_3^4\alpha_1\alpha_2 + \alpha_1^3\alpha_2^3 + \alpha_2^3\alpha_3^3 + \alpha_3^3\alpha_1^3 + 3\alpha_1^2\alpha_2^2\alpha_3^2 \\ &= s_3(\alpha_1^3 + \alpha_2^3 + \alpha_3^3) + 3s_3^2 + \sum_{i<j} \alpha_i^3\alpha_j^3\end{aligned}$$

Now

$$\begin{aligned} s_1^3 &= (\alpha_1+\alpha_2+\alpha_3)^3 \\ &= (\alpha_1^3+\alpha_2^3+\alpha_3^3)+3\sum_{i\neq j}\alpha_i^2\alpha_j+6\alpha_1\alpha_2\alpha_3 \end{aligned}$$

so that

$$\alpha_1^3+\alpha_2^3+\alpha_3^3 = s_1^3-6s_3-3(s_1s_2-3s_3)$$

Moreover,

$$\begin{aligned} s_2^3 &= (\alpha_1\alpha_2+\alpha_2\alpha_3+\alpha_3\alpha_1)^3 \\ &= \sum_{i<j}\alpha_i^3\alpha_j^3+3\sum_{i,j,k}\alpha_i^3\alpha_j^2\alpha_k+6\alpha_1^2\alpha_2^2\alpha_3^2 \\ &= \sum_{i<j}\alpha_i^3\alpha_j^3+3s_3\left(\sum_{i\neq j}\alpha_i^2\alpha_j\right)+6s_3^2 \end{aligned}$$

Therefore

$$\begin{aligned} \sum_{i<j}\alpha_i^3\alpha_j^3 &= s_2^3-3s_3(s_1s_2-3s_3)-6s_3^2 \\ &= s_2^3-3s_1s_2s_3+3s_3^2 \end{aligned}$$

Putting all these together,

$$\begin{aligned} \phi\psi &= s_3(s_1^3-3s_1s_2+3s_3)+s_2^3+3s_3^2-3s_1s_2s_3+3s_3^2 \\ &= s_1^3s_3+9s_3^2-6s_1s_2s_3+s_2^3 \end{aligned}$$

Hence ϕ and ψ are the roots of the quadratic equation

$$t^2\ \ at+b=0$$

where

$$\begin{aligned} a &= s_1s_2-3s_3 \\ b &= s_3(s_1^3-3s_1s_2+3s_3)+s_2^3+3s_3^2-3s_1s_2s_3+3s_3^2 \end{aligned}$$

By the formula for quadratics, this equation has rational zeros if and only if $\sqrt{a^2-4b}\in\mathbb{Q}$. Direct calculation shows that

$$a^2-4b = s_1^2s_2^2+18s_1s_2s_3-27s_3^2-4s_1^3s_3-4s_2^3$$

We denote this expression by Δ, because it turns out to be the discriminant of f. Thus we have proved:

Proposition 22.4. *Let* $f(t)=t^3-s_1t^2+s_2t-s_3\ \in\ \mathbb{Q}[t]$ *be irreducible over* $\mathbb{Q}$. *Then its Galois group is* $\mathbb{A}_3$ *if*

$$\Delta = s_1^2s_2^2+18s_1s_2s_3-27s_3^2-4s_1^3s_3-4s_2^3$$

is a perfect square in $\mathbb{Q}$, *and is* $\mathbb{S}_3$ *otherwise.*

Examples 22.5. (1) Let $f(t) = t^3 + 3t + 1$. This is irreducible, and

$$s_1 = 0 \qquad s_2 = 3 \qquad s_3 = -1$$

We find that $\Delta = -27 - 4.27 = -135$, which is not a square. Hence the Galois group is $\mathbb{S}_3$.
(2) Let $f(t) = t^3 - 3t - 1$. This is irreducible, and

$$s_1 = 0 \qquad s_2 = -3 \qquad s_3 = 1$$

Now $\Delta = 81$, which is a square. Hence the Galois group is $\mathbb{A}_3$.

22.3 The Discriminant

More elaborate versions of the above method can be used to treat quartics or quintics, but in this form the calculations are very unstructured. See Exercise 22.6 for quartics. In this section we provide an interpretation of the expression Δ above, and show that a generalisation of it distinguishes between polynomials of degree n whose Galois groups are, or are not, contained in $\mathbb{A}_n$.

The definition of the discriminant generalises to any field:

Definition 22.6. Suppose that $f(t) \in K(t)$ and let its zeros in a splitting field be $\alpha_1, \ldots, \alpha_n$. Let

$$\delta = \prod_{i<j}(\alpha_i - \alpha_j)$$

Then the *discriminant* $\Delta(f)$ of f is

$$\Delta(f) = \delta^2$$

Theorem 22.7. *Let $f \in K[t]$, where the characteristic of K is not 2. Then*

(1) $\Delta(f) \in K$.

(2) $\Delta(f) = 0$ *if and only if f has a multiple zero.*

(3) If $\Delta(f) \neq 0$ then $\Delta(f)$ *is a perfect square in K if and only if the Galois group of f, interpreted as a group of permutations of the zeros of f, is contained in the alternating group $\mathbb{A}_n$.*

Proof. Let $\sigma \in \mathbb{S}_n$, acting by permutations of the α_j. It is easy to check that if σ is applied to δ then it changes it to $\pm\delta$, the sign being $+$ if σ is an even permutation and $-$ if σ is odd. (Indeed in many algebra texts the sign of a permutation is defined in this manner.) Therefore $\delta \in \mathbb{A}_n^{\dagger}$. Further, $\Delta(f) = \delta^2$ is unchanged by any permutation in $\mathbb{S}_n$, hence lies in K. This proves (1).

Part (2) follows from the definition of $\Delta(f)$.

Let G be the Galois group of f, considered as a subgroup of $\mathbb{S}_n$. If $\Delta(f)$ is a perfect square in K then $\delta \in K$, so δ is fixed by G. Now odd permutations change δ to $-\delta$, and since $\mathrm{char}(K) \neq 2$ we have $\delta \neq -\delta$. Therefore all permutations in G are even, that is, $G \subseteq \mathbb{A}_n$. Conversely, if $G \subseteq \mathbb{A}_n$ then $\delta \in G^\dagger = K$. Therefore $\Delta(f)$ is a perfect square in K. □

In order to apply Theorem 22.7, we must calculate $\Delta(f)$ explicitly. Because it is a symmetric polynomial in the zeros α_j, it must be given by some polynomial in the elementary symmetric polynomials s_k. Brute force calculations show that if f is a cubic polynomial then

$$\Delta(f) = s_1^2 s_2^2 + 18 s_1 s_2 s_3 - 27 s_3^2 - 4 s_1^3 s_3 - 4 s_2^3$$

which is precisely the expression Δ obtained in Proposition 22.4. Proposition 22.4 is thus a corollary of Theorem 22.7.

22.4 General Algorithm for the Galois Group

We now describe a method which, in principle, will compute the Galois group of any polynomial. The practical obstacles involved in carrying it out are considerable for equations of even modestly high degree, but it does have the virtue of showing that the problem possesses an algorithmic solution. More efficient algorithms have been invented, but to describe them would take us too far afield: see previous references in this chapter.

Suppose that

$$f(t) = t^n - s_1 t^{n-1} + \cdots + (-1)^n s_n$$

is a monic irreducible polynomial over a field K, having distinct zeros $\alpha_1, \ldots, \alpha_n$ in a splitting field Σ. That is, we assume f is separable. The s_k are the elementary symmetric polynomials in the α_j. The idea is to consider not just how an element γ of the Galois group G of f acts on $\alpha_1, \ldots, \alpha_n$, but how γ acts on arbitrary 'linear combinations'

$$\beta = x_1 \alpha_1 + \cdots + x_n \alpha_n$$

To make this action computable we form polynomials having zeros $\gamma(\beta)$ as γ runs through G. To do so, let $x_1, \ldots, x_n$ be independent indeterminates, let β be defined as above, and for every $\sigma \in \mathbb{S}_n$ define

$$\begin{aligned}\sigma_x(\beta) &= x_{\sigma(1)} \alpha_1 + \cdots + x_{\sigma(n)} \alpha_n \\ \sigma_\alpha(\beta) &= x_1 \alpha_{\sigma(1)} + \cdots + x_n \alpha_{\sigma(n)}\end{aligned}$$

By rearranging terms, we see that $\sigma_\alpha(\beta) = \sigma_x^{-1}(\beta)$.

(The notation here reminds us that σ_x acts on the x_j, whereas σ_α acts on the α_j.)

Since f has distinct zeros, $\sigma_x(\beta) \neq \tau_x(\beta)$ if $s \neq \tau$. Define the polynomial

$$Q = \prod_{\sigma \in \mathbb{S}_n} (t - \sigma_x(\beta)) = \prod_{\sigma \in \mathbb{S}_n} (t - \sigma_\alpha(\beta))$$

If we use the second expression for Q, expand in powers of t, collect like terms, and write all symmetric polynomials in the α_j as polynomials in the s_k, we find that

$$Q = \sum_{j=0}^{n!} \left(\sum_i g_i(s_1, \ldots, s_n) x_1^{i_1} \ldots x_n^{i_n} \right) t^j$$

where the g_i are explicitly computable functions of $s_1, \ldots, s_n$. In particular $Q \in K[t, x_1, \ldots, x_n]$. (In the second sum above, i ranges over all n-tuples of nonnegative integers $(i_1, \ldots, i_n)$ with $i_1 + \cdots + i_n + j = n$)

Next we split Q into a product of irreducibles,

$$Q = Q_1 \ldots Q_k$$

in $K[t, x_1, \ldots, x_n]$. In the ring $\Sigma[t, x_1, \ldots, x_n]$ we can write

$$Q_j = \prod_{\sigma \in S_j} (t - \sigma_x(\beta))$$

where $\mathbb{S}_n$ is the disjoint union of the subsets S_j. We choose the labels so that the identity of $\mathbb{S}_n$ is contained in S_1, and then $t - \beta$ divides Q_1 in $\Sigma[t, x_1, \ldots, x_n]$.

If $\sigma \in \mathbb{S}_n$ then

$$Q = \sigma_x Q = (\sigma_x Q_1) \cdots (\sigma_x Q_k)$$

Hence σ_x permutes the irreducible factors Q_j of Q. Define

$$\mathbf{G} = \{\sigma \in \mathbb{S}_n : \sigma_x Q_1 = Q_1\}$$

a subgroup of $\mathbb{S}_n$. Then we have the following characterisation of the Galois group of f:

Theorem 22.8. *The Galois group G of f is isomorphic to the group* $\mathbf{G}$.

Proof. The subset S_1 of $\mathbb{S}_n$ is in fact equal to $\mathbf{G}$, because

$$\begin{aligned} S_1 &= \{\sigma : t - \sigma_x\beta \text{ divides } Q_1 \text{ in } \Sigma[t, x_1, \ldots, x_n]\} \\ &= \{\sigma : t - \beta \text{ divides } \sigma_x^{-1} Q_1 \text{ in } \Sigma[t, x_1, \ldots, x_n]\} \\ &= \{\sigma : \sigma_x^{-1} Q_1 = Q_1\} \\ &= \mathbf{G} \end{aligned}$$

Define

$$H = \prod_{\sigma \in G} (t - \sigma_\alpha(\beta)) = \prod_{\sigma \in G} (t - \sigma_x(\beta))$$

Clearly $H \in K[t, x_1, \ldots, x_n]$. Now H divides Q in $\Sigma[t, x_1, \ldots, x_n]$ so H divides Q

in $\Sigma(x_1,\ldots,x_n)[t]$. Therefore H divides Q in $K(x_1,\ldots,x_n)[t]$ so that H divides Q in $K[t,x_1,\ldots,x_n]$ by the analogue of Gauss's Lemma for $K(x_1,\ldots,x_n)[t]$, which can be proved in a similar manner to Lemma 3.17.

Thus H is a product of some of the irreducible factors Q_j of Q. Because $y-\beta$ divides H we know that Q_1 is one of these factors. Therefore Q_1 divides H in $K[t,x_1,\ldots,x_n]$ so $\mathbf{G}\subseteq G$.

Conversely, let $\gamma\in G$ and apply the automorphism γ to the relation $(t-\beta)|Q_1$. Since Q_1 has coefficients in K, we get $(t-\gamma_\alpha(\beta))|Q_1$. Now $t-\gamma_\alpha(\beta)=t-\gamma_x^{-1}(\beta)=\gamma_x^{-1}(t-\beta)$, so $\gamma_x^{-1}(t-\beta)|Q_1$. Equivalently, $(t-\beta)|\gamma_x(Q_1)$. But Q_1 is the unique irreducible factor of Q that is divisible by $t-\beta$, so $\gamma_x(Q_1)=Q_1$, so $\gamma\in\mathbf{G}$.

□

Example 22.9. Suppose that α, β are the zeros of a quadratic polynomial $t^2-At+B=0$, where $A=\alpha+\beta$ and $B=\alpha\beta$. The polynomial Q takes the form

$$\begin{aligned} Q &= (t-\alpha x-\beta y)(t-\alpha y-\beta x) \\ &= t^2-t(\alpha x+\beta y+\alpha y+\beta x)+[(\alpha^2+\beta^2)xy+\alpha\beta(x^2+y^2)] \\ &= t^2-t(Ax+Ay)+[(A^2-2B)xy+B(x^2+y^2)] \end{aligned}$$

This is either irreducible or has two linear factors. The condition for irreducibility is that

$$A^2(x+y)^2-4[(A^2-2B)xy+B(x^2+y^2)]$$

is not a perfect square. But this is equal to

$$(A^2-4B)(x-y)^2$$

which is a perfect square if and only if A^2-4B is a perfect square. Thus the Galois group G is trivial if A^2-4B is a perfect square, and is cyclic of order 2 if A^2-4B is not a perfect square.

It is of course much simpler to prove this directly, but the calculation illustrates how the theorem works.

EXERCISES

22.1 Let $f\in K[t]$ where char $(K)\neq 2$. If $\Delta(f)$ is not a perfect square in K and G is the Galois group of f, show that $G\cap\mathbb{A}_n$ has fixed field $K(\delta)$.

22.2* Find an expression for the discriminant of a quartic polynomial. [*Hint*: You may assume without proof that this is the same as the discriminant of its resolvent cubic.]

22.3 In the notation of Proposition 22.4, show that $\mathbb{A}_3^\dagger=\mathbb{Q}(\phi,\psi)$.

22.4 Show that δ or $-\delta$ in Definition 22.6 is given by the Vandermonde determinant (see Exercise 2.5)

$$\begin{vmatrix} 1 & 1 & \cdots & 1 \\ \alpha_1 & \alpha_2 & \cdots & \alpha_n \\ \alpha_1^2 & \alpha_2^2 & \cdots & \alpha_n^2 \\ \vdots & \vdots & \ddots & \vdots \\ \alpha_1^{n-1} & \alpha_2^{n-1} & \cdots & \alpha_n^{n-1} \end{vmatrix}$$

Multiply this matrix by its transpose and take the determinant to show that $\Delta(f)$ is equal to

$$\begin{vmatrix} \lambda_0 & \lambda_1 & \cdots & \lambda_{n-1} \\ \lambda_1 & \lambda_2 & \cdots & \lambda_n \\ \vdots & \vdots & \ddots & \vdots \\ \lambda_{n-1} & \lambda_n & \cdots & \lambda_{2n-2} \end{vmatrix}$$

where $\lambda_k = \alpha_1^k + \cdots + \alpha_n^k$. Hence, using Exercise 18.17, compute $\Delta(f)$ when f is of degree 2, 3, or 4. Check your result is the same as that obtained previously.

22.5* If $f(t) = t^n + at + b$, show that

$$\Delta(f) = \mu_{n+1} n^n b^{n-1} - \mu_n (n-1)^{n-1} a^n$$

where μ_n is 1 if n is a multiple of 4 and is -1 otherwise.

22.6* Show that any transitive subgroup of $\mathbb{S}_4$ is conjugate to one of $\mathbb{S}_4, \mathbb{A}_4, \mathbb{D}_4, \mathbb{V}$, or $\mathbb{Z}_4$, defined as follows:

$$\begin{aligned} \mathbb{A}_4 &= \text{alternating group of degree 4} \\ \mathbb{V} &= \{1, (12)(34), (13)(24), (14)(23)\} \\ \mathbb{D}_4 &= \text{group generated by } \mathbb{V} \text{ and } (12) \\ \mathbb{Z}_4 &= \text{group generated by } (1234) \end{aligned}$$

22.7* Let f be a monic irreducible quartic polynomial over a field K of characteristic $\neq 2, 3$ with discriminant Δ. Let g be its resolvent cubic, defined by the same formula that we derived for the general quartic, and let M be a splitting field for g. Show that:

(a) $\Gamma(f) \cong \mathbb{S}_4$ if and only if Δ is not a square in K and g is irreducible over K.

(b) $\Gamma(f) \cong \mathbb{A}_4$ if and only if Δ is a square in K and g is irreducible over K.

(c) $\Gamma(f) \cong \mathbb{D}_4$ if and only if Δ is not a square in K, g is reducible over K, and f is irreducible over M.

(d) $\Gamma(f) \cong \mathbb{V}$ if and only if Δ is a square in K and g is reducible over K.

(e) $\Gamma(f) \cong \mathbb{Z}_4$ if and only if Δ is not a square in K, g is reducible over K, and f is reducible over M.

22.8 Prove that $\{(123), (456), (14)\}$ generates a transitive subgroup of $\mathbb{S}_6$.

22.9 Mark the following true or false.

(a) Every nontrivial normal subgroup of $\mathbb{S}_n$ is transitive.

(b) Every nontrivial subgroup of $\mathbb{S}_n$ is transitive.

(c) Every transitive subgroup of $\mathbb{S}_n$ is normal.

(d) Every transitive subgroup of $\mathbb{S}_n$ has order divisible by n.

(e) The Galois group of any irreducible cubic polynomial over a field of characteristic zero is isomorphic either to $\mathbb{S}_3$ or to $\mathbb{A}_3$.

(f) If K is a field of characteristic zero in which every element is a perfect square, then the Galois group of any irreducible cubic polynomial over K is isomorphic to $\mathbb{A}_3$.

Chapter 23

Algebraically Closed Fields

Back to square one.

In Chapter 2 we proved the Fundamental Theorem of Algebra, Theorem 2.4, using some basic point-set topology and simple estimates. It is also possible to give an 'almost' algebraic proof, in which the only extraneous information required is that every polynomial of odd degree over $\mathbb{R}$ has a real zero. This follows immediately from the continuity of polynomials over $\mathbb{R}$ and the fact that an odd degree polynomial changes sign somewhere between $-\infty$ and $+\infty$.

We now present this almost-algebraic proof, which applies to a slight generalisation. The main property of $\mathbb{R}$ that we require is that $\mathbb{R}$ is an ordered field, with a relation $\leq$ that satisfies the usual properties. So we start by defining an ordered field. Then we develop some group theory, a far-reaching generalisation of Cauchy's Theorem due to the Norwegian mathematician Ludwig Sylow, about the existence of certain subgroups of prime power order in any finite group. Finally, we combine Sylow's Theorem with the Galois correspondence to prove the main theorem, which we set in the general context of an 'algebraically closed' field.

23.1 Ordered Fields and Their Extensions

As remarked in Chapter 2, the first proof of the Fundamental Theorem of Algebra was given by Gauss in his doctoral dissertation of 1799. His title (in Latin) was *A New Proof that Every Rational Integral Function of One Variable can be Resolved into Real Factors of the First or Second Degree*. Gauss was being polite in using the word 'new', because his was the first genuine proof. Even his proof, from the modern viewpoint, has gaps; but these are topological in nature and not hard to fill. In Gauss's day they were not considered to be gaps at all. Gauss came up with several different proofs of the Fundamental Theorem of Algebra; among them is a topological proof that can be found in Hardy (1960 page 492).

As discussed in Chapter 2, many other proofs are now known. Several of them use complex analysis. The one in Titchmarsh (1960 page 118) is probably the proof most commonly encountered in an undergraduate course.

Less well known is a proof by Clifford (1968 page 20) which is almost entirely algebraic. His idea is to show that any irreducible polynomial over $\mathbb{R}$ is of degree 1 or 2. The proof we give here is essentially due to Legendre, but his original proof had gaps which we fill using Galois theory.

It is unreasonable to ask for a *purely* algebraic proof of the theorem, since the real numbers (and hence the complex numbers) are defined in terms of analytic concepts such as Cauchy sequences, Dedekind cuts, or completeness in an ordering.

We begin by abstracting some properties of the reals.

Definition 23.1. An *ordered field* is a field K with a relation $\leq$ such that:

(1) $k \leq k$ for all $k \in K$.

(2) $k \leq l$ and $l \leq m$ implies $k \leq m$ for all $k, l, m \in K$.

(3) $k \leq l$ and $l \leq k$ implies $k = l$ for all $k, l \in K$.

(4) If $k, l \in K$ then either $k \leq l$ or $l \leq k$.

(5) If $k, l, m \in K$ and $k \leq l$ then $k + m \leq l + m$.

(6) If $k, l, m \in K$ and $k \leq l$ and $0 \leq m$ then $km \leq lm$.

The relation $\leq$ is an ordering on K. The associated relations $<$, $\geq$, $>$ are defined in terms of $\leq$ in the obvious way, as are the concepts 'positive' and 'negative'.

Examples of ordered fields are $\mathbb{Q}$ and $\mathbb{R}$. We need two simple consequences of the definition of an ordered field.

Lemma 23.2. *Let K be an ordered field. Then for any $k \in K$ we have $k^2 \geq 0$. Further, the characteristic of K is zero.*

Proof. If $k \geq 0$ then $k^2 \geq 0$ by (6). So by (3) and (4) we may assume $k < 0$. If now we had $-k < 0$ it would follow that

$$0 = k + (-k) < k + 0 = k$$

a contradiction. So $-k \geq 0$, whence $k^2 = (-k)^2 \geq 0$. This proves the first statement.

We now know that $1 = 1^2 > 0$, so for any finite n the number

$$n \cdot 1 = 1 + \cdots + 1 > 0$$

implying that $n \cdot 1 \neq 0$ and K must have characteristic 0. □

We quote the following properties of $\mathbb{R}$.

Lemma 23.3. $\mathbb{R}$, *with the usual ordering, is an ordered field. Every positive element of $\mathbb{R}$ has a square root in $\mathbb{R}$. Every odd degree polynomial over $\mathbb{R}$ has a zero in $\mathbb{R}$.*

These are all proved in any course in analysis, and depend on the fact that a polynomial function on $\mathbb{R}$ is continuous.

23.2 Sylow's Theorem

Next, we set up the necessary group theory. Sylow's Theorem is based on the concept of a p-group:

Definition 23.4. Let p be a prime. A finite group G is a *p-group* if its order is a power of p.

For example, the dihedral group $\mathbb{D}_4$ is a 2-group. If $n \geq 3$, then the symmetric group $\mathbb{S}_n$ is never a p-group for any prime p.

The p-groups have many pleasant properties (and many unpleasant ones, but we shall not dwell on their Dark Side). One is:

Theorem 23.5. *If $G \neq 1$ is a finite p-group, then G has non-trivial centre.*

Proof. The class equation (14.2) of G reads

$$p^n = |G| = 1 + |C_2| + \cdots + |C_r|$$

and Corollary14.12 implies that $|C_j| = p^{n_j}$ for some $n_j \geq 0$. Now p divides the right-hand side of the class equation, so that at least $p-1$ values of $|C_j|$ must be equal to 1. But if x lies in a conjugacy class with only one element, then $g^{-1}xg = x$ for all $g \in G$, that is, $gx = xg$. Hence $x \in Z(G)$. Therefore $Z(G) \neq 1$. □

From this we easily deduce:

Lemma 23.6. *If G is a finite p-group of order p^n, then G has a series of normal subgroups*

$$1 = G_0 \subseteq G_1 \subseteq \ldots \subseteq G_n = G$$

such that $|G_j| = p^j$ for all $j = 0, \ldots, n$.

Proof. Use induction on n. If $n = 0$ all is clear. If not, let $Z = Z(G) \neq 1$ by Theorem 23.5. Since Z is an abelian group of order p^m it has an element of order p. The cyclic subgroup K generated by such an element has order p and is normal in G since $K \subseteq Z$. Now G/K is a p-group of order p^{n-1}, and by induction there is a series of normal subgroups

$$K/K = G_1/K \subseteq \ldots \subseteq G_n/K$$

where $|G_j/K| = p^{j-1}$. But then $|G_j| = p^j$ and $G_j \lhd G$. If we let $G_0 = 1$, the result follows. □

Corollary 23.7. *Every finite p-group is soluble.*

Proof. The quotients G_{j+1}/G_j of the series afforded by Lemma 23.6 are of order p, hence cyclic and in particular abelian. □

In 1872 Sylow discovered some fundamental theorems about the existence of p-groups inside given finite groups. We shall need one of his results in this chapter. We state all of his results, though we shall prove only the one that we require, statement (1).

Theorem 23.8 (Sylow's Theorem). *Let G be a finite group of order $p^a r$ where p is prime and does not divide r. Then*

(1) *G possesses at least one subgroup of order p^a.*

(2) *All such subgroups are conjugate in G.*

(3) *Any p-subgroup of G is contained in one of order p^a.*

(4) *The number of subgroups of G of order p^a leaves remainder* 1 *on division by p.*

This result motivates:

Definition 23.9. If G is a finite group of order $p^a r$ where p is prime and does not divide r, then a *Sylow p-subgroup* of G is a subgroup of G of order p^a.

In this terminology Theorem 23.8 says that for finite groups Sylow p-subgroups exist for all primes p, are all conjugate, are the maximal p-subgroups of G, and occur in numbers restricted by condition (4).

Proof of Theorem 23.8(1). Use induction on $|G|$. The theorem is obviously true for $|G| = 1$ or 2. Let $C_1, \ldots, C_s$ be the conjugacy classes of G, and let $c_j = |C_j|$. The class equation of G is

$$p^a r = c_1 + \cdots + c_s \tag{23.1}$$

Let Z_j denote the centraliser in G of some element $x_j \in C_j$, and let $n_j = |Z_j|$. By Lemma 14.11

$$n_j = p^a r / c_j \tag{23.2}$$

Suppose first that some c_j is greater than 1 and not divisible by p. Then by (23.2) $n_j < p^a r$ and is divisible by p^a. Hence by induction Z_j contains a subgroup of order p^a. Therefore we may assume that for all $j = 1, \ldots, s$ either $c_j = 1$ or $p | c_j$. Let $z = |Z(G)|$. As in Theorem 23.5, z is the number of values of i such that $c_j = 1$. So $p^a r = z + kp$ for some integer k. Hence p divides z, and G has a non-trivial centre Z such that p divides $|Z|$. By Lemma 14.14, the group Z has an element of order p, which generates a subgroup P of G of order p. Since $P \subseteq Z$ it follows that $P \lhd G$. By induction G/P contains a subgroup S/P of order p^{a-1}, whence S is a subgroup of G of order p^a and the theorem is proved. □

Example 23.10. Let $G = \mathbb{S}_4$, so that $|G| = 24$. According to Sylow's theorem G must have subgroups of orders 3 and 8. Subgroups of order 3 are easy to find: any 3-cycle, such as (123) or (134) or (234), generates such a group. We shall find a subgroup of order 8. Let $\mathbb{V}$ be the Klein four-group, which is normal in G. Let τ be any 2-cycle, generating a subgroup T of order 2. Then $\mathbb{V} \cap T = 1$, and $\mathbb{V}T$ is a subgroup of order 8. (It is isomorphic to $\mathbb{D}_4$.)

Analogues of Sylow's theorem do not work as soon as we go beyond prime powers. Exercise 23.1 illustrates this point.

23.3 The Algebraic Proof

With Sylow's Theorem under our belt, all that remains is to set up a little more Galois-theoretic machinery.

Lemma 23.11. *Let K be a field of characteristic zero, such that for some prime p every finite extension M of K with $M \neq K$ has $[M:K]$ divisible by p. Then every finite extension of K has degree a power of p.*

Proof. Let N be a finite extension of K. The characteristic is zero so $N:K$ is separable. By passing to a normal closure we may assume $N:K$ is also normal, so that the Galois correspondence is bijective. Let G be the Galois group of $N:K$, and let P be a Sylow p-subgroup of G. The fixed field $P^\dagger$ has degree $[P^\dagger : K]$ equal to the index of P in G (Theorem 12.2(3)), but this is prime to p. By hypothesis, $P^\dagger = K$, so $P = G$. Then $[N:K] = |G| = p^n$ for some n. □

Theorem 23.12. *Let K be an ordered field in which every positive element has a square root and every odd-degree polynomial has a zero. Then $K(i)$ is algebraically closed, where $i^2 = -1$.*

Proof. K cannot have any extensions of finite odd degree greater than 1. For suppose $[M:K] = r > 1$ where r is odd. Let $\alpha \in M \backslash K$ have minimal polynomial m. Then ∂m divides r, so is odd. By hypothesis m has a zero in K, so is reducible, contradicting Lemma 5.6. Hence every finite extension of K has even degree over K. The characteristic of K is 0 by Lemma 23.2, so by Lemma 23.11 every finite extension of K has 2-power degree.

Let $M \neq K(i)$ be any finite extension of $K(i)$ where $i^2 = -1$. By taking a normal closure we may assume $M:K$ is normal, so the Galois group of $M:K$ is a 2-group. Using Lemma 23.6 and the Galois correspondence, we can find an extension N of $K(i)$ of degree $[N:K(i)] = 2$. By the formula for solving quadratic equations, $N = K(i)(\alpha)$ where $\alpha^2 \in K(i)$. But if $a, b \in K$ then recall (2.5):

$$\sqrt{a+bi} = \sqrt{\frac{a+\sqrt{a^2+b^2}}{2}} + i\sqrt{\frac{-a+\sqrt{a^2+b^2}}{2}}$$

where the square root of a^2+b^2 is the positive one, and the signs of the other two square roots are chosen to make their product equal to b. The square roots exist in K since the elements inside them are positive, as is easily checked.

Therefore $\alpha \in K(i)$, so that $N = K(i)$, which contradicts our assumption on N. Therefore $M = K(i)$, and $K(i)$ has no finite extensions of degree > 1. Hence any irreducible polynomial over $K(i)$ has degree 1, otherwise a splitting field would have finite degree > 1 over $K(i)$. Therefore $K(i)$ is algebraically closed. □

Corollary 23.13 (Fundamental Theorem of Algebra). *The field $\mathbb{C}$ of complex numbers is algebraically closed.*

Proof. Put $\mathbb{R} = K$ in Theorem 23.12 and use Lemma 23.3. □

EXERCISES

23.1 Show that $\mathbb{A}_5$ has no subgroup of order 15.

23.2 Show that a subgroup or a quotient of a p-group is again a p-group. Show that an extension of a p-group by a p-group is a p-group.

23.3 Show that $\mathbb{S}_n$ has trivial centre if $n \geq 3$.

23.4 Prove that every group of order p^2(with p prime) is abelian. Hence show that there are exactly two non-isomorphic groups of order p^2 for any prime number p.

23.5 Show that a field K is algebraically closed if and only if $L : K$ algebraic implies $L = K$.

23.6 Show that every algebraic extension of $\mathbb{R}$ is isomorphic to $\mathbb{R} : \mathbb{R}$ or $\mathbb{C} : \mathbb{R}$.

23.7 Show that $\mathbb{C}$, with the traditional field operations, cannot be given the structure of an ordered field. If we allow different field operations, can the set $\mathbb{C}$ be given the structure of an ordered field?

23.8 Prove the theorem whose statement is the title of Gauss's doctoral dissertation mentioned at the beginning of the chapter. ('Rational integral function' was his term for 'polynomial'.)

23.9 Suppose that $K : \mathbb{Q}$ is a finitely generated extension. Prove that there exists a $\mathbb{Q}$-monomorphism $K \to \mathbb{C}$. (*Hint:* Use cardinality considerations to adjoin transcendental elements, and algebraic closure of $\mathbb{C}$ to adjoin algebraic elements.) Is the theorem true for $\mathbb{R}$ rather than $\mathbb{C}$?

23.10 Mark the following true or false.

(a) Every soluble group is a p-group.

(b) Every Sylow subgroup of a finite group is soluble.

(c) Every simple p-group is abelian.

(d) The field $\mathbb{A}$ of algebraic numbers defined in Example 17.4 is algebraically closed.

(e) There is no ordering on $\mathbb{C}$ making it into an ordered field.

(f) Every ordered field has characteristic zero.

(g) Every field of characteristic zero can be ordered.

(h) In an ordered field, every square is positive.

(i) In an ordered field, every positive element is a square.

Chapter 24

Transcendental Numbers

Our discussion of the three geometric problems of antiquity—trisecting the angle, duplicating the cube, and squaring the circle—left one key fact unproved. To complete the proof of the impossibility of squaring the circle by a ruler-and-compass construction, crowning three thousand years of mathematical effort, we must prove that π is transcendental over $\mathbb{Q}$. (In this chapter the word 'transcendental' will be understood to mean transcendental over $\mathbb{Q}$.) The proof we give is analytic, which should not really be surprising since π is best defined analytically. The techniques involve symmetric polynomials, integration, differentiation, and some manipulation of inequalities, together with a healthy lack of respect for apparently complicated expressions.

It is not at all obvious that transcendental real (or complex) numbers exist. That they do was first proved by Liouville in 1844, by considering the approximation of reals by rationals. It transpires that algebraic numbers cannot be approximated by rationals with more than a certain 'speed' (see Exercises 24.5–24.7). To find a transcendental number reduces to finding a number that can be approximated more rapidly than the known bound for algebraic numbers. Liouville showed that this is the case for the real number

$$\xi = \sum_{n=1}^{\infty} 10^{-n!}$$

but no 'naturally occurring' number was proved transcendental until Charles Hermite, in 1873, proved that e, the 'base of natural logarithms', is. Using similar methods, Ferdinand Lindemann demonstrated the transcendence of π in 1882.

Meanwhile Georg Cantor, in 1874, had produced a revolutionary proof of the existence of transcendental numbers, without actually constructing any. His proof (see Exercises 24.1–24.4) used set-theoretic methods, and was one of the earliest triumphs of Cantor's theory of infinite cardinals. When it first appeared, the mathematical world viewed it with great suspicion, but nowadays it scarcely raises an eyebrow.

We shall prove four theorems in this chapter. In each case the proof proceeds by contradiction, and the final blow is dealt by the following simple result:

Lemma 24.1. *Let $f : \mathbb{Z} \to \mathbb{Z}$ be a function such that $f(n) \to 0$ as $n \to +\infty$. Then there exists $N \in \mathbb{Z}$ such that $f(n) = 0$ for all $n \geq N$.*

Proof. Since $f(n) \to 0$ as $n \to +\infty$, there exists $N \in \mathbb{Z}$ such that $|f(n) - 0| < \frac{1}{2}$ whenever $n \geq N$, for some integer N. Since $f(n)$ is an integer, this implies that $f(n) = 0$ for $n \geq N$. □

24.1 Irrationality

Lindemann's proof is ingenious and intricate. To prepare the way we first prove some simpler theorems of the same general type. These results are not needed for Lindemann's proof, but familiarity with the ideas is. The first theorem was initially proved by Johann Heinrich Lambert in 1770 using continued fractions, although it is often credited to Legendre.

Theorem 24.2. *The real number π is irrational.*

Proof. Consider the integral

$$I_n = \int_{-1}^{1} (1-x^2)^n \cos(\alpha x)\mathrm{d}x$$

Integrating by parts, twice, and performing some fairly routine calculations, this leads to a recurrence relation

$$\alpha^2 I_n = 2n(2n-1)I_{n-1} - 4n(n-1)I_{n-2} \tag{24.1}$$

if $n \geq 2$. After evaluating the cases $n = 0, 1$, induction on n yields

$$\alpha^{2n+1} I_n = n!(P_n \sin(\alpha) + Q_n \cos(\alpha)) \tag{24.2}$$

where P_n and Q_n are polynomials in α of degree $< 2n+1$ with integer coefficients. The term $n!$ comes from the factor $2n(2n-1)$ of (24.1).

Assume, for a contradiction, that π is rational, so that $\pi = a/b$ where $a, b \in \mathbb{Z}$ and $b \neq 0$. Let $\alpha = \pi/2$ in (24.2). Then

$$J_n = a^{2n+1} I_n / n!$$

is an integer. By the definition of I_n,

$$J_n = \frac{a^{2n+1}}{n!} \int_{-1}^{1} (1-x^2)^n \cos\frac{\pi}{2}x\mathrm{d}x$$

The integrand is > 0 for $-1 < x < 1$, so $J_n > 0$. Hence $J_n \neq 0$ for all n. But

$$\begin{aligned} |J_n| &\leq \frac{|a|^{2n+1}}{n!} \int_{-1}^{1} \cos\frac{\pi}{2}x\mathrm{d}x \\ &\leq 2|a|^{2n+1}/n! \end{aligned}$$

Hence $J_n \to 0$ as $n \to +\infty$. This contradicts Lemma 24.1, so the assumption that π is rational is false. □

The next, slightly stronger, result was proved by Legendre in his *Éléments de Géométrie* of 1794, which, as we remarked in the Historical Introduction, greatly influenced the young Galois.

Theorem 24.3. *The real number π^2 is irrational.*

Proof. Assume if possible that $\pi^2 = a/b$ where $a, b \in \mathbb{Z}$ and $b \neq 0$. Define

$$f(x) = x^n(1-x)^n/n!$$

and

$$G(x) = b^n\left(\pi^{2n}f(x) - \pi^{2n-2}f''(x) + \cdots + (-1)^n\pi^0 f^{(2n)}(x)\right)$$

where the superscripts on f indicate derivatives. We claim that any derivative of f takes integer values at 0 and 1. Recall Leibniz's rule for differentiating a product:

$$\frac{\mathrm{d}^m}{\mathrm{d}x^m}(uv) = \sum \binom{m}{r}\frac{\mathrm{d}^r u}{\mathrm{d}x^r}\frac{\mathrm{d}^{m-r}v}{\mathrm{d}x^{m-r}}$$

If both factors x^n or $(1-x)^n$ are differentiated fewer than n times, then the value of the corresponding term is 0 whenever $x = 0$ or 1. If one factor is differentiated n or more times, then the denominator $n!$ is cancelled out. Hence $G(0)$ and $G(1)$ are integers. Now

$$\begin{aligned}\frac{\mathrm{d}}{\mathrm{d}x}\left[G'(x)\sin(\pi x) - \pi G(x)\cos(\pi x)\right] &= \left[G''(x) + \pi^2 G(x)\right]\sin(\pi x)\\ &= b^n\pi^{2n+2}f(x)\sin(\pi x)\end{aligned}$$

since $f(x)$ is a polynomial in x of degree $2n$, so that $f^{(2n+2)}(x) = 0$. And this expression is equal to

$$\pi^2 a^n \sin(\pi x) f(x)$$

Therefore

$$\begin{aligned}\pi\int_0^1 a^n\sin(\pi x)f(x)\mathrm{d}x &= \left[\frac{G'(x)\sin(\pi x)}{\pi} - G(x)\cos(\pi x)\right]_0^1\\ &= G(0) + G(1)\end{aligned}$$

which is an integer. As before the integral is not zero. But

$$\begin{aligned}\left|\int_0^1 a^n\sin(\pi x)f(x)\mathrm{d}x\right| &\leq |a|^n\int_0^1|\sin(\pi x)||f(x)|\mathrm{d}x\\ &\leq |a|^n\int_0^1\frac{|x^n(1-x)^n|}{n!}\mathrm{d}x\\ &\leq \frac{1}{n!}\int_0^1|(ax)^n(1-x)^n|\mathrm{d}x\end{aligned}$$

which tends to 0 as n tends to $+\infty$. The usual contradiction completes the proof. □

24.2 Transcendence of e

We move from irrationality to the far more elusive transcendence. Hermite's original proof was simplified by Karl Weierstrass, Hilbert, Adolf Hurwitz, and Paul Gordan, and it is the simplified proof that we give here. The same holds for the proof of Lindemann's theorem in the next section.

Theorem 24.4 (Hermite). *The real number* e *is transcendental.*

Proof. Assume that e is not transcendental. Then

$$a_m \mathrm{e}^m + \cdots + a_1 \mathrm{e} + a_0 = 0$$

where without loss of generality we may suppose that $a_j \in \mathbb{Z}$ for all j and $a_0 \neq 0$. Define

$$f(x) = \frac{x^{p-1}(x-1)^p(x-2)^p \ldots (x-m)^p}{(p-1)!}$$

where p is an arbitrary prime number. Then f is a polynomial in x of degree $mp + p - 1$. Put

$$F(x) = f(x) + f'(x) + \cdots + f^{(mp+p-1)}(x)$$

and note that $f^{(mp+p)}(x) = 0$. Calculate:

$$\frac{\mathrm{d}}{\mathrm{d}x}(\mathrm{e}^{-x}F(x)) = \mathrm{e}^{-x}(F'(x) - F(x)) = -\mathrm{e}^{-x}f(x)$$

Hence for any j

$$\begin{aligned} a_j \int_0^j \mathrm{e}^{-x} f(x)\mathrm{d}x &= a_j\left[-\mathrm{e}^{-x}F(x)\right]_0^j \\ &= a_j F(0) - a_j \mathrm{e}^{-j} F(j)x \end{aligned}$$

Multiply by e^j and sum over j to get

$$\begin{aligned} \sum_{j=0}^{m}\left(a_j \mathrm{e}^j \int_0^j \mathrm{e}^{-x} f(x)\mathrm{d}x\right) &= F(0)\sum_{j=0}^{m} a_j \mathrm{e}^j - \sum_{j=0}^{m} a_j F(j) \\ &= -\sum_{j=0}^{m}\sum_{i=0}^{mp+p-1} a_j f^{(i)}(j) \end{aligned} \qquad (24.3)$$

from the equation supposedly satisfied by e.

We claim that each $f^{(i)}(j)$ is an integer, and that this integer is divisible by p unless $j = 0$ and $i = p - 1$. To establish the claim we use Leibniz's rule again; the only non-zero terms arising when $j \neq 0$ come from the factor $(x - j)^p$ being differentiated exactly p times. Since $p!/(p-1)! = p$, all such terms are integers divisible by p. In the exceptional case $j = 0$, the first non-zero term occurs when $i = p - 1$, and then

$$f^{(p-1)}(0) = (-1)^p \ldots (-m)^p$$

Subsequent non-zero terms are all multiples of p. The value of equation (24.3) is therefore

$$K_p + a_0(-1)^p \dots (-m)^p$$

for some $K \in \mathbb{Z}$. If $p > \max(m, |a_0|)$, then the integer $a_0(-1)^p \dots (-m)^p$ is not divisible by p. So for sufficiently large primes p the value of equation (6.3) is an integer not divisible by p, hence not zero.

Now we estimate the integral. If $0 \leq x \leq m$ then

$$|f(x)| \leq m^{mp+p-1}/(p-1)!$$

so

$$\begin{aligned} \left| \sum_{j=0}^{m} a_j e^j \int_0^j e^{-x} f(x) dx \right| &\leq \sum_{j=0}^{m} |a_j e^j| \int_0^j \frac{m^{mp+p-1}}{(p-1)!} dx \\ &\leq \sum_{j=0}^{m} |a_j e^j| j \frac{m^{mp+p-1}}{(p-1)!} \end{aligned}$$

which tends to 0 as p tends to $+\infty$.

This is the usual contradiction. Therefore e is transcendental. □

24.3 Transcendence of π

The proof that π is transcendental involves the same sort of trickery as the previous results, but is far more elaborate. At several points in the proof we use properties of symmetric polynomials from Chapter 18.

Theorem 24.5 (Lindemann). *The real number π is transcendental.*

Proof. Suppose for a contradiction that π is a zero of some non-zero polynomial over $\mathbb{Q}$. Then so is $i\pi$, where $i = \sqrt{-1}$. Let $\theta_1(x) \in \mathbb{Q}[x]$ be a polynomial with zeros $\alpha_1 = i\pi, \alpha_2, \dots, \alpha_n$. By a famous theorem of Euler,

$$e^{i\pi} + 1 = 0$$

so

$$(e^{\alpha_1} + 1)(e^{\alpha_2} + 1) \dots (e^{\alpha_n} + 1) = 0 \tag{24.4}$$

□

We now construct a polynomial with integer coefficients whose zeros are the exponents $\alpha_{i_1} + \cdots + \alpha_{j_r}$ of e that appear in the expansion of the product in (24.4). For example, terms of the form

$$e^{\alpha_s} \cdot e^{\alpha_t} \cdot 1 \cdot 1 \cdot 1 \cdots 1$$

give rise to exponents $\alpha_s + \alpha_t$. Taken over all pairs s, t we get exponents of the form $\alpha_1 + \alpha_2, \dots, \alpha_{n-1} + \alpha_n$. The elementary symmetric polynomials of these are symmetric in $\alpha_1, \dots, \alpha_n$, so by Theorem 18.10 they can be expressed as polynomials in the elementary symmetric polynomials of $\alpha_1, \dots, \alpha_n$. These in turn are expressible in terms of the coefficients of the polynomial θ_1 whose zeros are $\alpha_1, \dots, \alpha_n$. Hence the pairs $\alpha_s + \alpha_t$ satisfy a polynomial equation $\theta_2(x) = 0$ where θ_2 has rational coefficients. Similarly the sums of k of the α's are zeros of a polynomial $\theta_k(x)$ over $\mathbb{Q}$. Then

$$\theta_1(x)\theta_2(x)\dots\theta_n(x)$$

is a polynomial over $\mathbb{Q}$ whose zeros are the exponents of e in the expansion of equation (24.4). Dividing by a suitable power of x and multiplying by a suitable integer we obtain a polynomial $\theta(x)$ over $\mathbb{Z}$, whose zeros are the non-zero exponents $\beta_1, \dots, \beta_r$ of e in the expansion of equation (24.4).

Now (24.4) takes the form

$$\mathrm{e}^{\beta_1} + \dots + \mathrm{e}^{\beta_r} + \mathrm{e}^0 + \dots + \mathrm{e}^0 = 0$$

that is,

$$\mathrm{e}^{\beta_1} + \dots + \mathrm{e}^{\beta_r} + k = 0 \tag{24.5}$$

where $k \in \mathbb{Z}$. The term $1 \cdot 1 \cdots 1$ occurs in the expansion, so $k > 0$.

Suppose that

$$\theta(x) = cx^r + c_1 x^{r-1} + \dots + c_r$$

We know that $c_r \neq 0$ since 0 is not a zero of θ. Define

$$f(x) = \frac{c^s x^{p-1}[\theta(x)]^p}{(p-1)!}$$

where $s = rp - 1$ and p is any prime number. Define also

$$F(x) = f(x) + f'(x) + \dots + f^{(s+p+r-1)}(x)$$

and note that $f^{(s+p+r)}(x) = 0$. As before

$$\frac{\mathrm{d}}{\mathrm{d}x}[\mathrm{e}^{-x}F(x)] = -\mathrm{e}^{-x}f(x)$$

Hence

$$\mathrm{e}^{-x}F(x) - F(0) = -\int_0^x \mathrm{e}^{-y}f(y)\mathrm{d}y$$

Putting $y = \lambda x$ we get

$$F(x) - \mathrm{e}^x F(0) = -x\int_0^1 \exp[(1-\lambda)x]f(\lambda x)\mathrm{d}\lambda$$

Let x range over $\beta_1, \dots, \beta_r$ and sum: by (24.5)

$$\sum_{j=1}^r F(\beta_j) + kF(0) = -\sum_{j=1}^r \beta_j \int_0^1 \exp[(1-\lambda)\beta_j]f(\lambda\beta_j)\mathrm{d}\lambda \tag{24.6}$$

We claim that for all sufficiently large p the left-hand side of (24.6) is a non-zero integer. To prove the claim, observe that

$$\sum_{j=1}^{r} f^{(t)}(\beta_j) = 0$$

if $0 < t < p$. Each derivative $f^{(t)}(\beta_j)$ with $t \geq p$ has a factor p, since we must differentiate $[\theta(x)]^p$ at least p times to obtain a non-zero term. For any such t,

$$\sum_{j=1}^{r} f^{(t)}(\beta_j)$$

is a symmetric polynomial in the β_j of degree $\leq s$. Thus by Theorem 18.10 it is a polynomial of degree $\leq s$ in the coefficients c_i/c. The factor c^s in the definition of $f(x)$ makes this into an integer. So for $t \geq p$

$$\sum_{j=1}^{r} f^{(t)}(\beta_j) = pk_t$$

for suitable $k_t \in \mathbb{Z}$.

Now we look at $F(0)$. Computations show that

$$f^{(t)}(0) = \begin{cases} 0 & (t \leq p-2) \\ c^s c_r^p & (t = p-1) \\ l_t p & (t \geq p) \end{cases}$$

for suitable $l_t \in \mathbb{Z}$. Consequently the left-hand side of (24.6) is

$$mp + kc^s c_r^p$$

for some $m \in \mathbb{Z}$. Now $k \neq 0$, $c \neq 0$, and $c_r \neq 0$. If we take

$$p > \max(k, |c|, |c_r|)$$

then the left-hand side of (24.6) is an integer not divisible by p, so is non-zero.

The last part of the proof is routine: we estimate the size of the right-hand side of (24.6). Now

$$|f(\lambda\beta_j)| \leq \frac{|c|^s |\beta_j|^{p-1} (m(j))^p}{(p-1)!}$$

where

$$m(j) = \sup_{0 \leq \lambda \leq 1} |\theta(\lambda\beta_j)|$$

Therefore

$$\left| -\sum_{j=1}^{r} \beta_j \int_0^1 \exp[(1-\lambda)\beta_j] f(\lambda\beta_j) \mathrm{d}\lambda \right| \leq \sum_{j=1}^{r} \frac{|\beta_j|^p |c^s| |m(j)|^p B}{(p-1)!}$$

where

$$B = \left| \max_j \int_0^1 \exp[(1-\lambda)\beta_j] \mathrm{d}\lambda \right|$$

Thus the expression tends to 0 as p tends to $+\infty$. By the standard contradiction, π is transcendental.

EXERCISES

The first four exercises outline Cantor's proof of the existence of transcendental numbers, using what are now standard results on infinite cardinals.

24.1 Prove that $\mathbb{R}$ is uncountable, that is, there is no bijection $\mathbb{Z} \to \mathbb{R}$.

24.2 Define the *height* of a polynomial

$$f(t) = a_0 + \cdots + a_n t^n \in \mathbb{Z}[t]$$

to be

$$h(f) = n + |a_0| + \cdots + |a_n|$$

Prove that there is only a finite number of polynomials over $\mathbb{Z}$ of given height h.

24.3 Show that any algebraic number satisfies a polynomial equation over $\mathbb{Z}$. Using Exercise 24.2 show that the algebraic numbers form a countable set.

24.4 Combine Exercises 24.1 and 24.3 to show that transcendental numbers exist.

The next three exercises give Liouville's proof of the existence of transcendental numbers.

24.5* Suppose that x is irrational and that

$$f(x) = a_n x^n + \cdots + a_0 = 0$$

where $a_0, \ldots, a_n \in \mathbb{Z}$. Show that if $p, q \in \mathbb{Z}$ and $q \neq 0$, and $f(p/q) \neq 0$, then

$$|f(p/q)| \geq 1/q^n$$

24.6* Now suppose that $x - 1 < p/q < x + 1$ and p/q is nearer to x than any other zero of f. There exists M such that $|f'(y)| < M$ if $x - 1 < y < x + 1$. Use the mean value theorem to show that

$$|p/q - x| \geq M^{-1} q^{-n}$$

Hence show that for any $r > n$ and $K > 0$ there exist only finitely many p and q such that

$$|p/q - x| < K q^{-r}$$

24.7 Use this result to prove that $\sum_{n=1}^{\infty} 10^{-n!}$ is transcendental.

24.8 Prove that $z \in \mathbb{C}$ is transcendental if and only if its real part is transcendental or its imaginary part is transcendental.

24.9 Mark the following true or false.

(a) π is irrational.

(b) All irrational numbers are transcendental.

(c) Any nonzero rational multiple of π is transcendental.

(d) $\pi + i\sqrt{5}$ is transcendental.

(e) e is irrational.

(f) If α and β are real and transcendental then so is $\alpha + \beta$.

(g) If α and β are real and transcendental then so is $\alpha + i\beta$.

(h) Transcendental numbers form a subring of $\mathbb{C}$.

(i) The field $\mathbb{Q}(\pi)$ is isomorphic to $\mathbb{Q}(t)$ for any indeterminate t.

(j) $\mathbb{Q}(\pi)$ and $\mathbb{Q}(\mathrm{e})$ are non-isomorphic fields.

(k) $\mathbb{Q}(\pi)$ is isomorphic to $\mathbb{Q}(\pi^2)$.

Chapter 25

What Did Galois Do or Know?

This is not a scholarly book on the history of mathematics, but it does contain a substantial amount of historical material, intended to locate the topic in its context and to motivate Galois theory as currently taught at undergraduate level. (At the research frontiers, the entire subject is even more general and more abstract.)

There is a danger in this approach: it can mix up history as it actually happened with how we reformulate the ideas now. This can easily be misinterpreted, distorting our view of the past and propagating historical myths. Peter Neumann makes this point very effectively in his admirable English translation of Galois's writings, Neumann (2011). The book covers both Galois's published papers and those of his unpublished manuscripts that have survived—very few, even when brief scraps are included.

To set the record straight, we now take a look at what this material tells us about what Galois actually did, what he knew, and what he might have been able to prove. Placing the material at the end of this book allows us to refer back to all of the historical and mathematical material.

The folklore story is: Galois proved that $\mathbb{A}_5$ is simple, indeed, the smallest simple group other than cyclic groups of prime order. From this he deduced that the quintic is not soluble by radicals. However, as Neumann states, the first statement is claimed without proof (and it is questionable whether Galois possessed one), while the link to the second does not appear explicitly anywhere in the extant manuscripts. The central issue, and our main focus here, is the relation between solving the quintic by radicals and the alternating group $\mathbb{A}_5$. It would be easy to imagine, and has often been asserted, that Galois viewed these topics in the same way as they have been presented in earlier chapters, and that in particular that the key issue, for him, was to prove that $\mathbb{A}_5$ is simple.

Not so.

However, history is seldom straightforward, especially when sources are fragmentary and limited. Closely related statements do appear, enough to justify Galois's stellar reputation among mathematicians and to credit him with the most penetrating insights of his period into the solution of equations by radicals and its relation to groups of permutations. As Neumann writes: 'The [First] memoir on the conditions for solubility of equations by radicals is undoubtedly Galois's most important work. It is here that he presented his original approach to the theory of equations that has now become known as Galois Theory.'

25.1 List of the Relevant Material

Galois's published papers are five in number, and only one, 'Analysis of a memoir on the algebraic solution of equations', is relevant here. After Galois died, his manuscripts went to a literary executor, his friend Auguste Chevalier. Chevalier passed them on to Liouville, who brought Galois's work to the attention of the mathematical community, probably encouraged by the brother, Alfred Galois. Liouville's daughter Mme de Blignières gave them to the French Academy of Sciences in 1905 or 1906, where they were organised into 25 'dossiers' and bound into a single volume. Parts were published or analysed by Chevalier, Liouville, Jules Tannery, and Émile Picard. Bourgne and Azra (1962) published a complete edition. The first and currently the only complete English translation is Neumann (2011). This also contains a printed version of the French originals, in parallel with the translation for ease of comparison. Scans of the manuscripts are available on the internet at

www.bibliotheque-institutdefrance.fr/numerisation/

The documents referred to below (the dossier numbers are those assigned by the Academy) are:

Analysis of a memoir on the algebraic solution of equations, *Bulletin des Sciences Mathématiques, Physiques et Chimiques* **13** (April 1830) 271–272.

Testamentary Letter, 29 May 1832, to Chevalier.

First Memoir, sent to the Academy.

Second Memoir, sent to the Academy.

Dossier 8: Torn fragment related to the First Memoir.

Dossier 10: Publication project and note on Abel.

Dossier 15: Fragments on permutations and equations.

Several other documents refer to groups and algebraic equations, and there are some on other topics altogether.

25.2 The First Memoir

The document called the First Memoir is the one that Galois sent to the Academy on 17 January 1831; it is actually his third submission, the other two having been lost. In the opening paragraph to the First Memoir, which functions as an abstract of the contents, Galois states that he will present

> … a general *condition satisfied by every equation that is soluble by radicals*, and which conversely ensures their solubility. An application is made just to equations of which the degree is a prime number. Here is the theorem given by our analysis:
>
> In order that an equation of prime degree … be soluble by radicals, it is *necessary* and it is *sufficient* that all the roots be rational functions of any two of them.

He adds that his theory has other applications, but 'we reserve them for another occasion.'

In this abstract, there is no mention of the quintic as such, although its degree 5 is prime, so his main theorem obviously applies to it. It is not mentioned in the rest of the paper either. There is also no mention of the concept of a group. It is hard not to have some sympathy for Poisson and Lacroix, the referees: it looks like they did a professional job, and spotted a key weakness in the theorem upon which Galois places so much emphasis. (Admittedly, this is not difficult.) Namely: although Galois's condition 'all the roots be rational functions of any two of them' is indeed necessary and sufficient for solubility by radicals, it is hard to think of any practical way to verify it for any specific equation.

The Historical Introduction mentioned the referees' statement that 'one could not derive from [Galois's condition] any good way of deciding whether a given equation of prime degree is soluble or not by radicals,' and the remark by Tignol (1988) that Galois's memoir 'did not yield any workable criterion to determine whether an equation is solvable by radicals.' I also wrote: 'What the referees wanted was some kind of condition on the *coefficients* that determined solubility; what Galois gave them was a condition on the *roots*.' But I think that a stronger criticism is in order: apparently, there is no algorithmic procedure to check whether the condition on the roots is valid. Or to prove that it is not. How, for example, would we use it to prove the quintic insoluble?

It turns out that this judgement is not entirely correct, but further work is needed to see why. It is implicit in a table that Galois includes titled 'Example of Theorem VII', and I'll come back to that shortly. But *he does not make the connection explicit.*

25.3 What Galois Proved

Before discussing possible reasons for the (to our eyes) curious omission of the application to quintics, we review the results that Galois does include in the First Memoir. These alone would establish his reputation.

The work is short, succinct, and clearly written. A modern reader will have no difficulty in following the reasoning, once they get used to the terminology. He develops several key ideas needed to prove his necessary and sufficient condition for solubility by radicals, which we *now* recognise as the core concepts of Galois Theory. It is clear that Galois recognised the importance of these ideas, but, once again, *he does not say so in the paper*.

After a few preliminaries, which would have been familiar to anyone working in the area, Galois presents his first key theorem:

Proposition 25.1. *Let an equation be given of which the m roots are $a, b, c, \ldots$. There will always be a group of permutations of the letters $a, b, c, \ldots$ which will enjoy the following property:*

> *That every function of the roots invariant* [a footnote explains this term] *under the substitutions of this group will be rationally known;*
>
> *Conversely, that every function of the roots that is rationally determinable will be invariant under the substitutions.*

This is his definition of what we now call the Galois group. It also makes the central point about the Galois correspondence, expressed in terms of the roots rather than the modern interpretation in terms of the subfield they generate.

Next, he studies how the group can be decomposed by adjoining the roots of auxiliary equations; that is, extending the field. He deduces that when a pth root is extracted, for (without loss of generality) prime p, the group must have what we would now express as a normal subgroup of index p. This leads to the next big result, initially posed as a question:

Proposition 25.2. *Under what circumstances is an equation soluble by radicals?*

Galois writes '... to solve an equation it is necessary to reduce its group successively to the point where it does not contain more than a single permutation.' He analyses what happens when the reduction is performed by adjoining 'radical quantities'. He concludes, slightly obscurely, that the group of the equation must have a normal subgroup of prime index, which in turn has a normal subgroup of prime index, and so on, until we reach the group with a single element. In short: the equation is soluble by radicals if and only if its group is soluble. But he fails to state this as an explicit proposition.

Galois goes on to illustrate the result for the general quartic equation, obtaining essentially what we found in Section 18.5 of Chapter 18. This of course was a known result, and Lagrange had already related it to permutation groups in his *Traité de la Résolution des Équations Numériques de Tous les Degrés*. But instead of continuing to the quintic, and proving that the group is not soluble, Galois does something that is in some ways more interesting, but answers another (closely related) question instead:

Proposition 25.3. *What is the group of an equation of prime degree n that is soluble by radicals?*

His answer is that if the roots are suitably numbered, the group of the equation can contain only substitutions of the form

$$x_k \mapsto x_{ak+b} \tag{25.1}$$

where the roots are the x_k, the symbols a, b denote constants, and $ak+b$ is to be computed modulo n.

To modern eyes, what he *should* have remarked at this point is that when $n = 5$ the group of all such substitutions has 4.5 = 20 elements (we need $0 \neq a \in \mathbb{Z}_5$ and $b \in \mathbb{Z}_5$), so it cannot equal $\mathbb{S}_5$, the group of the general quintic. Moreover, Galois definitely *knew* that for any m the group of the general equation of degree m is the symmetric group $\mathbb{S}_m$. He states as much in the discussion of his Proposition I:

> In the case of algebraic equations, this group is nothing other than the collection of the $1.2.3 \ldots m$ possible permutations on the m letters, because in this case, only the symmetric functions are rationally determinable.

By 'algebraic equation' he meant what we now call the 'general polynomial equation'. Galois distinguished 'numerical' and 'literal' equations: those in which the coefficients are specific numbers, and those in which they are arbitrary symbols. He is clearly thinking of literal equations here. But to a casual reader this statement is somewhat confusing.

Anyway, Galois does no such thing. Instead, he in effect observes that once you have two numbers of the form $ak+b, a'k+b'$, you can generate all numbers of this form. Whence the criterion that given any two roots, the others are all rationally expressible.

25.4 What is Galois Up To?

Taking inspiration and historical information from Neumann (2011), I now think there is a sensible explanation of what at first sight seems to be a strange series of omissions and obscurities, in which Galois wanders all round a key idea without ever putting his finger on it. Namely: Galois wasn't interested in discussing the quintic. He was after something quite different.

We know that he had taken on board the work of Ruffini and Abel, because Dossier 10 refers to Abel's proof that the quintic is insoluble, and Dossier 8 states:

> It is today a commonly known truth that general equations of degree greater than the 4th cannot be solved by radicals.
>
> This truth has become commonly known to some extent by hearsay and even though most geometers do not know the proofs of it given by Ruffini, Abel, etc., proofs founded upon the fact that such a solution is already impossible for the fifth degree.

This being so, why should Galois place any emphasis on the quintic? I think he had his sights set on something more ambitious: to say something *new* about solutions by radicals.

The first piece of evidence is the continuation of the above quotation: 'In the first instance it would seem that the [theory] of solution of equations by radicals would end there.' Unfortunately the text on that side of the paper ends at this point, and the other side merely lists titles of four memoirs.

Another is Dossier 9, which includes:

> The proposed goal is to determine the characteristics for the solubility of equations by radicals... that is the question to which we offer a complete solution.

He then acknowledges that in practice 'the calculations are impracticable,' but attempts to justify the importance of the result nonetheless:

> ... most of the time in algebraic analysis one is led to equations all of whose properties one knows beforehand: properties by means of which it will always be easy to answer the question by the rules we shall expound ... I will cite, for example, the equations which give the division of elliptic functions and which the celebrated Abel has solved ...

Galois refers to these 'modular equations' from the theory of elliptic functions elsewhere, and they presumably played a major role in his thinking.

Dossier 10 states:

> ... Abel did not know the particular circumstances of solution by radicals ... he has left nothing on the general discussion of the problem which has occupied us. Once and for all, what is remarkable in our theory [is to be able to answer yes or no in all cases, *crossed out*].

Over and over again Galois places emphasis not on proving equations such as the general quintic insoluble, but on *finding equations that* are *soluble*. The title of the First Memoir says it all: 'Memoir on the conditions for solubility of equations by radicals.' So does that of the Second Memoir: 'On primitive equations which are soluble by radicals.' Galois is not interested in impossibility proofs. To him, they are old hat; they do not lead anywhere new. This, I suspect, is why he does not use the quintic as an example in the First Memoir; it is most definitely why his main general result is Proposition VII. In modern terms, he is telling us that an equation is soluble by radicals if and only if its Galois group is conjugate to a subgroup of the affine general linear group $\mathbb{AGL}(1,n)$, which consists of the transformations (25.1). These are the equations that Galois considers interesting; this is the theorem of which he is justly proud, since it constitutes a major advance and characterises soluble equations.

It is also worth remarking that the form in which Galois states Proposition VII does not involve the notion of a group. It would be immediately comprehensible to any algebraist of the period, without having to explain to them the new—and rather unorthodox—concept of a group. This is reminiscent of the way that Isaac Newton used classical geometry rather than calculus to prove many statements in his *Principia Mathematica*, even though he probably used calculus to derive them in the first place. Ironically, by trying—for once—to make his ideas more accessible, Galois obscured their importance.

25.5 Alternating Groups, Especially $\mathbb{A}_5$

Neumann (2011) discusses several myths about Galois. Prominent among them is the claim that he proved the alternating group $\mathbb{A}_n$ is simple when $n \geq 5$. However, these groups are not mentioned in any of the works of Galois published by Liouville in 1846, which was the main source for professional mathematicians. There is no mention even of $\mathbb{A}_5$, and even the symmetric groups are mentioned only to illustrate Proposition I of the First Memoir (see the quotation in Section 25.3) and as an example for Proposition V when the degree is 4.

One reason why Galois did not mention the simplicity of $\mathbb{A}_n$ or even of $\mathbb{A}_5$ is that he didn't need it. His necessary and sufficient condition for solubility—having a group conjugate to a subgroup of $\mathbb{AGL}(1,n)$—was all he needed. We can prove that $\mathbb{A}_5$ cannot occur rather easily: its order is 60 while that of $\mathbb{AGL}(1,5)$ is only 20. Simplicity is not the issue. However, Galois doesn't even say that: insolubility is also not the issue, for him.

But...

As Neumann recognises, Galois does give brief mention to alternating groups in a few manuscripts. One is Dossier 15, which consists of a series of short headings. It looks suspiciously like the outline of a lecture course. Could it be the one on advanced algebra that he offered on 13 January 1831? It might be a plan for a memoir, or even a book, for all we know. Crossed out, we find the words:

Example. Alternate groups (Two similar groups). Properties of the alternate groups.

By 'two similar groups' Galois is referring to two cosets with the same structure: this was his way to say 'normal subgroup of index 2', no doubt in $\mathbb{S}_n$. The same text appears slightly later, also crossed out. Later still we find 'New proof of the theorem relative to the alternate groups', not crossed out. This is followed shortly by 'One may suppose that the group contains only even substitutions', which I take to be a 'without loss of generality we may assume the group is contained in the alternating group'.

There is a simple way to set this up, which was known to every algebraist, and Galois would have learned it at his mother's knee. It uses the quantity δ defined in (1.13). This changes sign if any two roots are interchanged; that is, it is invariant under $\mathbb{A}_n$ but not $\mathbb{S}_n$. However, its square $\Delta = \delta^2$ is a symmetric function of the roots and therefore can be expressed as a function of the coefficients. It is the discriminant of the equation, so named because its traditional role is to provide a computable algebraic test for the existence of a multiple root. Indeed, $\Delta = 0$ if and only if the equation has a multiple root.

Since Δ is a rational function of the coefficients, we can adjoin δ by taking a square root. As far as solving equations by radicals goes, this is harmless, and it reduces the group to its intersection with $\mathbb{A}_n$. Probably Galois had something like this in mind.

The same document includes a reference to Cauchy's work on permutations, including

> Theorem. If a function on m indeterminates is given by an equation of degree m all of whose coefficients [are symmetric functions, permanent or alternating, of these indeterminates], this function will be symmetric, permanent or alternating, with respect to all letters or at least with respect to $m-1$ among them.
>
> Theorem. No algebraic equation of degree higher than 4 may be solved or reduced.

So there is no doubt that Galois was *aware* of the link between $\mathbb{S}_5, \mathbb{A}_5$, and the quintic.

25.6 Simple Groups Known to Galois

What about simple groups? Neumann points out that Galois definitely knew about simple groups (his term is 'indecomposable'). But the examples he cites are the projective special linear groups $\mathbb{PSL}(2,p)$ for prime p. His Second Memoir was clearly heading in that direction, and this fact is stated explicitly in the letter to Chevalier: '[this group] is not further decomposable unless $p=2$ or $p=3$.'

This bring us to another statement in the letter to Chevalier, which Neumann reasonably considers a 'mysterious assertion'. Namely:

> The smallest number of permutations which can have an indecomposable [simple] group, when this number is not prime, is 5.4.3.

That is, the smallest order for a simple group is 60. Neumann argues persuasively that Galois was thinking of $\mathbb{PSL}(2,5)$, not $\mathbb{A}_5$. Agreed, these groups are isomorphic, but Galois writes extensively about what we now call $\mathbb{PSL}(2,p)$, and says virtually nothing about $\mathbb{A}_n$.

Neumann also provides a fascinating discussion of whether Galois actually possessed a proof that the smallest order for a simple group is 60.

> He was so insightful that, perhaps, yes, he could have known it. Nevertheless, I very much doubt it. How could he have excluded orders such as 30, 32, 36, 40, 48, 56? With Sylow's theorems and some calculation, such orders can be excluded... but... it seems unlikely that Galois had Sylow's theorems available to him. Besides, there is no hint in any of the extant manuscripts and scraps of the kind of case-by-case analysis that is needed...

It is of course *conceivable* that Galois knew the results we now call Sylow's Theorem. He was very clever, and his known insights into group theory are impressive. However, even granting that, the viewpoint needed to prove Sylow's Theorem seems too sophisticated for the period. The biggest problem is that it is difficult to imagine him failing to tell anyone about such discoveries, and some hint ought to have survived among his papers. In their absence, Neumann's last point is especially telling. On the other hand, and grasping at straws, Galois's affairs were somewhat chaotic. Like most mathematicians, he probably threw a lot of scraps away, especially 'rough work'. In the Historical Introduction we saw that when at school he did a lot of work in his head, instead of on paper—and was criticised for it. So the absence of evidence is not evidence of absence.

25.7 Speculations about Proofs

It is worth examining just what a mathematician of the period would have needed to prove Galois's statement about the smallest order for a simple group. What follows illustrates what might have been possible given a little ingenuity. We use only a few basic theorems in group theory, all of which have easy proofs, well within Galois's capabilities. We also make no claim that he was aware of any of this material.

He knew about subgroups, cosets, conjugacy, and normal subgroups. He read Lagrange and must have known Lagrange's theorem: the order of a subgroup (or element) divides the order of the group.

He could have defined the normaliser $N_G(H)$ of a subgroup of G, which is the set of all $g \in G$ such that $g^{-1}Hg = H$. This is obviously a subgroup, and $H \lhd N_G(H)$. Moreover, it is evident that the number of distinct conjugates of H is equal to the index $|G : N_G(H)|$. The index of a subgroup $K \subseteq G$, usually denoted $|G : H|$, is equal to $|G|/|H|$ for finite groups, and is the number of distinct cosets (left or right) of H in G. Galois knew about cosets (though he called them 'groups'.)

Galois would also have been aware of what we now call the centraliser $C_G(g)$ of an element $g \in G$: the set of all $h \in G$ such that $h^{-1}gh = g$. This too is a subgroup, and the number of distinct conjugates of g is equal to the index $|G : C_G(H)|$. This line of thinking leads inevitably to the *class equation* discussed in Chapter 14 (14.2). We rewrite it in the form:

$$|G| = 1 + \sum_{g_i} |G : C_G(g_i)| \tag{25.2}$$

where $\{g_i\}$ is a set of representatives of the non-identity conjugacy classes of G. The extra 1 takes care of the identity. As we will see, the class equation is a surprisingly powerful tool when investigating simple groups of small order.

Indeed, using the class equation, Galois would easily have been able to prove Theorem 14.15, published in 1845 by Cauchy. This is a limited converse to Lagrange's theorem: if a prime number p divides the order of a finite group, the group has an element of order p. The class equation is the key to the proof, as we saw in Chapter 14.

It turns out that for putative simple groups of small order, Cauchy's Theorem works fairly well as a substitute for Sylow's theorem(s). Some systematic counting of elements then goes a long way. However, it is a bit of a scramble. The main results we need are:

Lemma 25.4. *Let G be a non-cyclic finite simple group. Then:*

(1) *The normaliser of any proper subgroup of G is a proper subgroup.*

(2) *The centraliser of any element of G is a proper subgroup of G.*

(3) *No prime p can divide the indices of all proper subgroups of G.*

(4) *There cannot exist a unique proper subgroup of G of given order $k > 1$.*

Proof. (1) If not, the subgroup is normal.

(2) If not, the element generates a cyclic normal subgroup.

(3) If such a p exists, the class equation takes the form

$$1 + c_1 + \cdots + c_k = |G|$$

where the c_j are the indices of centralisers of non-identity elements, which by (2) are proper subgroups. Therefore $p|c_j$ for all j. Also p divides $|G|$ since p divides c_1, which divides $|G|$. So the class equation taken (mod p) implies that $1 \equiv 0 \pmod{p}$, a contradiction.

(4) Suppose that H is the unique subgroup of order k. The order of any conjugate $g^{-1}Hg$ is also k, so $g^{-1}Hg = H$ for all $g \in G$. Therefore $H \triangleleft G$, a contradiction. □

We need one further idea. Galois's definition of 'normal' immediately implies that a subgroup of index 2 is normal. More generally, a little thought about the conjugates of a subgroup leads to a useful generalisation:

Lemma 25.5. *Let G be a finite group and let H be a non-normal subgroup of index m. Then G has a proper normal subgroup of index dividing $m!$ In particular, G cannot be simple if $|G| > m!$*

Proof. The subgroup H has m conjugates $H_i = g_i^{-1}Hg_i$ for $1 \leq i \leq m$. For any $g \in G$ the conjugate $g^{-1}Hg$ is one of the H_i. The map $\phi : G \to \mathbb{S}_m$ defined by $\phi(g) = g_i$ is a homomorphism. Its kernel K is a normal subgroup of G of index at most $|\mathbb{S}_m| = m!$. If $k \in K$ then $k^{-1}Hk = H$, so $K \subseteq N_G(H) \neq G$, and K is proper. □

Armed with these weapons, Galois would easily have been able to prove:

Theorem 25.6. *Let p, q be distinct primes and $k \geq 2$. A finite non-cyclic simple group cannot have order p^k, pq, $2p^k$, $3p^k$, $4p^k$, or $4p$ for $p \geq 7$.*

Proof. (1) Order p^k is ruled out by Lemma 25.4, since p divides the index of any proper subgroup. This is how we proved Theorem 23.5, but there we obtained a further consequence: the group has non-trivial centre.

(2) Suppose G is simple of order pq. By Cauchy's Theorem it has subgroups H of order p and K of order q. All nontrivial proper subgroups have order p or q. Each of H, K must equal its normaliser, otherwise it would be a normal subgroup. Therefore

H has q conjugates, which intersect pairwise in the identity, and K has p conjugates, which intersect pairwise in the identity. Therefore G has 1 element of order 1, at least $(p-1)q$ elements of order p, and at least $p(q-1)$ elements of order q. These total $2pq-p-q+1=pq+(p-1)(q-1)$ elements, a contradiction since $p,q>1$.
(3) Suppose G is simple of order $2p^k$. There is no subgroup of index 2, so every proper subgroup has index divisible by p, contrary to Lemma 25.4(3).
(4) Suppose G is simple of order $3p^k$. Since $3p^k \geq 8$, Lemma 25.5 implies that there is no subgroup of index ≤ 3. Therefore every proper subgroup has index divisible by p, contrary to Lemma 25.4(3).
(5) Suppose G is simple of order $4p^k$. If $p=2$ apply part (1). Otherwise $4p^k \geq 36$. By Lemma 25.5 there is no subgroup of index ≤ 4, so every proper subgroup has index divisible by p, contrary to Lemma 25.4(3).
(6) Suppose G is simple of order $4p$. Since $p \geq 7$ we have $|G| > 24$, so by Lemma 25.5 there is no proper subgroup of index ≤ 4. In particular there is no subgroup of order p, contrary to Cauchy's Theorem. □

We now present a proof, using nothing that could not easily have been known to Galois, of his mysterious statement:

Theorem 25.7. *There is no non-cyclic simple group of order less than* 60.

Proof. Let G be a non-cyclic simple group of order less than 60. This rules out groups of prime order, and Theorem 25.6 rules out many other orders. Only six orders survive:

$$20 \quad 30 \quad 40 \quad 42 \quad 45 \quad 56$$

and we dispose of these in turn.

Throughout, we apply Lemma 25.4(1, 2) without further comment.

Order 20

By Lemma 25.5 G has no subgroups of index ≤ 3. Therefore the possible orders of nontrivial proper subgroups are 2, 4, 5 only. By Cauchy's Theorem there exist elements of orders 2 and 5.

The class equation does not lead directly to a contradiction, so we argue as follows. Let N be the normaliser of any order-5 subgroup H. This is a proper subgroup. Since all proper subgroups have order 1, 2, 4, or 5, we have $|N|=5$. Therefore H has $20/5=4$ distinct conjugates. Since 5 is prime, these conjugates intersect only in the identity. Each non-identity element of $\mathbb{Z}_5$ has order 5, so there are 4 elements of order 5 in each order-5 subgroup. Therefore together these conjugates contain 4.4 = 16 elements of order 5.

There is also at least one element of order 2. Its normaliser has order 2 or 4, so cannot contain an element of order 5. It therefore has 5 distinct conjugates by any order-5 element. Therefore G has at least 1+16+5 = 22 elements, contradiction.

Order 30

Since $30 > 4!$, Lemma 25.5 implies that G has no subgroups of index ≤ 4. Therefore the possible orders of nontrivial proper subgroups are 2, 3, 5, 6 only. By Cauchy's Theorem there exist elements of orders 2, 3, and 5.

The class equation can be used here, but there is a simpler argument. The normaliser of any $\mathbb{Z}_5$ subgroup has order 5, hence index 6. Thus there are at least $6.4 = 24$ elements of order 5. The normaliser of any $\mathbb{Z}_3$ subgroup has order 3 or 6, hence index 10 or 5. Thus there are at least $5.2 = 10$ elements of order 3. But $24 + 10 = 34 > 30$, a contradiction.

Order 40

Lemma 25.5 implies that G has no subgroups of index ≤ 4. Therefore the possible orders of nontrivial proper subgroups are 2, 4, 5, 8 only. By Cauchy's Theorem there exist elements of orders 2 and 5.

The normaliser of any $\mathbb{Z}_5$ subgroup has order 5, hence index 8. Thus there are at least $8.4 = 32$ elements of order 5. Each has centraliser of order 5, so its conjugacy class has 8 elements. Any further order-5 element gives rise to 32 more elements for the same reason, not conjugate to the above, which is impossible. So we have found all order-5 elements and their conjugacy classes.

The centraliser of any element of order 2^k has order 2, 4, or 8, hence index 20, 10, or 5.

The class equation therefore becomes

$$40 = 1 + 32 + 5a + 10b + 20c$$

so

$$7 = 5a + 10b + 20c$$

which is impossible since $5 \nmid 7$.

Order 42

Lemma 25.5 implies that G has no subgroups of index ≤ 4. Therefore the possible orders of nontrivial proper subgroups are 2, 3, 6, 7 only. Their indices are 21, 14, 7, and 6. The class equation takes the form

$$42 = 1 + 6a + 7b + 14c + 21d$$

where a arises from elements of order 7. Consider this (mod 7) to deduce that $a \equiv 1 \pmod 7$. If $a = 1$ then there is a unique $\mathbb{Z}_7$ subgroup. But this contradicts Lemma 25.4(4). Otherwise $a \geq 8$, which yields at least $6.8 = 48$ elements of order 7, contradiction.

Order 45

Lemma 25.5 implies that G has no subgroups of index ≤ 4. Therefore the possible orders of nontrivial proper subgroups are 3, 5, 9 only. Their indices are 15, 9, and 5.

The centraliser of any order-5 element has order 5, index 9. So there are at least $9.4 = 36$ elements of order 5.

The centraliser of any order-3 element has order 3 or 9, index 15 or 5. So there are at least $2.5 = 10$ elements of order 3, giving at least $36 + 10 = 46$ elements, contradiction.

Order 56

Lemma 25.5 implies that G has no subgroups of index ≤ 4. Therefore the possible orders of nontrivial proper subgroups are 2, 4, 7, 8 only. Their indices are 28, 14, 8, and 7.

The normaliser of any $\mathbb{Z}_7$ subgroup has order 7, index 8, yielding at least $6.8 = 48$ elements of order 7.

The normaliser of any $\mathbb{Z}_2$ subgroup has order 2, 4, or 8, index 28, 14, or 7, yielding at least 7 elements of order 2.

Together with the identity, these give all 56 elements. Therefore there are exactly 48 order-7 elements and 7 order-2 elements.

The centraliser of any order-7 element must have order 7, index 8. So there are 6 conjugacy classes of order-7 elements.

The centraliser of any order-2 element must have order 2, 4, or 8, index 28, 14, or 7.

The class equation takes the form

$$56 = 1 + 48 + 7a + 14b + 28c$$

so $a = 1, b = c = 0$ and there are precisely 7 order-2 elements, all conjugate to each other. Their centralisers have order 8, so do not contain any order-7 element; therefore each has the same centraliser. This is the unique order-8 subgroup, contradicting Lemma 25.4(4). □

Galois would have had little difficulty with these orders. If he needed scrap paper calculations, they would have been short, and easily lost or thrown away. However, history relies on written evidence, and there is no documentary evidence that Galois ever proved Theorem 25.7. However, the above proof makes it plausible that Galois *could have* known how to prove that the smallest non-cyclic simple group has order 60.

EXERCISES

25.1 Prove, using the methods of this chapter, that a simple group cannot have order $5p^k$ where $k \geq 2$ and $p \geq 5$ is prime.

25.2 Using the methods of this chapter, extend the list of impossible orders for non-cyclic simple groups from 61 upwards, as far as you can using the methods of this chapter.

(Using more advanced methods it can be proved that the next possible order is 168, so there are plenty of orders to try. Orders $72, 80, 84, 90$ seem to require new ideas and may be beyond the methods of this chapter.)

References

GALOIS THEORY

Artin, E. (1948) *Galois Theory*, Notre Dame University Press, Notre Dame.

Bastida, J.R. (1984) *Field Extensions and Galois Theory*, Addison-Wesley, Menlo Park.

Berndt, B.C., Spearman, B.K., and Williams, K.S. (2002) Commentary on a unpublished lecture by G.N. Watson on solving the quintic, *Mathematical Intelligencer* **24** number 4, 15–33.

Bewersdorrff, J. *Galois Theory for Beginners: A Historical Perspective*, American Mathematical Society, Providence.

Cox, D.A. (2012) *Galois Theory*, 2nd ed., Wiley-Blackwell, Hoboken.

Edwards, H.M. (1984) *Galois Theory*, Springer, New York.

Fenrick, M.H. (1992) *Introduction to the Galois Correspondence*, Birkhäuser, Boston.

Garling, D.J.H. (1960) *A Course in Galois Theory*, Cambridge University Press, Cambridge.

Hadlock, C.R. (1978) *Field Theory and its Classical Problems*, Carus Mathematical Monographs **19**, Mathematical Assocation of America, Washington.

Howie, J.M. (2005) *Fields and Galois Theory*, Springer, Berlin.

Isaacs, M. (1985) Solution of polynomials by real radicals, *Amer. Math. Monthly* **92** 571–575.

Jacobson, N. (1964) *Theory of Fields and Galois Theory*, Van Nostrand, Princeton.

Kaplansky, I. (1969) *Fields and Rings*, University of Chicago Press, Chicago.

King, R.B. (1996) *Beyond the Quartic Equation*, Birkhäuser, Boston.

Kuga, M. (2013) *Galois' Dream: Group Theory and Differential Equations*, Birkhäuser, Basel.

Lidl, R. and Niederreiter, H. (1986) *Introduction to Finite Fields and Their Applications*, Cambridge University Press, Cambridge.

Lorenz, F. and Levy, S. (2005) *Algebra Volume 1: Fields and Galois Theory*, Springer, Berlin.

Morandi, P. (1996) *Field and Galois Theory*, Graduate Texts in Mathematics **167**, Springer, Berlin.

Newman, S.C. (2012) *A Classical Introduction to Galois Theory*, Wiley-Blackwell, Hoboken.

Postnikov, M.M. (2004) *Foundations of Galois Theory*, Dover, Mineola.

Rotman, J. (2013) *Galois Theory*, Springer, Berlin.

Tignol, J.-P. (1988) *Galois' Theory of Algebraic Equations*, Longman, London.

Van der Waerden, B.L. (1953) *Modern Algebra* (2 vols), Ungar, New York.

ADDITIONAL MATHEMATICAL MATERIAL

Adams, J.F. (1969) *Lectures on Lie Groups*, University of Chicago Press, Chigago.

Anton, H. (1987) *Elementary Linear Algebra* (5th ed.), Wiley, New York.

Braden, H., Brown, J.D., Whiting, B.F., and York, J.W. (1990) *Physical Review* **42** 3376–3385.

Chang, W.D. and Gordon, R.A. (2014) Trisecting angles in Pythagorean triangles, *Amer. Math. Monthly* **121** 625–631.

Conway, J.H. (1985) The weird and wonderful chemistry of audioactive decay, *Eureka* **45** 5–18.

Hardy, G.H. (1960) *A Course of Pure Mathematics*, Cambridge University Press, Cambridge.

Dudley, U. (1987) *A Budget of Trisections*, Springer, New York.

Fraleigh, J.B. (1989) *A First Course in Abstract Algebra*, Addison-Wesley, Reading.

Gleason, A.M. (1988) Angle trisection, the heptagon, and the triskaidecagon, *American Mathematical Monthly* **95** 185–194.

Hardy, G.H. and Wright, E.M. (1962) *The Theory of Numbers*, Oxford University Press, Oxford.

Heath, T.L. (1956) *The Thirteen Books of Euclid's Elements* (3 vols) (2nd ed.), Dover, New York.

Herz-Fischler, R. (1998) *A Mathematical History of the Golden Number* (2nd ed.), Dover, Mineola.

Hulke, A. (1996) *Konstruktion transitiver Permutationsgruppen*, Dissertation, Rheinisch Westflische Technische Hochschule, Aachen.

Humphreys, J.F. (1996) *A Course in Group Theory*, Oxford University Press, Oxford.

Livio, M. (2002) *The Golden Ratio*, Broadway Books, New York.

Neumann, P.M., Stoy, G.A., and Thompson, E.C. (1994) *Groups and Geometry*, Oxford University Press, Oxford.

Oldroyd, J.C. (1955) Approximate constructions for 7, 9, 11 and 13-sided polygons, *Eureka* **18**, 20.

Ramanujan, S. (1962) *Collected Papers of Srinivasa Ramanujan*, Chelsea, New York.

Salmon, G. (1885) *Lessons Introductory to the Modern Higher Algebra*, Hodges, Figgis, Dublin.

Sharpe, D. (1987) *Rings and Factorization*, Cambridge University Press, Cambridge.

Soicher, L. and McKay, J. (1985) Computing Galois groups over the rationals, *Journal of Number Theory* **20** 273–281.

Stewart. I. (1977) Gauss, *Scientific American* **237** 122–131.

Stewart, I. and Tall, D. (1983) *Complex Analysis*, Cambridge University Press, Cambridge.

Stewart, I. and Tall, D. (2002) *Algebraic Numbers and Fermat's Last Theorem* (3rd ed.), A. K. Peters, Natick MA.

Thompson, T.T. (1983) *From Error-Correcting Codes Through Sphere-Packings to Simple Groups*, Carus Mathematical Monographs **21**, Mathematical Assocation of America, Washington DC.

Titchmarsh, E.C. (1960) *The Theory of Functions*, Oxford University Press, Oxford.

HISTORICAL MATERIAL

Bell, E.T. (1965) *Men of Mathematics* (2 vols), Penguin, Harmondsworth, Middlesex.

Bertrand, J. (1899) La vie d'Évariste Galois, par P. Dupuy, *Bulletin des Sciences Mathématiques*, **23** , 198–212.

Bortolotti, E. (1925) L'algebra nella scuola matematica bolognese del secolo XVI, *Periodico di Matematica*, **5**(4), 147–84.

Bourbaki, N. (1969) *Éléments d'Histoire des Mathématiques*, Hermann, Paris.

Bourgne, R. and Azra, J.-P. (1962) *Écrits et Mémoires Mathématiques d'Évariste Galois*, Gauthier-Villars, Paris.

Cardano, G. (1931) *The Book of my Life*, Dent, London.

Clifford, W.K. (1968) *Mathematical Papers*, Chelsea, New York.

Coolidge, J.L. (1963) *The Mathematics of Great Amateurs*, Dover, New York.

Dalmas, A. (1956) *Évariste Galois, Révolutionnaire et Géomètre*, Fasquelle, Paris.

Dumas, A. (1967) *Mes Memoirs* (volume 4 chapter 204), Editions Gallimard, Paris.

Dupuy, P. (1896) La vie d'Évariste Galois, *Annales de l'École Normale*, **13**(3), 197–266.

Galois, E. (1897) *Oeuvres Mathématiques d'Évariste Galois*, Gauthier-Villars, Paris.

Gauss, C.F. (1966) *Disquisitiones Arithmeticae*, Yale University Press, New Haven.

Henry, C. (1879) Manuscrits de Sophie Germain, *Revue Philosophique* **631**.

Huntingdon, E.V. (1905) *Trans. Amer. Math. Soc.* **6**, 181.

Infantozzi, C.A. (1968) Sur l'a mort d'Évariste Galois, *Revue d'Histoire des Sciences* **2**, 157.

Joseph, G.G. (2000). *The Crest of the Peacock*, Penguin, Harmondsworth.

Klein, F. (1913) *Lectures on the Icosahedron and the Solution of Equations of the Fifth Degree*, Kegan Paul, London.

Klein, F. (1962) *Famous Problems and other Monographs*, Chelsea, New York.

Kollros, L. (1949) *Évariste Galois*, Birkhäuser, Basel.

La Nave, F., and Mazur, B. (2002) Reading Bombelli, *Mathematical Intelligencer* **24** number 1, 12–21.

Midonick, H. (1965) *The Treasury of Mathematics* (2 vols), Penguin, Harmondsworth, Middlesex.

Neumann, P.M. (2011) *The Mathematical Writings of Évariste Galois*, European Mathematical Society, Zürich.

Richelot, F.J. (1832) De resolutione algebraica aequationis $x^{257} = 1$, sive de divisione circuli per bisectionam anguli septies repetitam in partes 257 inter se aequales commentatio coronata, *Journal für die Reine and Angewandte Mathematik* **9**, 1–26, 146–61, 209–30, 337–56.

Richmond, H.W. (1893) *Quarterly Journal of Mathematics* **26**, 206–7; and *Mathematische Annalen* **67** (1909), 459–61.

Rothman, A. (1982a) The short life of Évariste Galois, *Scientific American*, April, 112–20.

Rothman, A. (1982b) Genius and Biographers: The Fictionalization of Évariste Galois, *Amer. Math.Monthly* **89** 84–106.

Tannery, J. (1908) (ed.) *Manuscrits d'Évariste Galois*, Gauthier-Villars, Paris.

Taton, R. (1947) Les relations d'Évariste Galois avec les mathématiciens de son temps. Cercle International de Synthèse, *Revue d'Histoire des Sciences* **1**, 114.

Taton, R. (1971) Sur les relations scientifiques d'Augustin Cauchy et d'Évariste Galois, *Revue d'Histoire des Sciences* **24**, 123.

THE INTERNET

Websites come and go, and there is no guarantee that any of the following will still be in existence when you try to access them. Try entering 'Galois' in a search engine, and look him up in Wikipedia.

Scans of the manuscripts:

www.bibliotheque-institutdefrance.fr/numerisation/

The Évariste Galois archive.

http://www.galois-group.net/

Évariste Galois.

http://www-gap.dcs.st-and.ac.uk/~history/Mathematicians/Galois.html

Évariste Galois postage stamp.

http://perso.club-internet.fr/orochoir/Timbres/tgalois.htm

Bright, C. Computing the Galois group of a polynomial.

https://cs.uwaterloo.ca/~cbright/reports/pmath641-proj.pdf

GAP data library containing all transitive subgroups of $\mathbb{S}_n$ for $n \leq 30$:

http://www.gap-system.org/Datalib/trans.html

Hulpke, A. Determining the Galois group of a rational polynomial.

http://www.math.colostate.edu/~hulpke/talks/galoistalk.pdf

Hulpke, A. Techniques for the computation of Galois groups.

http://www.math.colostate.edu/~hulpke/paper/gov.pdf

Fermat numbers

http://http://www.fermatsearch.org/stat/stats.php

Mersenne primes

http://www.isthe.com/chongo/tech/math/prime/mersenne.html

Pierpont primes

http://en.wikipedia.org/wiki/Pierpont_prime

---

iPAD APP

Stewart, I. (2014) *Professor Stewart's Incredible Numbers*, TouchPress.

Index

◎编辑手记

本书是一部英文版的数学专著，中文书名可译为《伽罗瓦理论》(第4版). 本书的作者是伊恩·斯图尔特(Ian Stewart)博士，他是英国华威大学的教授.

伽罗瓦理论是学术界和科普界的一个非常热门的话题. 对于这种专家与大众都感兴趣的东西一定要慎重，因为大众可能更需要学术. 对于"大众是否需要学术"的讨论，一位专家的答案是：大众一定需要学术. 只不过，很多时候大众没有意识到这一点，或者说他们没有意识到自己在享受学术的成果. 举个简单的例子：《唐诗三百首》大家都看过，现在市面上有成百上千个版本，价格往往也很低廉，平时随手买一本翻翻，这跟学术有什么关系？但事实上，最初这本书必然也是通过古籍整理，从繁体转为简体，加上标点，很多版本还要加上注释，这些都是整理者受过学术训练才能做的. 类似的情况还有很多，读者们往往习焉不察罢了. 反过来说，如果整理者和编辑没有经过合格的学术训练，即便是这样一本大众化程度已经极高的图书，也可能错误百出. 这么多年来国家通报批评的不合格图书中有很多属于此类情形，我想，大家都不希望自己教孩子读唐诗，结果孩子背了一个错误的版本吧！

同样，伽罗瓦理论也有许多普及性的介绍，作者鱼龙混杂，作品泥沙俱下，令人目不暇接、真伪不辨. 而本书无疑是一个较优秀的解读，因为作者靠谱，而且已出到第4版了. 正如作者在第4版前言中所描述的那样：

又一个十年,又一个版本……

这一次,我克制住了想修改补充本书基本结构的冲动.非常感谢George Bergman, David Derbes, Peter Mulligan, Gerry Myerson, Jean Pierre Ortolland, F. Javier Trigos-Arrieta, Hemza Yagoub,以及Carlo Wood的指正与建议.这个版本很大程度上得益于他们的建议.已知的印刷错误已经改正了,不过无疑还会有新的错误产生.像参考文献之类的材料已经更新,阐述也进行了微小的改进.

主要改变处如下:

第2章中,我用一个不需要太多复杂背景的证明替换了代数基本定理的拓扑证明,即从点集拓扑和估计中得到的一个简单而合理的结果,这对任何上过分析学入门课程的人来说都是熟悉的.

第7章被重新整理了,用复平面$\mathbb{C}$说明了欧几里得平面$\mathbb{R}^2$.这使得讨论一个点$x+iy=z\in\mathbb{C}$可以用尺规作图来构造,而不是单独考虑它的x和y坐标.这个结果理论更加简洁,一些证明也更加简单,注意力可以放在有理数$\mathbb{Q}$的毕达哥拉斯闭包$\mathbb{Q}^{py}$上面,该闭包可以精确地由来自$\{0,1\}$的点组成.为了保持一致性,我对第20章中的正多边形做了类似但不是很大的修改.我在第21章添加了一个简短的章节,该章节是关于构造中可以使用角三等分线的内容,因为这是针对我们所发展出来的方法的一个有趣而直接的应用.

我阅读了并对彼得·诺伊曼(Peter Neumann)翻译的艾瓦利斯特·伽罗瓦(Évariste Galois)的英文出版物与手稿非常深刻.我已经把他的建议记在心里了,同时添加了一个最后一章第25章.与假设伽罗瓦做了多少工作相比,最后一章可以让我们对他究竟做了什么有一个回顾,并且明白这本书的意义是什么.人们很容易认为,今天的演讲内容只是伽罗瓦工作成果的一个简化和概括的版本.然而,数学史很少遵循现在看起来很明显的路径,伽罗瓦的情况也是这样.

当我们积累了必要的术语并理解了所需的思想时,在本书的结尾讨论这个主题更加容易.主要问题是伽罗瓦在多大程度上依靠证明交换群$\mathbb{A}_5$是简单的,或者说至少是不可解的.或许令人惊讶的回答是"一点也没有".他最大的贡献是引入了伽罗瓦对应,并证明了(用我们的语言)当且仅当一个方程的伽罗瓦群是可解时,这个方程也可以通过根式解出来.他肯定知道一般五次群是对称群$\mathbb{S}_5$,并且该群是不可解的,但是他没有强调这一点.他的主要目的是刻画能被根式解的素数方程. 他是通过推导相关的伽罗瓦群的结构来做到这一点的,但加罗瓦群是不对称的,因为在其他特征之中它的阶更小. 然而,他并没有明确地指出这一点.

诺伊曼还讨论了两个谬论:伽罗瓦证明了当 $n \geqslant 5$ 时,交换群 $\mathbb{A}_n$ 是简单的,且除了素数阶的循环群之外,$\mathbb{A}_5$是最小的简单群. 因为诺伊曼指出第一个谬论(几乎没有证据证明伽罗瓦关心交换群)肯定没有证据. 第二个谬论的唯一的证据是一个随意的声明,即在一场致命决斗的前一夜,伽罗瓦给他的朋友奥古斯特·舍瓦利叶(Auguste Chevalier) 写的信. 伽罗瓦让人捉摸不透地指出,最小非循环简单群有"5.4.3" 个元素. 诺伊曼举了一个非常好的例子反驳了这个谬论, 即伽罗瓦当时想的不是$\mathbb{A}_5$这种群, 而是同构群$\mathbb{PSL}(2,5)$. 伽罗瓦肯定知道$\mathbb{PSL}(2,5)$ 是简单的,但是在他现有的著作中甚至没有任何证据表明,非循环简单群可能有更小的阶. 我与诺伊曼在伽罗瓦是否证明过这个理论上面有不同的观点. 我相信伽罗瓦有可能证明过这个理论,尽管我同意由于缺乏证据支撑而可能性不大. 在证明方面,我仅使用伽罗瓦轻松发现和证明的思想给出证明. 无论如何仅用经典思想和一些原创性工具来证明是可能的——而且比我们期望的可能更简单.

本书较之普通伽罗瓦理论介绍书的一大特点是作者读原著、悟思想、新思路、新解读. 而国内诸多著作者没听说哪位读过伽罗瓦的原著,而大多是二手、三手、……、n 手的转述,所以以讹传讹不可避免.

本书的版权编辑李丹女士为我们翻译了本书的目录如下:

11 场同构
12 伽罗瓦对应
13 一个生动的例子
14 可解性与简单性
15 通过根式得到的解
16 抽象环与域
17 抽象域扩张
18 一般多项式方程
19 有限域
20 正多边形
21 分圆问题
22 计算伽罗瓦群
23 代数闭域
24 超越数
25 伽罗瓦做了什么或知道什么?

20 世纪 80 年代搞文艺理论的有一个热门叫比较文学研究,以北京大学乐黛云教授为旗手,很是热闹了一阵子,这里笔者也来凑个热闹,搞一个中西伽罗瓦理论学习方式比较案例. 中方是举一位搞微分方程的老先生自学伽罗瓦理论的心得体会为例,看中西方对这一重大理论的学习和理解路径与机制的不同.

第一章　排列与置换

有 n 个(互不相同)元素的集 N,不妨用各元素的序号记此集的元素,即

$$N=\{1,2,\cdots,n\}$$

设集 N 的任一排列为

$$(a_1,a_2,a_3,a_4,\cdots,a_n)$$

在行列式论中已定义此排列的反序数,现记为

$$\sigma(a_1,a_2,\cdots,a_n)$$

又定义了奇(偶) 排列,并证明了:

引理 1.1　排列中对换任二元素必改变排列的奇、偶性.

引理1.2　任一排列必可经若干次(含0次,即不必对换)对换变成标准排列

$$(1,2,\cdots,n)$$

反之亦然,且对换次数的奇偶性与此排列的奇偶性相同.

现再证:

引理1.3　集N的所有奇、偶排列个数相同,均为$\frac{n!}{2}$.

证　设N的所有偶排列集

$$\{(a_1,a_2,a_3,a_4,\cdots,a_n):2\mid\sigma(a_1,a_2,\cdots,a_n)\}$$

(其中记号$p\mid q$表整数p能整除整数q).

在每个偶排列中对换开头二元素a_1,a_2,由引理1.1知得奇排列集

$$\{(a_2,a_1,a_3,a_4,\cdots,a_n):2\nmid\sigma(a_1,a_2,\cdots,a_n)\}$$

(其中记号$p\nmid q$表整数p不能整除整数q).

显然不同的偶排列经此对换得不同的奇排列;又任一奇排列

$$(b_1,b_2,b_3,b_4,\cdots,b_n)$$

必可由一个偶排列$(b_2,b_1,b_3,b_4,\cdots,b_n)$对换开头二元素$b_2,b_1$而得,故所有偶排列与所有奇排列有一一对应关系,其总个数相同,显然都等于$\frac{n!}{2}$个.

设

$$N=\{a_1,a_2,\cdots,a_n\}$$

($a_1,a_2,\cdots,a_n$为$1,2,\cdots,n$的一排列),作一一映射

$$a_i\to b_i\quad(i=1,2,\cdots,n)$$

即把每个元素a_i变成N中某一元素b_i(b_i可等于a_i,即a_i不变),$a_{i_1}\neq a_{i_2}$时$b_{i_1}\neq b_{i_2}$(不同的a_i变成不同的b_i).则称此映射为N的一个置换,记为

$$\begin{pmatrix}a_1,a_2,\cdots,a_n\\b_1,b_2,\cdots,b_n\end{pmatrix}$$

显然$(b_1,b_2,\cdots,b_n)$也是$1,2,\cdots,n$的一个排列.

实际不妨取N为标准排列$(1,2,\cdots,n)$,则任一置换必可唯一地写成

$$\begin{pmatrix}1,2,\cdots,n\\c_1,c_2,\cdots,c_n\end{pmatrix}$$

例如

$$\begin{pmatrix}2,3,1,4\\3,1,4,2\end{pmatrix}=\begin{pmatrix}1,2,3,4\\4,3,1,2\end{pmatrix}$$

其中元素个数 n 称此置换次数,此置换称 n 次置换.

显然任一 n 次置换对应唯一的(n 个元素)排列$(c_1,c_2,\cdots,c_n)$,反之任一(n 个元素)排列$(c_1,c_2,\cdots,c_n)$对应唯一的 n 次置换,故 n 次置换总数为 $n!$.

若 N 的一个置换中有 $n-k(n-2\leqslant k\leqslant n)$ 个元素(不妨设为 $a_{k+1},a_{k+2},\cdots,a_n$)不变①,而其余元素有循环变换关系,即(不妨设)把 a_1 变成 a_2,a_2 变成 a_3,a_3 变成 a_4,$\cdots$,a_{k-1} 变成 a_k,最后的 a_k 变成 a_1,亦即

$$\begin{pmatrix}a_1,a_2,a_3,\cdots,a_{k-1},a_k,a_{k+1},a_{k+2},\cdots,a_n\\a_2,a_3,a_4,\cdots,a_k,\ \ a_1,a_{k+1},a_{k+2},\cdots,a_n\end{pmatrix}$$

则称此置换为 $a_1,a_2,\cdots,a_k$ 的轮换,特记为

$$(a_1,a_2,\cdots,a_k)$$

称 k 为此轮换阶数,此轮换称为 k 阶轮换. 当然此轮换亦可记为$(a_2,a_3,\cdots,a_k,a_1)$,$(a_3,\cdots,a_k,a_1,a_2)$,$\cdots$.

元素 a_i 与 a_j 的对换是其特例,即二阶轮换(a_i,a_j).

对排列$(a_1,a_2,\cdots,a_n)$进行置换$\begin{pmatrix}1, 2, \cdots,n\\c_1,c_2,\cdots,c_n\end{pmatrix}$所得排列,记为

$$(a_1,a_2,\cdots,a_n)\begin{pmatrix}1, 2, \cdots,n\\c_1,c_2,\cdots,c_n\end{pmatrix}$$

如

$$(2,3,4,1)\begin{pmatrix}1,2,3,4\\4,2,1,3\end{pmatrix}=(2,1,3,4)$$

对 N 先进行置换 σ_1,再进行置换 σ_2 所得结果(又得一个置换)称之为置换 σ_1 与置换 σ_2 之积,记为 $\sigma_1\sigma_2$.

但有些书记为 $\sigma_2\sigma_1$. 相应把排列$(a_1,a_2,\cdots,a_n)$进行置换$\begin{pmatrix}1, 2, \cdots,n\\c_1,c_2,\cdots,c_n\end{pmatrix}$,所得排列记为

$$\begin{pmatrix}1, 2, \cdots,n\\c_1,c_2,\cdots,c_n\end{pmatrix}(a_1,a_2,\cdots,a_n)$$

①当 $k=n$ 时无元素不变.

如

$$\begin{pmatrix}1,2,3,4\\2,4,3,1\end{pmatrix}\begin{pmatrix}1,2,3,4\\3,4,1,2\end{pmatrix}=\begin{pmatrix}1,2,3,4\\2,4,3,1\end{pmatrix}\begin{pmatrix}2,4,3,1\\4,2,1,3\end{pmatrix}=\begin{pmatrix}1,2,3,4\\4,2,1,3\end{pmatrix}$$

(实际可直接写出结果).

注意 置换的乘法不适合交换律.如

$$\begin{pmatrix}1,2,3\\2,1,3\end{pmatrix}\begin{pmatrix}1,2,3\\1,3,2\end{pmatrix}=\begin{pmatrix}1,2,3\\3,1,2\end{pmatrix}$$

但

$$\begin{pmatrix}1,2,3\\1,3,2\end{pmatrix}\begin{pmatrix}1,2,3\\2,1,3\end{pmatrix}=\begin{pmatrix}1,2,3\\2,3,1\end{pmatrix}$$

定理 1.1 除恒等置换$\begin{pmatrix}1,2,\cdots,n\\1,2,\cdots,n\end{pmatrix}$(即各元素都不变)外,任一置换必可分解为若干($k$)个无公共元素的轮换(即任二轮换中的元素互不相同,当 $k=1$ 时此置换就是一个轮换)之积,且此分解式是唯一的(不计各轮换因式的次序).

其道理很简单,可用例子说明:

设有置换

$$\begin{pmatrix}1,2,3,4,5,6,7,8,\ 9,10\\3,5,4,1,2,6,8,10,7,9\end{pmatrix}$$

从上行第一个数1开始,由1变成3,3变成4,4变成1,得一个轮换(1,3,4);

再从上行未取的第一数2开始,由2变成5,5变成2,又得一个轮换(2,5);

上行未取的第一数6不变;

上行随后未取第一数7,由7变成8,8变成10,10变成9,9变成7,又得一轮换(7,8,10,9).

上行10个数已取尽,于是原置换分解为

$$(1,3,4)(2,5)(7,8,10,9)$$

显然三个轮换因式的次序可任意调换,即两两无公共元素的轮换适合乘法交换律.

此定理提供了表示置换的简式.

定义 1.1 在置换

$$\begin{pmatrix}1,\ 2,\ \cdots,n\\c_1,c_2,\cdots,c_n\end{pmatrix}$$

中,若排列

$$(c_1,c_2,\cdots,c_n)$$

为偶(奇)排列,则称此置换为偶(奇)置换.

恒等置换是偶置换.

由引理1.3知,n次偶、奇置换总数均为$\frac{n!}{2}$个.

引理1.4 偶(奇)置换必可分解为偶(奇)数个对换之积.

证 从引理1.2知,标准排列$(1,2,\cdots,n)$经偶(奇)数次对换变成偶(奇)排列$(c_1,c_2,\cdots,c_n)$. 于是此偶(奇)置换为此偶(奇)数个对换之积.

推论1 奇(偶)置换改变(不改变)排列的奇偶性,反之亦然.

推论2 两偶置换之积,两奇置换之积为偶置换;奇置换与偶置换之积,偶置换与奇置换之积为奇置换.

推论3 若干个置换之积,若其中奇置换个数为偶(奇)数,则此积为偶(奇)置换.

引理1.5 偶(奇)阶轮换是奇(偶)置换.

证 易见

$$(a_1,a_2,\cdots,a_k)=(a_1,a_2)(a_1,a_3)\cdots(a_1,a_k)$$

即为$k-1$个对换之积,于是偶(奇)数k阶轮换为奇(偶)数次对换之积,但由引理1.1知对换改变排列之奇偶性,故为奇置换,再据上述推论3知所述轮换是奇(偶)置换.

推论 在置换用轮换因式之积的表示式中,若偶阶轮换因式个数为偶(奇)数,则此置换为偶(奇)置换,反之亦然.

第二章 置换群

定理2.1 n次置换的乘法适合结合律:即对任何三个n次置换$\sigma_1,\sigma_2,\sigma_3$有

$$(\sigma_1\sigma_2)\sigma_3=\sigma_1(\sigma_2\sigma_3)$$

证 若σ_1把元素a_{n_1}变成元素a_{m_1},σ_2把元素a_{m_1}变成元素a_{p_1},σ_3把元素a_{p_1}变成元素a_{q_1},则$\sigma_1\sigma_2$把元素a_{n_1}变成元素a_{p_1},再由上述知$(\sigma_1\sigma_2)\sigma_3$把元素$a_{n_1}$变成元素$a_{q_1}$;又由上述知$\sigma_2\sigma_3$把元素$a_{m_1}$变成元素$a_{q_1}$,于是$\sigma_1(\sigma_2\sigma_3)$亦把元素$a_{n_1}$变成元素$a_{q_1}$. 因$a_{n_1}$可代表$N$中任一元素,故定理结论得证.

同样置换的乘法结合律可推广到任一个置换之积(即可任意添加括号,但不能改变各置换因式的次序).

称置换

$$\begin{pmatrix}1,2,\cdots,n\\1,2,\cdots,n\end{pmatrix}$$

为单位置换(或恒等置换),即使各元素不变,特记为I.

一般

$$\begin{pmatrix}a_1,a_2,\cdots,a_n\\a_1,a_2,\cdots,a_n\end{pmatrix}=I$$

显然

$$\sigma I=I\sigma=\sigma$$

易见I为左(右)乘任一置换使之不变的唯一置换.

设n次置换

$$\sigma=\begin{pmatrix}1,2,\cdots,n\\c_1,c_2,\cdots,c_n\end{pmatrix}$$

称置换

$$\sigma'=\begin{pmatrix}c_1,c_2,\cdots,c_n\\1,2,\cdots,n\end{pmatrix}$$

为σ的逆置换,记为σ^{-1},则有$\sigma^{-1}=\sigma'$.

显然σ为σ'的逆置换,称二置换为互逆置换.

一般

$$\begin{pmatrix}a_1,a_2,\cdots,a_n\\b_1,b_2,\cdots,b_n\end{pmatrix}^{-1}=\begin{pmatrix}b_1,b_2,\cdots,b_n\\a_1,a_2,\cdots,a_n\end{pmatrix}$$

显然

$$\sigma\sigma^{-1}=\sigma^{-1}\sigma=I$$

且σ^{-1}为左乘(右乘)σ使积为I的唯一置换.

引理2.1 σ与σ^{-1}有相同奇偶性.

证 显然I是偶置换,据$\sigma\sigma^{-1}=I$及引理1.4的推论2知本引理成立(用反证法).

由于在$N=\{1,2,\cdots,n\}$的所有置换中可定义乘法运算,它适合结合律,又存在单位置换I(使对任一置换σ适合$\sigma I=I\sigma=\sigma$),对任一置换σ存在唯一逆置换σ^{-1}(使$\sigma\sigma^{-1}=\sigma^{-1}\sigma=I$).即对所有$n$次置换的乘法运算适合群的定义,故所有$n$次置换之集对其乘法运算为一个群,称之为$n$次对称群,记之为$S_n$.

因单位置换 I 为偶置换,任两偶置换之积为偶置换,任何偶置换之逆置换亦为偶置换.故所有 n 次偶置换组成之集对其乘法运算为一个群,称之为 n 次交代群,记之为 A_n.

只含(n 次)单位置换之集 $\{I\}$,易见对其乘法运算为一个群($I\cdot I=I$, $I^{-1}=I$),称之为 n 次单位置换群,或 n 次恒等置换群.

因二次置换只有 I 及 $(1,2)$,易见二次群只有

$$\{I\},S_2=\{I,(1,2)\}$$

一般情况,若由一些(部分或全部) n 次置换所组成的置换集对置换乘法符合群的定义(即集的任二置换之积为集的置换,置换积适合乘法结合律,集含单位置换,集的任一置换之逆置换亦为集的置换),则称此置换集为一个(n 次)置换群,群的所有置换个数 q 称为此群的阶数,又称此群为 q 阶群.

如 $\{I\}$ 的阶数为 1;对称群 S_n 及交代群 A_n 的阶数分别为 $n!$ 及 $\frac{n!}{2}$.

同样记(置换 σ 的 k 次幂)

$$\underbrace{\sigma\sigma\cdots\sigma}_{k个}=\sigma^k$$

同样定义任何置换 σ 的零次幂为单位置换 I

$$\sigma^0=I$$

易见对任何 $k_1,k_2\in\mathbf{N}$($\mathbf{N}=\mathbf{Z}^+\cup\{0\}$——按现行中学课本规定)有

$$\sigma^{k_1}\sigma^{k_2}=\sigma^{k_1+k_2}$$

$$(\sigma^{k_1})^{k_2}=\sigma^{k_1k_2}$$

从而 σ 的幂适合乘法交换律.

又对任何自然数 k(含 0)有

$$I^k=I$$

显然对轮换的幂有

$$(a_1,a_2,\cdots,a_k)^2=\begin{pmatrix}a_1,a_2,\cdots,a_{k-2},a_{k-1},a_k\\a_3,a_4,\cdots,a_k,\quad a_1,\quad a_2\end{pmatrix}$$

$$(a_1,a_2,\cdots,a_k)^3=\begin{pmatrix}a_1,a_2,\cdots,a_{k-3},a_{k-2},a_{k-1},a_k\\a_4,a_5,\cdots,a_k,\quad a_1,\quad a_2,\quad a_3\end{pmatrix}$$

$$\vdots$$

$$(a_1,a_2,\cdots,a_k)^{k-2}=\begin{pmatrix}a_1,\quad a_2,a_3,a_4,\cdots,a_k\\a_{k-1},a_k,a_1,a_2,\cdots,a_{k-2}\end{pmatrix}$$

$$(a_1,a_2,\cdots,a_k)^{k-1}=\begin{pmatrix}a_1,a_2,a_3,\cdots,a_k\\a_k,a_1,a_2,\cdots,a_{k-1}\end{pmatrix}=(a_1,a_k,a_{k-1},\cdots,a_3,a_2)$$

$$(a_1,a_2,\cdots,a_k)^k=I$$

$$(a_1,a_2,\cdots,a_k)^{mk}=I\quad(m\in\mathbf{Z}^+)$$

$$(a_1,a_2,\cdots,a_k)^{mk+r}=(a_1,a_2,\cdots,a_k)^r$$

$$(m\in\mathbf{Z}^+,r=0,1,2,\cdots,k-1)$$

易见

$$(a_1,a_2,\cdots,a_k)^r\quad(r=0,1,2,\cdots,k-1)$$

的逆元

$$((a_1,a_2,\cdots,a_k)^r)^{-1}=(a_1,a_2,\cdots,a_k)^{k-r}$$

定理 2.2 对任何置换 σ,必存在唯一正整数 m,使

$$\sigma^m=I$$

且 m 为使"σ 的幂等于 I"的最小次数.

证 当 $\sigma=I$,取 $m=1$ 即可. 当 $\sigma\neq I$,把 σ 分解为若干个两两无公共元素的轮换因式之积,取各轮换因式阶数之最小公倍数 m(若 σ 就是一个轮换,则取 m 为此轮换阶数). 易见 $\sigma^m=I$.

再证 m 为使 σ 的幂等于 I 的最小次数:对小于 m 的正整数 m',因 m' 必然至少不是某轮换因式阶数 p 的倍数,由上述各轮换幂的公式知,$\sigma^{m'}$(导至此轮换的 m' 次幂)必不能使此轮换中各元素不变,于是 $\sigma^{m'}\neq I$.

定理 2.3 设定理 2.2 中所述 n 次置换 σ 及正整数 m,则 σ 的所有自然数次幂只有如下 m 个互不相同的置换

$$\sigma,\sigma^2,\sigma^3,\cdots,\sigma^{m-1},\sigma^m=\sigma^0=I\tag{2.1}$$

它们组成一个置换群,称为由 σ 生成的(n 次 m 阶)循环置换群(或简称循环群),称 σ 为此循环置换群的生成元.

证 对任何自然数 p,对除数 m 作带余除法得

$$p=qm+r\quad(q\in\mathbf{N},r=0,1,2,\cdots,m-1)$$

从而

$$\sigma^p=\sigma^{qm+r}=(\sigma^m)^q\sigma^r=I^q\sigma^r=I\sigma^r=\sigma^r$$

即 σ 的所有幂都在式(2.1)内.

再证式(2.1)内任二置换不等:

否则有 $r_1,r_2\in\{0,1,2,\cdots,m-1\}$,不妨设 $r_1<r_2$,使

$$\sigma^{r_1}=\sigma^{r_2}$$

即

$$\sigma^{r_1}=\sigma^{r_1}\sigma^{r_2-r_1}$$

两边左乘 σ^{m-r_1} 得

$$I=\sigma^{r_2-r_1}$$

但

$$0<r_2-r_1\leqslant r_2<m$$

这与 m 为"使 σ 的幂等于 I"的最小次数矛盾.

又因式(2.1)中有单位置换 I,其内任二置换 σ^{r_1} 与 σ^{r_2} 之积 $\sigma^{r_1+r_2}$ 仍(可化)为式(2.1)内的置换,对其内任一置换 $\sigma^{r}(r=0,1,2,\cdots,m-1)$,易见有逆置换 σ^{m-r} 在式(2.1)内,故式(2.1)组成一个置换群.

特例:k 阶轮换 σ 为 k 阶循环置换群的生成元,且 $\sigma^{k}=I$.

类似有:

定理2.4 若一 n 次置换集内可进行乘法运算,则此置换集为一个置换群.

证 只要再证此集含单位置换且对集内任一置换 σ,在此集存在逆置换 σ^{-1}:由定理2.2知存在最小正整数 m,使 $\sigma^{m}=I$,由假设知此集含 $\sigma^{m}=I$;又易见 $\sigma\sigma^{m-1}=\sigma^{m-1}\sigma=I$,于是此集又含 $\sigma^{m-1}=\sigma^{-1}$.故此置换集为一置换群.

例1 设5次置换

$$\sigma=(1,2,3)(4,5)$$

则因3,2之最小公倍数为6,σ 生成6阶循环群

$$\sigma=(1,2,3)(4,5)$$
$$\sigma^{2}=(1,3,2)$$
$$\sigma^{3}=(4,5)$$
$$\sigma^{4}=(1,2,3)$$
$$\sigma^{5}=(1,3,2)(4,5)$$
$$\sigma^{6}=I$$

例2 设6阶轮换

$$\sigma=(1,2,3,4,5,6)$$

则

$$\sigma=(1,2,3,4,5,6)$$
$$\sigma^{2}=(1,3,5)(2,4,6)$$
$$\sigma^{3}=(1,4)(2,5)(3,6)$$
$$\sigma^{4}=(1,5,3)(2,6,4)$$

$$\sigma^5=(1,6,5,4,3,2)$$
$$\sigma^6=I$$

这6个置换组成6阶循环群.

例3 设5阶轮换

$$\sigma=(1,2,3,4,5)$$

则由σ产生5阶循环群

$$\{(1,2,3,4,5),(1,3,5,2,4),(1,4,2,5,3),(1,5,4,3,2),I\}$$

定义2.1 若两个n次置换群S_1,S_2有(作为置换集)$S_1\subset S_2$,则称S_1是S_2的子群.

易见任何n次置换群是S_n的子群,$\{I\}$是任何n次置换群的子群.

定理2.5 置换群G的阶数p必为其子群G'的阶数p'的整数倍.此倍数记为$[G:G']$,$[G:G']=\frac{p}{p'}$.

证 若$p'=p$,结论显然,下设$p'<p$,又设

$$G'=\{\sigma_1,\sigma_2,\cdots,\sigma_{p'}\}$$

因$p'<p$,群G内必有不属于G'的置换τ_2,从而含置换

$$\sigma_1\tau_2,\sigma_2\tau_2,\cdots,\sigma_{p'}\tau_2 \tag{2.2}$$

上述p'个置换互不相同(否则如$\sigma_1\tau_2=\sigma_2\tau_2$,两边右乘$\tau_2^{-1}$得$\sigma_1=\sigma_2$,矛盾).且其中任一置换不与$G'$中任一置换相同(否则如$\sigma_1\tau_2=\sigma_2$,则$\tau_2=\sigma_1^{-1}\sigma_2$,因群$G'$含$\sigma_1,\sigma_2$,必含逆置换$\sigma_1^{-1}$,从而又含$\sigma_1^{-1}\sigma_2=\tau_2$,与前述矛盾).

若除G'及式(2.2)中的置换外G再无其他置换,则$p=2p'$,结论得证.若G尚有其他置换,在其中取置换τ_3,从而G又含置换

$$\sigma_1\tau_3,\sigma_2\tau_3,\cdots,\sigma_{p'}\tau_3 \tag{2.3}$$

同样式(2.3)中各置换互不相同,且其任一置换不与G'中的置换相同.再证式(2.3)中任一置换不与式(2.2)中任一置换相同:否则如$\sigma_3\tau_3=\sigma_2\tau_2$,则

$$\tau_3=\sigma_3^{-1}(\sigma_2\tau_2)$$

即

$$\tau_3=(\sigma_3^{-1}\sigma_2)\tau_2$$

但$\sigma_3^{-1}\sigma_2$为G'的置换,故τ_3为式(2.2)中的置换,这与假设G尚有在G'及式(2.2)中的置换之外的置换τ_3矛盾.

若除G'及式(2.2),式(2.3)中的置换外G不含其他置换,则

$p=3p'$,命题得证.否则继续做同样论证,因 G 的置换个数 p 有限,故必到某一步取尽 G 的所有置换,即 G 的所有置换排成下表

$$\begin{array}{c}\sigma_1,\quad \sigma_2,\quad \cdots,\sigma_{p'}\\ \sigma_1\tau_2,\sigma_2\tau_2,\cdots,\sigma_{p'}\tau_2\\ \sigma_1\tau_3,\sigma_2\tau_3,\cdots,\sigma_{p'}\tau_3\\ \vdots\\ \sigma_1\tau_m,\sigma_2\tau_m,\cdots,\sigma_{p'}\tau_m\end{array}$$

从而 $p=mp'$,定理证毕.

定理 2.6 设有 k 阶循环置换群

$$G=\{\sigma,\sigma^2,\cdots,\sigma^{k-1},I\}\quad(\sigma^k=I)$$

又设正整数 m 与 k 的最大公约数(记为)$(k,m)=d$,设 $\frac{k}{d}=k'$,则置换集

$$G'=\{\sigma^m,(\sigma^m)^2,\cdots,(\sigma^m)^{k'}\}=\{\sigma^m,\sigma^{2m},\cdots,\sigma^{k'm}\}$$

为 k' 阶循环置换群,且为 G 的子群.

证 设

$$\frac{m}{d}=m'$$

求得

$$(\sigma^m)^{k'}=\sigma^{\frac{mk}{d}}=\sigma^{m'k}=(\sigma^k)^{m'}=I^{m'}=I$$

再对任何小于 k' 的正整数 k'',证 $(\sigma^m)^{k''}\neq I$:否则若 $(\sigma^m)^{k''}=I$,即 $\sigma^{mk''}=I$,因 k 为以 σ 为生成元的循环群之阶数,故 mk'' 除以 k 所得余数 $r=0$(否则 $\sigma^{mk''}=\sigma^r\neq I$).于是

$$mk''=pk\quad(p\in\mathbf{Z}^+)$$

两边除以 d 得

$$m'k''=pk'$$

但 k' 与 m' 分别为 k 与 m 除以其最大公约数 d 所得之商,故 k' 与 m' 互质,于是再从上式知 $k'\mid k''$,这与 $k''<k'$ 矛盾.

于是 σ^m 及 k' 适合定理 2.2 对 σ 及 k 的条件,故从定理 2.3 知由 σ^m 生成 k' 阶循环群,又显然是 G 的子群.

推论 1 当 k,m 互质(或称互素,即 $d=1$)时,上述 σ^m 亦为 G 的生成元(这时 G 与其子群 G' 有同样阶数 k,故二群相同).

推论 2 当 k 为素数,则上述 k 阶循环群 G 中除 I 外任一置换 σ^m 均为 G 的生成元(这时 m 非 k 的整数倍数,k 与 m 互素).

例4　设例1中由 $\sigma=(1,2,3)(4,5)$ 产生6阶循环群 G.

取 $m=4$,4与6之最大公约数为2,$\frac{6}{2}=3$,由 $\sigma^4=(1,2,3)$ 产生3阶循环群(G之子群)为

$$\{(1,2,3),(1,3,2),I\}$$

取 $m=3$,3与6之最大公约数为3,$\frac{6}{3}=2$,由 $\sigma^3=(4,5)$ 产生2阶循环群(G之子群)为

$$\{(4,5),I\}$$

例5　设例2中由 $\sigma=(1,2,3,4,5,6)$ 产生6阶循环群 G. 分别取 $m=4,3,6$,则由

$$\sigma^4=(1,5,3)(2,6,4)$$
$$\sigma^3=(1,4)(2,5)(3,6)$$
$$\sigma^5=(1,6,5,4,3,2)$$

产生的循环群分别为

$$\{(1,5,3)(2,6,4),(1,3,5)(2,4,6),I\}$$
$$\{(1,4)(2,5)(3,6),I\}$$
$$\{(1,6,5,4,3,2),(1,5,3)(2,6,4),(1,4)(2,5)(3,6),(1,3,5)(2,4,6),(1,2,3,4,5,6),I\}$$

它们都是 G 之子群,第三群即 G.

例6　设例3中由 $\sigma=(1,2,3,4,5)$ 产生(素数)5阶循环群 G. 取 $m=2$,由 σ^2 产生的循环群(为 G)

$$\{(1,3,5,2,4),(1,5,4,3,2),(1,2,3,4,5),(1,4,2,5,3),I\}$$

不变子群(或称自共轭子群,标准子群)的定义:

定义2.2　设群 G 的子群 G',若对任何 $\sigma\in G$ 均有

$$\sigma^{-1}G'\sigma\triangleq\{\sigma^{-1}\tau\sigma:\tau\in G'\}=G'$$

("$\triangleq$"表"定义"),则称 G' 为 G 的不变子群.

定义2.3　设群 G 的子群 G',若对任何 $\sigma\in G,\tau\in G'$,有 $\sigma^{-1}\tau\sigma\in G'$,则称 G' 为 G 的不变子群.

现证上两定义等价:

当 G 与 G' 符合定义2.2的条件时,显然也符合定义2.3的条件;

当 G 与 G' 符合定义2.3的条件时,显然对任何 $\sigma\in G'$ 有 $\sigma^{-1}G\sigma\subset G'$,再证 $G'\subset\sigma^{-1}G\sigma$ 即可. 对任何 $\sigma''\in G'\subset G$ 及任何 $\sigma\in G$,令

$$\tau = \sigma\sigma''\sigma^{-1}$$

则易见 $\tau \in G$ 且

$$\sigma'' = \sigma^{-1}\tau\sigma \in \sigma^{-1}G\sigma$$

得证

$$G' \subset \sigma^{-1}G\sigma$$

不变子群的例子(可按定义2.3判定):

A_n 是 S_n 的不变子群;

设 n 次单位置换 I,则恒等置换群 $\{I\}$ 是任何 n 次置换群的不变子群.

定理2.7 设置换 σ 及置换

$$\tau = (x_{m_1}, x_{m_2}, \cdots, x_{m_k})(x_{m_{k+1}}, \cdots)\cdots$$

置换 σ 使 x_{m_i} 变成 y_{m_i},记为

$$x_{m_i}\sigma = y_{m_i} \quad (i = 1, 2, \cdots, k, k+1, \cdots)$$

则

$$\sigma^{-1}\tau\sigma = (y_{m_1}, y_{m_2}, \cdots, y_{m_k})(y_{m_{k+1}}, \cdots)\cdots$$

证 由 $x_{m_1}\sigma = y_{m_1}$ 得 $y_{m_1}\sigma^{-1} = x_{m_1}$,又 $x_{m_1}\tau = x_{m_2}$ 故

$$y_{m_1}\sigma^{-1}\tau\sigma = x_{m_1}\tau\sigma = x_{m_2}\sigma = y_{m_2}$$

同样 y_{m_2} 经置换 $\sigma^{-1}\tau\sigma$ 变成 y_{m_3}(当 $k \geqslant 3$ 时),$\cdots$,y_{m_k} 经 $\sigma^{-1}\tau\sigma$ 变成 y_{m_1},故得轮换 $(y_{m_1}, y_{m_2}, \cdots, y_{m_k})$. 同理得轮换 $(y_{m_{k+1}}, \cdots)$,$\cdots$,证毕.

本定理表明,把 τ 的轮换表示式中的任一 x_{m_i} 改成 $x_{m_i}\sigma = y_{m_i}$,便得 $\sigma^{-1}\tau\sigma$ 的表示式. 于是不变子群 G' 的意义为:G' 的所有置换 τ 的轮换表示式对其中元素作由 G 的任何置换 σ 产生的上述变换后所得轮换表示式仍属 G'(定义2.3);G' 作为置换集经上述变换后仍得 G'(定义2.2,即 G' 不变).

第三章 数域,代数扩域

定义3.1 一个包含0与1的复数集 Ω,如其中任二数之和、差、积、商(除数不为0)亦为此集 Ω 中的数,则称 Ω 为一数域(或称数体).

因 Ω 含有1,故必含所有正整数(几个1之和);又因 Ω 含有0,故必含所有负整数(0与正整数之差);从而 Ω 必须含有所有有理数(两整数之商(除数非0)).

数域的例子:

(所有)有理数域 $\mathbf{Q}$;

(所有)实数域 $\mathbf{R}$;

(所有)复数域 $\mathbf{C}$.

易证 $\{a+b\sqrt{2}:a,b\in\mathbf{Q}\}$ 为一数域,它由部分实数组成;

易证 $\{a+b\mathrm{i}:a,b\in\mathbf{Q}\}$ 为一数域,它由部分复数组成.

$\{a+b\sqrt{2}+c\sqrt{3}:a,b,c\in\mathbf{Q}\}$ 不是数域,因它不含其中二数 $\sqrt{2}$ 与 $\sqrt{3}$ 之积 $\sqrt{6}$.

若两数域 $\Omega_1,\Omega_2,\Omega_1\subset\Omega_2$,则称 Ω_1 为 Ω_2 的子域,Ω_2 为 Ω_1 的扩域.

复数域 $\mathbf{C}$ 是实数域 $\mathbf{R}$ 的扩域. 由上述知有理数域 $\mathbf{Q}$ 是任一数域的子域,即有理数域是最小的数域,是所有数域的交集.

由部分或全部实(复)数组成的数域叫实(复)域.

注意实(复)域不一定是实数域 $\mathbf{R}$(复数域 $\mathbf{C}$),后者由所有实(复)数组成. 如 $\{a+b\sqrt{2}:a,b\in\mathbf{Q}\}$ 是实域但不是实数域,$\{a+b\mathrm{i}:a,b\in\mathbf{Q}\}$ 是复域但不是复数域. 而有理数域 $\mathbf{Q}$ 的任何真子集不是数域.

设数域 Δ 的扩域 Ω,因对 Ω 中任何数 a 及任何数 $\lambda\in\Delta\subset\Omega$,$\Omega$ 必须含 $0-a=-a$ 及 λa,于是 Ω 可看成域 Δ 上的线性空间(视 Ω 中的数为空间的向量),若此线性空间是有限 n 维的,则称域 Ω 为域 Δ 的有限扩域,n 称 Ω 对 Δ 的维数,记为

$$(\Omega:\Delta)=n$$

例如 $\Omega=\{a+b\sqrt{2}:a,b\in\mathbf{Q}\}$ 是 $\mathbf{Q}$ 的扩域,易见 1 与 $\sqrt{2}$ 在 $\mathbf{Q}$ 上线性无关(因仅当有理数 $a=b=0$ 时 $a+b\sqrt{2}=0$),故为 $\mathbf{Q}$ 上线性空间 Ω 的基底,Ω 是 $\mathbf{Q}$ 的有限扩域,$(\Omega:\mathbf{Q})=2$.

$\Omega=\{a+b\mathrm{i},a,b\in\mathbf{Q}\}$ 也是 $\mathbf{Q}$ 的有限扩域,$(\Omega:\mathbf{Q})=2$(基底为 $1,\mathrm{i}$).

复数域 $\mathbf{C}$ 是实数域 $\mathbf{R}$ 的有限扩域,$(\mathbf{C}:\mathbf{R})=2$(基底为 $1,\mathrm{i}$).

实数域 $\mathbf{R}$ 是有理数域 $\mathbf{Q}$ 的扩域,但不是有限扩域,因可证对任意多个素数 $p_1,p_2,\cdots,p_n$,实数 $\sqrt{p_1},\sqrt{p_2},\cdots,\sqrt{p_n}$ 在 $\mathbf{Q}$ 上线性无关.

定义 3.2 代数方程

$$x^n+a_1x^{n-1}+a_2x^{n-2}+\cdots+a_{n-1}x+a_n=0$$

的有理域 Δ:在有理数域 $\mathbf{Q}$ 添加 $a_1,a_2,\cdots,a_n$ 所成的最小数域,即一切形如($a_1,a_2,\cdots,a_n$ 在 $\mathbf{Q}$ 上的有理式)

$$\frac{\sum\limits_{p_1,p_2,\cdots p_n\in\mathbf{N}} c_{p_1p_2\cdots p_n} a_1^{p_1} a_2^{p_2}\cdots a_n^{p_n}}{\sum\limits_{p_1,p_2,\cdots,p_n\in\mathbf{N}} c'_{p_1p_2\cdots p_n} a_1^{p_1} a_2^{p_2}\cdots a_n^{p_n}} \tag{3.1}$$

(其中 $c_{p_1p_2\cdots p_n}, c'_{p_1p_2\cdots p_n} \in \mathbf{Q}$,分子、分母均取有限个项,分母不为0)的数组成的数域(易见这些数符合数域的定义且包含任何有理数及 $a_1,a_2,\cdots,a_n$). 原方程可看成定义在 Δ 上的代数方程.

当 $a_1,a_2,\cdots,a_n$ 均为有理数时,Δ 即有理数域 $\mathbf{Q}$. 当 $a_1,a_2,\cdots,a_n$ 全为实数但非全为有理数时,则 Δ 非域 $\mathbf{Q}$,而是一个实域(当然也是复域),但不是实数域 $\mathbf{R}$. 当 $a_1,a_2,\cdots,a_n$ 中含有虚数,则 Δ 不是实域而是复域,但不是复数域 $\mathbf{C}$.

举例如下:

方程

$$x^3 + \sqrt{2}x^2 + 3x - 1 = 0$$

的有理域

$$\Delta = \{a + b\sqrt{2} : a,b \in \mathbf{Q}\}$$

多项式

$$x^2 + \mathrm{i}x - 3$$

的复域

$$\Delta = \{a + b\mathrm{i} : a,b \in \mathbf{Q}\}$$

定理 3.1 设数域 F 的有限扩域为 F',数域 F' 的有限扩域为 F'',则数域 F'' 为 F 的有限扩域,且

$$(F'':F) = (F'':F')\cdot(F':F)$$

证 设

$$(F':F) = m$$
$$(F'':F') = n$$

则域 F' 有 m 个数 $u_1,u_2,\cdots,u_m$ 为域 F 上线性空间 F' 的基底,域 F'' 有 n 个数 $v_1,v_2,\cdots,v_n$ 为域 F' 上线性空间 F'' 的基底. 域 F'' 上任一数 v 可唯一地用 $v_1,v_2,\cdots,v_n$ 在域 F' 上的线性组合表示,即

$$v = \sum_{i=1}^{n} c_i v_i$$

其中 $c_i \in F'$,它们可唯一地用 $u_1,u_2,\cdots,u_m$ 在域 F 上的线性组合表示,即

$$v_i = \sum_{j=1}^{m} c_{ji} u_j$$

其中 $c_{ji} \in F$,于是

$$v = \sum_{i=1}^{n} \sum_{j=1}^{m} c_{ji}(u_j v_i)$$

即 v 为域 F'' 上 mn 个数 $u_j v_i (j = 1,2,\cdots,m; i = 1,2,\cdots,n)$ 在域 F 上的线性组合,故 F'' 为域 F 的有限扩域,再证这些 $u_j v_i$ 在 F 上线性无关(从而知 $(F'':F) = nm = (F'':F') \cdot (F':F)$) 即可.

若

$$\sum_{i=1}^{n} \sum_{j=1}^{m} c'_{ji}(u_j v_i) = 0 \quad (c'_{ji} \in F)$$

即

$$\sum_{i=1}^{n} \left(\sum_{j=1}^{m} c'_{ji} u_j \right) v_i = 0$$

因 $c'_{ji} \in F \subset F'$,$u_j \in F'$,故 $\sum_{j=1}^{m} c'_{ji} u_j \in F'$,而 $v_1, v_2, \cdots, v_n$ 为域 F' 上线性空间 F'' 的基底,它们在域 F' 上线性无关,从而有

$$\sum_{j=1}^{m} c'_{ji} u_j = 0 \quad (i = 1,2,\cdots,n)$$

又因 $c'_{ji} \in F$,而 $u_1, u_2, \cdots, u_m$ 为域 F 上线性空间 F' 的基底,它们在域 F 上线性无关,从而又有

$$c'_{ji} = 0 \quad (i = 1,2,\cdots,n; j = 1,2,\cdots,m)$$

得证这些 $u_j v_i$ 在 F 上线性无关.

定理 3.2 设数域 Δ 的有限扩域 Ω,又 Δ 的扩域 Σ 是 Ω 的子域,则 Σ 是 Δ 的有限扩域,且 $(\Omega:\Delta)$ 是 $(\Sigma:\Delta)$ 的整数倍.

证 设 $(\Omega:\Delta) = m$,则 Ω 内对 Δ 线性无关的元素个数不大于 m,Ω 的子域 Σ 内对 Δ 线性无关的元素个数更不大于 m,故 Σ 为 Δ 的有限扩张. 设域 Δ 上线性空间 Ω 的基底为 $v_1, v_2, \cdots, v_m$,它们可在域 Δ 线性表示 Ω 内任一数,从而更可在域 Δ 线性表示 Σ 内任一数,故它们中的在 Σ 上线性无关的最大组是域 Σ 上线性空间 Ω 的基底,Ω 是 Σ 的有限扩张. 从而据定理 3.1 知 $(\Omega:\Delta) = (\Omega:\Sigma) \cdot (\Sigma:\Delta)$ 为 $(\Sigma:\Delta)$ 的整数倍.

定义 3.3 数域 Δ 上的代数扩域:在 Δ 上添加在其上不可约的代数方程 $P(x) = 0$ 的一根 α 所成的扩域,即 Δ 的含有 α 的最小扩域,记之为 $\Delta(\alpha)$.

引理 3.1 设数域 Δ 上两多项式 $P_1(x)$,$P_2(x)$ 有公共根 α,$P_2(x)$ 在 Δ 上不可约,则 $P_2(x)$ 为 $P_1(x)$ 的因式,$P_2(x)$ 的任一根都

是 $P_1(x)$ 的根.

证 因 $P_2(x)$ 在 Δ 上不可约,故 $P_1(x)$ 与 $P_2(x)$ 在域 Δ 上的最高公因式为域 Δ 的非 0 常数(不妨设为 1)或 $P_2(x)$(这时 $P_2(x)$ 为 $P_1(x)$ 的因式),证前者不成立即可:否则 $P_1(x)$ 与 $P_2(x)$ 的最高公因式为 1,可求出域 Δ 上的多项式 $Q_1(x)$ 与 $Q_2(x)$ 使

$$P_1(x)Q_1(x)+P_2(x)Q_2(x)\equiv 1$$

令 $x=\alpha$ 代入,因 $P_1(\alpha)=P_2(\alpha)=0$,故代入后得 $0=1$,矛盾. 于是引理得证.

推论 1 设 α 为域 Δ 上的"代数数"(即 α 为域 Δ 上一个代数方程的根),则在 Δ 上不可约且以 α 为其一根的"首一"(最高次项系数为 1)多项式唯一.

证 设两多项式 $P_1(x)$,$P_2(x)$ 为所求多项式,因二者均在 Δ 上不可约且有公共根 α,故二者互为因式,于是二者必为同次多项式且对应项系数成比例,但因二者最高次项系数同为 1,故二者对应项系数相等,为同一多项式.

推论 2 设域 Δ 上不可约的 n 次多项式 $P(x)$ 的一根为 α,则 $1,\alpha,\alpha^2,\cdots,\alpha^{n-1}$ 在域 Δ 上线性无关.

证 否则 $1,\alpha,\alpha^2,\cdots,\alpha^{n-1}$ 在 Δ 上线性相关,存在不全为 0 的 $b_0,b_1,b_2,\cdots,b_{n-1}\in\Delta$ 使

$$b_0+b_1\alpha+b_2\alpha^2+\cdots+b_{n-1}\alpha^{n-1}=0$$

于是域 Δ 上的至多 $n-1$ 次多项式 $b_0+b_1x+b_2x^2+\cdots+b_{n-1}x^{n-1}$ 与域 Δ 上不可约多项式 $P(x)$ 有公共根 α,从而 $P(x)$ 为 $b_0+b_1x+b_2x^2+\cdots+b_{n-1}x^{n-1}$ 的因式,但 $P(x)$ 的次数 n 高于 $b_0+b_1x+b_2x^2+\cdots+b_{n-1}x^{n-1}$ 的次数,矛盾.

定理 3.3 设上述代数扩域 $\Delta(\alpha)$ 定义中 $P(x)$ 次数为 n,则域 $\Delta(\alpha)$ 上任一数可唯一地表示为

$$c_0+c_1\alpha+c_2\alpha^2+\cdots+c_{n-1}\alpha^{n-1}\quad(c_0,c_1,c_2,\cdots,c_{n-1}\in\Delta)\tag{3.2}$$

证 易见 $\Delta(\alpha)$ 上任一数形如 $\dfrac{P_1(\alpha)}{P_2(\alpha)}$,其中 $P_1(x)$,$P_2(x)$ 为在 Δ 上互素的多项式,$P_2(\alpha)\neq 0$. 因 $P(x)$ 在 Δ 上不可约,$P_2(x)$ 与 $P(x)$ 在 Δ 上互素,可取 Δ 上多项式 $Q(x)$,$Q_2(x)$ 使

$$P(x)Q(x)+P_2(x)Q_2(x)\equiv 1$$

以 $x=\alpha$ 代入,因 $P(\alpha)=0$,故

$$P_2(\alpha)Q_2(\alpha)=1$$

$$\frac{P_1(\alpha)}{P_2(\alpha)}=P_1(\alpha)Q_2(\alpha)$$

用 $P(x)$ 除 $P_1(x)Q_2(x)$,设商为 $F(x)$,余式为 $R(x)$,即

$$P_1(x)Q_2(x)=F(x)P(x)+R(x)$$

注意 $P(\alpha)=0$,于是

$$\frac{P_1(\alpha)}{P_2(\alpha)}=P_1(\alpha)Q_2(\alpha)=F(\alpha)P(\alpha)+R(\alpha)=R(\alpha)$$

而 $R(x)$ 的次数低于 $P(x)$ 的次数 n,故式(3.2)成立.

若又有

$$\frac{P_1(\alpha)}{P_2(\alpha)}=c'_0+c'_1\alpha+c'_2\alpha^2+\cdots+c'_{n-1}\alpha^{n-1}$$

$$(c'_0,c'_1,c'_2,\cdots,c'_{n-1}\in\Delta)$$

且 $c'_0,c'_1,c'_2,\cdots,c'_{n-1}$ 与 $c_0,c_1,c_2,\cdots,c_{n-1}$ 不全对应相等,则

$$\sum_{j=0}^{n-1}c_j\alpha^j=\sum_{j=0}^{n-1}c'_j\alpha^j$$

$$\sum_{j=0}^{n-1}(c_j-c'_j)\alpha^j=0$$

而 $c_j-c'_j(j=0,1,2,\cdots,n-1)$ 不全为0,且属于 Δ,于是 $1,\alpha,\alpha^2,\cdots,\alpha^{n-1}$ 在 Δ 上线性相关,与引理3.1推论2矛盾.得证表示式(3.2)的唯一性.

推论 $\Delta(\alpha)$ 为 Δ 的有限扩域,$(\Delta(\alpha):\Delta)=n$,$\Delta(\alpha)$ 在 Δ 上的基底为 $1,\alpha,\alpha^2,\cdots,\alpha^{n-1}$.

本定理说明,有理式 $\frac{P_1(\alpha)}{P_2(\alpha)}$ 必可有理化成 α 的多项式.

有理数域 **Q** 添加超越数 π(不是任何有理系数代数方程的根)所得扩域 $Q(\pi)$ 不是 **Q** 的有限扩域:因由 π 为超越数知,对任何正整数 m

$$1,\pi,\pi^2,\cdots,\pi^m$$

在 **R** 上线性无关,故域 $Q(\pi)$ 上不存在有限个数可作为域 **Q** 在 **R** 上的基底.实际上,域 $Q(\pi)$ 中的数也不一定能用 π 的幂在 **R** 上的线性组合表示,即不能如同代数扩域 $\Delta(\alpha)$,必可把 $\frac{P_1(\alpha)}{P_2(\alpha)}$ 进行“分母有理化”.

以下只讨论代数扩域.

第四章 代数方程的根域

在现行高等代数课本的多项式理论中已述,对多项式$f(x)$,先求$f(x)$与其导数$f'(x)$的最高公因式$(f(x),f'(x))$,再求$\frac{f(x)}{(f(x),f'(x))}$,则得一个没有重根的多项式,且其所有根为$f(x)$所有互不相同的根.书中以下均假定$n$次多项式$f(x)$无重根.又如无特别说明,均设$f(x)$为其有理域$\Delta$(见上节定义)上的多项式(即不同于一般课本设$f(x)$为整个有理数域$\mathbf{Q}$或整个实数域$\mathbf{R}$或整个复数域$\mathbf{C}$上的多项式).又设$f(x)$的所有根为$x_1,x_2,\cdots,x_n$(互不相同).

定义4.1 在域Δ上添加$x_1,x_2,\cdots,x_n$所成的最小数域,记之为$\Delta(x_1,x_2,\cdots,x_n)$,称之为域$\Delta$上代数方程$f(x)=0$(或多项式$f(x)$)的根域,以下记之为$\Omega$.

Ω中所有数的一般通式可仿式(3.1)写出:把$a_0,a_1,a_2,\cdots,a_n$改为$x_1,x_2,\cdots,x_n$,把域$\mathbf{Q}$改为域Δ.又实际上Ω是在有理数域$\mathbf{Q}$上添加系数$a_0,a_1,a_2,\cdots,a_n$及$x_1,x_2,\cdots,x_n$所成的最小数域,只是复数域$\mathbf{C}$的子域,或实数域$\mathbf{R}$的子域(当所有系数及根全为实数).

注意域Ω与上节所述代数扩域$\Delta(\alpha)$不同:$\Delta(\alpha)$只是在域Δ添加在其上不可约多项式的一个根α所成最小数域;而域Ω则要在Δ添加(不一定在Δ上可约)多项式$f(x)$的所有根.但我们必可求出域Δ上的另一个多项式$F(t)$,使此Ω成为在Δ上添加$F(t)$的任一个根ρ所得代数扩域,即$\Omega=\Delta(\rho)$.即有:

定理4.1 对域Δ上多项式$f(x)$,存在Δ上的多项式$F(t)$,其所有根为$x_1,x_2,\cdots,x_n$在Δ上的线性组合,且对$F(t)$的任一根ρ有:域Δ上多项式$f(x)$的根域$\Omega=\Delta(\rho)$.

先证:

引理4.1 设多项式$\varphi(y)$无重根,其所有根为$y_1,y_2,\cdots,y_m$,令多项式

$$\Psi(y)=\sum_{j=1}^{m}N_j\frac{\varphi(y)}{y-y_j}\quad(N_j\in\mathbf{C})\tag{4.1}$$

(其中$\frac{\varphi(y)}{y-y_j}$表示$\varphi(y)$约去其因式$y-y_j$所得结果).则

$$N_{j_0}=\frac{\Psi(y_{j_0})}{\varphi'(y_{j_0})}\quad (j_0=1,2,\cdots,m) \tag{4.2}$$

(φ' 表 φ 之导数).

证 当 $y=y_{j_0}$ 时,式(4.1) 右边除第 j_0 项中 $\varphi(y)$ 被约去 $y-y_{j_0}$ 外,其余各项为0,故

$$\Psi(y_{j_0})=N_{j_0}\frac{\varphi(y)}{y-y_{j_0}}\Bigg|_{y=y_{j_0}}=N_{j_0}\lim_{y\to y_{j_0}}\frac{\varphi(y)-\varphi(y_{j_0})}{y-y_{j_0}}=N_{j_0}\varphi'(y_{j_0})$$

因 y_{j_0} 非重根,故 $\varphi'(y_{j_0})\neq 0$,式(4.2) 成立.

引理 4.2 设域 Ω 上任意 m 个 p 元齐次线性函数

$$L_h(y_1,y_2,\cdots,y_p)=\sum_{j=1}^{p}c_{h_j}y_j\quad (h=1,2,\cdots,m;c_{h_j}\in\Omega,y_j\in\Omega)$$

又有无限集 $\Delta'\subset\Omega$,则在 Δ' 内可取一组值

$$y_1^*,y_2^*,\cdots,y_p^*$$

使当 $h',h''\in\{1,2,\cdots,m\},h'\neq h''$ 时

$$L_{h'}(y_1^*,y_2^*,\cdots,y_p^*)\neq L_{h''}(y_1^*,y_2^*,\cdots,y_p^*)$$

证 对 p 用归纳法:

当 $p=1$ 时,这时 $c_{h'1}\neq c_{h''1}$ 对 $h'\neq h''$,只要取 $y_1^*\neq 0$ 即可使 $L_{h'}(y_1^*)\neq L_{h''}(y_1^*)$(即 $c_{h'1}y_1^*\neq c_{h''1}y_1^*$).

设命题对 p 成立,则对任意 m 个互不恒等的 $p+1$ 元线性函数

$$L_h(y_1,y_2,\cdots,y_{p+1})=\sum_{j=1}^{p}c_{hj}y_j+c_{h,p+1}y_{p+1}\quad (h=1,2,\cdots,m)$$

它们的前 p 项之和为 p 元齐次线性函数,按归纳法设在 Δ' 取 $y_1=y_1^*,y_2=y_2^*,\cdots,y_p=y_p^*$,使任两个不恒等的上述 p 元齐次线性函数取不同的值.现对 $y_1^*,y_2^*,\cdots,y_p^*$ 取定这些值,则:① 当上述两 $p+1$ 元齐次线性函数的前 p 项恒等时,其第 $p+1$ 项系数不等(否则两个 $p+1$ 元齐次线性函数恒等,违反假设),于是只当再取 $y_{p+1}=0$ 时,两个 $p+1$ 元齐次线性函数取相等的值;② 当上述两个 $p+1$ 元齐次线性函数(设为 $L_{h'}$ 与 $L_{h''}$)的前 p 项不恒等,若它们的第 $p+1$ 项系数相等,则任取 y_{p+1} 之值 y_{p+1}^* 不能使 $L_{h'}$ 与 $L_{h''}$ 在 $y_1^*,y_2^*,\cdots,y_p^*,y_{p+1}^*$ 之值相等;若它们的第 $p+1$ 项系数 $c_{h',p+1}$ 与 $c_{h'',p+1}$ 不等,则只当 y_{p+1} 取一个值 y'_{p+1} 时 $L_{h'}$ 与 $L_{h''}$ 在 $y_1^*,y_2^*,\cdots,y_p^*,y'_{p+1}$ 之值相等(因求 y_{p+1} 时得到一次方程),于是在无限集 Δ' 中再取 y_{p+1}^* 非0且非上述最后的 y'_{p+1},则任两个 $p+1$ 元齐次线性函数在 $y_1^*,y_2^*,\cdots,y_p^*,y_{p+1}^*$ 之值

不等，即推出命题对 $p+1$ 亦成立.

证定理4.1，分解 $f(x)$ 为在 Δ 上不可约因式之积

$$f(x)=f_1(x)f_2(x)\cdots f_k(x)$$

设其中 $f_j(x)(j=1,2,\cdots,k)$ 为 n_j 次多项式，其根为 $x_{j1},x_{j2},\cdots,x_{jn_j}$. 为书写简便取 $k=2,n_1=2,n_2=3$（一般情况完全类似可证）.

设1，2的所有排列 (i_1,i_2) 构成集 M_1，1，2，3的所有排列 (j_1,j_2,j_3) 构成集 M_2. 取域 Ω 上齐次线性函数

$$\rho_{i_1i_2j_1j_2j_3}(y_{11},y_{12},y_{21},y_{22},y_{23})=(x_{1i_1}y_{11}+x_{1i_2}y_{12})+(x_{2j_1}y_{21}+x_{2j_2}y_{22}+x_{2j_3}y_{23})$$

$$(i_1,i_2)\in M_1,(j_1,j_2,j_3)\in M_2$$

按引理4.2可在（无限集）Δ 上取变元 $y_{11},y_{12},y_{21},y_{22},y_{23}$ 的一组值，使得所有齐次线性函数 $\rho_{i_1i_2j_1j_2j_3}$ 在其上之值互不相等，以下把所得之值简记为 $\rho^*_{i_1i_2j_1j_2j_3}$，取多项式（无重根）

$$F(t)=\prod_{\substack{(i_1,i_2)\in M_1\\(j_1,j_2,j_3)\in M_2}}(t-\rho^*_{i_1i_2j_1j_2j_3})$$

其 t 的各次幂的系数对 x_{11},x_{12} 对称，故可用 $f_1(x)$（不妨设 $f(x)$ 及其各因式的最高次项系数为1）的各项系数（x_{11},x_{12} 的基本对称多项式，为域 Δ 的数）及 x_{21},x_{22},x_{23} 的有理式表示，即用 x_{21},x_{22},x_{23} 在域 Δ 的有理式表示. 又易见这些有理式又对 x_{21},x_{22},x_{23} 对称，故同样可用 $f_2(x)$ 的各项系数（属 Δ）的有理式表示. 于是 t 的各次幂系数属 Δ，$F(t)$ 为域 Δ 上的多项式，它的各根为 $x_1,x_2,\cdots,x_n$ 在 Δ 上的齐次线性函数，属于根域 Ω.

对 $F(t)$ 的任一根 $\rho^*_{i_1i_2j_1j_2j_3}$，因它属于 Ω，易见 $\Delta(\rho^*_{i_1i_2j_1j_2j_3})\subset\Omega$. 现再证 Ω 上任一数 N 可用 $\rho^*_{i_1i_2j_1j_2j_3}$ 在 Δ 上有理式表示（从而 $\Omega\subset\Delta(\rho^*_{i_1i_2j_1j_2j_3})$）即可：不妨设 (i_1,i_2,j_1,j_2,j_3) 为 $(1,2,1,2,3)$（因 $f_1(x)$，$f_2(x)$ 各根的序号可任意定）. 因

$$N=\Psi(x_{11},x_{12},x_{21},x_{22},x_{23})\triangleq N_{12123}\quad（域\ \Delta\ 上有理式）$$

一般记

$$\Psi(x_{1i_1},x_{1i_2},x_{2j_1},x_{2j_2},x_{2j_3})=N_{i_1i_2j_1j_2j_3}$$

取多项式

$$G(t)\equiv\sum_{\substack{(i_1,i_2)\in M_1\\(j_1,j_2,j_3)\in M_2}}N_{i_1i_2j_1j_2j_3}\cdot\frac{F(t)}{t-\rho^*_{i_1i_2j_1j_2j_3}}$$

其 t 的各次幂系数对 x_{11},x_{12} 对称,也对 x_{21},x_{22},x_{23} 对称,故亦为域 Δ 上的多项式,又 $F(t)$ 无重根,故由引理 4.1 知

$$N=N_{12123}=\frac{G(\rho_{12123}^{*})}{f'(\rho_{12123}^{*})}$$

为 ρ_{12123}^{*} 的有理式,证毕.

例如有理数域 $\mathbf{Q}$ 上多项式

$$f(x)=(x^2-2)(x^2-3)$$

的各根为 $\pm\sqrt{2}$, $\pm\sqrt{3}$,以此四根为系数取 4 个齐次线性函数

$$\sqrt{2}y_{11}-\sqrt{2}y_{12}+\sqrt{3}y_{21}-\sqrt{3}y_{22}$$
$$-\sqrt{2}y_{11}+\sqrt{2}y_{12}+\sqrt{3}y_{21}-\sqrt{3}y_{22}$$
$$\sqrt{2}y_{11}-\sqrt{2}y_{12}-\sqrt{3}y_{21}+\sqrt{3}y_{22}$$
$$-\sqrt{2}y_{11}+\sqrt{2}y_{12}-\sqrt{3}y_{21}+\sqrt{3}y_{22}$$

在域 $\mathbf{Q}$ 上取 $y_{11}=2,y_{12}=1,y_{21}=2,y_{22}=1$ 得互不相同的 4 个数 $\pm\sqrt{2}\ \pm\sqrt{3}$(两项的正、负号任意配搭,下同),以此 4 个数为根取多项式

$$(t-\sqrt{2}-\sqrt{3})(t-\sqrt{2}+\sqrt{3})(t+\sqrt{2}-\sqrt{3})(t+\sqrt{2}+\sqrt{3})=$$
$$(t^2-2\sqrt{2}t-1)(t^2+2\sqrt{2}t-1)=t^4-10t^2+1 \tag{4.3}$$

用其任一根(如$\sqrt{2}+\sqrt{3}$)在 $\mathbf{Q}$ 上的有理式可表示 $f(x)$ 的各根

$$\sqrt{2}=\frac{1}{2}[(\sqrt{2}+\sqrt{3})+(\sqrt{2}-\sqrt{3})]=$$
$$\frac{1}{2}(\sqrt{2}+\sqrt{3})-\frac{1}{2}\frac{1}{\sqrt{2}+\sqrt{3}}$$
$$-\sqrt{2}=-\frac{1}{2}(\sqrt{2}+\sqrt{3})+\frac{1}{2}\frac{1}{\sqrt{2}+\sqrt{3}}$$
$$\sqrt{3}=\frac{1}{2}[(\sqrt{2}+\sqrt{3})-(\sqrt{2}-\sqrt{3})]=$$
$$\frac{1}{2}(\sqrt{2}+\sqrt{3})+\frac{1}{2}\frac{1}{\sqrt{2}+\sqrt{3}}$$
$$-\sqrt{3}=-\frac{1}{2}(\sqrt{2}+\sqrt{3})-\frac{1}{2}\frac{1}{\sqrt{2}+\sqrt{3}}$$

故 $\mathbf{Q}$ 上多项式 $f(x)$ 的根域 $\Omega=Q(\sqrt{2}+\sqrt{3})$.

这时求得的多项式 $F(t)=t^4-10t^2+1$ 在 $\mathbf{Q}$ 上不可约[①]. 但定理4.1的证明中未证 $F(t)$ 在 $\mathbf{Q}$ 上不可约. 实际上 $F(t)$ 可能在 Δ 上可约. 举例如下:取

$$f(x)=(x^2-2)(x^2-8)(x^2-3)$$

它有6个根(分成三组,每组两个根): $\pm\sqrt{2}$, $\pm2\sqrt{2}$, $\pm\sqrt{3}$. 相应取8个$(2!\times2!\times2!)$齐次线性函数

$$\sqrt{2}y_{11}-\sqrt{2}y_{12}+2\sqrt{2}y_{21}-2\sqrt{2}y_{22}+\sqrt{3}y_{31}-\sqrt{3}y_{32}$$
$$\sqrt{2}y_{11}-\sqrt{2}y_{12}+2\sqrt{2}y_{21}-2\sqrt{2}y_{22}-\sqrt{3}y_{31}+\sqrt{3}y_{32}$$
$$\sqrt{2}y_{11}-\sqrt{2}y_{12}-2\sqrt{2}y_{21}+2\sqrt{2}y_{22}+\sqrt{3}y_{31}-\sqrt{3}y_{32}$$
$$\sqrt{2}y_{11}-\sqrt{2}y_{12}-2\sqrt{2}y_{21}+2\sqrt{2}y_{22}-\sqrt{3}y_{31}+\sqrt{3}y_{32}$$
$$-\sqrt{2}y_{11}+\sqrt{2}y_{12}+2\sqrt{2}y_{21}-2\sqrt{2}y_{22}+\sqrt{3}y_{31}-\sqrt{3}y_{32}$$
$$-\sqrt{2}y_{11}+\sqrt{2}y_{12}+2\sqrt{2}y_{21}-2\sqrt{2}y_{22}-\sqrt{3}y_{31}+\sqrt{3}y_{32}$$
$$-\sqrt{2}y_{11}+\sqrt{2}y_{12}-2\sqrt{2}y_{21}+2\sqrt{2}y_{22}+\sqrt{3}y_{31}-\sqrt{3}y_{32}$$
$$-\sqrt{2}y_{11}+\sqrt{2}y_{12}-2\sqrt{2}y_{21}+2\sqrt{2}y_{22}-\sqrt{3}y_{31}+\sqrt{3}y_{32}$$

取 $y_{11}=2,y_{12}=1,y_{21}=2,y_{22}=1,y_{31}=2,y_{32}=1$ 得8个互不相同的数 $\pm\sqrt{2}\ \pm\sqrt{3}$, $\pm3\sqrt{2}\ \pm\sqrt{3}$. 同样易见对以此8个数为根的多项式 $F(t)$ 的任一根(如$\sqrt{2}+\sqrt{3}$)有:$f(x)$ 的根域 $\Omega=Q(\sqrt{2}+\sqrt{3})$. 但这时 $F(t)$(八次多项式)在 $\mathbf{Q}$ 上可约,它在 $\mathbf{Q}$ 上有四次因式

$$(t-\sqrt{2}-\sqrt{3})(t-\sqrt{2}+\sqrt{3})(t+\sqrt{2}-\sqrt{3})(t+\sqrt{2}+\sqrt{3})=t^4-10t^2+1$$

从以上论证知多项式 $F(t)$ 并非唯一.

对定理所述之 ρ,设它为 $F(t)$ 在域 Δ 上不可约因式 $F^*(t)$(设为 m 次)之根,又 Ω 显然为 Δ 之扩域,从本定理及定理3.3推论知域 Ω 在 Δ 上的基底为 $1,\rho,\rho^2,\cdots,\rho^{m-1}$,$(\Omega:\Delta)=m$,此维数由 Δ 及多项式 $f(x)$ 而定(当 Δ 及 $f(x)$ 已定,则域 Ω 亦定),与定理证明中所选取的 ρ 无关.

定理4.1推论　设多项式 $f(x)$ 如上所述,则在 Δ 上的任何有理式 $H(x_1,x_2,\cdots,x_n)$,即根域 $\Omega=\Delta(x_1,x_2,\cdots,x_n)$ 任何数必可化成

---

①否则它可分解为域 $\mathbf{Q}$ 上一次因式与三次因式之积,或域 $\mathbf{Q}$ 上两个二次因式之积,但由因式分解唯一性,$F(t)$ 的一、二次因式必为式(4.3)中的一次因式或两个一次因式之积,易见它们都非域 $\mathbf{Q}$ 上多项式.

域Δ上$x_1,x_2,\cdots,x_n$的(多元)多项式①.

证 因每一根$x_j(j=1,2,\cdots,n)$属于$\Omega=\Delta(\rho)$,故

$$x_j=\xi_j(\rho)\quad(\text{域}\Delta\text{上有理式})$$

从而

$$H(x_1,x_2,\cdots,x_n)=H(\xi_1(\rho),\xi_2(\rho),\cdots,\xi_n(\rho))$$

为域Δ上ρ的有理式,即属于$\Delta(\rho)$,故由定理3.3知它可化成域Δ上ρ的多项式. 又ρ为$x_1,x_2,\cdots,x_n$在Δ上的线性组合更属$x_1,x_2,\cdots,x_n$在Δ上的(多元)多项式. 于是$H(x_1,x_2,\cdots,x_n)$可化成$x_1,x_2,\cdots,x_n$在Δ上的多项式.

本推论其实可作为定理3.3的直接推论:逐次取在域Δ上含根x_1的不可约多项式,在域$\Delta(x_1)$上含根x_2的不可约多项式,在域$\Delta(x_1,x_2)$上含根x_3的不可约多项式……推导出$\Delta(x_1)$的所有数,$\Delta(x_1,x_2)$的所有数,$\Delta(x_1,x_2,x_3)$的所有数……分别可表为x_1,x_1及x_2,x_1,x_2及$x_3\cdots$在Δ上的多项式,最后知,$\Delta(x_1,x_2,\cdots,x_n)$的任何数可表为$x_1,x_2,\cdots,x_n$在$\Delta$上的多项式.

定义4.2 在域Δ上不可约方程的各根称之为在Δ上互相共轭. 以Δ上代数数α(域Δ上代数方程之根)为一根且在Δ上不可约的方程(不妨设最高次项系数为1,据定理3.2推论1知此"首一"方程唯一,从而其所有根亦唯一确定)的其余各根称为α在域Δ上的共轭数.

例如$\alpha=1+\sqrt{2}$,以其为一根且在有理数域$\mathbf{Q}$上不可约的方程为

$$x^2-2x-1=0$$

(可从$x=1+\sqrt{2}$,$x-1=\sqrt{2}$,再两边平方求得),它尚有根$1-\sqrt{2}$. 故$1+\sqrt{2}$在$\mathbf{Q}$上的共轭数为$1-\sqrt{2}$,反之亦然,$1+\sqrt{2}$与$1-\sqrt{2}$在$\mathbf{Q}$上互相共轭.

又如$\alpha=\sqrt[3]{2}$,以其为一根且在有理数域$\mathbf{Q}$上不可约的方程为

$$x^3=2$$

它尚有两根$\sqrt[3]{2}\,\mathrm{e}^{2\pi\mathrm{i}/3}$,$\sqrt[3]{2}\,\mathrm{e}^{4\pi\mathrm{i}/3}$,故$\sqrt[3]{2}$在$\mathbf{Q}$上的共轭数为上述二数,此三数在$\mathbf{Q}$上互相共轭,任一数在$\mathbf{Q}$上的共轭数为其余二数.

定义4.3 设域Δ的扩域$\mathscr{D}$,若$\mathscr{D}$内任一数的所有在Δ的共轭

①实际应为一个n元多项式在$x_1,x_2,\cdots,x_n$之值,下文亦常把一元或多元多项式之值称为多项式.

数亦属于$\mathscr{D}$,则称$\mathscr{D}$为Δ的正规扩域.

下文将证明,域Δ上多项式$f(x)$的根域Ω必是Δ的正规扩域.

但取有理数域$\mathbf{Q}$上不可约多项式x^3-2,在$\mathbf{Q}$上添加此多项式一根$\sqrt[3]{2}$所得代数扩域$Q(\sqrt[3]{2})$不是$\mathbf{Q}$的正规扩域:因$Q(\sqrt[3]{2})$是实域,包含$\sqrt[3]{2}$但不包含$\sqrt[3]{2}$在$\mathbf{Q}$的共轭数$\sqrt[3]{2}\mathrm{e}^{2\pi i/3}$.

定理4.2 在定理4.1中所述$f(x)$的根域$\Omega=\Delta(\rho)$(m维)中任一数α必为Δ上一个m次"首一"多项式$\Psi(y)$的根,且$\Psi(y)$的所有根全属于Ω;又设以α为一根且在Δ不可约的"首一"多项式为$\psi(y)$,则

$$\Psi(y)=(\psi(y))^k \quad (k\in\mathbf{Z}^+)$$

证 由$\alpha\in\Delta(p)$,据定理3.3可设

$$\alpha=g(\rho) \quad (\text{域}\Delta\text{上多项式})$$

又设上述$F^*(t)$的m个根为

$$\rho,\rho_2,\cdots,\rho_m$$

因它们亦为$F(t)$的根,故为$f(x)$的根$x_1,x_2,\cdots,x_n$在域Δ上的齐次线性式,从而属于其根域Ω,于是域Δ上有理式

$$g(\rho),g(\rho_2),g(\rho_3),\cdots,g(\rho_m)$$

亦然.作多项式

$$\Psi(y)\equiv(y-g(\rho))(y-g(\rho_2))(y-g(\rho_3))\cdots(y-g(\rho_m))$$

其y的各次幂系数对$\rho,\rho_2,\rho_3,\cdots,\rho_m$对称,故为$F^*(t)$各系数的有理式属于$\Delta$,即$\Psi(y)$为$\Delta$上多项式,它有根$g(\rho)=\alpha$,所有根属于$\Omega$.设$\Psi(y)$在$\Delta$上不可约的"首一"因式为$\psi(y)$,则它的任一根(单重,因不可约多项式必无重根)必为$\Psi(y)$的某一根$g(\rho_j)$,即

$$\psi(g(\rho_j))=0$$

于是Δ上变元t的多项式$\psi(g(t))$与不可约多项式$F^*(t)$有公共根ρ_j,故由引理3.1知,$F^*(t)$的每个根$\rho,\rho_2,\rho_3,\cdots,\rho_m$为$\psi(g(t))$的根,即$g(\rho)=\alpha,g(\rho_2),g(\rho_3),\cdots,g(\rho_m)$都是$\Delta$上不可约多项式$\psi(y)$的根(前已述为单重),于是$\psi(y)$合本定理要求.又$\psi(y)$为$\Psi(y)$的因式,故$\psi(y)$无($\Psi(y)$的根)$g(\rho),g(\rho_2),g(\rho_3),\cdots,g(\rho_m)$以外的根.于是$\psi(y)$的根由$g(\rho),g(\rho_2),g(\rho_3),\cdots,g(\rho_m)$中互不相同的根组成,故$\Psi(y)$的任何两个在$\Delta$不可约的"首一"因式都有完全相同的(单)根集,于是这些不可约因式(设有k个)恒等,得证

$$\Psi(y)=(\psi(y))^k$$

推论1 Δ上多项式$f(x)$的根域Ω为Δ的正规扩域.

(因Ω中任一数α的所有在Δ上共轭数为$\psi(y)$的根,更是$\Psi(y)$的根,全属于Ω).

推论2 以Ω上任一数为一根且在Δ上不可约的多项式的次数为$(\Omega:\Delta)=m$的约数.

(由$\Psi(y)=(\psi(y))^k$知$\Psi(y)$的次数m为$\psi(y)$的次数的k倍).

定义4.4 设域Δ上多项式$f(x)$的根域Ω上一数α,若以α为一根且在Δ上不可约的方程的次数为$(\Omega:\Delta)=m$,则称α为Δ上多项式$f(x)$的根域Ω的本原数.

显然这时α的任一共轭数亦是所述本原数,定理4.1所述的多项式的含有根ρ且在域Δ上不可约的因式$F^*(t)$的任一根都是所述本原数.如定理4.1后之例$f(x)=(x^2-2)(x^2-3)$,相应的$F(t)=t^4-10t^2+1$在域$\mathbf{Q}$不可约,即$F^*(t)=F(t)$,故$F(t)$的四根$\sqrt{2}\pm\sqrt{3}$,$-\sqrt{2}\pm\sqrt{3}$为共轭本原数.

有理数域$\mathbf{Q}$上二次多项式

$$x^2+px+q$$

当$\sqrt{p^2-4q}\notin\mathbf{Q}$(可为虚数)时它在$\mathbf{Q}$上不可约,其二根

$$x_1=\frac{-p+\sqrt{p^2-4q}}{2}$$

$$x_2=\frac{-p-\sqrt{p^2-4q}}{2}$$

易见x_1,x_2在$\mathbf{Q}$上的多项式可表为形如下式之数

$$a+b\sqrt{p^2-4q}\quad(a,b\in\mathbf{Q})$$

此形任二数的和、差、积、商(除数不为0)亦为此形之数.又

$$\sqrt{p^2-4q}=x_1-x_2$$

属于此多项式的根域Ω.于是Ω为此形的所有数,其维数等于此多项式次数2,故x_1,x_2均是根域Ω的本原数.

下定理反映本原数的特征.

定理4.3 设域Δ上多项式$f(x)$的根域Ω(m维)有一数α,则如下命题等价:

①α为根域Ω的本原数$\Leftrightarrow$②$\Omega=\Delta(\alpha)\Leftrightarrow$③$f(x)$的各根可用$\alpha$在$\Delta$的多项式表示.

显然②$\Leftrightarrow$③(注意由定理3.3知$\Delta(\alpha)$的数可表为α的多项

式).

证 ①⇒②:这时α为Δ上不可约m次多项式之根,据定理3.3知$1,\alpha,\alpha^2,\cdots,\alpha^{m-1}$为$\Delta$上$m$维线性空间$\Delta(\alpha)$之基底,但这$m$个数属$m$维空间$\Omega$(根域)且在$\Delta$上线性无关,故亦为域$\Delta$上线性空间$\Omega$的基底,于是$\Delta(\alpha)=\Omega$.

②⇒①:若α不是根域Ω的本原数,因由上定理4.2推论2知以α为一根且在Δ上不可约的多项式次数为m的约数,从α不是本原数知此次数小于m,再从定理3.3推论知Δ上线性空间$\Delta(\alpha)$的维数小于m——线性空间Ω的维数,于是$\Delta(\alpha)$与Ω是Δ上不同的线性空间,与②矛盾.

推论 设α为域Δ上多项式根域Ω(m维)的本原数,则Ω的任一数可唯一表示为α的不高于m次的多项式.

(由$\Omega=\Delta(\alpha)$再据定理3.3可证).

虽然前述有理数域$\mathbf{Q}$上不可约的二次多项式之根为其根域Ω的本原数,但域$\mathbf{Q}$上不可约的三次多项式x^3-2的根$\sqrt[3]{2}$不是其根域Ω的本原数,因域$Q(\sqrt[3]{2})$是实域,不同于Ω(包含复数$\sqrt[3]{2}\mathrm{e}^{2\pi i/3}$),不适合上定理中的②.

第五章 代数方程的 Galois 群

设域Δ上n次方程$f(x)=0$如同第四章.其根域Ω对Δ的维数为m,则有m个互相共轭的本原数$\rho_j(j=1,2,\cdots,m)$.据定理4.3知,$\Omega=\Delta(\rho_j)$,$f(x)$的n个根(属Ω)都可用任一ρ_j的多项式表示(据定理3.3).域Δ上方程$f(x)=0$的Galois群实际是这n个根的一个置换群,群中的置换是由上述表示式中作本原数的变换而导致这n个根的置换.本原数的变换又导致根域Ω在Δ上的所有自同构映射(见定理5.1的4°,5°).

定理5.1 设$f(x)$各根用根域Ω的本原数ρ_1的多项式表示为

$$x_1=\xi_1(\rho_1),x_2=\xi_2(\rho_1),\cdots,x_n=\xi_n(\rho_1) \qquad (5.1_1)$$

共轭本原数为$\rho_1,\rho_2,\cdots,\rho_m$,则

1° 对$j=1,2,\cdots,m$,数列

$$\xi_1(\rho_j),\xi_2(\rho_j),\cdots,\xi_n(\rho_j) \qquad (5.1_j)$$

为根$x_1,x_2,\cdots,x_n$的一个排列(即变换$\rho_1\to\rho_j$导致n个根的一个置换);

2° 当 $j', j'' \in \{1,2,\cdots,m\}, j' \neq j''$ 时数列 $(5.1_{j'})$ 与 $(5.1_{j''})$ 不全同（不同的变换 $\rho_1 \to \rho_{j'}, \rho_1 \to \rho_{j''}$ 导致不同的置换）；

3° 设 1° 中

$$\xi_1(\rho_j) = x_{\rho_{j1}}, \xi_2(\rho_j) = x_{\rho_{j2}}, \cdots, \xi_n(\rho_j) = x_{\rho_{jn}}$$

即变换 $\rho_1 \to \rho_j$ 引起 n 个根的置换为

$$\begin{pmatrix} x_1, & x_2, & \cdots, & x_n \\ x_{\rho_{j1}}, & x_{\rho_{j2}}, & \cdots, & x_{\rho_{jn}} \end{pmatrix}$$

则对 $j = 1,2,\cdots,m$，所有这些置换构成一个群；

4° 对任一 $j = 1,2,\cdots,m$，设根域 Ω 任一数

$$\alpha = \eta(x_1, x_2, \cdots, x_n) = \eta(\xi_1(\rho_1), \xi_2(\rho_1), \cdots, \xi_n(\rho_1)) \triangleq \xi(\rho_1)$$

$$(\eta(x_1, x_2, \cdots, x_n) \text{ 为域 } \Delta \text{ 上有理式})$$

作与变换 $\rho_1 \to \rho_j$ 相应的映射

$$T_j\alpha = \eta(\xi_1(\rho_j), \xi_2(\rho_j), \cdots, \xi_n(\rho_j)) = \xi(\rho_j) = \eta(x_{\rho_{j1}}, x_{\rho_{j2}}, \cdots, x_{\rho_{jn}})$$

则此映射为 Ω 在域 Δ 上的自同构映射，即：

① 映射 T_j 把 Ω 的任一数 α 映射为 Ω 的一个确定的数 β，$T_j\alpha = \beta$；

② 对任何 $\beta \in \Omega$，在 Ω 必存在唯一的数 α，使 $T_j\alpha = \beta$；

③ 对任何 $c \in \Delta$ 有 $T_jc = c$（T_j 使 Δ 中的数不变）；

④ 对任何 $\alpha, \alpha' \in \Omega$ 有

$$T_j(\alpha + \alpha') = T_j\alpha + T_j\alpha'$$

$$T_j(\alpha\alpha') = T_j\alpha \cdot T_j\alpha'$$

5° 上述映射为 Ω 在 Δ 上的所有自同构映射.

证 1° 设 $\rho_1, \rho_2, \cdots, \rho_m$ 为 Δ 上不可约多项式 $\psi(t)$ 的根，由

$$f(\xi_s(\rho_1)) = f(x_s) = 0 \quad (s = 1,2,\cdots,m)$$

知 Δ 上两多项式（变元 t）$f(\xi_s(t))$ 与 $\psi(t)$ 有公共根 ρ_1，而 $\psi(t)$ 不可约，故由引理 3.1 知其任一根 ρ_j 也是 $f(\xi_s(t))$ 的根，故 $f(\xi_s(\rho_j)) = 0$，即 $\xi_s(\rho_j)$ 是 $f(x)$ 的某个根 ρ_h：$\xi_s(\rho_j) = \rho_h$.

下证 $s, s' \in \{1,2,\cdots,m\}$，当 $s \neq s'$ 时 $\xi_s(\rho_j) \neq \xi_{s'}(\rho_j)$：否则 Δ 上两方程 $\xi_s(t) = \xi_{s'}(t)$ 与 $\psi(t) = 0$ 有公共根 ρ_j，而 $\psi(t)$ 不可约，故其根 ρ_1 亦是前者的根，于是 $\xi_s(\rho_1) = \xi_{s'}(\rho_1)$，即 $x_s = x_{s'}$，与设 $f(x)$ 无重根矛盾.

于是序列 (5.1_j) 由 $f(x)$ 的根组成，其中没有相同的根，故为 $x_1, x_2, \cdots, x_m$ 的一个排列.

证 2°　由假设及定理 4.1 推论知

$$\rho_1=\eta(x_1,x_2,\cdots,x_n)=\eta(\xi_1(\rho_1),\xi_2(\rho_1),\cdots,\xi_n(\rho_1))$$

（n 元多项式）

于是域 Δ 上两方程

$$t=\eta(\xi_1(t),\xi_2(t),\cdots,\xi_n(t))$$

与 $\psi(t)=0$ 有公共根 ρ_1，故在 Δ 上不可约的方程 $\psi(t)=0$ 的任二根 $\rho_{j'},\rho_{j''}$ 亦是前者的根

$$\rho_{j'}=\eta(\xi_1(\rho_{j'}),\xi_2(\rho_{j'}),\cdots,\xi_n(\rho_{j'}))$$

$$\rho_{j''}=\eta(\xi_1(\rho_{j''}),\xi_2(\rho_{j''}),\cdots,\xi_n(\rho_{j''}))$$

若数列 $(5.1_{j'})$ 与数列 $(5.1_{j''})$ 全相同，则上两式右边相等，从而 $\rho_{j'}=\rho_{j''}$，$\psi(t)$ 有重根与其在 Δ 不可约矛盾.

证 3°　因 ρ_1 为本原数，任何 $\rho_k\in\Omega=\Delta(\rho_1)$（据定理 4.3），又由定理3.3 知

$$\rho_k=\lambda_k(\rho_1)\ (\text{多项式})\quad(k=1,2,\cdots,m)$$

对任何 $k_1,k_2\in\{1,2,\cdots,m\}$，变换 $\rho_1\to\rho_{k_1}$ 使数列

$$x_j=\xi_j(\rho_1)\quad(j=1,2,\cdots,m)$$

化成

$$\xi_j(\rho_{k_1})=\xi_j(\lambda_{k_1}(\rho_1))\quad(j=1,2,\cdots,m)$$

再变换 $\rho_1\to\rho_{k_2}$，则上序列又化成

$$\xi_j(\lambda_{k_1}(\rho_{k_2}))\quad(j=1,2,\cdots,m)$$

下证 $\lambda_{k_1}(\rho_{k_2})$ 等于某个 ρ_s：因 $\psi(\lambda_{k_1}(\rho_1))=\psi(\rho_{k_1})=0$，故在域 Δ 上两多项式 $\psi(\lambda_{k_1}(t))$ 与（不可约）$\psi(t)$ 有公共根 ρ_1，故 $\psi(t)$ 的根 ρ_{k_2} 亦是前者之根：$\psi(\lambda_{k_1}(\rho_{k_2}))=0$，于是 $\lambda_{k_1}(\rho_{k_2})$ 为 $\psi(t)$ 的某一根 ρ_s. 从而上数列又化成

$$\xi_j(\rho_s)\quad(j=1,2,\cdots,m)$$

即与两变换 $\rho_1\to\rho_{k_1},\rho_1\to\rho_{k_2}$ 相应的两置换之积为与 $\rho_1\to\rho_s$ 相应的置换，即 3° 中所述 m 个置换所组成的集内可进行乘法运算，由定理 2.4 知它们构成一个置换群.

证 4°　① 显然成立.

② 因 ρ_j 也是域 Δ 上多项式 $f(x)$ 根域 Ω 的本原数，按定理 4.3 及定理 3.3 可先把 β 唯一地表示成

$$\beta=c'_0+c'_1\rho_j+c'_2\rho_j^2+\cdots+c'_{m-1}\rho_j^{m-1}$$

因根据 Ω 任一数 α 可唯一地表示为

$$\alpha=c_0+c_1\rho_1+c_2\rho_1^2+\cdots+c_{m-1}\rho_1^{m-1}\qquad(5.2)$$

从而

$$T_j\alpha = c_0 + c_1\rho_j + c_2 p_j^2 + \cdots + c_{m-1}\rho_j^{m-1}$$

要使 $T_j\alpha = \beta$，由 β 表示式唯一性知，唯有

$$c_0 = c'_0, c_1 = c'_1, c_2 = c'_2, \cdots, c_{m-1} = c'_{m-1}$$

于是求得适合 $T_j\alpha = \beta$ 的唯一的

$$\alpha = c'_0 + c'_1\rho_1 + c'_2\rho_1^2 + \cdots + c'_{m-1}\rho_1^{m-1}$$

③c 唯一地表成

$$c + 0\rho_1 + 0\rho_1^2 + \cdots + 0\rho_1^{m-1}$$

于是

$$T_j c = c + 0\rho_j + 0\rho_j^2 + \cdots + 0\rho_j^{m-1} = c$$

④ 把 α 与 $T_j\alpha$ 分别用 ρ_1 与 ρ_j 表示的多项式（两多项式对应系数相同，下称为同一多项式）及把 α'，$T_j\alpha'$ 分别用 ρ_1，ρ_j 表示的同一多项式合并，得 $\alpha + \alpha'$ 与 $T_j\alpha + T_j\alpha'$ 分别用 ρ_1 与 ρ_j 表示的同一多项式（仍为至多 $m-1$ 次），由于对应系数相同，故 $T_j(\alpha + \alpha') = T_j\alpha + T_j\alpha'$.

再证 $T_j(\alpha\alpha') = T_j\alpha \cdot T_j\alpha'$：把 α 及 α' 用 ρ_1 表示的多项式相乘化简合并，又把 $T_j\alpha$ 及 $T_j\alpha'$ 用 ρ_j 表示的多项式（与上用 ρ_1 表示的两多项式分别相同）相乘化简合并，得 $\alpha\alpha'$ 及 $T_j\alpha \cdot T_j\alpha'$ 分别用 ρ_1 及 ρ_j 表示的同一多项式的表示式，但表示式会出现超过 $m-1$ 次的项. 设有最高为 $m+p$ 次项（$p \geqslant 0$），由 ρ_1 及 ρ_j 为 m 次多项式 $\psi(t)$ 的根，知 $m+p$ 次多项式

$$\rho_1^p\psi(\rho_1) = \rho_j^p\psi(\rho_j) = 0$$

于是可把二者的 $m+p$ 次项分别化成 ρ_1，ρ_j 的至多为 $m+p-1$ 次的相同多项式，再把它们与 $\alpha\alpha'$，$T_j\alpha \cdot T_j\alpha'$ 的表示式的其余低次项合并，把 $\alpha\alpha'$ 及 $T_j\alpha \cdot T_j\alpha'$ 分别化成用 ρ_1 及 ρ_j 表示的至多 $m+p-1$ 次相同的多项式. 继续用同样方法逐步降低此两相同多项式的次数，最后都可降低为分别用 ρ_1 及 ρ_j 表示的至多 $m-1$ 次的相同多项式，于是得证 $T_j(\alpha\alpha') = T_j\alpha \cdot T_j\alpha'$.

证 5° 共轭本原数 $\rho_1, \rho_2, \cdots, \rho_m$ 为域 Δ 上不可约多项式

$$\psi(t) \equiv b_0 + b_1 t + b_2 t^2 + \cdots + b_m t^m$$

的全部根，从

$$b_0 + b_1\rho_1 + b_2\rho_1^2 + \cdots + b_m\rho_1^m = 0$$

由于 Ω 在 Δ 上的任一自同构映射保持 Ω 中的数的和、积关系并使 Δ 中的数不变，故

$$b_0+b_1(T\rho_1)+b_2(T\rho_1)^2+\cdots+b_m(T\rho_1)^m=0$$

即 $T\rho_1$ 为 $\psi(t)$ 的某一根 ρ_j. 再据 Ω 在 Δ 上自同构定义知对 Ω 中的任何 α(用式(5.2)表示)有

$$T\alpha=c_0+c_1(T\rho_1)+c_2(T\rho_1)^2+\cdots+c_{m-1}(T\rho_1)^{m-1}=$$

$$c_0+c_1\rho_j+c_2\rho_j^2+\cdots+c_{m-1}\rho_j^{m-1}$$

即映射 T 实必为某个映射 T_j. 证毕.

由于上述置换群所有置换(由 $\rho_1\to\rho_j(j=1,2,\cdots,m)$ 导致的)相应于根域 Ω 在 Δ 上的所有自同构映射,且不同的置换(由 ρ_1 变为不同的 ρ_j 所导致的)相应于不同的自同构映射,故组成此置换群的置换集与在证明 1° ~ 3° 时所取的在 Δ 上不可约的 m 次多项式 $\psi(t)$ 及所选的根 ρ_1(实即 $\psi(t)$ 的任一根)无关. 我们称此置换群为域 Δ 上方程 $f(x)=0$(或多项式 $f(x)$)的 Galois 置换群,或简称 Galois 群.

例1 第四章例中多项式 $(x^2-2)(x^2-3)$ 在域 $\mathbf{Q}$ 的共轭本原数为 $\sqrt2+\sqrt3$, $-\sqrt2-\sqrt3$, $\sqrt2-\sqrt3$, $-\sqrt2+\sqrt3$(即含有根 $\sqrt2+\sqrt3$ 且在域 Δ 不可约的多项式 t^4-10t^2+1 的所有根)在用 $\sqrt2+\sqrt3$ 表示四根

$$x_1=\sqrt2$$

$$x_2=-\sqrt2$$

$$x_3=\sqrt3$$

$$x_4=-\sqrt3$$

的表示式中把本原数 $\sqrt2+\sqrt3$ 改为本原数 $-\sqrt2-\sqrt3$,则

$$x_1=\sqrt2=\frac12(\sqrt2+\sqrt3)-\frac12\frac{1}{\sqrt2+\sqrt3}$$

改成

$$\frac12(-\sqrt2-\sqrt3)-\frac12\frac{1}{-\sqrt2-\sqrt3}=-\sqrt2=x_2$$

类似知 x_2 改成 x_1, x_3 改成 x_4, x_4 改成 x_3,故得各根相应的置换

$$(x_1,x_2)(x_3,x_4)$$

把本原数 $\sqrt2+\sqrt3$ 改为本原数 $\sqrt2-\sqrt3$,则 x_1 改成

$$\frac12(\sqrt2-\sqrt3)-\frac12\frac{1}{\sqrt2-\sqrt3}=$$

$$\frac{1}{2}(\sqrt{2}-\sqrt{3})+\frac{1}{2}(\sqrt{2}+\sqrt{3})=\sqrt{2}=x_1 \quad (\text{不变})$$

从而 $x_2=-x_1$ 亦不变，x_3 改为

$$\frac{1}{2}(\sqrt{2}-\sqrt{3})+\frac{1}{2}\frac{1}{\sqrt{2}-\sqrt{3}}=$$

$$\frac{1}{2}(\sqrt{2}-\sqrt{3})-\frac{1}{2}(\sqrt{2}+\sqrt{3})=x_4$$

从而 $x_4=-x_3$ 改成与上结果相反的

$$-x_4=x_3$$

故得相应的置换为

$$(x_3,x_4)$$

把本原数 $\sqrt{2}+\sqrt{3}$ 改为本原数 $-\sqrt{2}+\sqrt{3}$（与上述 $\sqrt{2}-\sqrt{3}$ 相反，故结果亦相反），则：x_1 改为 $-x_1=x_2$，x_2 改为 $-x_2=x_1$，x_3 改为 $-x_4=x_3$，x_4 改为 $-x_3=x_4$，故得相应置换为

$$(x_1,x_2)$$

当本原数 $\sqrt{2}+\sqrt{3}$ 不变，则 $f(x)$ 四根亦不变，得相应置换为恒等置换 I.

故由本原数 $\sqrt{2}+\sqrt{3}$ 变成与其共轭的所有本原数（含不变）相应得出 $f(x)$ 各根的四置换

$$(x_1,x_2)(x_3,x_4),(x_3,x_4),(x_1,x_2),I$$

易验证此四置换组成一个群，即域 $\mathbf{Q}$ 上多项式 $(x^2-2)(x^2-3)$ 的 Galois 群.

上述根域 Ω 的任一数当然是域 Δ 上 $x_1,x_2,\cdots,x_n$ 的有理式（即式中各系数属于 Δ），但这有理式表示并非唯一. 如上例中根域 Ω 的数 $-\sqrt{2}$ 可表示为

$$x_2,\ -x_1,2x_2+x_1,\frac{x_1^2}{x_2}$$

这 4 个表示式作（不属 Galois 群的）置换 (x_2,x_3) 分别变成

$$x_3,\ -x_1,2x_3+x_1,\frac{x_1^2}{x_3}$$

其值分别为

$$\sqrt{3},\ -\sqrt{2},2\sqrt{3}+\sqrt{2},\frac{2}{\sqrt{3}}$$

得出不同的值.

但若取 Galois 群的置换(x_1,x_2),则原4个表示式分别变成

$$x_1,\ -x_2, 2x_1+x_2, \frac{x_2^2}{x_1}$$

它们之值都是$\sqrt{2}$,实际上有:

定理5.1推论 域Δ上多项式$f(x)$的根域Ω的任一数ξ用$f(x)$诸根$x_1,x_2,\cdots,x_n$表示的所有有理式经域Δ上多项式$f(x)$的 Galois 群的一置换s所变成的有理式之值相同.

证 因置换s是由定理5.1证明中的一个本原数ρ_1变成与其共轭的本原数ρ_j所导致的. 由此$\rho_1\to\rho_j$相应导致(按此定理中4°)根域Ω在Δ上的自同构映射把ξ变成一个确定的数. 又因为此自同构映射T使Δ中的数不变且保持和、积关系,从而亦保持商的关系(设$\frac{a}{b}=c$,则$a=bc, Ta=Tb\cdot Tc, T\frac{a}{b}=Tc=\frac{Ta}{Tb}$). 于是对$\xi$用$x_1, x_2,\cdots,x_n$表示的任一有理式$\xi=R(x_1,x_2,\cdots,x_n)$有

$$T\xi=R(Tx_1,Tx_2,\cdots,Tx_n)$$

右边即$R(x_1,x_2,\cdots,x_n)$经置换s所得结果,故所得结果之值等于确定的数$T\xi$.

以后对域Δ上多项式$f(x)$根域的数ξ用$f(x)$诸根$x_1,x_2,\cdots, x_n$表示的所有有理式经域Δ上多项式$f(x)$的 Galois 群中的置换s所变成的有理式的值记为ξs,或按$\xi=R(x_1,x_2,\cdots,x_n)$记为$R(x_1, x_2,\cdots,x_n)s$.

以下讨论 Galois 群的性质.

定理5.2 设$M(x_1,x_2,\cdots,x_n)$为Δ上的有理式,其值为μ,则有等价关系:$\mu\in\Delta\Leftrightarrow$经$\Delta$上多项式$f(x)$的 Galois 群任一置换$M(x_1,x_2,\cdots,x_n)$不变.

证 $\Rightarrow$:设$\mu\in\Delta$. 因 Galois 群任一置换为本原数ρ_1变成ρ_j时导致n个根排列的一个置换——把$(x_1,x_2,\cdots,x_n)$变成$(x_{\rho_{j1}}, x_{\rho_{j2}},\cdots,x_{\rho_{jn}})$,这时亦导致根域$\Omega$在$\Delta$上的一个自同构$T_j$

$$T_jM(x_1,x_2,\cdots,x_n)=M(x_{\rho_{j1}},x_{\rho_{j2}},\cdots,x_{\rho_{jn}})$$

因$M(x_1,x_2,\cdots,x_n)=\mu\in\Delta$,故

$$T_jM(x_1,x_2,\cdots,x_n)=T_j\mu=\mu$$

不变,即$M(x_1,x_2,\cdots,x_n)$经上述置换后不变

$$M(x_{p_{j1}},x_{p_{j2}},\cdots,x_{p_{jn}})=\mu$$

$\Leftarrow$:用本原数ρ_1表示

$$M(x_1,x_2,\cdots,x_n)=\mu=c_0+c_1\rho_1+c_2\rho_1^2+\cdots+c_{m-1}\rho_1^{m-1}$$
$$(c_0,c_1,c_2,\cdots,c_{m-1}\in\Delta)$$

经 Galois 群任一置换(设与 $\rho_1\to\rho_j$ 相应),由假设知上式右边值 μ 不变,故

$$\mu=c_0+c_1\rho_j+c_2\rho_j^2+\cdots+c_{m-1}\rho_j^{m-1}$$

于是 Δ 上至多 $m-1$ 次方程

$$\mu=c_0+c_1t+c_2t^2+\cdots+c_{m-1}t^{m-1}$$

有 m 个不同根 $\rho_1,\rho_2,\cdots,\rho_m$,故必须 $\mu-c_0=c_1=c_2=\cdots=c_{m-1}=0$,$\mu=c_0\in\Delta$.

例2 第四章例中域 $\mathbf{Q}$ 上多项式 $(x^2-2)(x^2-3)$,取有理式 x_1+x_2,其值为 $0\in\Delta$,易见它经 Galois 群 4 个置换 $I,(x_1,x_2)(x_3,x_4),(x_1,x_2),(x_3,x_4)$ 不变值;取有理式 x_1-x_2,其值为 $2\sqrt{2}\notin\mathbf{Q}$,它经 Galois 群中的置换 (x_1,x_2) 成为 $x_2-x_1=-2\sqrt{2}$,变值.

再叙述 Galois 群的迷向子群的概念.

设 Δ 上多项式 $f(x)$ 的 Galois 群为 G,域 Δ 上各根的有理式 $M(x_1,x_2,\cdots,x_n)$,易见 G 中使 $M(x_1,x_2,\cdots,x_n)$ 不变值的所有置换组成的集 G' 为一个群①,称之为 G 对 $M(x_1,x_2,\cdots,x_n)$ 的迷向子群,记之为

$$I_G(M)=G'$$

反之称 $M(x_1,x_2,\cdots,x_n)$ 为 G 的子群 G' 的不变式.

例3 仍取上例多项式,取有理式 $M=x_1=\sqrt{2}$,则 Galois 群 G 中使 x_1 不变值的置换为 $I,(x_3,x_4)$,显然它们构成一个群,即

$$I_G(x_1)=\{I,(x_3,x_4)\}$$

注意置换 (x_2,x_3,x_4) 虽使 x_1 不变,但此置换不属 G,故不属 $I_G(x_1)$.

现讨论迷向子群的性质及 Lagrange 预解方程的概念.

定理 5.3 设有上述 Galois 群 G(m 阶),G 对有理式 $M(x_1,x_2,\cdots,x_n)$ 的迷向子群 $I_G(M)$(q 阶),则 $M(x_1,x_2,\cdots,x_n)$ 经 G 的所有置换得出 $[G:I_G(M)]=\dfrac{m}{q}=r$ 个不同的值,它们是 Δ 上一个不可约方程的全部的根(此方程称为方程 $f(x)$ 对 $M(x_1,x_2,\cdots,x_n)$ 的 Lagrange 预解方程).

---

①易见:$I\in G'$,当 $s,s'\in G'$ 时,$ss'\in G'$ 且 $s^{-1}\in G'$(从 $Ms=M$ 两边取置换 s^{-1} 得 $M=Ms^{-1}$).

推论 以根域Ω上任一数$M(x_1,x_2,\cdots,x_n)$为一根且在Δ上不可约的方程的次数为Ω维数m之约数(注意$M(x_1,x_2,\cdots,x_n)$含于上述r个值中,为上述不可约方程(r次)之根).

证 由定理2.5之证明知,Galois群G的$m=qr$个置换可分为r组

$$G_1=I_G(M)=\{\sigma_1,\sigma_2,\cdots,\sigma_q\}\quad(\sigma_1=I)$$
$$G_2=\{\sigma_1\tau_2,\sigma_2\tau_2,\cdots,\sigma_q\tau_2\}\quad(\tau_2\in G\backslash G_1)$$
$$G_3=\{\sigma_1\tau_3,\sigma_2\tau_3,\cdots,\sigma_q\tau_3\}\quad(\tau_3\in G\backslash(G_1\cup G_2))$$
$$\vdots$$
$$G_r=\{\sigma_1\tau_r,\sigma_2\tau_r,\cdots,\sigma_q\tau_r\}\quad(\tau_r\in G\backslash(G_1\cup G_2\cup\cdots\cup G_{r-1}))$$

而

$$M(x_1,x_2,\cdots,x_n)\sigma_j=M(x_1,x_2,\cdots,x_n)\triangleq M_1\quad(j=1,2,\cdots,q)$$
$$M(x_1,x_2,\cdots,x_n)\sigma_j\tau_k=$$
$$M(x_1,x_2,\cdots,x_n)\tau_k\triangleq M_k\quad(k=1,2,3,\cdots,r)(\tau_1=I)$$

再证$k\neq k'$(不妨设$k'>k$)时$M_{k'}\neq M_k$,否则

$$M(x_1,x_2,\cdots,x_n)\tau_{k'}=M(x_1,x_2,\cdots,x_n)\tau_k$$
$$M(x_1,x_2,\cdots,x_n)\tau_{k'}\tau_k^{-1}=M(x_1,x_2,\cdots,x_n)$$

从而

$$u\triangleq\tau_{k'}\tau_k^{-1}\in I_G(M)$$
$$\tau_{k'}=u\tau_k\in G_k\subset G_1\cup G_2\cup\cdots\cup G_{k'-1}$$

与前述$\tau_{k'}\notin G_1\cup G_2\cup\cdots\cup G_{k'-1}$矛盾.得证第一结论.

对任一个M_k(用$x_1,x_2,\cdots,x_n$的表示式)进行G中任一置换σ的结果,实际是先对M_1先进行G中某置换得M_k,再对M_k进行G中的置换σ,结果是对M_1进行这两置换之积,仍是群G中的置换,于是把每个M_k变成某个M_{p_k},易证当$k\neq k'$时$M_k\sigma\neq M_{k'}\sigma$(否则两边右乘$\sigma^{-1}$得$M_k=M_{k'}$,矛盾),故对排列$(M_1,M_2,\cdots,M_r)$各数进行$G$中的置换得出$M_1,M_2,\cdots,M_r$的某个排列.取多项式

$$\Phi(y)\equiv(y-M_1)(y-M_2)\cdots(y-M_r)$$

它的各次项系数对$M_1,M_2,\cdots,M_r$对称,从上述论证知它们对Galois群G的任一置换不变值,据定理5.2知它们属于Δ,$\Phi(y)$为域Δ上的多项式.

又设以M_1为一根且在Δ上不可约的"首一"多项式为$\Phi^*(y)$,故

$$\Phi^*(M(x_1,x_2,\cdots,x_n))=\Phi^*(M_1)=0\in\Delta$$

于是按定理5.2知左边经G中任一置换不变值，但这时$M(x_1, x_2, \cdots, x_n)$可变成任一M_k，即

$$\Phi^*(M_k)=0$$

$\Phi^*(y)$有根$M_1, M_2, \cdots, M_r$，于是$\Phi^*(y)$有因式$\Phi(y)$（Δ上多项式，最高次项系数亦为1），但$\Phi^*(y)$在Δ上不可约，故

$$\Phi^*(y) \equiv \Phi(y)$$

$\Phi(y)$在Δ上不可约，符合定理结论要求，证毕.

例4　仍取第四章中域$\mathbf{Q}$上多项式$(x^2-2)(x^2-3)$. 取有理式x_1，它经Galois群G的4个置换（见第五章）分别得x_2, x_1, x_2, x_1. 只得两个不同的值$x_1=\sqrt{2}, x_2=-\sqrt{2}$. 其个数2正是Galois群阶数4除以迷向子群$I_G(x_1)$阶数2的商. 此两个不同的值$\sqrt{2}, -\sqrt{2}$正是$\mathbf{Q}$上不可约方程$x^2-2=0$的全部根.

定理5.4　设Δ上多项式$f(x)$的Galois群为G，根域Ω有数ξ，则$I_G(\xi)$为$\Delta(\xi)$上$f(x)$的Galois群G'.

证　把$f(x)$看成Δ的扩域$\Delta(\xi)$上的方程，易见其根域亦为Ω，故G'的任一置换σ'导致Ω在$\Delta(\xi)$上的自同构映射，即Ω的使ξ及Δ中所有数不变的自同构映射. 从它不改变Δ中所有数知它是Δ上Galois群G中置换相应的映射，它使ξ（按$x_1, x_2, \cdots, x_n$表示式）不变，故σ'属于$I_G(\xi)$中的置换，得证$G' \subset I_G(\xi)$.

再证$I_G(\xi) \subset G'$即可：$I_G(\xi)$中任一置换σ是Δ上$f(x)$的Galois群G中使ξ（用$x_1, x_2, \cdots, x_n$的表示式）不变的置换，相应于根域Ω使ξ及Δ中所有数不变的自同构映射，从而使$\Delta(\xi)$的数不变，故亦可看成Ω在$\Delta(\xi)$上的自同构映射，即$\Delta(\xi)$上$f(x)$的Galois群G'的置换相应的映射，故σ为G'的置换，得证$I_G(\xi) \subset G'$.

推论　设有上述Galois群G及数ξ，则$\Delta(\xi)$为Ω中被$I_G(\xi)$中任何置换不变值的所有数组成的域.

证　因$I_G(\xi)$中所有置换即$\Delta(\xi)$上$f(x)$的Galois群中所有置换，故据定理5.2知$\Delta(\xi)$为Ω中被$I_G(\xi)$中任何置换不变值的所有数组成的域.

例5　仍取例4中的多项式及有理式x_1，则经所有$I_G(x_1)$的置换$(I,(x_3, x_4))$不变的有理式M^*不能含x_3, x_4，即不能含$\sqrt{3}$（包括$\sqrt{6}=\sqrt{2}\times\sqrt{3}$），即只能含$\sqrt{2}$，故$M^* \in Q(\sqrt{2})$，反之$Q(\sqrt{2})$的任一数显然经$I$及$(x_3, x_4)$不变值.

第六章 用 Galois 群的不变式导出 Lagrange 预解方程,从而推出三、四次方程的求根公式

域 Δ 上三(四) 次方程的 Galois 群未必是对称群 $S_3(S_4)$,在较特殊情况下可能是 $S_3(S_4)$ 的真子群. 虽然本章在 Galois 群为对称群的较一般情况下推导求根公式,但在下文可见,当求出 Lagrange 预解方程各根用原三、四次方程的根的表示式后的计算适用于任何三(四) 次方程,故求得的公式实际适用于所有三(四) 次方程.

对任何 n 次方程(不妨设最高次项系数为 1)

$$x^n + a_1x^{n-1} + a_2x^{n-2} + \cdots + a_{n-1}x + a_n = 0$$

作变换 $x = x' - \frac{a_1}{n}$,得对未知数 x' 的方程

$$(x' - \frac{a_1}{n})^n + a_1(x' - \frac{a_1}{n})^{n-1} + \cdots + a_n = 0$$

易见经化简后得

$$x'^n + a_2'x'^{n-2} + \cdots + a_n' = 0$$

故不妨设原 n 次方程无 $n-1$ 次项①.

(1) 解三次方程(复系数)

$$x^3 + px + q = 0 \tag{6.1}$$

按附录 Ⅰ 求得的 S_3 的真子群

$$G = A_3$$

$$[S_3 : A_3] = 2$$

S_3 的子群 A_3 有不变式

$$\alpha^3 = (x_1 + \omega_3x_2 + \omega_3^2x_3)^3$$

其中

$$\alpha = x_1 + \omega_3x_2 + \omega_3^2x_3$$

$$\omega_3 = e^{\frac{2}{3}\pi i}$$

因 α^3 经 A_3 中轮换 (x_1, x_2, x_3) 得

$$(x_2 + \omega_3x_3 + \omega_3^2x_1)^3 =$$

$$\omega_3^3(x_2 + \omega_3x_3 + \omega_3^2x_1)^3 =$$

①对二次方程进行此变换(实即进行配方) 后,即可推得其求根公式.

$$[\omega_3(x_2+\omega_3x_3+\omega_3^2x_1)]^3=$$
$$(\omega_3x_2+\omega_3^2x_3+x_1)^3=\alpha^3\quad(因\ \omega_3^3=1)$$

从而 α^3 经置换 $(x_1,x_2,x_3)=(x_1,x_2,x_3)^2$ 仍不变,但经 A_3 以外的置换(对换)变值.

于是 S_3 对 α^3 的迷向子群为 A_3. 再对 α^3 进行 S_3(偶群)以外的奇置换——A_3 的各置换乘 S_3 中奇置换 (x_2,x_3) 之积(据附录Ⅰ中引理),实际上是先对 A_3 进行 A_3 各置换再进行置换 (x_2,x_3) 的结果,因第一次置换使 α^3 不变,故最后结果都得

$$(x_1+\omega_3x_3+\omega_3^2x_2)^3\triangleq\beta^3$$

其中

$$\beta=x_1+\omega_3x_3+\omega_3^2x_2$$

从而 α^3 经 S_3 所有置换得两个不同的值 α^3 及 β^3. 现求以 α^3 及 β^3 为两根之二次方程(Lagrange 预解方程)

$$\alpha^3+\beta^3=(x_1+\omega_3x_2+\omega_3^2x_3)^3+(x_1+\omega_3x_3+\omega_3^2x_2)^3=$$
$$2(x_1^3+x_2^3+x_3^3)+3(\omega_3+\omega_3^2)(x_1^2x_2+x_2^2x_1+x_2^2x_3+x_3^2x_2+x_3^2x_1+x_1^2x_3)+12x_1x_2x_3=$$
$$2\sum x_1^3-3\sum x_1^2x_2+12x_1x_2x_3$$

(其中"$\sum$"表示对 x_1,x_2,x_3 的所有同型的项之和,其后只写出一个"代表项". 因 $\omega_3^3-1=0$,即 $(\omega_3-1)(\omega_3^2+\omega_3+1)=0$,又 $\omega_3-1\neq0$,故 $\omega_3^2+\omega_3+1=0$,$\omega_3^2+\omega_3=-1$).

继续求得(先展开 $(\sum x_1)^3$,再把 $\sum x_1^3$ 用 $(\sum x_1)^3,\cdots$ 表示).

$$\alpha^3+\beta^3=2[(\sum x_1)^3-3\sum x_1^2x_2-6x_1x_2x_3]-$$
$$3\sum x_1^2x_2+12x_1x_2x_3=$$
$$2(\sum x_1)^3-9\sum x_1^2x_2=$$
$$2(\sum x_1)^3-9(\sum x_1x_2\cdot\sum x_1-3x_1x_2x_3)=-27q$$
$$(注意\sum x_1=0,\sum x_1x_2=p,x_1x_2x_3=-q)$$

$$\alpha\beta=(x_1+\omega_3x_2+\omega_3^2x_3)(x_1+\omega_3x_3+\omega_3^2x_2)=$$
$$(x_1^2+x_2^2+x_3^2)+(\omega_3+\omega_3^2)(x_1x_2+x_2x_3+x_3x_1)=$$
$$(x_1+x_2+x_3)^2-2(x_1x_2+x_2x_3+x_3x_1)-$$
$$(x_1x_2+x_2x_3+x_3x_1)=-3p$$

$$\alpha^3\beta^3 = -27p^3$$

故 α^3,β^3 为下述方程之根

$$t^2 + 27qt - 27p^3 = 0$$

从而

$$\alpha^3,\beta^3 = \frac{-27q \pm \sqrt{27^2q^2 + 4\cdot 27p^3}}{2}$$

$$\alpha,\beta = 3\sqrt[3]{-\frac{q}{2} \pm \sqrt{\frac{q^2}{4} + \frac{p^3}{27}}}$$

对复数 p,q,上式表(3 个立方根中) 某个立方根,但由 $\alpha\beta = -3p$,故必须要求对 α,β 分别所取的立方根之积为 $-\frac{p}{3}$. 特别当 p,q 为实数,$\frac{q^2}{4} + \frac{p^3}{27} \geqslant 0$,则可设平方根号表算术根(非负),立方根可取实根①. 取对 α,β 的一组解得

$$\begin{cases} x_1 + \omega_3 x_2 + \omega_3^2 x_3 = 3\sqrt[3]{-\frac{q}{2} + \sqrt{\frac{q^2}{4} + \frac{p^3}{27}}} \\ x_1 + \omega_3^2 x_2 + \omega_3 x_3 = 3\sqrt[3]{-\frac{q}{2} - \sqrt{\frac{q^2}{4} + \frac{p^3}{27}}} \end{cases} \tag{6.2}$$

(另一组解为右边对调,实际相当左边 x_2,x_3 对调,因各根序号是随意定的,故可认为只有上述一组解).

又

$$x_1 + x_2 + x_3 = 0$$

以上三式相加得

$$x_1 = \sqrt[3]{-\frac{q}{2} + \sqrt{\frac{q^2}{4} + \frac{p^3}{27}}} + \sqrt[3]{-\frac{q}{2} - \sqrt{\frac{q^2}{4} + \frac{p^3}{27}}} \tag{6.3}$$

第 1,2,3 式分别乘 $\omega_3^2,\omega_3,1$ 或分别乘 $\omega_3,\omega_3^2,1$ 后相加得

$$x_2 = \omega_3^2\sqrt[3]{-\frac{q}{2} + \sqrt{\frac{q^2}{4} + \frac{p^3}{27}}} + \omega_3\sqrt[3]{-\frac{q}{2} - \sqrt{\frac{q^2}{4} + \frac{p^3}{27}}} \tag{6.4}$$

---

①易见当 $a,b\in\mathbf{R}$ 且取立方根为实根时 $\sqrt[3]{a}\sqrt[3]{b} = \sqrt[3]{ab}$(两边均为 ab 的唯一实数立方根),故当 $p,q\in\mathbf{R}$,且 $\frac{q^2}{4} + \frac{p^3}{27} \geqslant 0$ 按上述取算术平方根及实数立方根,则有

$$\sqrt[3]{-\frac{q}{2} + \sqrt{\frac{q^2}{4} + \frac{p^3}{27}}}\cdot\sqrt[3]{-\frac{q}{2} - \sqrt{\frac{q^2}{4} + \frac{p^3}{27}}} = \sqrt[3]{\frac{q^2}{4} - (\frac{q^2}{4} + \frac{p^3}{27})} = \sqrt[3]{-\frac{p^3}{27}} = -\frac{p}{3}$$

$$x_3=\omega_3\sqrt[3]{-\frac{q}{2}+\sqrt{\frac{q^2}{4}+\frac{p^3}{27}}}+\omega_3^2\sqrt[3]{-\frac{q}{2}-\sqrt{\frac{q^2}{4}+\frac{p^3}{27}}} \quad (6.5)$$

其中要求两(多值)立方根之积为 $-\frac{p}{3}$. 各式中同一数的立(平)方根要相同.

这就是 Cardan(卡当)公式.

(2) 解四次方程

$$x^4+px^2+qx+r=0$$

解法一(Euler(欧拉))①

按附录 Ⅰ 求得 S_4 的真子群($G_8^{(4)}$ 型)

$$G=\{I,(x_1,x_2)(x_3,x_4),(x_1,x_3)(x_2,x_4),(x_1,x_4)(x_2,x_3),(x_1,x_3),(x_2,x_4),(x_1,x_2,x_3,x_4),(x_1,x_4,x_3,x_2)\}$$

$$[S_4:G]=3 \quad (6.6)$$

取

$$\alpha=\frac{1}{4}(x_1+x_3-x_2-x_4)$$

$$\alpha^2=\frac{1}{16}(x_1+x_3-x_2-x_4)^2$$

易见 α^2 为 S_4 的子群 G 的不变式, S_4 对 α^2 的迷向子群为 G, α^2 经 S_4 的一切置换除得 α^2 不变外还得出

$$\beta^2=\frac{1}{16}(x_2+x_3-x_1-x_4)^2$$

$$\gamma^2=\frac{1}{16}(x_4+x_3-x_2-x_1)^2$$

其中

$$\beta=\frac{1}{4}(x_2+x_3-x_1-x_4)$$

$$\gamma=\frac{1}{4}(x_4+x_3-x_2-x_1)$$

现求以 $\alpha^2,\beta^2,\gamma^2$ 为根的(Lagrange 预解)方程. 先求出(其中"$\sum$"表示对 x_1,x_2,x_3,x_4 的所有同型项之和)

$$\alpha^2+\beta^2+\gamma^2=\frac{3}{16}\sum x_1^2-\frac{1}{8}\sum x_1x_2=$$

---

①用 Galois 群的不变式求得 Lagrange 预解方程推导出的解四次方程过程与 Euler 解法相同,但推导与 Euler 不同,下文解法二(Ferrari)亦然.

$$\frac{3}{16}\left[\left(\sum x_1\right)^2-2\sum x_1x_2\right]-\frac{1}{8}\sum x_1x_2=$$

$$-\frac{1}{2}\sum x_1x_2=-\frac{1}{2}p\quad(\text{注意}\sum x_1=0)$$

$$\alpha^2\beta^2+\beta^2\gamma^2+\gamma^2\alpha^2=\frac{1}{256}\{[(x_1-x_2)^2-(x_3-x_4)^2]^2+[(x_3-x_1)^2-(x_2-x_4)^2]^2+[(x_1-x_4)^2-(x_2-x_3)^2]^2\}\qquad(6.7)$$

现改用待定系数法求它用 p,q,r 的表示式. 易验证它对各根 x_1,x_2,x_3,x_4 的任何对换不变,为 x_1,x_2,x_3,x_4 的对称多项式(或据定理5.3,Lagrange 预解方程为域 Δ 上的方程,上式为其系数属 Δ,再据定理5.2知它对Galois群 S_n 任何置换不变,必为对称多项式),故可用基本对称多项式

$$\sigma_1=x_1+x_2+x_3+x_4$$

$$\sigma_2=x_1x_2+x_1x_3+x_1x_4+x_2x_3+x_2x_4+x_3x_4$$

$$\sigma_3=x_1x_2x_3+x_2x_3x_4+x_3x_4x_1+x_4x_1x_2$$

$$\sigma_4=x_1x_2x_3x_4$$

的多项式表示. 但式(6.7)为四次对称式,要使各项为 x_1,x_2,x_3,x_4 的四次项只能表示为

$$A\sigma_4+B\sigma_3\sigma_1+C\sigma_2^2+D\sigma_2\sigma_1^2+E\sigma_1^4$$

但 $\sigma_1=0$,故实际只有两项

$$A\sigma_4+C\sigma_2^2=Ax_1x_2x_3x_4+C(x_1x_2+x_1x_3+x_1x_4+x_2x_3+x_2x_4+x_3x_4)^2\qquad(6.8)$$

取 $x_1=1,x_2=-1,x_3=x_4=0$(要使和为0)代入式(6.7)及式(6.8),其右边应相等,从而得

$$\frac{1}{16}=C$$

再取 $x_1=x_3=1,x_2=x_4=-1$(和为0)代入得

$$0=A+4C$$

$$A=-4C=-\frac{1}{4}$$

于是得

$$\alpha^2\beta^2+\beta^2\gamma^2+\gamma^2\alpha^2=-\frac{1}{4}x_1x_2x_3x_4+\frac{1}{16}\left(\sum x_1x_2\right)^2=\frac{1}{16}p^2-\frac{1}{4}r$$

再求

$$
\begin{aligned}
\alpha\beta\gamma = & \frac{1}{64}\left(\sum x_1^3 - \sum x_1^2 x_2 + 2\sum x_1 x_2 x_3\right) = \\
& \frac{1}{64}\left[\left(\sum x_1\right)^3 - 3\sum x_1^2 x_2 - 6\sum x_1 x_2 x_3 - \right. \\
& \left. \sum x_1^2 x_2 + 2\sum x_1 x_2 x_3\right] = \\
& \frac{1}{64}\left(-4\sum x_1^2 x_2 - 4\sum x_1 x_2 x_3\right) = \\
& \frac{1}{64}\left[-4\left(\sum x_1 \cdot \sum x_1 x_2 - 3\sum x_1 x_2 x_3\right) - 4\sum x_1 x_2 x_3\right] = \\
& \frac{1}{8}\sum x_1 x_2 x_3 = -\frac{1}{8}q
\end{aligned}
$$

$$
\alpha^2\beta^2\gamma^2 = (\alpha\beta\gamma)^2 = \frac{1}{64}q^2
$$

故 $\alpha^2,\beta^2,\gamma^2$ 为下面三次方程的根

$$
t^3 + \frac{1}{2}pt^2 + \left(\frac{1}{16}p^2 - \frac{1}{4}r\right)t - \frac{1}{64}q^2 = 0 \tag{6.9}
$$

解此三次方根求得其三根，再开平方得 α,β,γ 之值(均为两解)，但应要求三个平方根之积 $\alpha\beta\gamma = -\frac{1}{8}q$，故取定 α,β(共四组解) 则 γ 随之确定，于是得 α,β,γ 的四组解，随意取定其中一组解，从而得

$$
\begin{aligned}
\frac{1}{4}(x_1 + x_3 - x_2 - x_4) = \alpha \\
\frac{1}{4}(x_2 + x_3 - x_1 - x_4) = \beta \\
\frac{1}{4}(x_4 + x_3 - x_1 - x_2) = \gamma
\end{aligned} \tag{6.10}
$$

此外还有

$$
\frac{1}{4}(x_1 + x_2 + x_3 + x_4) = 0
$$

解得

$$
\begin{cases}
x_3 = \alpha + \beta + \gamma \\
x_2 = -\alpha + \beta - \gamma \\
x_1 = \alpha - \beta - \gamma \\
x_4 = -\alpha - \beta + \gamma
\end{cases} \tag{6.11}
$$

(要使 $\alpha\beta\gamma = -\frac{1}{8}q$)

其中右边四式中任两式含 α,β 的前两项,若有一(二)项变号则含 γ 的第三项变号(不变).易见当 $\alpha(\beta)$ 变号,γ 随之变号时,式(6.11)各式右边分别改成 x_2,x_3,x_4,x_1(分别改成 x_1,x_4,x_3,x_2)之值;而当 α,β 变号,γ 不变时,式(6.11)各式右边分别表示 x_4,x_1,x_2,x_3 之值.即 α,β 可随意选定一组解求得四根都有相同答案(但各根序号不同).

解法二(Ferrari(费拉里))

解四次方程

$$x^4+ax^3+px^2+qx+r=0$$

同样取 S_4 的真子群 G 如(6.6),改取

$$\alpha=x_1x_3+x_2x_4$$

它也是 S_4 对子群 G 的不变式,S_4 对 α 的迷向子群为 G,α 经 S_4 的一切置换除得 α 不变外还得出

$$\beta=x_2x_3+x_1x_4$$

$$\gamma=x_4x_3+x_2x_1$$

现求以 α,β,γ 为根的 Lagrange 预解方程,先求

$$\alpha+\beta+\gamma=\sum x_1x_2=p$$

$$\alpha\beta+\beta\gamma+\gamma\alpha=\sum x_1^2x_2x_3=\sum x_1x_2x_3\cdot\sum x_1-4x_1x_2x_3x_4=qa-4r$$

$$\begin{aligned}\alpha\beta\gamma&=x_1x_2x_3x_4\cdot\sum x_1^2+\sum x_1^2x_2^2x_3^2=\\&x_1x_2x_3x_4\left[\left(\sum x_1\right)^2-2\sum x_1x_2\right]+\\&\left[\left(\sum x_1x_2x_3\right)^2-2x_1x_2x_3x_4\cdot\sum x_1x_2\right]=\\&r(a^2-2p)+(q^2-2rp)=\\&ra^2+q^2-4rp\end{aligned}$$

故 α,β,γ 为下述三次方程之根

$$t^3-pt^2+(aq-4r)t+(4rp-q^2-ra^2)=0\qquad(6.12)$$

任取一根 t^*,不妨设为 γ 之值(因各根序号任定),于是

$$x_1x_2+x_3x_4=t^*$$

$$x_1x_2\cdot x_3x_4=r$$

故 x_1x_2 与 x_3x_4 为下述二次方程的两根

$$u^2-t^*u+r=0$$

即

$$\frac{t^*}{2}\pm\sqrt{\frac{t^{*2}}{4}-r}$$

同样不妨认为

$$\begin{cases}x_1x_2=\frac{t^*}{2}+\sqrt{\frac{t^{*2}}{4}-r}\\ x_3x_4=\frac{t^*}{2}-\sqrt{\frac{t^{*2}}{4}-r}\end{cases}\tag{6.13}$$

注意

$$x_1+x_2+x_3+x_4=-a$$

$$x_1+x_2+a=-(x_3+x_4)$$

于是

$$q=-(x_1+x_2)x_3x_4-(x_3+x_4)x_1x_2=$$

$$-(x_1+x_2)x_3x_4+(x_1+x_2+a)x_1x_2=$$

$$(x_1+x_2)(x_1x_2-x_3x_4)+ax_1x_2$$

以上述求得之 x_1x_2, x_3x_4 之值代入得

$$q=(x_1+x_2)\cdot 2\sqrt{\frac{t^{*2}}{4}-r}+a\left(\frac{t^*}{2}+\sqrt{\frac{t^{*2}}{4}-r}\right)$$

从而

$$\left(x_1+x_2+\frac{a}{2}\right)\sqrt{\frac{t^{*2}}{4}-r}=\frac{1}{2}\left(q-\frac{at^*}{2}\right)\tag{6.14}$$

又

$$(x_1+x_2)(x_3+x_4)=p-(x_1x_2+x_3x_4)=p-t^*$$

$$(x_1+x_2)+(x_3+x_4)=-a$$

故 x_1+x_2, x_3+x_4 为下述方程(未知数 u)之根

$$u^2+au+(p-t^*)=0$$

即

$$-\frac{a}{2}\pm\sqrt{\frac{a^2}{4}-p+t^*}$$

再从式(6.14),对 x_1+x_2 应取上式根号前的符号,使

$$\pm\sqrt{\frac{a^2}{4}-p+t^*}\sqrt{\frac{t^{*2}}{4}-r}=\frac{1}{2}\left(q-\frac{at^*}{2}\right)\tag{6.15}$$

于是

$$x_1+x_2=-\frac{a}{2}\pm\sqrt{\frac{a^2}{4}-p+t^*}\tag{6.16}$$

$$x_3 + x_4 = -\frac{a}{2} \mp \sqrt{\frac{a^2}{4} - p + t^*} \tag{6.17}$$

(式(6.16)中的"±"选择要适合式(6.15),式(6.17)与式(6.16)的"±"相反).

再由式(6.13)知x_1与x_2,x_3与x_4分别是下述方程之根

$$x^2 + \left(\frac{a}{2} \mp \sqrt{\frac{a^2}{4} + t^* - p}\right)x + \left(\frac{t^*}{2} + \sqrt{\frac{t^{*2}}{4} - r}\right) = 0 \tag{6.18}$$

$$x^2 + \left(\frac{a}{2} \pm \sqrt{\frac{a^2}{4} + t^* - p}\right)x + \left(\frac{t^*}{2} - \sqrt{\frac{t^{*2}}{4} - r}\right) = 0 \tag{6.19}$$

(根号前的"±"选择要适合式(6.15)).

实际上,当$\sqrt{\frac{t^{*2}}{4} - r} \neq 0$时,两方程一次项系数中(据式(6.15))

$$\mp \sqrt{\frac{a^2}{4} + t^* - p} = -\left(q - \frac{at^*}{2}\right)/\left(2\sqrt{\frac{t^{*2}}{4} - r}\right) \tag{6.20}$$

$$\pm \sqrt{\frac{a^2}{4} + t^* - p} = \left(q - \frac{at^*}{2}\right)/\left(2\sqrt{\frac{t^{*2}}{4} - r}\right) \tag{6.21}$$

而当$\sqrt{\frac{t^{*2}}{4} - r} = 0$时,从式(6.15)应有$\frac{1}{2}\left(q - \frac{at^*}{2}\right) = 0$,这时由式(6.13)知$x_1x_2 = x_3x_4 = \frac{t^*}{2}$,$x_1$与$x_2$,$x_3$与$x_4$两组根尚可对调,即上述两一次项系数中的"±"可任取,只要两个根号前的符号相反即可.

易见上述两二次方程与一般高等代数课本中所述由Ferrari求得的二次方程实质相同.

第七章 循环方程

定义7.1 设域Δ上方程$f(x)=0$如同第四章,若其Galois群为循环群,则称$f(x)=0$为循环方程.

最常见的循环方程是与素数次分圆多项式相应的分圆方程(见下文定理7.4).

现讨论循环方程性质,先引入可迁(非可迁)群定义.

定义 7.2 n 个元素 $x_1, x_2, \cdots, x_n$ 的置换群 P，若任一元素能经 P 中的置换变成其余任一元素，则称 P 为可迁群；否则 P 称为非可迁群——即 n 个元素可分为至少两组，任一组的任一元素可经 P 中的置换变成该组其余任一元素，而不能经 P 中的置换变成另一组的任一元素.

引理 7.1 设域 Δ 上 n 次多项式 $f(x)$，其 p 个根 $x_1, x_2, \cdots, x_p$. 令

$$F(x) \equiv (x - x_1)(x - x_2)\cdots(x - x_p) \tag{7.1}$$

则有等价关系：$F(x)$ 为域 Δ 上多项式 $\Leftrightarrow$ 域 Δ 上 $f(x)$ 的 Galois 群把 $x_1, x_2, \cdots, x_p$ 中任一根变成 $x_1, x_2, \cdots, x_p$ 中某一根.

证 $\Leftarrow$：易见 $F(x)$ 展开式中 x 的各次幂的系数对 $x_1, x_2, \cdots, x_p$ 对称，而这时 Galois 群中任一置换把排列 $x_1, x_2, \cdots, x_p$ 变成这 p 个根的某一排列，故上述系数对 Galois 群任一置换不变，从而据定理5.2知上述系数属于 Δ，故 $F(x)$ 为 $f(x)$ 在域 Δ 上的因式.

$\Rightarrow$：由 Vieta（韦达）定理知

$$F(x) \equiv x^p - x^{p-1}\sum_{j=1}^{p} x_j + x^{p-2}\sum_{1 \leqslant j_1 < j_2 \leqslant p} x_{j_1}x_{j_2} + \cdots + (-1)^p x_1 x_2 \cdots x_p$$

因 $x_s(s = 1, 2, \cdots, p)$ 为其根，故

$$x_s^p - x_s^{p-1}\sum_{j=1}^{p} x_j + x_s^{p-2}\sum_{1 \leqslant j_1 < j_2 \leqslant p} x_{j_1}x_{j_2} + \cdots + (-1)^p x_1 x_2 \cdots x_p = 0$$

此式为 $x_1, x_2, \cdots, x_p$ 之恒等式（因用任何 p 个数 $x_1, x_2, \cdots, x_p$ 可构造多项式 $F(x)$，从而上式成立）. 设域 Δ 上 $f(x) = 0$ 的 Galois 群的任一置换 σ 把 x_j 变成 x_{σ_j}，则在上述恒等式中换元得

$$x_{\sigma_s}^p - x_{\sigma_s}^{p-1}\sum_{j=1}^{p} x_{\sigma_j} + x_{\sigma_s}^{p-2}\sum_{1 \leqslant j_1 < j_2 \leqslant p} x_{\sigma_{j_1}}x_{\sigma_{j_2}} + \cdots + (-1)^p x_{\sigma_1}x_{\sigma_2}\cdots x_{\sigma_p} = 0$$

但域 Δ 上多项式 $f(x)$ 的系数 $-\sum_{j=1}^{p} x_j$，$\sum_{1 \leqslant j_1 < j_2 \leqslant p} x_{j_1}x_{j_2}$，$\cdots$，$(-1)^p x_1 x_2 \cdots x_p$ 属于 Δ. 故由定理5.2知经 Galois 群的置换 σ 不变，于是

$$x_{\sigma_s}^p - x_{\sigma_s}^{p-1}\sum_{j=1}^{p} x_j + x_{\sigma_s}^{p-2}\sum_{1 \leqslant j_1 < j_2 \leqslant p} x_{j_1}x_{j_2} + \cdots + (-1)^p x_1 x_2 \cdots x_p = 0$$

即 x_{σ_s} 为 $f(x)$ 的某一根，$x_1, x_2, \cdots, x_p$ 中任一根 x_s 经 Galois 群任一置换 σ 变成 $x_1, x_2, \cdots, x_p$ 中的一根.

推论 域 Δ 上多项式 $f(x)$ 在 Δ 上可约（不可约）的充要条件为：Δ 上 $f(x)$ 的 Galois 群为非可迁群（可迁群）.

(因多项式 $f(x)$ 在域 Δ 可约,即有部分 $p(<n)$ 个根 $x_1,x_2,\cdots,x_p$ 相应的多项式 $F(x)$ 为域 Δ 上的多项式).

定理 7.1 设域 Δ 上循环方程 $f(x)=0$,其 Galois(循环)群生成元为 σ,则有等价关系:

①$f(x)$ 在域 Δ 不可约;$\Leftrightarrow$②σ 为 $f(x)$ 所有根的一个轮换.

证 由上推论知只要证:②$\Leftrightarrow$ 所述 Galois 群为可迁群.

把 σ 表示为无公共元素的轮换之积

$$(x_{a_1},x_{a_2},\cdots,x_{a_{n'}})\cdots$$

易见 $n'=n(n'<n)$ 时所述 Galois 群为非可迁群(可迁群),从而所述得证——证"$\Leftarrow$"时用反证法.

定理 7.2 域 Δ 上 n 次不可约循环方程 $f(x)=0$ 的所有根可在 Δ 上循环有理表示,即(不妨必要时改变各根的序号)

$$x_2=\varphi(x_1),x_3=\varphi(x_2),\cdots,x_n=\varphi(x_{n-1})$$

$$x_1=\varphi(x_n)\quad(\varphi\text{ 为 }\Delta\text{ 上有理式})\tag{7.2}$$

证 由上定理可不妨设域 Δ 上方程 $f(x)=0$ 的 Galois(循环)群的生成元为

$$s=(x_1,x_2,\cdots,x_n)$$

取多项式

$$\Psi(x)=x_2\frac{f(x)}{x-x_1}+x_3\frac{f(x)}{x-x_2}+\cdots+x_n\frac{f(x)}{x-x_{n-1}}+x_1\frac{f(x)}{x-x_n}$$

易见其中 x 的各次幂系数对 Galois 群任一置换(s 的幂)不变,故按定理 5.2 知这些系数属于域 Δ,$\Psi(x)$ 为域 Δ 上多项式,按引理 4.1 知

$$x_2=\frac{\Psi(x)}{f'(x_1)},x_3=\frac{\Psi(x)}{f'(x_2)},\cdots,x_n=\frac{\Psi(x)}{f'(x_{n-1})},x_1=\frac{\Psi(x)}{f'(x_n)}$$

故定理结论成立,其中 $\varphi(x)=\dfrac{\Psi(x)}{f'(x)}$.

推论 域 Δ 上不可约循环方程任一根必是其根域的本原数.

定理 7.3 若域 Δ 上素数 p 次方程 $f(x)=0$ 所有各根可在 Δ 上循环有理表示如式(7.2),则它是域 Δ 上不可约循环方程. 相应 Galois 群为 p 阶循环群,以轮换 $s=(x_1,x_2,\cdots,x_n)$ 为其生成元.

证 由式(7.2) 对 $j=2,3,\cdots,p$ 有

$$x_j=\varphi\underbrace{(\varphi(\cdots\varphi(x_1)\cdots))}_{j-1\text{层括号}}\triangleq\varphi_{j-1}(x)$$

而

$$\varphi_p(x_1)=\varphi(\varphi_{p-1}(x_1))=\varphi(x_p)=x_1$$

即 p 层括号相当于无括号.

在根列用 x_1 的表示式

$$x_1=x_1,x_2=\varphi_1(x_1),x_3=\varphi_2(x_1),\cdots,x_p=\varphi_{p-1}(x_1) \quad (7.3)$$

中各式右边把 x_1 改成 x_2,实际相当于增加一层括号,上根列变成

$$x_2=\varphi_1(x_1),x_3=\varphi_2(x_1),x_4=\varphi_3(x_1),\cdots,$$
$$x_p=\varphi_{p-1}(x_1),x_1=\varphi_p(x_1) \quad (7.4)$$

即根列进行了轮换

$$s=(x_1,x_2,\cdots,x_p)$$

若在式(7.3)中各式右边把 x_1 改为 x_3,实际即在式(7.4)中再加一层括号,上述根列变成

$$x_3,x_4,x_5,\cdots,x_p,x_1,x_2$$

即对式(7.3)进行置换 s^2.

一般在式(7.3)各式右边把 x_1 改成 $x_j(j=1,2,\cdots,p)$,相当于对式(7.3)进行置换 $s^{j-1}(j-1=0,1,2,\cdots,p-1)$,它们组成以轮换 s 为生成元的循环群 G.

由于 $f(x)$ 各根可用根域中 x_1 的有理式表示,于是据定理4.3知 x_1 为域 Δ 上 $f(x)$ 根域的本原数,设以 x_1 为一根且在 Δ 不可约的多项式 $g(x)$ 的各根(含 x_1)为域 Δ 上 $f(x)$ 根域的共轭本原数,Δ 上多项式 $f(x)$ 与不可约多项式 $g(x)$ 有公共根 x_1,故据引理3.1知 $g(x)$ 的所有根为 $f(x)$ 的根,可设 $g(x)$ 的所有根为

$$x_1,x_{a_2},\cdots,x_{a_m} \quad (m\leqslant p)$$

它们为 Δ 上 $f(x)$ 根域的共轭本原数. 于是域 Δ 上多项式 $f(x)$ 的 Galois 群为与本原数的变换

$$x_1\to x_1,x_1\to x_{a_2},x_1\to x_{a_3},\cdots,x_1\to x_{a_m}$$

导致 $f(x)$ 各根的置换所组成的群,它显然是上述(素数)p 阶循环群 G 之子群,由定理2.5知此子群的阶数 m 为 G 之阶数 p 的约数 p 或1,但当 $m=1$ 时 $x-x_1$ 为 $f(x)$ 在 Δ 上的不可约因式,与假设 $f(x)$ 在 Δ 上不可约矛盾. 故只有 $m=p$,于是上述 Galois 群为 G 之同阶子群,实际上,Galois 群即循环群 G,又 $f(x)$ 与其因式 $g(x)$ 同次,故 $f(x)$ 与 $g(x)$ 一样在域 Δ 上不可约,$f(x)=0$ 为 Δ 上不可约循环方程.

下文讨论素数 p 次循环方程与1的 p 次方根(称 p 次单位根)的

关系,记

$$\omega_p = e^{2\pi i/p}$$

定理7.4 设p为奇素数,r为模p之原根(见附录Ⅱ),则有理数域$\mathbf{Q}$上分圆方程(左边称分圆多项式)

$$x^{p-1}+x^{p-2}+\cdots+x+1=0 \tag{7.5}$$

为域$\mathbf{Q}$上不可约循环方程,其Galois(循环)群生成元为轮换①

$$(\omega_p^{r},\omega_p^{r^2},\omega_p^{r^3},\cdots,\omega_p^{r^{p-1}})$$

(实际上$\omega_p^{r^{p-1}}=\omega_p$,因$r$非$p$的倍数,由Fermat定理(附录Ⅱ定理2)$r^{p-1}\equiv 1(\bmod p)$).

证 式(7.5)实为由所有p次单位根所适合的方程

$$x^p-1=0$$

约去因式$x-1$,即除去p次单位根1所得的方程,故其根为

$$\omega_p^{r},\omega_p^{r^2},\omega_p^{r^3},\cdots,\omega_p^{r^{p-1}}=\omega_p$$

(因由附录Ⅱ定理4,上述ω_p的幂的指数与$1,2,\cdots,p-1$对模p同余(次序不相同),注意当$a\equiv b(\bmod r)$时$\omega_r^a=\omega_r^b$,即上述ω_p的各幂实际为$\omega_p,\omega_p^2,\omega_p^3,\cdots,\omega_p^{p-1}$——除1以外的所有$p$次单位根,即式(7.5)的所有根).

再由

$$\begin{aligned}\omega_p^{r^2}&=(\omega_p^{r})^{r}\\ \omega_p^{r^3}&=(\omega_p^{r^2})^{r}\\ &\vdots\\ \omega_p=\omega_p^{r^{p-1}}&=(\omega_p^{r^{p-2}})^{r}\\ \omega_p^{r}&=(\omega_p)^{r}\end{aligned}$$

知式(7.5)各根可在域$\mathbf{Q}$上循环有理表示,于是据定理7.3知式(7.5)为域$\mathbf{Q}$上循环方程,其Galois(循环)群的生成元为式(7.5)所有根的轮换$(\omega_p^{r},\omega_p^{r^2},\omega_p^{r^3},\cdots,\omega_p^{r^{p-1}})$,再由定理7.1知方程(7.5)在域$\mathbf{Q}$不可约.

例1 验证域$\mathbf{Q}$上方程(7.5)当$p=7$时的Galois群为循环群.

由附录Ⅱ定理3例中素数模7有原根3. $3,3^2,3^3,3^4,3^5,3^6$除以7分别得余数3,2,6,4,5,1. 故

$$\omega_7^{3},\omega_7^{3^2},\omega_7^{3^3},\omega_7^{3^4},\omega_7^{3^5},\omega_7^{3^6} \tag{7.6}$$

①不可取轮换$(\omega_p,\omega_p^2,\omega_p^3,\cdots,\omega_p^{p-1})$.

分别等于(当 $p=7$ 时方程(7.5)的所有根)

$$\omega_7^3,\omega_7^2,\omega_7^6,\omega_7^4,\omega_7^5,\omega_7$$

ω_7 属域 $\mathbf{Q}$ 上方程(7.5)(当 $p=7$)的根域 Ω,且可在域 $\mathbf{Q}$ 上有理表示方程(7.5)各根,从而 $\Omega=Q(\omega_7)$,据定理4.3知 ω_7 为根域的本原数.以 ω_7 为一根且在域 $\mathbf{Q}$ 不可约的方程就是方程(7.5),故方程(7.5)各根即式(7.6)中各数就是共轭本原数集.

取本原数的变换:ω_7 变成 $\omega_7^{3^r}(r=0,1,2,3,4,5)$,则数列(7.6)各数分别变为

$$\omega_7^{3^{r+1}},\omega_7^{3^{r+2}},\cdots,\omega_7^{3^{r+(6-r)}}=\omega_7,\omega_7^{3^{r+(6-r)+1}}=\omega_7^3,\cdots,\omega_7^{3^r}$$

即方程(7.5)的上述根列进行了置换 s^r,其中

$$s=(\omega_7^3,\omega_7^{3^2},\omega_7^{3^3},\omega_7^{3^4},\omega_7^{3^5},\omega_7^{3^6})$$

于是域 $\mathbf{Q}$ 上方程(7.5)(当 $p=7$ 时)的Galois群为以 s 为生成元的循环群.

引理7.2(Abel) 设 p 为素数,$a\in\Delta$,则有等价关系:

① 方程

$$x^p=a \tag{7.7}$$

在 Δ 上可约(不可约)⇔② 方程(7.7)在域 Δ 上有根(无根).

显然只需证可约及有根等价性后用反证法即可推出无根及不可约的等价性.又显然从②(有根)可推出①(可约).现只证①(可约)⇒②(有根):

设方程(7.7)在复数域一根为 ξ,则其所有根为

$$\xi\omega_p^j\quad(j=0,1,2,\cdots,p-1)$$

故

$$x^p-a=\prod_{j=0}^{p-1}(x-\xi\omega_p^j)^{①}$$

因方程(7.7)在 Δ 上可约,故 x^p-a 在 Δ 上有一 $p'(0<p'<p)$ 次因式,它是 p' 个 $x-\xi\omega_p^j$ 型因式之积,据Vieta定理,其常数项(属 Δ)

$$c=(-1)^{p'}\xi^{p'}\omega_p^s\quad(s\in\mathbf{Z}^+)$$

$$c^p=(-1)^{p'p}\xi^{p'p}\omega_p^{ps}=$$
$$[(-1)^p\xi^p]^{p'}=$$
$$(-a)^{p'}$$

---

①连乘号 $\prod_{k=1}^{n}a_k=a_1a_2\cdots a_n$.

(因 $\xi^p = a, p$ 为奇数, $(-1)^p = -1$). 从而

$$a^{p'} = (-1)^{p'} c^p$$

又因 p 为素数, $0 < p' < p$, p 与 p' 互素, 故存在 $u, u' \in \mathbf{Z}$, 使

$$u'p' + up = 1$$

于是

$$\begin{aligned} a = a^{u'p'+up} &= \\ (a^{p'})^{u'} \cdot (a^u)^p &= \\ [(-1)^{p'} c^p]^{u'} (a^u)^p &= \\ (-1)^{p'u'} (c^{u'} a^u)^p \end{aligned}$$

故

$$a = \begin{cases} (c^{u'} a^u)^p & \text{当 } p'u' \text{ 为偶数} \\ (-c^{u'} a^u)^p & \text{当 } p'u' \text{ 为奇数} \end{cases}$$

即存在 $c^{u'} a^u$ 或 $-c^{u'} a^u$ 属 Δ 为方程(7.7) 之根. 证毕.

推论 若引理 7.2 中域 Δ 含有 ω_p, 则有等价关系:

① 方程(7.7) 在 Δ 上可约(不可约)$\Leftrightarrow$② 方程(7.7) 在 Δ 上有根(无根)$\Leftrightarrow$③ 方程(7.7) 的所有根属于 Δ(至少有一根不属于 Δ).

证 只要证 ②$\Leftrightarrow$③, 实际亦只要证: Δ 上有根 $\Leftrightarrow$ 所有根属 Δ, 但"$\Leftarrow$"显然, 只要证"$\Rightarrow$":

设方程(7.7) 在 Δ 有根 α, 即 $\alpha^p = a$, 易见方程(7.7) 所有根为 $\alpha\omega_p^j (j = 0, 1, 2, \cdots, p-1)$, 因 $\omega_p \in \Delta, \alpha \in \Delta$, 于是方程(7.7) 的任一根 $\alpha\omega_p^j \in \Delta$.

定理 7.5 设 p 为素数, $a \in \Delta, \omega_p \in \Delta$, 方程(7.7) 至少有一根不属于 Δ, 则方程(7.7) 为域 Δ 上不可约循环方程.

证 由上推论知方程(7.7) 在域 Δ 不可约且无根. 又其所有根

$$x_j = x_1 \omega_p^{j-1} \quad (j = 1, 2, 3, \cdots, p)$$

故它们可在 Δ 上循环有理表示为

$$x_2 = x_1 \omega_p, x_3 = x_2 \omega_p, \cdots, x_p = x_{p-1} \omega_p, x_1 = x_1 \omega_p^p = x_p \omega_p$$

故再从定理 7.3 知方程(7.7) 为域 Δ 上不可约循环方程.

定理 7.6 域 Δ 上素数 p 次不可约循环方程 $f(x) = 0$ 的各根可用域 $\Delta(\omega_p)$ 中的数的 p 次方根在域 $\Delta(\omega_p)$ 上的有理式表示.

证 由定理 7.1 知域 Δ 上方程 $f(x) = 0$ 的 Galois(循环) 群 G 的生成元为所有各根的轮换(不妨必要时改变轮换中各根的序号)

$$s = (x_1, x_2, \cdots, x_p)$$

对 $\lambda = 0, 1, 2, \cdots, p-1$, 记

$$\alpha_\lambda = x_1\omega_p^{\lambda} + x_2\omega_p^{2\lambda} + \cdots + x_{p-1}\omega_p^{(p-1)\lambda} + x_p = \sum_{k=1}^{p} x_k\omega_p^{k\lambda} \quad (7.8)$$

先证 $\alpha_\lambda^p \in \Delta(\omega_p)$

$$\alpha_\lambda^p = (\omega_p^\lambda)^p(x_1 + x_2\omega_p^\lambda + x_3\omega_p^{2\lambda} + \cdots + x_{p-1}\omega_p^{(p-2)\lambda} + x_p\omega_p^{(p-1)\lambda})^p =$$
$$(x_1\omega_p^\lambda + x_2\omega_p^{2\lambda} + \cdots + x_{p-2}\omega_p^{(p-2)\lambda} + x_{p-1}\omega_p^{(p-1)\lambda} + x_p)^p$$

即域 $\Delta(\omega_p)$ 上有理式 α_λ^p 经轮换 s 不变值，从而经所述 Galois 群 G 的所有置换（s 的幂）不变值. 又据定理 5.4 知域 $\Delta(\omega_p)$ 上 $f(x)=0$ 的 Galois 群为 G 的迷向子群 $I_G(\omega_p)$，故 α_λ^p 经此子群 $I_G(\omega_p)$ 的所有置换不变值，再据定理 5.2 知 $\alpha_\lambda^p \in \Delta(\omega_p)$，即 α_λ 为域 $\Delta(\omega_p)$ 中的数的 p 次方根.

由式(7.8)，对 $j=1,2,\cdots,p$ 有

$$\sum_{\lambda=0}^{p-1}\alpha_\lambda\omega_p^{-j\lambda} = \sum_{\lambda=0}^{p-1}\left(\sum_{k=1}^{p}x_k\omega_p^{k\lambda}\right)\omega_p^{-j\lambda} =$$
$$\sum_{\lambda=0}^{p-1}\sum_{k=1}^{p}x_k\omega_p^{(k-j)\lambda} =$$
$$\sum_{k=1}^{p}\sum_{\lambda=0}^{p-1}x_k\omega_p^{(k-j)\lambda} =$$
$$\sum_{k=1}^{p}\left(\sum_{\lambda=0}^{p-1}\omega_p^{(k-j)\lambda}\right)x_k$$

当 $k=j$ 时

$$\sum_{\lambda=0}^{p-1}\omega_p^{(k-j)\lambda} = \sum_{\lambda=0}^{p-1}1 = p$$

当 $k\neq j$ 时，由 $1\leqslant j\leqslant p, 1\leqslant k\leqslant p$，故 $-(p-1)\leqslant k-j\leqslant p-1$，于是 $k-j$ 非 p 的倍数，由附录 Ⅱ 定理 1 知 $(k-j)\lambda(\lambda=0,1,\cdots,p-1)$ 遍取对模 p 的所有 p 个剩余类，即 $\omega_p^{(k-j)\lambda}$ 遍取 1 的所有 p 个 p 次方根，从而

$$\sum_{\lambda=0}^{p-1}\omega_p^{(k-j)\lambda} = 0$$

所以

$$\sum_{\lambda=0}^{p-1}\alpha_\lambda\omega_p^{-j\lambda} = px_j$$

$$x_j = \frac{1}{p}\sum_{\lambda=0}^{p-1}\alpha_\lambda\omega_p^{-j\lambda} \quad (7.9)$$

即 $f(x)$ 的任一根 x_j 可用域 $\Delta(\omega_p)$ 中的数的 p 次方根 $\alpha_0,\alpha_1,\alpha_2,\cdots,\alpha_{p-1}$ 在 $\Delta(\omega_p)$ 的有理式表示.

表述方程可根式解性的 Galois 理论基本定理及(用圆规直尺)等分圆周需要研究分圆方程(7.5)(p 为奇素数)的根的表示式:多次以循环方程的根扩展系数域,用不可约素数次循环方程的根迭代表示分圆方程的(单位)根.

设

$$p-1=p'q' \quad (p' \text{为(正)素数}, q' \in \mathbf{Z}^+)$$

取模 p 的原根 r,由定理 7.4,有理数域 $\mathbf{Q}$ 上方程(7.5)的 Galois(循环)群 G^* 生成元为

$$s=(\omega_p^r, \omega_p^{r^2}, \cdots, \omega_p^{r^{p-2}}, \omega_p)$$

把上轮换中 $p-1$ 个(方程(7.5)的根)元素排成 $p' \times q'$ 矩阵(先排第一列,再排第二列……),再取和,令

$$\eta_1=\omega_p^r+\omega_p^{r^{1+p'}}+\omega_p^{r^{1+2p'}}+\cdots+\omega_p^{r^{1+(q'-1)p'}}$$

$$\eta_2=\omega_p^{r^2}+\omega_p^{r^{2+p'}}+\omega_p^{r^{2+2p'}}+\cdots+\omega_p^{r^{2+(q'-1)p'}}$$

$$\vdots$$

$$\eta_{p'}=\omega_p^{r^{p'}}+\omega_p^{r^{2p'}}+\omega_p^{r^{3p'}}+\cdots+\omega_p$$

(由 Fermat 定理最后一式末项 $\omega_p^{r^{q'p'}}=\omega_p^{r^{p-1}}=\omega_p^{r^0}=\omega_p$). 作方程

$$(t-\eta_1)(t-\eta_2)\cdots(t-\eta_{p'})=0 \tag{7.10}$$

易见(把 $f(x)$ 各根 $x_1, x_2, \cdots, x_n$ 的有理式 $M(x_1, x_2, \cdots, x_n)$ 经(Galois 群中)置换 s 所得结果记为 $M(x_1, x_2, \cdots, x_n)s$ 或 Ms)

$$\eta_1 s=\eta_2, \eta_2 s=\eta_3, \cdots, \eta_{p'-1}s=\eta_{p'}, \eta_{p'}s=\eta_1 \tag{7.11}$$

故式(7.10)中 t 的各次幂系数对域 $\mathbf{Q}$ 上方程(7.5)的 Galois 群的任一置换(s 的幂)不变,故由定理 5.2 知这些系数属于域 $\mathbf{Q}$,式(7.10)为域 $\mathbf{Q}$ 上的方程,且其 p' 个根经轮换 s 得出 p' 元的轮换

$$\sigma=(\eta_1, \eta_2, \cdots, \eta_{p'})$$

下证域 $\mathbf{Q}$ 上方程(7.10)的 Galois 群 G 是以 σ 为生成元的循环群 G':

由域 $\mathbf{Q}$ 上分圆方程(7.5)根域 $\Omega=Q(\omega_p)$ 的本原数 ω_p 变为本原数 ω_p^r(见定理 7.2 推论),导致方程(7.5)根域 Ω 在域 Δ 上的一个自同构映射 T,亦导致方程(7.5)所有各根的轮换 s,且导致方程(7.10)各根的轮换 σ. 于是:

① 域 $Q(\eta_1, \eta_2, \cdots, \eta_{p'})$($\Omega$ 的子域)的任一数 α 变成仍属此域的一个确定的数 β.

② 域 $Q(\eta_1, \eta_2, \cdots, \eta_{p'})$ 的任一数 β 可由此域的一个确定的数 α 经映射 T 而得.

(用 $\eta_1,\eta_2,\cdots,\eta_{p'}$) 在域 $\mathbf{Q}$ 上的有理式表示

$$\beta=M(\eta_1,\eta_2,\cdots,\eta_{p'})\in Q(\eta_1,\eta_2,\cdots,\eta_{p'})\subset\Omega$$

取

$$\alpha=M(\eta_{p'},\eta_1,\cdots,\eta_{p'-1})\in Q(\eta_1,\eta_2,\cdots,\eta_{p'})$$

易见 $T\alpha=\beta$,由根域 Ω 的映射 T 的逆映射唯一性知 α 是唯一的).

③ 域 $\mathbf{Q}$ 上任一有理式 $P(\eta_1,\eta_2,\cdots,\eta_{p'})$ 与群 G' 有等价关系

$$P(\eta_1,\eta_2,\cdots,\eta_{p'})=\mu\in\mathbf{Q}$$

$\Leftrightarrow P(\eta_1,\eta_2,\cdots,\eta_{p'})$ 经群 G' 的任一置换不变值

因域 $\mathbf{Q}$ 上的有理式 $P(\eta_1,\eta_2,\cdots,\eta_{p'})$ 实际是域 $\mathbf{Q}$ 上方程(7.5) 各根 $\omega_p,\omega_p^r,\omega_p^{r^2},\cdots,\omega_p^{r^{p-2}}$ 的有理式,据定理 5.2 知

$$P(\eta_1,\eta_2,\cdots,\eta_{p'})=\mu\in\mathbf{Q}$$

$\Leftrightarrow P(\eta_1,\eta_2,\cdots,\eta_{p'})$ 经 Galois 群 G^*(由 s 的幂组成)

任一置换不变值

$\Leftrightarrow P(\eta_1,\eta_2,\cdots,\eta_{p'})s=P(\eta_1,\eta_2,\cdots,\eta_{p'})$

$\Leftrightarrow P(\eta_1,\eta_2,\cdots,\eta_{p'})\sigma=P(\eta_1,\eta_2,\cdots,\eta_{p'})$

$\Leftrightarrow P(\eta_1,\eta_2,\cdots,\eta_{p'})$ 经群 G' 中任一置换不变值

④ 由映射 T 在域 Ω 保持和、积关系不变知映射 T 限制在 Ω 的子域 $Q(\eta_1,\eta_2,\cdots,\eta_{p'})$ 亦然.

于是限制在域 $Q(\eta_1,\eta_2,\cdots,\eta_{p'})$ 上的映射 T 为域 $Q(\eta_1,\eta_2,\cdots,\eta_{p'})$ 在域 $\mathbf{Q}$ 上的一个自同构映射,它导致方程(7.10) 各根的置换 σ 为域 $\mathbf{Q}$ 上方程(7.10) 的 Galois 群 G 的一个置换,从而以 σ 为生成元的循环群 G'(由 σ 的幂组成) 的所有置换亦然,于是 G' 为 G 之子群.

再证 G 之子群 G' 与 G 同阶(从而二群相同) 即可:设 G 为 m 阶,G' 为 m' 阶,若 $m'<m$. Galois 群 G 之子群 G' 中 m' 个置换相应于根域本原数的 m' 个变换,把本原数 α 变成其部分共轭数(不妨设)

$$\alpha_j\quad(j=1,2,\cdots,m')$$

$$\alpha_1=\alpha\tag{7.12}$$

再进行 G' 中一置换,即进行群 G' 二置换之积(仍为 G' 中置换),故相应根域中两次变换之积把 α 变成上述共轭数之一. 于是可证 G' 中任一置换 s 把上述任一共轭数仍变成这部分共轭数之一(否则若 G' 中相应变换把 α 变成 $\alpha_{j'}$ 使某个 α_j 变成 α^* 非上述部分共轭数,则进行根域与 G' 的二置换相应的变换 $\alpha\to\alpha_j,\alpha\to\alpha_{j'}$(使 $\alpha_j\to\alpha^*$) 之积(仍为与 G' 相应的变换) 使 $\alpha\to\alpha^*$,与前述矛盾). 且

$\alpha_1,\alpha_2,\cdots,\alpha_{m'}$ 中不同的数变成不同的数(否则与 s^{-1} 相应的变换把同一数变成不同的数). 故上述排列(7.12) 经与 s 相应的变换后仍得这部分共轭数的一排列,由此可知,方程

$$(t-\alpha_1)(t-\alpha_2)\cdots(t-\alpha_{m'})=0$$

中 t 的各次幂的系数($\alpha_1,\alpha_2,\cdots,\alpha_{m'}$ 的对称式) 经 G' 中任一置换不变值. 于是按上述 ③ 中的"$\Leftarrow$"知这些系数属于 $\mathbf{Q}$,上述方程为 $\mathbf{Q}$ 上的方程它与 $\mathbf{Q}$ 上不可约方程有公共根 α,故后者的 m 个根[①]都是前者的根,这与它只有 $m'(<m)$ 矛盾. 故反证假设 $m'<m$ 不成立,必有 $m'=m$.

于是域 $\mathbf{Q}$ 上方程(7.10) 的 Galois 群为以 σ 为生成元的循环群,它是可迁群,再由引理 7.1 推论,知方程(7.10) 在域 $\mathbf{Q}$ 不可约,故方程(7.10) 为域 $\mathbf{Q}$ 上素数 p' 次不可约循环方程.

再对 $j'=1,2,\cdots,p'$,以 $\eta_{j'}$ 的各项为根,作 p' 个方程

$$(y-\omega_p^{r^{j'}})(y-\omega_p^{r^{j'+p'}})(y-\omega_p^{r^{j'+2p'}})\cdots(y-\omega_p^{r^{j'+(q'-1)p'}})=0 \tag{7.13}_{j'}$$

证其为域 $Q(\eta_{j'})$(即 $Q(\eta_{p'})$[②]) 上 q' 次不可约循环方程:

由已证方程(7.10) 在域 $\mathbf{Q}$ 不可约故无重根,其各根 $\eta_{j'}(j'=1,2,\cdots,p')$ 互不相等. 于是由上设 $\eta_{j'}$ 的表示式知,对任一 $\eta_{j'}$(属域 $\mathbf{Q}$ 上方程(7.5) 的根域),当且仅当 m 为 p' 的倍数时 $\eta_{j'}s^m=\eta_{j'}$. 故对群 G' 的任一置换 s^m,m 为 p' 的倍数,有 $\eta_{j'}s^m=\eta_{j'}$;而对 $G\backslash G'$ 的任一置换 $s^{m'}$,m' 非 p' 的倍数,有 $\eta_{j'}s^{m'}\neq\eta_{j'}$. 故 Galois 群 G^* 对 $\eta_{j'}$ 的迷向子群为 G'. 而方程$(7.13)_{j'}$ 中 y 的各次幂的系数对 $\eta_{j'}$ 的各项对称,故它对 G' 的任一置换不变,据定理 5.4 推论知这些系数属于域 $Q(\eta_{j'})$,于是方程$(7.13)_{j'}$ 为域 $Q(\eta_{j'})$ 上的方程. 又因

$$\omega_p^{r^{j'+p'}}=(\omega_p^{r^{j'}})^{r^{p'}}$$
$$\omega_p^{r^{j'+2p'}}=(\omega_p^{r^{j'+p'}})^{r^{p'}}$$
$$\vdots$$
$$\omega_p^{r^{j'+(q'-1)p'}}=(\omega_p^{r^{j'+(q'-2)p'}})^{r^{p'}}$$
$$\omega_p^{r^{j'}}=(\omega_p^{r^{p'q'}})^{r^{j'}}=(\omega_p^{r^{j'+(q'-1)p'}})^{r^{p'}}$$

即方程$(7.13)_{j'}$ 各根可在域 $\mathbf{Q}$ 循环有理表示,故由定理 7.3 知方程

①因域 $\mathbf{Q}$ 上方程(7.10) 的 Galois 群为 m 阶,α 为域 $\mathbf{Q}$ 上方程(7.10) 根域的本原数,故以 α 为一根在域 $\mathbf{Q}$ 上不可约的方程为 m 次,有 m 个互异之根;

②因方程(7.10) 为域 $\mathbf{Q}$ 上不可约循环方程,各根可相互有理表示(据定理 7.2).

$(7.13)_{j'}$ 为域 $Q(\eta_{j'})$ 上 q' 次不可约循环方程,其 Galois(循环)群为

$$\{s^{p'},s^{p'2},s^{p'3},\cdots,(s^{p'})^{q'}\}\quad (其中(s^{p'})^{q'}=s^{p'q'}=s^{p-1}=I)$$

若 q' 非奇素数,再分解

$$q'=p''q''\quad (p''为奇素数,q''\in\mathbf{Z}^+)$$

类似于方程(7.10),把方程 $(7.13)_{j'}$ 的根列排列 $p''\times q''$ 矩阵后取和

$$\eta_{j'1}=\omega_p^{r^{j'}}+\omega_p^{r^{j'+p''p'}}+\omega_p^{r^{j'+2p''p'}}+\cdots+\omega_p^{r^{j'+(q''-1)p''p'}}$$

$$\eta_{j'2}=\omega_p^{r^{j'+p'}}+\omega_p^{r^{j'+(p''+1)p'}}+\omega_p^{r^{j'+(2p''+1)p'}}+\cdots+\omega_p^{r^{j'+((q''-1)p''+1)p'}}$$

$$\vdots$$

$$\eta_{j'j''}=\omega_p^{r^{j'+(j''-1)p'}}+\omega_p^{r^{j'+(p''+j''-1)p'}}+\omega_p^{r^{j'+(2p''+j''-1)p'}}+\cdots+\omega_p^{r^{j'+((q''-1)p''+j''-1)p'}}$$

$$\vdots$$

$$\eta_{j'p''}=\omega_p^{r^{j'+(p''-1)p'}}+\omega_p^{r^{j'+(2p''-1)p'}}+\omega_p^{r^{j'+(3p''-1)p'}}+\cdots+\omega_p^{r^{j'+(q'p''-1)p'}}$$

(最后一式末项实际为 $\omega_p^{r^{j'+p-1-p'}}=(\omega_p^{r^{p-1}})^{r^{j'-p'}}=\omega_p^{r^{j'-p'}}$).类似于方程(7.10)、方程 $(7.13)_{j'}$ 作方程

$$(t-\eta_{j'1})(t-\eta_{j'2})\cdots(t-\eta_{j'p''})=0\qquad (7.14)_{j'}$$

$$\prod_{k=0}^{q''-1}(y-\omega_p^{r^{j'+(kp''+j''-1)p'}})=0\qquad (7.15)_{j'j''}$$

以轮换

$$s'=(\omega_p^{r^{j'}},\omega_p^{r^{j'+p'}},\omega_p^{r^{j'+2p'}},\cdots,\omega_p^{r^{j'+(q''p''-1)p'}})$$

取代轮换 s,同样可证式 $(7.14)_{j'}$ 为域 $Q(\eta_{j'})=Q(\eta_{p'})$ 上素数 p' 次不可约循环方程,其 Galois(循环)群生成元为轮换

$$\sigma_{j'}=(\eta_{j'1},\eta_{j'2},\cdots,\eta_{j'p''})$$

而方程 $(7.15)_{j'j''}$ 为域 $Q(\eta_{j'j''})$ 上 q'' 次不可约循环方程,其 Galois(循环)群为

$$G''=\{s^{p'p''},(s^{p'p''})^2,(s^{p'p''})^3,\cdots,(s^{p'p''})^{q'}\}$$

$((s^{p'p''})^{q''}=s^{p'p''q''}=s^{p'q'}=s^{p-1}=I)$.

现证任一域

$$Q(\eta_{j'j''})=Q(\eta_{p'p''})$$

只要证 $\eta_{j'j''}$ 与 $\eta_{p'p''}$ 可相互有理表示:

考虑方程

$$\prod_{\substack{1\leqslant j'\leqslant p'\\1\leqslant j''\leqslant p''}}(t-\eta_{j'j''})=0\qquad (7.16)$$

它有 $p'p''$ 个根,每个根 $\eta_{j'j''}$ 为 ω_p 的 q'' 个不同幂之和,且任两个不同根中的各项亦互不相同,故所有这些根的各项遍取 ω_p 的 $p'p''q''=$

$p-1$ 个不同的幂,即方程(7.5)的所有根,且方程(7.16)的每个根的 q'' 个项中 r 的指数成等差数列,其公差为 $p''p'$,即这 $p'p''$ 个项实际为

$$t_1=\omega_p^{r}+\omega_p^{r^{1+p'p''}}+\omega_p^{r^{1+2p'p''}}+\cdots+\omega_p^{r^{1+(q'-1)p'p''}}$$

$$t_2=\omega_p^{r^2}+\omega_p^{r^{2+p'p''}}+\omega_p^{r^{2+2p'p''}}+\cdots+\omega_p^{r^{2+(q'-1)p'p''}}$$

$$\vdots$$

$$t_{p'p''}=\omega_p^{r^{p'p''}}+\omega_p^{r^{2p'p''}}+\cdots+\omega_p^{r^{q'p'p''}}$$

因

$$\omega_p^{r^{q'p'p''}}=\omega_p^{r^{p-1}}=\omega_p^{1}=\omega_p$$

故又有

$$t_{p'p''}=\omega_p+\omega_p^{r^{p'p''}}+\omega_p^{r^{2p'p''}}+\cdots+\omega_p^{r^{(q'-1)p'p''}}$$

于是易见它们经置换 s,有

$$t_1s=t_2,t_2s=t_3,t_3s=t_4,\cdots,t_{p'p''-1}s=t_{p'p''},t_{p'p''}s=t_1$$

故类似于方程(7.10)、方程(7.16)为域 $\mathbf{Q}$ 上不可约循环方程,各根可相互有理表示.

若 q'' 非素数,再分解

$$q''=p'''q'''\quad(p'''\text{ 为素数},q'''\in\mathbf{Z}^{+})$$

类似把所有方程$(7.15)_{j'j''}$ 的各根排成 $p'''\times q'''$ 矩阵再取和 …… 直到最后得出 $q^{(l)}$ 是素数为止,于是得出最后结论:

定理 7.7 设 p 为奇素数,分解 $p-1$ 为素因数之积

$$p-1=p'p''p'''\cdots p^{(l)}q^{(l)}$$

则方程(7.5)的各根为

$$\eta_{j'j''\cdots j^{(l)}j^{(l+1)}}$$

$$(j'=1,2,\cdots,p';j''=1,2,\cdots,p'';\cdots;j^{(l)}=1,2,\cdots,p^{(l)};j^{(l+1)}=1,2,\cdots,q^{(l)})$$

它们可迭代表示为:

域 $\mathbf{Q}$ 上素数 p' 次不可约循环方程(7.10)之根 $\eta_{j'}$;

域 $Q(\eta_{p'})$ 上素数 p'' 次不可约循环方程$(7.14)_{j'}$ 之根 $\eta_{j'j''}$;

域 $Q(\eta_{p'p''})$ 上素数 p''' 次不可约循环方程之根 $\eta_{j'j''j'''}$;

$$\vdots$$

域 $Q(\eta_{p'p''\cdots p^{(l)}})$ 上素数 $q^{(l)}$ 次不可约循环方程之根 $\eta_{j'j''j'''\cdots j^{(l)}j^{(l+1)}}$.

其中 p' 次方程(7.10)的所有根 $\eta_{j'}(j'=1,2,\cdots,p')$ 为 $q'=\dfrac{p-1}{p'}$ 项之和,各项为 ω_p 之幂,它们的(ω_p 的)指数为等比数列,公

比为 $r^{p'}$，且所有 p' 个根的各项的(ω_p 的)指数取尽集$\{1,2,\cdots,p-1\}$.

p'' 次方程$(7.14)_{j'}$ 的所有根 $\eta_{j'j''}(j''=1,2,\cdots,p'')$ 为 $q''=\frac{q'}{p''}=\frac{p-1}{p'p''}$ 项之和，各项为 ω_p 之幂，它们的指数为等比数列，公比为 $r^{p'p''}$，且所有 p'' 个根的各项的指数取尽 $\eta_{j'}$ 中各项的指数.

一般，上述素数 $p^{(s)}(s=2,3,\cdots,l)$ 次方程的所有根 $\eta_{j'j''\cdots j^{(s)}}(j^{(s)}=1,2,\cdots,p^{(s)})$ 为 $q^{(s)}=\frac{q^{(s-1)}}{p^{(s)}}=\frac{p-1}{p'p''\cdots p^{(s)}}$ 项之和，各项为 ω_p 之幂，它们的指数为等比数列，公比为 $r^{p'p''\cdots p^{(s)}}$，且所有 $p^{(s)}$ 个根的各项的指数取尽 $\eta_{j'j''\cdots j^{(s-1)}}$ 各项的指数.

最后，上述素数 $q^{(l)}$ 次方程的所有根 $\eta_{j'j''\cdots j^{(l)}j^{(l+1)}}(j^{(l+1)}=1,2,\cdots,p^{(l)})$ 为 $q^{(l)}$ 项之和，各项为 ω_p 之幂，它们的指数为等比数列，公比为 $r^{p'p''\cdots p^{(l)}}$，且所有 $q^{(l)}$ 个根的各项取尽 $\eta_{j'j''\cdots j^{(l)}}$ 中各项的指数.

第八章　用不可约方根表示单位根，用直尺、圆规把圆分为 Fermat(费马)素数等份

由引理7.2知，当 a 不是任何有理数的 p 次幂，即方程 $x^p=a$ 在有理数域 $\mathbf{Q}$ 上无根时此方程在域 $\mathbf{Q}$ 上不可约，这时我们称 a 的所有 p 次方根(即 $x^p=a$ 的所有根)为不可约方根.

即使 p 为素数，但方程 $x^p=1$ 并非在域 $\mathbf{Q}$ 的不可约方程，故其根 $1,\omega_p^j(j=1,2,\cdots,p-1)$ 并非不可约方根.

定理7.7说明，当 p 为奇素数时，可把分圆方程

$$x^{p-1}+x^{p-2}+\cdots+x+1=0$$

的各根 ω_p^j 用其中所述素数 $p',p'',\cdots,p^{(l)},q^{(l)}$ 次不可约循环方程之根迭代表示，再由定理7.6这些循环方程之根可用 $p',p'',\cdots,p^{(l)},q^{(l)}$ 次不可约方根表示，但这两定理未讲述如何求出这些循环方程. 本章将详述如何求出这些循环方程各项系数. 若 $p',p'',\cdots,p^{(l)},q^{(l)}$ 都是2,3,4，即这些方程都是2,3,4次方程，则可求出其根，从而求出所有 ω_p^j 用上述不可约方根的表示式. 特别当素数 p 为 Fermat

素数时(当素数形如$2^{2^n}+1(n\in\mathbf{Z}^+)$①),则得$\omega_p$用多层平方根表示式,从而可得(用圆规、直尺)$p$等分圆的作法.

引理8.1 定理7.7中域$\mathbf{Q}$上方程(7.10)的根域为域$\mathbf{Q}$上以$\eta_1,\eta_2,\cdots,\eta_{p'}$为基底之线性空间;域$Q(\eta_{p'})$上方程$(7.13)_{j'}$的根域为域$Q(\eta_{p'})$上以$\eta_{j'1},\eta_{j'2},\cdots,\eta_{j'p''}$为基底的线性空间……

证 因域$\mathbf{Q}$上不可约循环方程(7.10)的各根可表示为一根$\eta_{p'}$在域$\mathbf{Q}$的有理式,故根域为$Q(\eta_{p'})$,再据定理3.3推论知$Q(\eta_{p'})$为域$\mathbf{Q}$上p'维线性空间,而$\eta_1,\eta_2,\cdots,\eta_{p'}$的在$\mathbf{Q}$上任何"系数不全为0"的线性组合必为$\omega_p^r,\omega_p^{r^2},\cdots,\omega_p^{r^{p'q'}}(=\omega_p^{r^0}=\omega_p)$在$\mathbf{Q}$上的"系数不全为0"的线性组合(因$\eta_1,\eta_2,\cdots,\eta_{p'}$用$\omega_p$的幂的表示式中各项的$\omega_p$的次数全不相同).再对域$\mathbf{Q}$上不可约方程(7.5)用上述推论知域$Q(\omega_{p'})$为域$\mathbf{Q}$上$p-1$维线性空间,从其基底为1,$\omega_p,\omega_p^2,\cdots,\omega_p^{p-2}$,可改取基底为$\omega_p,\omega_p^2,\omega_p^3,\cdots,\omega_p^{p-1}$(因$\sum\limits_{k=1}^{p-1}c_k\omega_p^k=0\Leftrightarrow\sum\limits_{k=1}^{p-1}c_k\omega_p^{k-1}=0$即$\sum\limits_{k'=0}^{p-2}c_{k'+1}\omega_p^{k'}=0\Leftrightarrow c_1=c_2=\cdots=c_{p-1}=0$,即$\omega_p$,$\omega_p^2,\cdots,\omega_p^{p-1}$在域$\mathbf{Q}$上线性无关),故从上述两线性组合的系数不全为0知线性组合不等于0,得证$\eta_1,\eta_2,\cdots,\eta_{p'}$在域$\mathbf{Q}$上线性无关,为所述根域的基底.类似可证域$Q(\eta_{p'})$上方程$(7.13)_{j'}$的根域为域$Q(\eta_{p'})$上以$\eta_{j'1},\eta_{j'2},\cdots,\eta_{j'p}$为基底的线性空间……

定理8.1 定理7.7中域$\mathbf{Q}$上方程(7.10)的各项系数必为整数;第二项为$t^{p'-1}$.又该定理所述域$Q(\eta_{p'p''\cdots p^{(s)}})(s=1,2,\cdots,l)$上素数$p^{(s+1)}$次($s=l$时为$q^{(l)}$次)不可约循环方程各项系数为所有$\eta_{j'j''\cdots j^{(s)}}(j'=1,2,\cdots,p';j''=1,2,\cdots,p'';\cdots;j^{(s)}=1,2,\cdots,p^{(s)})$的整系数线性组合.以$\eta_{j'j''\cdots j^{(s-1)}j^{(s)}m}(m=1,2,\cdots,p^{(s+1)})$为根的方程的第二项为$-\eta_{j'j''\cdots j^{(s-1)}j^{(s)}}t^{p^{(s+1)}-1}$.

证 易见方程(7.10)中t的各次幂系数为$\eta_1,\eta_2,\cdots,\eta_{p'}$的整系数对称多项式,各$\eta_{j'}$用$\omega_p$表示后化简便把上述系数化成$\omega_p$的整系数多项式,而由于(定理7.7已证)方程(7.10)为域$\mathbf{Q}$上方程,故这些系数为有理数.又上述多项式(之值)实际等于它除以$\omega_p^{p-1}+\omega_p^{p-2}+\cdots+\omega_p+1$(值为0,因$\omega_p$为方程(7.5)之根)所得余式(次数小于$p$),故这些系数形如

①$2^{2^n}+1$型整数不一定是素数,如$2^{2^5}+1=641\times6\ 700\ 417$非素数.

$$m + k_1\omega_p + k_2\omega_p^2 + \cdots + k_{p-1}\omega_p^{p-1} = b \in \mathbf{Q}$$
$$(m, k_1, k_2, \cdots, k_{p-1} \in \mathbf{Z})$$①

从而

$$(m - b) + k_1\omega_p + k_2\omega_p^2 + \cdots + k_{p-1}\omega_p^{p-1} = 0$$

可证

$$m - b = k_1 = k_2 = \cdots = k_{p-1} \in \mathbf{Z}$$

否则与 p 为素数,ω_p 在 $\mathbf{Q}$ 上适合的不可约方程为(7.5)——它是 ω_p 所适合的(最高次项系数为1)在域 $\mathbf{Q}$ 上唯一的不超过 $p-1$ 次方程(据引理3.1推论)矛盾.又因 $m \in \mathbf{Z}$,故知 t 的各次幂系数形如

$$b = m - k_1 \in \mathbf{Z}$$

因

$$\eta_1 + \eta_2 + \cdots + \eta_{p'} = \sum_{m=1}^{p-1} \omega_p^{r^m} = \sum_{b=1}^{p-1} \omega_p^b = -1$$②

故 $t^{p'-1}$ 的系数为1.

式$(7.13)_{j'}$中 t 的各次幂系数同样可化成 ω_p 的低于 p 次的整系数多项式,但据定理7.7的推导过程知它们属于域 $Q(\eta_{p'})$,更属于在域 $\mathbf{Q}$ 上方程(7.10)的根域,据引理8.1后者实为域 $\mathbf{Q}$ 上以 η_1,η_2,…,$\eta_{p'}$ 为基底的线性空间,故 ω_p 的这些整系数多项式形如

$$\begin{aligned}
&m + (k_{11}\omega_p^{r} + k_{12}\omega_p^{r^{1+p'}} + k_{13}\omega_p^{r^{1+2p'}} + \cdots + k_{1q'}\omega_p^{r^{1+(q'-1)p'}}) + \\
&(k_{21}\omega_p^{r^2} + k_{22}\omega_p^{r^{2+p'}} + k_{23}\omega_p^{r^{2+2p'}} + \cdots + k_{2q'}\omega_p^{r^{2+(q'-1)p'}}) + \cdots + \\
&(k_{p'1}\omega_p^{r^{p'}} + k_{p'2}\omega_p^{r^{2p'}} + \cdots + k_{p'q'}\omega_p^{r^{q'p'}}) = l_1\eta_1 + l_2\eta_2 + \cdots + l_{p'}\eta_{p'} \\
&(m, k_{11}, k_{12}, \cdots, k_{1q'}, k_{21}, k_{22}, \cdots, k_{2q'}, \cdots, k_{p'1}, k_{p'2}, \cdots, k_{p'q'} \in \mathbf{Z}) \\
&\qquad (l_1, l_2, \cdots, l_{p'} \in \mathbf{Q})
\end{aligned} \tag{8.1}$$

以 $\eta_1,\eta_2,\cdots,\eta_{p'}$ 用 ω_p 的表示式代入化简得

$$\begin{aligned}
&m + [(k_{11} - l_1)\omega_p^{r} + (k_{12} - l_1)\omega_p^{r^{1+p'}} + (k_{13} - l_1)\omega_p^{r^{1+2p'}} + \cdots + \\
&(k_{1q'} - l_1)\omega_p^{r^{1+(q'-1)p'}}] + [(k_{21} - l_2)\omega_p^{r^2} + \\
&(k_{22} - l_2)\omega_p^{r^{2+p'}} + (k_{23} - l_2)\omega_p^{r^{2+2p'}} + \cdots + (k_{2q'} - l_2)\omega_p^{r^{2+(q'-1)p'}}] + \cdots + \\
&[(k_{p'1} - l_{p'})\omega_p^{r^{p'}} + (k_{p'2} - l_{p'})\omega_p^{r^{2p'}} + (k_{p'3} - l_{p'})\omega_p^{r^{3p'}} + \cdots + \\
&(k_{p'q'} - l_{p'})\omega_p^{r^{p'q'}}] = 0
\end{aligned}$$

同样由 ω_p 适合域 $\mathbf{Q}$ 上不可约方程(7.5)$(p-1=p'q')$及上式各项

---

①实际进行竖式除法,易见所得不完全商及余式的各项系数全为整数.

②据附录Ⅱ定理4.

ω_p 的指数遍取$0,r,r^2,\cdots,r^{p-1}$,即遍取$0,1,2,\cdots,p-1$且不重复(见附录 Ⅱ 定理4,注意r为模p之原根),故得

$$m=k_{11}-l_1=k_{12}-l_1=\cdots=k_{1q'}-l_1$$
$$m=k_{21}-l_2=k_{22}-l_2=\cdots=k_{2q'}-l_2$$
$$\vdots$$
$$m=k_{p'1}-l_{p'}=k_{p'2}-l_{p'}=\cdots=k_{p'q'}-l_{p'}$$

于是

$$k_{11}=k_{12}=\cdots=k_{1q'}=m+l_1\in\mathbf{Z}$$
$$k_{21}=k_{22}=\cdots=k_{2q'}=m+l_2\in\mathbf{Z}$$
$$\vdots$$
$$k_{p'1}=k_{p'2}=\cdots=k_{p'q'}=m+l_{p'}\in\mathbf{Z} \tag{8.2}$$

故

$$l_1=k_{11}-m\in\mathbf{Z},l_2=k_{21}-m\in\mathbf{Z},\cdots,l_{p'}=k_{p'1}-m\in\mathbf{Z}$$

得证$s=1$时本定理结论.

$s\geqslant 2$时的结论类似可证(只是与式(8.1)相应的等式右边用ω_p的表示式代入后未能取尽左边ω_p的各次幂,从而这些幂之系数都等于m).

证最后结论:由所有根之和

$$\sum_{m=1}^{p(s+1)}\eta_{j'j''\cdots j(s-1)j(s)_m}=\eta_{j'j''\cdots j(s-1)j(s)}$$

及 Vieta 定理知结论成立.

上述证明过程已得出求上述整系数多项式,所有$\eta_{j'j''\cdots j(s)}$的整系数线性组合的方法:把所有$\eta_1,\eta_2,\cdots,\eta_{p'}$或所有$\eta_{j'j''\cdots j(s)}$用$\omega_p$的表示式代入$t$的各次幂的系数表示式再化简.由$k_1=k_2=\cdots=k_{p-1}$便得方程(7.10)中$t$的各次幂的(不一定相同)系数为

$$m+(k_1\omega_p+k_1\omega_p^2+\cdots+k_1\omega_p^{p-1})=m-k_1$$

(因$1+(\omega_p+\omega_p^2+\cdots+\omega_p^{p-1})=0$).由式(8.2)得式(8.1)左边为

$$m\cdot 1+\sum_{s=1}^{p'}(\sum_{u=0}^{q'-1}k_{su}\omega_p^{r^{s+up'}})=$$
$$m\sum_{s=1}^{p'}[\sum_{u=0}^{q'-1}(-\omega_p^{r^{s+up'}})]+\sum_{s=1}^{p'}(\sum_{u=0}^{q'-1}k_{s1}\omega_p^{r^{s+up'}})=$$
$$\sum_{s=1}^{p'}[(k_{s1}-m)\sum_{u=0}^{q'-1}\omega_p^{r^{s+up'}}]=\sum_{s=1}^{p'}(k_{s1}-m)\eta_s$$

$s\geqslant 2$时亦有类似结果(上述左边中与右边无相应项之各幂的系数为m,这些幂的项正好与由$m\cdot 1$化出的相应项——$-m$乘ω的幂

抵消).

但实际计算时如灵活应用上定理的结论及对称性可更直接求出各整数系数.

如求方程(7.10)中 t^{p-s} 的系数 $(-1)^s\sigma_s$, σ_s 为 Vieta 定理中 $\eta_1,\eta_2,\cdots,\eta_{p'}$ 的 s 次基本对称多项式. 把 $\eta_1,\eta_2,\cdots,\eta_{p'}$ 用 ω_p 的表示式代入 σ_s 展开后设 σ_s 的表示式共有 h 项($h=C_{p'}^s q'^s$), 先观察这 h 项中有 n 项(设 m 项)等于 1(即 $\omega_p^p,\omega_p^{2p},\cdots$ 型项), 余下 $h-m$ 项应为 $p-1$ 项之和 $\omega_p+\omega_p^2+\cdots+\omega_p^{p-1}(=-1)$ 的整数(设为 k)倍, 即余下应有 $k(p-1)$ 项, 故

$$k(p-1)=h-m$$

$$k=\frac{h-m}{p-1}$$

于是

$$\sigma_s=m+k\cdot(-1)=m-k=m-\frac{h-m}{p-1} \tag{8.3}$$

从而 t^{p-s} 的系数为

$$(-1)^s\left(m-\frac{h-m}{p-1}\right)$$

求方程 $(7.13)_{j'}$ 及定理 7.7 中其余方程各项系数(各个 $\eta_{j'j''\cdots j^{(s)}}$ 的(整系数)线性组合)的简便方法可见后面例子.

下面再推导求 $\eta_{j_1}\cdot\eta_{j_2}$ 的简便公式

$$\begin{aligned}
\eta_{j_1}\cdot\eta_{j_2}=&(\omega_p^{r^{j_1}}+\omega_p^{r^{j_1+p'}}+\omega_p^{r^{j_1+2p'}}+\cdots+\omega_p^{r^{j_1+(q'-1)p'}})\cdot\\
&(\omega_p^{r^{j_2}}+\omega_p^{r^{j_2+p'}}+\omega_p^{r^{j_2+2p'}}+\cdots+\omega_p^{r^{j_2+(q'-1)p'}})=\\
&(\omega_p^{r^{j_1}}+\omega_p^{r^{j_1+p'}}+\omega_p^{r^{j_1+2p'}}+\cdots+\omega_p^{r^{j_1+(q'-1)p'}})\cdot\omega_p^{r^{j_2}}+\\
&(\omega_p^{r^{j_1+p'}}+\omega_p^{r^{j_1+2p'}}+\omega_p^{r^{j_1+3p'}}+\cdots+\omega_p^{r^{j_1+q'p'}}①)\cdot\omega_p^{r^{j_2+p'}}+\\
&(\omega_p^{r^{j_1+2p'}}+\omega_p^{r^{j_1+3p'}}+\omega_p^{r^{j_1+4p'}}+\cdots+\omega_p^{r^{j_1+(q'+1)p'}})\cdot\omega_p^{r^{j_2+2p'}}+\cdots+\\
&(\omega_p^{r^{j_1'+(q'-1)p'}}+\omega_p^{r^{j_1+q'p'}}+\omega_p^{r^{j_1+(q'+1)p'}}+\cdots+\omega_p^{r^{j_1+(2q'-2)p'}})\omega_p^{r^{j_2+(q'-1)p'}}
\end{aligned}$$

各括号内第一项与括号外因式乘积之和 s_0 为

$$\omega_p^{r^{j_1}+r^{j_2}}+\omega_p^{(r^{j_1}+r^{j_2})r^{p'}}+\omega_p^{(r^{j_1}+r^{j_2})r^{2p'}}+\cdots+\omega_p^{(r^{j_1}+r^{j_2})r^{(q'-1)p'}} \tag{8.4}$$

当 $p\mid r^{j_1}+r^{j_2}$ 时

① $\omega_p^{r^{j_1+q'p'}}=(\omega_p^{r^{q'p'}})^{r^{j_1}}=(\omega_p^{r^{p-1}})^{r^{j_1}}=(\omega_p^{r^0})^{r^{j_1}}=\omega_p^{r^{j_1}}$, 同样下行括号内末项为 $\omega_p^{r^{j_1+p'}}\cdots$

$$\omega_p^{(r^{j_1}+r^{j_2})r^{sp'}} = 1^{r^{sp'}} = 1 \quad (s = 0,1,2,\cdots,q'-1)$$

式(8.4)等于

$$1 + 1 + \cdots + 1 = q'$$

当 $p \nmid r^{j_1} + r^{j_2}$ 时,取 $r^{j_1} + r^{j_2}$ 除以 p 所得余数 $m_0 \in \{1,2,\cdots,p-1\}$,因 r 为模 p 的原根,由附录Ⅱ定理4知存在 $b_0 \in \{1,2,\cdots,p-1\}$,使

$$r^{b_0} \equiv m_0 \equiv r^{j_1} + r^{j_2} \pmod p$$

故

$$\omega_p^{r^{j_1}+r^{j_2}} = \omega_p^{r^{b_0}}$$

$$\omega_p^{(r^{j_1}+r^{j_2})r^{sp'}} = (\omega_0^{r^{b_0}})^{r^{sp'}} = \omega_p^{r^{b_0}r^{sp'}} = \omega_p^{r^{b_0+sp'}}$$

式(8.4)等于

$$\omega_p^{r^{b_0}} + \omega_p^{r^{b_0+p'}} + \omega_p^{r^{b_0+2p'}} + \cdots + \omega_p^{r^{b_0+(q'-1)p'}} \tag{8.5}$$

$\omega^{r^{b_0}}$ 必为 $\eta_1,\eta_2,\cdots,\eta_{p'}$ 中某个 η_{d_0} 的一项,与上述 $\eta_{j_1}\cdot\eta_{j_2}$ 表示式中 η_{j_1} 可改为从原表示式中任一项开始,其余 $q'-1$ 项中 r 的指数逐次增加 p' 同理. 式(8.5)等于 η_{d_0},故式(8.4)等于

$$\begin{cases} q' & \text{当 } p \mid r^{j_1} + r^{j_2} \text{ 时} \\ \eta_{d_0} & \text{当 } p \nmid r^{j_1} + r^{j_2} \text{ 时} \end{cases}$$

其中取 $r^{b_0} \equiv r^{j_1} + r^{j_2} \pmod p$,即 $\omega_p^{r^{j_1}} \cdot \omega_p^{r^{j_2}} = \omega_p^{r^{b_0}}$($b_0 \in \{1,2,\cdots,p-1\}$),$\omega_p^{r^{b_0}}$ 为 η_{d_0} 的一项.

同理 $\eta_{j_1}\cdot\eta_{j_2}$ 最后表示式中各括号内第二项与括号外因式乘积之和 s_1 为

$$\omega_p^{r^{j_1+p'}}\omega_p^{r^{j_2}} + \omega_p^{r^{j_1+2p'}}\omega_p^{r^{j_2+p'}} + \cdots + \omega_p^{r^{j_1+q'p'}}\omega_p^{r^{j_2+(q'-1)p'}} =$$

$$\begin{cases} q' & \text{当 } p \mid r^{j_1+p'} + r^{j_2} \text{ 时} \\ \eta_{d_1} & \text{当 } p \nmid r^{j_1+p'} + r^{j_2} \text{ 时} \end{cases}$$

其中取 $r^{b_1} \equiv r^{j_1+p'} + r^{j_2} \pmod p$,即 $\omega_p^{r^{j_1+p'}}\omega_p^{r^{j_2}} = \omega_p^{r^{b_1}}$($b_1 \in \{1,2,\cdots,p-1\}$),$\omega_p^{r^{b_1}}$ 为 η_{d_1} 的一项.

继续同样推证,即可得 q' 项之和,分别由

$$\omega_p^{r^{j_1+kp'}}\omega_p^{r^{j_2}} = \omega_p^{r^{j_1+kp'}+r^{j_2}} \quad (k = 0,1,2,\cdots,q'-1)$$

的指数 $r^{j_1+kp'} + r^{j_2}$ 而定,即 η_{j_1} 各项分别与 η_{j_2} 第一项乘积之指数而定. 于是

$$\eta_{j_1}\eta_{j_2} = \sum_{k=0}^{q'-1} s_k \tag{8.6}$$

$$s_k=\begin{cases} q' & \text{当 } p \mid r^{j_1+kp'}+r^{j_2} \text{ 时} \\ \eta_{d_k} & \text{当 } p \nmid r^{j_1+kp'}+r^{j_2} \text{ 时} \end{cases}$$

其中取 $r^{b_k}\equiv r^{j_1+kp'}+r^{j_2}(\bmod p)$,即 $\omega_p^{r^{j_1+kp}}\omega_p^{r^{j_2}}=\omega_p^{r^{b_k}}(b_k\in\{1,2,\cdots,p-1\})$,$\omega_p^{r^{b_k}}$ 为 η_{d_k} 的一项.

但实际上不必算出上述 r 的幂,求出这些幂除以 p 的余数,用余数代替这些幂即可.

求得任一 $\eta_{j_1}\eta_{j_2}$ 用 $\eta_1,\eta_2,\cdots,\eta_{p'}$ 的表示式后可用同法求 $\eta_{j_1}\eta_{j_2}\eta_{j_3},\cdots$

求 $\eta_{j'j''_1}\cdot\eta_{j'j''_2}$ 亦有类似方法.

由于 $p-1$ 为偶数,定理 7.7 中可取 $q^{(l)}=2$,这时有下列结论.

定理 8.2 设定理 7.7 中 $q^{(l)}=2$,则:

① 对 $s=1,2,\cdots,l$,当 $j_1^{(s)}\neq j_2^{(s)}$ 时把 $\eta_{j'j''\cdots j^{(s-1)}j_1^{(s)}}\cdot\eta_{j'j''\cdots j^{(s-1)}j_2^{(s)}}$ 用 ω_p 表示展开后所得的 $(q^{(s)})^2$ 个项都不等于 1;

② 任何 $\eta_{j'},\eta_{j'j''},\cdots,\eta_{j'j''\cdots j^{(l)}}(j'=1,2,\cdots,p';j''=1,2,\cdots,p'';\cdots;j^{(l)}=1,2,\cdots,p^{(l)})$ 都是实数,而所有 $\eta_{j'j''\cdots j^{(l)}}$ 为所有 $2\cos\frac{2\pi m}{p}(m=1,2,\cdots,\frac{p-1}{2})$ 之值;

③ $\eta_{j'j''\cdots j^{(l)}1}$ 与 $\eta_{j'j''\cdots j^{(l)}2}$ 为下述方程之根

$$t^2-\eta_{j'j''\cdots j^{(l)}}t+1=0 \tag{8.7}$$

先证二引理:

引理 8.2 设 r 为奇素数模 p 之原根,则在 $r,r^2,\cdots,r^{p-1}$ 中只有

$$r^{\frac{p-1}{2}}\equiv -1(\bmod p)$$

证 考察二次同余式方程

$$x^2\equiv 1(\bmod p) \tag{8.8}$$

即

$$x^2-1\equiv 0(\bmod p)$$

$$(x+1)(x-1)\equiv 0(\bmod p)$$

素数

$$p\mid (x+1)(x-1)$$

故

$$p\mid (x+1) \quad 或 \quad p\mid (x-1)$$

即

$$x+1\equiv 0(\bmod p) \quad 或 \quad x-1\equiv 0(\bmod p)$$

$$x \equiv -1 \equiv p-1 \pmod p \quad 或 \quad x \equiv 1 \pmod p$$

又由 Fermat 定理知

$$(r^{\frac{p-1}{2}})^2 \equiv r^{p-1} \equiv 1 \pmod p$$

$$(r^{p-1})^2 \equiv 1^2 \equiv 1 \pmod p$$

故整数 $r^{\frac{p-1}{2}}, r^{p-1}$ 是方程(8.8)的解,从而对模 p 与 1 或 $p-1$ 同余. 又由 r 为模 p 之原根,据附录 Ⅱ 定理4,$r, r^2, \cdots, r^{p-1}$ 除以 p 所得余数互不相同,故

$$r^{\frac{p-1}{2}} \not\equiv r^{p-1} \equiv 1 \pmod p$$

于是唯有

$$r^{\frac{p-1}{2}} \equiv p-1 \equiv -1 \pmod p$$

且在 $t=1,2,\cdots,p-1$ 中除 $t=\frac{p-1}{2}$ 外再无 t 可使 $r^t \equiv -1 \pmod p$.

引理 8.3 假设同上引理,则在

$$\omega_p^r, \omega_p^{r^2}, \cdots, \omega_p^{r^{p-2}}, \omega_p^{r^{p-1}} (= \omega_p^{r^0} = \omega_p) \tag{8.9}$$

中正好有 $\frac{p-1}{2}$ 对数,其乘积为 1,即

$$\omega_p^{r^b} \cdot \omega_p^{r^{b+\frac{p-1}{2}}} = 1 \quad (b=1,2,\cdots,\frac{p-1}{2}) \tag{8.10}$$

证 在式(8.9)中求乘积为 1 的一对数通式,设

$$\omega_p^{r^b} \cdot \omega_p^{r^{b+k}} = 1 = \omega_p^0$$

(k 为自然数,$1 \leqslant b \leqslant b+k \leqslant p-1$). 即

$$\omega_p^{r^b + r^{b+k}} = \omega_p^0$$

从而

$$r^b(1+r^k) = r^b + r^{b+k} \equiv 0 \pmod p$$

因 r 为模 p 原根,$1 \leqslant b \leqslant p-1 < p$,故

$$r^b \not\equiv 0 \pmod p$$

而 p 为素数,故必有

$$1 + r^k \equiv 0 \pmod p$$

$$r^k \equiv -1 \pmod p$$

由上引理知只有

$$k = \frac{p-1}{2}$$

再从

$$1 \leqslant b \leqslant b+k \leqslant p-1$$

即知

$$1 \leqslant b \leqslant b+\frac{p-1}{2} \leqslant p-1 \Leftrightarrow 1 \leqslant b \leqslant \frac{p-1}{2}$$

本引理结论得证.

证定理8.2.

证①:当 $s=1$ 时,把 ω_p 的幂排成

$$\omega_p^{r}, \omega_p^{r^2}, \omega_p^{r^3}, \cdots, \omega_p^{r^{p-1}}$$

其中 r 的指数成等差数列,公差为1;

当 $s \geqslant 2$ 时,任一 $\eta_{j'j''\cdots j^{(s-1)}}$ 为 $q^{(s-1)}$ 个项之和,其中各项中 r 的指数为等差数列,公差为 $p'p''\cdots p^{(s-1)}$.

因再对 $j^{(s)}=1,2,\cdots,p^{(s)}$,取 $\eta_{j'j''\cdots j^{(s-1)}j^{(s)}}$ 为上述第 $j^{(s)}$ 项,第 $j^{(s)}+p^{(s)}$ 项,第 $j^{(s)}+2p^{(s)}$ 项与第 $j^{(s)}+(q^{(s)}-1)p^{(s)}$ 项之和,故 $\eta_{j'j''\cdots j^{(s-1)}j^{(s)}}$ 中各项 r 之指数为等差数列,公差为 $p'p''\cdots p^{(s-1)}p^{(s)}$,即任两项中 r 的指数之差为 $p'p''\cdots p^{(s-1)}p^{(s)}$ 之倍数.但任二 $\eta_{j'j''\cdots j^{(s-1)}j_1^{(s)}}$ 与 $\eta_{j'j''\cdots j^{(s-1)}j_2^{(s)}}(j_1^{(s)} \neq j_2^{(s)})$ 首项中 r 的指数之差非 $p'p''\cdots p^{(s-1)}p^{(s)}$ 之倍数,于是 $\eta_{j'j''\cdots j^{(s-1)}j_1^{(s)}}$ 中任一项与 $\eta_{j'j''\cdots j^{(s-1)}j_2^{(s)}}$ 中任一项的 r 的指数差非 $p'p''\cdots p^{(s-1)}p^{(s)}$ 之倍数,故非 $\frac{p-1}{2}\left(=\frac{p-1}{q^{(l)}}=p'p''\cdots p^{(s)} \cdot p^{(s+1)}\cdots p^{(l)}\right)$,从而据引理8.3,①所述 $(q^{(s)})^2$ 个项(乘积)都不等于1.

证②③:任一 $\eta_{j'j''\cdots j^{(s)}}$ 中有 $q^{(s)}$ 个项,其中相邻两项 r 的指数递增 $p'p''\cdots p^{(s)}$,又 $q^{(s)}$ 为 $q^{(s+1)}, q^{(s+2)}, \cdots, q^{(l)}(=2)$ 的倍数,即 $q^{(s)}$ 为偶数,$\frac{1}{2}q^{(s)}$ 为正整数,每一项与其后第 $\frac{1}{2}q^{(s)}$ 项的 r 的指数差为

$$p'p''\cdots p^{(s)} \cdot \frac{1}{2}q^{(s)}=\frac{1}{2}p'p''\cdots p^{(s-1)}q^{(s-1)}=$$

$$\frac{1}{2}p'p''\cdots p^{(s-2)}q^{(s-2)}=\cdots=\frac{1}{2}p'q'=\frac{p-1}{2}$$

于是 $\eta_{j'j''\cdots j^{(s)}}$ 的 $q^{(s)}$ 个项中前 $\frac{1}{2}q^{(s)}$ 个项与后 $\frac{1}{2}q^{(s)}$ 个项正好配成(按原来的前后次序) $\frac{1}{2}q^{(s)}$ 对数,每对数之积为1(据引理8.3),而和形如

$$\omega_p^{r^b}+\omega_p^{r^{b+\frac{p-1}{2}}}=\omega_p^{r^b}+\left(\omega_p^{r^{\frac{p-1}{2}}}\right)^{r^b}=\omega_p^{r^b}+\left(\omega_p^{-1}\right)^{r^b}=\omega_p^{r^b}+\omega_p^{-r^b}=$$

$$e^{\frac{2\pi r^b}{p}i}+e^{-\frac{2\pi r^b}{p}i}=2\cos\frac{2\pi r^b}{p}\in\mathbf{R}$$

故上述$\frac{1}{2}q^{(s)}$对数之和即$\eta_{j'j''\cdots j^{(s)}}$为实数.

任一$\eta_{j'j''\cdots j^{(l)}}$有$q^{(l)}=2$个项,即$\eta_{j'j''\cdots j^{(l)}1}$与$\eta_{j'j''\cdots j^{(l)}2}$,按上述正好配成一对,其和为$\eta_{j'j''\cdots j^{(l)}}$,其积为1,据Vieta定理得证③.

现证②中最后结论:由上述知任一$\eta_{j'j''\cdots j^{(l)}}$形如$\omega_p^{r^b}+\omega_p^{r^{b+\frac{p-1}{2}}}$,其中$r$的指数$b,b+\frac{p-1}{2}\in\{1,2,\cdots,p\}$,而$r$为模$p$之原根,故$r^b$,$r^{b+\frac{p-1}{2}}\not\equiv 0(\bmod p)$,于是$r^b$,$r^{b+\frac{p-1}{2}}$除以$p$所得余数亦属于$\{1,2,\cdots,p\}$.又因式(8.10)即

$$r^b+r^{b+\frac{p-1}{2}}\equiv 0(\bmod p)$$

于是$\{1,2,\cdots,p-1\}$内的上述两余数之和(最小为2,最大为$2p-2$)为p之倍数,这只能等于p,故两余数之一为小于$\frac{p}{2}$的正整数,即属于集$\{1,2,\cdots,\frac{p-1}{2}\}$的数$m$,另一为$p-m\in\{\frac{p+1}{2},\frac{p+3}{2},\cdots,p-1\}$.从而

$$\omega_p^{r^b}+\omega_p^{r^{b+\frac{p-1}{2}}}=\omega_p^m+\omega_p^{p-m}=\omega_p^m+\omega_p^p\omega_p^{-m}=\omega_p^m+\omega_p^{-m}=$$
$$e^{\frac{2\pi m}{p}i}+e^{-\frac{2\pi m}{p}i}=2\cos\frac{2\pi m}{p}$$

因所有$p'p''\cdots p^{(l)}=\frac{p-1}{q^{(l)}}=\frac{p-1}{2}$个数$\eta_{j'j''\cdots j^{(l)}}$,每个数有2项,共$p-1$个项,其中$\omega_p$的指数取尽$r,r^2,r^3,\cdots,r^{p-1}$,因$r$为模$p$的原根,由附录Ⅱ定理4知$r$的上述各幂除以$p$所得余数取尽$1,2,\cdots,p-1$,即正好分成$\frac{p-1}{2}$对

$$(m,p-m)\quad(m=1,2,\cdots,\frac{p-1}{2})$$

每一对$(m,p-m)$正好是某个$\eta_{j'j''\cdots j^{(l)}}$的两项中$\omega_p$的指数$b,b+\frac{p-1}{2}$除以$p$所得余数.即相应的$2\cos\frac{2\pi m}{p}$正是此$\eta_{j'j''\cdots j^{(l)}}$之值.

例1 用域$\mathbf{Q}$上素数次不可约循环方程之根迭代表示$p=13$时分圆方程(7.5)的各根.

解 取模 13 的原根 $r=6$①，则方程(7.5)在域 $\mathbf{Q}$ 上的 Galois(循环)群生成元为轮换

$$s=(\omega_{13}^{6},\omega_{13}^{6^2},\omega_{13}^{6^3},\omega_{13}^{6^4},\omega_{13}^{6^5},\omega_{13}^{6^6},\omega_{13}^{6^7},\omega_{13}^{6^8},\omega_{13}^{6^9},\omega_{13}^{6^{10}},\omega_{13}^{6^{11}},\omega_{13}^{6^{12}})=$$
$$(\omega_{13}^{6},\omega_{13}^{10},\omega_{13}^{8},\omega_{13}^{9},\omega_{13}^{2},\omega_{13}^{12},\omega_{13}^{7},\omega_{13}^{3},\omega_{13}^{5},\omega_{13}^{4},\omega_{13}^{11},\omega_{13})$$

(由第一数的指数 6 起，逐次乘 6 取对模 13 的余数得下一个指数)分解 $13-1=12=3\times 4$，取 $p'=3,q'=4$，把轮换式 s 中各数排成 3×4 矩阵(先排第 1,2,… 列)再取和得

$$\eta_1=\omega_{13}^{6}+\omega_{13}^{9}+\omega_{13}^{7}+\omega_{13}^{4}$$
$$\eta_2=\omega_{13}^{10}+\omega_{13}^{2}+\omega_{13}^{3}+\omega_{13}^{11}$$
$$\eta_3=\omega_{13}^{8}+\omega_{13}^{12}+\omega_{13}^{5}+\omega_{13}$$

求 $\eta_1\eta_2$ 用 η_1,η_2,η_3 的线性组合表示式：用 η_2 表示式第一项 ω_{13}^{10} 乘 η_1 表示式，各项分别得积 $\omega_{13}^{16},\omega_{13}^{19},\omega_{13}^{17},\omega_{13}^{14}$，各指数除以 13 分别得余数 3,6,4,1，即上述积为 $\omega_{13}^{3},\omega_{13}^{6},\omega_{13}^{4},\omega_{13}$，它们分别为 $\eta_2,\eta_1,\eta_1,\eta_3$ 的项，故由式(8.6)得

$$\eta_1\eta_2=\eta_2+\eta_1+\eta_1+\eta_3=2\eta_1+\eta_2+\eta_3$$

用同样方法得(亦可在上式对 η_1,η_2,η_3 作轮换(η_1,η_2,η_3)得②)

$$\eta_2\eta_3=2\eta_2+\eta_3+\eta_1$$
$$\eta_3\eta_1=2\eta_3+\eta_1+\eta_2$$

故

$$\eta_1\eta_2+\eta_2\eta_3+\eta_3\eta_1=4\eta_1+4\eta_2+4\eta_3=-4$$

用同样方法求

$$\eta_1^2=\eta_1\eta_1=\eta_3+\eta_2+4+\eta_2=2\eta_2+\eta_3+4$$

(注意 $\omega^6\omega^7=\omega^{13}=1$，式(8.6)中相应项为 $q'=4$). 再求

$$\eta_1\eta_2\eta_3=\eta_1(2\eta_2+\eta_1+\eta_3)=2\eta_1\eta_2+\eta_1^2+\eta_1\eta_3=$$
$$2(2\eta_1+\eta_2+\eta_3)+(2\eta_2+\eta_3+4)+$$
$$(\eta_1+\eta_2+2\eta_3)=$$
$$5(\eta_1+\eta_2+\eta_3)+4=-5+4=-1$$

故由 Vieta 定理知 η_1,η_2,η_3 为下述方程(由定理 8.1 知第二项为 t^2)之根

$$t^3+t^2-4t+1=0$$

再分解 $q'=4=2\times 2$，取 $p''=2,q''=2$，分别把 η_1,η_2,η_3 用 ω_{13} 表

①在数论教材中有素数的原根表.

②据 η_1,η_2,η_3 有轮换对称性.

示式各项排成 2×2 矩阵(先排各列)取和得

$$\eta_{11}=\omega_{13}^{6}+\omega_{13}^{7}=2\cos\frac{12\pi}{13}$$

$$\eta_{12}=\omega_{13}^{9}+\omega_{13}^{4}=2\cos\frac{8\pi}{13}$$

$$\eta_{21}=\omega_{13}^{10}+\omega_{13}^{3}=2\cos\frac{6\pi}{13}$$

$$\eta_{22}=\omega_{13}^{2}+\omega_{13}^{11}=2\cos\frac{4\pi}{13}$$

$$\eta_{31}=\omega_{13}^{8}+\omega_{13}^{5}=2\cos\frac{10\pi}{13}$$

$$\eta_{32}=\omega_{13}^{12}+\omega_{13}=2\cos\frac{2\pi}{13}$$

则得(其中求 $\eta_{11}\eta_{12}$ 按类似式(8.6)的公式处理)

$$\eta_{11}+\eta_{12}=\eta_1$$

$$\eta_{11}\eta_{12}=\eta_{22}+\eta_{21}=\eta_2$$

故 η_{11},η_{12} 为下述方程之根

$$t^2-\eta_1 t+\eta_2=0$$

同理(按轮换)η_{21},η_{22} 为下述方程之根

$$t^2-\eta_2 t+\eta_3=0$$

η_{31},η_{32} 为下述方程之根

$$t^2-\eta_3 t+\eta_1=0$$

最后易见(或按定理 8.2 的 ③):

$\omega_{13}^{6},\omega_{13}^{7}$ 适合方程

$$t^2-\eta_{11}t+1=0$$

$\omega_{13}^{9},\omega_{13}^{4}$ 适合方程

$$t^2-\eta_{12}t+1=0$$

$\omega_{13}^{10},\omega_{13}^{3}$ 适合方程

$$t^2-\eta_{21}t+1=0$$

$\omega_{13}^{2},\omega_{13}^{11}$ 适合方程

$$t^2-\eta_{22}t+1=0$$

$\omega_{13}^{8},\omega_{13}^{5}$ 适合方程

$$t^2-\eta_{31}t+1=0$$

$\omega_{13}^{12},\omega_{13}$ 适合方程

$$t^2-\eta_{32}t+1=0$$

又解(按式(8.3)求$\eta_1\eta_2+\eta_2\eta_3+\eta_3\eta_1,\eta_1\eta_2\eta_3,\eta_{11}\eta_{12}$)用$\omega_{13}$的幂的表示式共有$3\times 4^2=48$项，据定理8.2无一项等于1，于是按式(8.3)得

$$\eta_1\eta_2+\eta_2\eta_3+\eta_3\eta_1=0-\frac{48-0}{13-1}=-4$$

$\eta_1\eta_2\eta_3$共$4^3=64$项，其中值为1者有$\omega_{13}^6\omega_{13}^2\omega_{13}^5,\omega_{13}^9\omega_{13}^3\omega_{13},\omega_{13}^7\omega_{13}^{11}\omega_{13}^8,\omega_{13}^4\omega_{13}^{10}\omega_{13}^{12}$共4项，按式(8.3)得

$$\eta_1\eta_2\eta_3=4-\frac{64-4}{13-1}=4-5=-1$$

$\eta_{11}\eta_{12}$共$2^2=4$项，而η_1,η_2,η_3均为4项，故$\eta_{11}\eta_{12}$等于η_1,η_2,η_3中之一(据定理8.1). 取$\eta_{11}\eta_{12}$展开式的一项$\omega_{13}^6\omega_{13}^9=\omega_{13}^{15}=\omega_{13}^2$为$\eta_2$之项，故

$$\eta_{11}\eta_{12}=\eta_2$$

最后举例论述如何(用圆规、直尺①)分圆为Fermat素数p等份，关键在于用多层二次根式表示$\cos\frac{2\pi}{p}$(从而可作出角$\frac{2\pi}{p}$).

例2 分圆为5等份($5=2^{2^1}+1$).

解 先用平方根表示$2\cos\frac{2\pi}{5}$:

考虑当$p=5$时分圆方程(7.5)，取模5原根$r=2$，则方程(7.5)在域$\mathbf{Q}$上的Galois(循环)群生成元

$$s=(\omega_5^2,\omega_5^{2^2},\omega_5^{2^3},\omega_5^{2^4})=(\omega_5^2,\omega_5^4,\omega_5^3,\omega_5)$$

分解

$$5-1=4=2\times 2$$

取$p'=2,q'=2$，把s中各元素排成2×2矩阵(先排各列)再取和得

$$\eta_1=\omega_5^2+\omega_5^3=2\cos\frac{4\pi}{5}<0$$

$$\eta_2=\omega_5^4+\omega_5=2\cos\frac{2\pi}{5}>0$$

易见(按式(8.3))

$$\eta_1\eta_2=0-\frac{2\times 2-0}{5-1}=-1$$

于是η_1,η_2为下述方程之根

①无刻度之直尺.

$$t^2 + t - 1 = 0$$

即
$$\frac{-1 \pm \sqrt{5}}{2}$$

因 $2\cos\frac{2\pi}{5} > 0$,故

$$2\cos\frac{2\pi}{5} = \frac{-1+\sqrt{5}}{2} = \sqrt{\frac{5}{4}} - \frac{1}{2} = \sqrt{1+(\frac{1}{2})^2} - \frac{1}{2}$$

$$\cos\frac{2\pi}{5} = \frac{1}{2}(\sqrt{1+(\frac{1}{2})^2} - \frac{1}{2})$$

不妨设所分圆 O 的半径为长度单位1,即圆 O 是单位圆. 于是先在圆 O 内作出长为 $\sqrt{1+(\frac{1}{2})^2} - \frac{1}{2}$ 的线段 ON,如图8.1所示①,再作 ON 的中垂线与圆 O 交于点 P,P',并得 ON 中点 Q,则 $OQ = \frac{1}{2}(\sqrt{1+(\frac{1}{2})^2} - \frac{1}{2})$,$\angle A'OP = \frac{2\pi}{5}$,…(或据 $2\sin\frac{\pi}{10} = 2\cos\frac{2\pi}{5} = ON$ 知 ON 为圆 O 内接正十边形边长……).

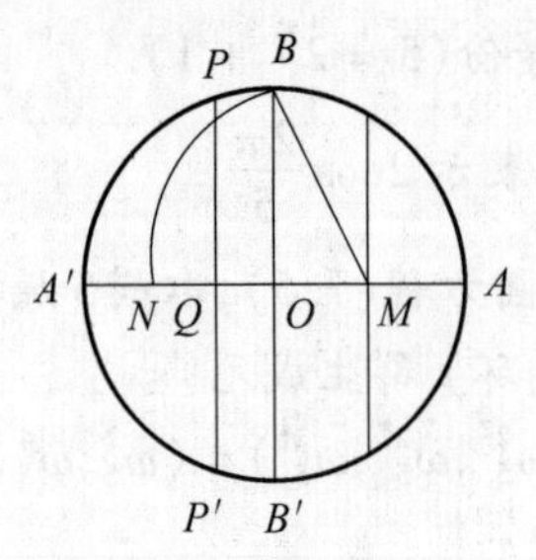

图8.1

此作法较一般书上(最简)作法(BN 即圆 O 内接正五边形的边长……)多出一步——作 ON 之中垂线,但此最简作法之推导(分析)较难. 固然上述作法需要以循环方程,分圆方程理论作基础.

例3 用多层二次根式表示 $2\cos\frac{2}{17}\pi(17 = 2^{2^2}+1)$.

解 取 $p = 17$ 时分圆方程(7.5). 取模17的原根 $r = 3$,3的1,2,3,…,16次幂除以17所得余数为

$$3,9,10,13,5,15,11,16,14,8,7,4,12,2,6,1$$

取

---

①作互相垂直的两直径 $A'A,B'B$,以 OA 中点 M 为圆心,MB 为半径作弧交 OA' 于点 N.

$$\eta_1=\omega_{17}^3+\omega_{17}^{10}+\omega_{17}^5+\omega_{17}^{11}+\omega_{17}^{14}+\omega_{17}^7+\omega_{17}^{12}+\omega_{17}^6$$

$$\eta_2=\omega_{17}^9+\omega_{17}^{13}+\omega_{17}^{15}+\omega_{17}^{16}+\omega_{17}^8+\omega_{17}^4+\omega_{17}^2+\omega_{17}$$

则

$$\eta_1\eta_2=0-\frac{8\times 8-0}{17-1}=-4$$

故 η_1,η_2 为下述方程之根

$$t^2+t-4=0$$

因(第1,5项合并,第2,6项合并,第3,7项合并,第4,8项合并)

$$\eta_1=2\cos\frac{6\pi}{17}+2\cos\frac{14\pi}{17}+2\cos\frac{10\pi}{17}+2\cos\frac{12\pi}{17}=$$

$$2\left(\cos\frac{6\pi}{17}-\cos\frac{3\pi}{17}-\cos\frac{7\pi}{17}-\cos\frac{5\pi}{17}\right)<0$$

又以 $\eta_1\eta_2=-4<0$ 知 $\eta_2>0$,于是解上述方程得

$$\eta_1=\frac{-1-\sqrt{17}}{2}(<0)$$

$$\eta_2=\frac{-1+\sqrt{17}}{2}(>0)$$

令

$$\eta_{11}=\omega_{17}^3+\omega_{17}^5+\omega_{17}^{14}+\omega_{17}^{12}=2\cos\frac{6\pi}{17}+2\cos\frac{10\pi}{17}$$

$$\eta_{12}=\omega_{17}^{10}+\omega_{17}^{11}+\omega_{17}^7+\omega_{17}^6=2\cos\frac{14\pi}{17}+2\cos\frac{12\pi}{17}$$

$$\eta_{21}=\omega_{17}^9+\omega_{17}^{15}+\omega_{17}^8+\omega_{17}^2=2\cos\frac{16\pi}{17}+2\cos\frac{4\pi}{17}$$

$$\eta_{22}=\omega_{17}^{13}+\omega_{17}^{16}+\omega_{17}^4+\omega_{17}=2\cos\frac{8\pi}{17}+2\cos\frac{2\pi}{17}$$

因 $2\cos\frac{2\pi}{17}$ 在 η_{22} 中,先求 η_{21},η_{22} 所适合的方程(按类似于式(8.6)求 $\eta_{21}\eta_{22}$)

$$\eta_{21}+\eta_{22}=\eta_2$$

$$\eta_{21}\eta_{22}=\eta_{11}+\eta_{12}+\eta_{22}+\eta_{21}=\eta_1+\eta_2=-1$$

故 η_{21},η_{22} 为下述方程之根

$$t^2-\eta_2 t-1=0 \tag{8.11}$$

易见 $\eta_{21}<0,\eta_{22}>0$,故解上述方程得

$$\eta_{21}=\frac{\eta_2-\sqrt{\eta_2^2+4}}{2},\eta_{22}=\frac{\eta_2+\sqrt{\eta_2^2+4}}{2} \tag{8.12}$$

类似按轮换(1,2)知$\eta_{11}(>0)$,$\eta_{12}(<0)$为下述方程之根

$$t^2-\eta_1 t-1=0$$

解得

$$\eta_{11}=\frac{\eta_1+\sqrt{\eta_1^2+4}}{2},\eta_{12}=\frac{\eta_1-\sqrt{\eta_1^2+4}}{2} \tag{8.13}$$

因$2\cos\frac{2\pi}{17}$在η_{22}中,取

$$\eta_{221}=\omega_{17}^{13}+\omega_{17}^{4}=2\cos\frac{8\pi}{17}$$

$$\eta_{222}=\omega_{17}^{16}+\omega_{17}=2\cos\frac{2\pi}{17}$$

$\eta_{221}\eta_{222}$有$2\times 2=4$项,应为$\eta_{11},\eta_{12},\eta_{21},\eta_{22}$的整系数线性组合,此四数每个均有4项,故$\eta_{221}\eta_{222}$应等于此四数之一,由$\omega_{17}^{13}\omega_{17}^{16}=\omega_{17}^{29}=\omega_{17}^{12}$为$\eta_{11}$的一项,故

$$\eta_{221}\eta_{222}=\eta_{11}$$

知η_{221},η_{222}为下述方程之根

$$t^2-\eta_{22}t+\eta_{11}=0 \tag{8.14}$$

易见$\eta_{222}>\eta_{221}$,故解上方程得

$$\eta_{222}=2\cos\frac{2\pi}{17}=\frac{\eta_{22}+\sqrt{\eta_{22}^2-4\eta_{11}}}{2}$$

再据上述结果讲述分圆17等份之方法.

因要三次求二次方程之根,先讲述(不先求这些根用平方根表示式)用作图方法直接在坐标平面作出表示有二实根的二次方程

$$x^2+ax+b=0 \tag{8.15}$$

的两根x_1,x_2的有向线段的方法:

取点$P(-a,b)$及$K(0,1)$,以PK为直径作圆必与x轴交于点X_1,X_2(可能重合),如图8.2所示,且

$$\overline{OX_1}=x_1$$

$$\overline{OX_2}=x_2$$

(以x轴正向为有向线段正向).

证 因方程(8.15)有实根,故判别式

$$\Delta=a^2-4b\geqslant 0$$

所作的圆的圆心

$$O'\left(-\frac{a}{2},\frac{b+1}{2}\right)$$

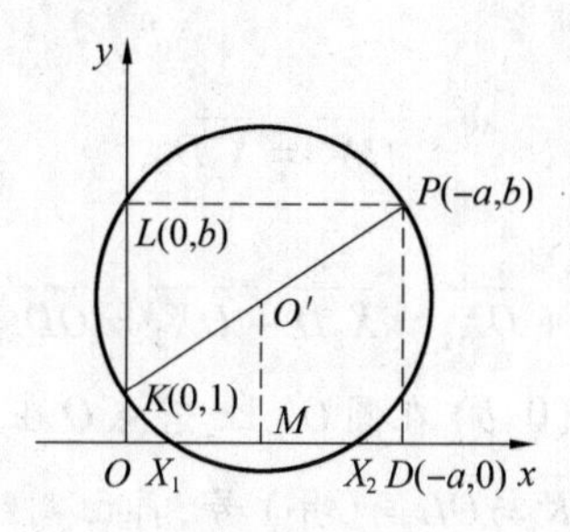

图 8.2

到 x 轴的距离

$$d=\left|\frac{b+1}{2}\right|$$

半径

$$r=\frac{1}{2}\sqrt{a^2+(b-1)^2}$$

先证圆 O' 与 x 轴相交或相切:

只要证

$$|d|\leqslant r$$

即证(可逆推)

$$d^2\leqslant r^2$$

即证

$$\frac{(b+1)^2}{4}\leqslant\frac{1}{4}[a^2+(b-1)^2]$$

即证

$$(b+1)^2\leqslant a^2+(b-1)^2$$

即证

$$4b\leqslant a^2$$

由上述 $a^2-4b\geqslant 0$,故结论成立.

再证 $\overline{OX_1}+\overline{OX_2}=-a$, $\overline{OX_1}\cdot\overline{OX_2}=b$(从而 $\overline{OX_1}$, $\overline{OX_2}$ 为方程(8.15)之两根)即可:

设点 O', P 在 x 轴射影分别为 M, D 两点. 因 $\overline{KO'}=\overline{O'P}$,故其在 x 轴射影

$$\overline{OM}=\overline{MD}$$

又 M 为 $\overline{X_1X_2}$ 中点,故

$$\overline{X_1M}=\overline{MX_2}$$

上两式相减得

$$\overline{OX_1} = \overline{X_2D}$$

于是

$$\overline{OX_1} + \overline{OX_2} = \overline{X_2D} + \overline{OX_2} = \overline{OD} = -a$$

易见P在y轴射影$L(0,b)$在圆O'上. 若点O在圆O'外(内),则$\overline{OX_1}$与$\overline{OX_2}$同(异)号,$\overline{OK}$与$\overline{OL}$同(异)号,再由割线定理,相交弦定理,或(特殊情况)切线割线定理,切线长定理知

$$\overline{OX_1} \cdot \overline{OX_2} = \overline{OK} \cdot \overline{OL} = 1 \cdot b = b$$

当原点O在圆O'上(即当$b=0$)时$\overline{OX_1}$或$\overline{OX_2}$为0,$\overline{OL}=0$,上式亦成立.

例4 作角$\frac{2\pi}{17}$.

解 即把单位圆(圆心未定)17等分,从上例中式(8.12)及式(8.13)得

$$\eta_{22} = \frac{\eta_2}{2} + \sqrt{\left(\frac{\eta_2}{2}\right)^2 + 1}$$

$$\eta_{11} = \frac{\eta_1}{2} + \sqrt{\left(\frac{\eta_1}{2}\right)^2 + 1}$$

现从η_1,η_2所适合的方程(8.10)求$\frac{\eta_1}{2},\frac{\eta_2}{2}$所适合的以$t'$为未知数的方程,应有$t' = \frac{t}{2}$,即$t = 2t'$代入方程(8.10)得

$$4t'^2 + 2t' - 4 = 0$$

$$t'^2 + \frac{1}{2}t' - 1 = 0$$

现要在坐标平面作出表示$\frac{\eta_2}{2},\frac{\eta_1}{2}$的有向线段,按上述方法,应取点$P(-\frac{1}{2},-1)$,$K(0,1)$,以$PK$为直径作圆,其圆心为$PK$中点$B(-\frac{1}{4},0)$,半径为$|KB|$,此圆交$x$轴于$E,D$,如图8.3所示,则

$$\overline{OE} = \frac{\eta_1}{2}$$

$$\overline{OD} = \frac{\eta_2}{2}$$

$$|EK| = \sqrt{(\frac{\eta_1}{2})^2 + 1}$$

$$|DK| = \sqrt{(\frac{\eta_2}{2})^2 + 1}$$

在 x 轴取 $\overline{EG} = |EK|$, $\overline{DF} = |DK|$, 则

$$\overline{OG} = \overline{OE} + \overline{EG} = \frac{\eta_1}{2} + \sqrt{(\frac{\eta_1}{2})^2 + 1} = \eta_{22}$$

$$\overline{OF} = \overline{OD} + \overline{DF} = \frac{\eta_2}{2} + \sqrt{(\frac{\eta_2}{2})^2 + 1} = \eta_{11}$$

取点 $H(\eta_{11}, \eta_{22})$——即过点 F 作 x 轴垂线,在其上于 x 轴上方取 $FH = OG$. 再以 KH 为直径作圆与 x 轴交于点 X_1, X_2,即作出表示方程(8.14)之两根 η_{221}, η_{222} 之有向线段

$$\overline{OX_1} = \eta_{222} = 2\cos\frac{2\pi}{17}$$

$$\overline{OX_2} = \eta_{221} = 2\cos\frac{8\pi}{17}$$

由此可得最后作法:以 O 为圆心作单位圆与 x 轴正向交于 A_0,过 OX_1 中点 M 作 x 轴垂线与此圆交于 A_1,则 $\angle A_0OA_1 = \frac{2\pi}{17}$;或由 $2\sin\frac{\pi}{34} = 2\cos\frac{8\pi}{17} = \overline{OX_2}$,知 $|OX_2|$ 为单位圆内接正34边形的边长……

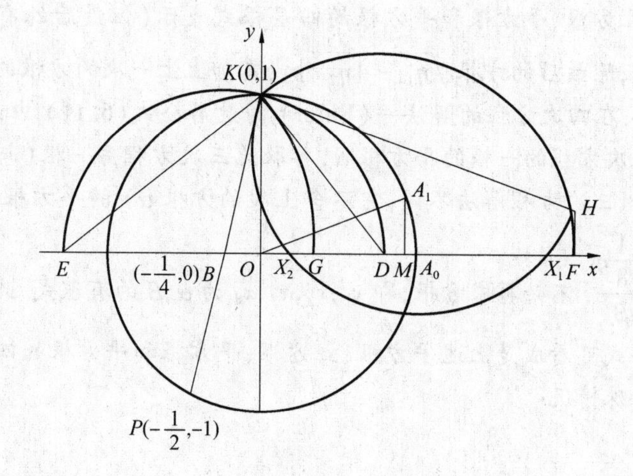

图 8.3

第九章　代数方程的多层根式解

显然域Δ上二次方程$x^2+px+q=0$的两根是域Δ的数p^2-4q的平方根$\sqrt{p^2-4q}$在域Δ上的有理式；在三次方程的根的公式(6.3)～(6.5)中的第一项实际是$\sqrt{\frac{q^2}{4}+\frac{p^3}{27}}$的有理式$-\frac{q}{2}+\sqrt{\frac{q^2}{4}+\frac{p^3}{27}}$的3个立方根，第二项是$-\frac{p}{3}$分别除以上述3个立方根所得之商①，即$x_1,x_2,x_3$都是这些立方根的有理式，我们称$x_1,x_2,x_3$可用方程的系数的平方根、立方根的两层根式表示；在四次方程的第二解法(Ferrari)的公式(6.18)或式(6.19)中两个平方根之积为$-(q-\frac{at^*}{2})/2$，故其一次项系数(记为η_1)及常数项(记为η_2)为t^*(三次方程之根，可用其系数之上述平方根、立方根的两层根式表示)之有理式$\frac{t^{*2}}{4}-r$的平方根的有理式，从而二次方程(6.18)(6.19)的各根可用η_1及$\sqrt{\eta_1^2-4\eta_2}$的有理式表示，实际上$\eta_1^2-4\eta_2$亦是$\frac{t^{*2}}{4}-r$(在域Δ上添加了前述平方根、立方根后所得数域之数)的平方根的有理式，故四次方程之各根可用其系数的平方根、立方根、平方根和平方根的四层根式表示(但最后的有理式的变元，除最后的方根$\sqrt{\eta_1^2-4\eta_2}$外尚需加上上一次的方根的有理式η_1)；在四次方程的解法一(Euler)的最后公式(6.11)中，先取一个三次方程的一根的平方根α，再取此三次方程另一根(与解法一及二、三次方程解法不同，它不含上次的方根α)的平方根β，而$\gamma=\frac{-\frac{1}{8}q}{\alpha\beta}$，不必再求方根，即$x_1,x_2,x_3,x_4$为$\alpha,\beta$的有理式. 此解法最后公式可看成是上述平方根、立方根、平方根和平方根的四层根式的特殊情况.

---

①注意式(6.3)中要求两立方根之积为$-\frac{p}{3}$.

如由域Δ上三次方根与五次方根(按顺序)构成的二层根式的一般形式为

$$\sum_{j=0}^{4}(a_{j0}+a_{j1}\sqrt[3]{A}+a_{j2}(\sqrt[3]{A})^2)(\sqrt[5]{b_0+b_1\sqrt[3]{A}+b_2(\sqrt[3]{A})^2})^j$$

$$(a_{j0},a_{j1},a_{j2}\in\Delta,b_0,b_1,b_2\in\Delta,A\in\Delta)$$

注意$b_1=b_2=0$的特殊情况,这时五次根号内无三次方根;又由于

$$\sqrt[p_1p_2]{A}=\sqrt[p_2]{\sqrt[p_1]{A}}\quad(p_1,p_2\in\mathbf{Z}^+)$$

不妨设根指数为素数.

现用记号准确阐述用域Δ上多层(含一层)根式表示代数方程一根之意义.

域Δ上多项式$f(x)$如第四章所述,其一根x^*可用域Δ上多层根式表示,其意义为:

存在(复数)$\xi^{(1)},\xi^{(2)},\cdots,\xi^{(k)}$及素数$n_1,n_2,\cdots,n_k$使

$$(\xi^{(1)})^{n_1}=x^{(0)}\in\Delta$$

(即$\xi^{(1)}$为域Δ上数$x^{(0)}$的n_1次方根)

$$(\xi^{(2)})^{n_2}=x^{(1)}=\varphi_1(\xi^{(1)})\text{——域}\Delta\text{上有理式}$$

(即$x^{(1)}\in\Delta(\xi^{(1)})$)

$$(\xi^{(3)})^{n_3}=x^{(2)}=\varphi_2(\xi^{(1)},\xi^{(2)})\text{——域}\Delta\text{上二元有理式}$$

(即$x^{(2)}\in\Delta(\xi^{(1)},\xi^{(2)})$)

$$\vdots$$

$$(\xi^{(k)})^{n_k}=x^{(k-1)}=\varphi_k(\xi^{(1)},\xi^{(2)},\cdots,\xi^{(k-1)})\text{——域}\Delta\text{上}k-1\text{元有理式}$$

$$x^*=\varphi_k(\xi^{(1)},\xi^{(2)},\cdots,\xi^{(k-1)},\xi^{(k)})\text{——域}\Delta\text{上}k\text{元有理式}$$

若对某些$j=1,2,\cdots,k$有

$$\xi^{(j)}\in\Delta(\xi^{(1)},\xi^{(2)},\cdots,\xi^{(j-1)})$$

则

$$\xi^{(j)}=\phi(\xi^{(1)},\xi^{(2)},\cdots,\xi^{(j-1)})\text{——域}\Delta\text{上}j-1\text{元有理式}$$

代入x^*的表示式知,实际上x^*的表示式内不含$\xi^{(j)}$,故可省去$\xi^{(j)}$,今后总假定有

$$\xi^{(j)}\notin\Delta(\xi^{(1)},\xi^{(2)},\cdots,\xi^{(j-1)})\quad(j=1,2,\cdots,k)$$

($j=1$时右边域为Δ).同样不妨设

$$x^{(j)}\notin\Delta(\xi^{(1)},\xi^{(2)},\cdots,\xi^{(j-1)})$$

$$x^*\notin\Delta(\xi^{(1)},\xi^{(2)},\cdots,\xi^{(k-1)})$$

但前面所述求二、三、四次方程的根的公式中,各根的迭代根

式表示式都是"同型"的，即各表示式中各次方根的根指数 n_1，$n_2, \cdots, n_k$ 相同，$x^{(0)}$ 相同，各有理式 $\varphi_1, \varphi_2, \cdots, \varphi_k$ 相同，但它们所取的变元可以不同，但只是由于在各次取不同的方根使变元得不同的值. 如方程(6.18)、方程(6.19) 中各项系数的有理式的平方根表示两方程的根 x_1, x_2, x_3, x_4 的式子中，因取$\frac{t^{*2}}{4}-r$的两不同平方根，使上述有理式的变元不同.

以后我们将证明，至少五次的代数方程在绝大多数情况下不存在多层根式解. 但只要有一个根可用多层根式解出，则方程的所有各根必定可用同型的多层根式解出. 下文即将证此结论，先引入一些记号表示上述同型的多层根式.

设方程 $y^{n_1}=x^{(0)} \in \Delta$ 的 n_1 个根为

$$\xi_{j'} \quad (j'=0,1,2,\cdots,n_1-1)$$

其中

$$\xi_0=\xi^{(1)}$$

令

$$x_{j'}=\varphi_1(\xi_{j'}) \text{—— 域 } \Delta \text{ 上有理式}$$

显然

$$x_0=x^{(1)}$$

设方程 $y^{n_2}=x_{j'}$ 的 n_2 个根为

$$\xi_{j'j''} \quad (j''=0,1,2,\cdots,n_2-1)$$

其中

$$\xi_{00}=\xi^{(1)}$$

令

$$x_{j'j''}=\varphi_2(\xi_{j'},\xi_{j'j''}) \text{—— 域 } \Delta \text{ 上二元有理式}$$

显然

$$x_{00}=x^{(2)}$$

设方程 $y^{n_3}=x_{j'j''}$ 的 n_3 个根为

$$\xi_{j'j''j'''} \quad (j'''=0,1,2,\cdots,n_3-1)$$

其中

$$\xi_{000}=\xi^{(3)}$$

令

$$x_{j'j''j'''}=\varphi_3(\xi_{j'},\xi_{j'j''},\xi_{j'j''j'''}) \text{——} \Delta \text{ 上三元有理式}$$

显然

$$x_{000}=x^{(3)}$$

$$\vdots$$

设方程 $y^{n_k}=x_{j'j''\cdots j(k-1)}$ 的 n_k 个根为

$$\xi_{j'j''\cdots j(k-1)j(k)}\quad (j^{(k)}=0,1,2,\cdots,n_k-1)$$

其中

$$\xi_{00\cdots 00}=\xi^{(k)}$$

令

$$x_{j'j''\cdots j(k-1)j(k)}=\varphi_k(\xi_{j'},\xi_{j'j''},\cdots,\xi_{j'j''\cdots j(k-1)j(k)}) \tag{9.1}$$

显然

$$x_{00\cdots 00}=x^*$$

$x_{j'j''\cdots j(k)}$ 就是与 x^* 同型的迭代根式.

再记多项式

$$F_{j'j''\cdots j(k)}(x)=x-x_{j'j''\cdots j(k)} \tag{9.2}$$

$$F_{j'j''\cdots j(l)}(x)=\prod_{j^{(l+1)}=0}^{n_{l+1}-1}\prod_{j^{(l+2)}=0}^{n_{l+2}-1}\cdots\prod_{j^{(k)}=0}^{n_k-1}(x-x_{j'j''\cdots j(l)j(l+1)j(l+2)\cdots j(k)})$$

$$(l=k-1,k-2,\cdots,1)$$

$$F(x)=\prod_{j'=0}^{n_1-1}\prod_{j''=0}^{n_2-1}\cdots\prod_{j^{(k)}=0}^{n_k-1}(x-x_{j'j''\cdots j(k)})$$

引理9.1 对 $l=k,k-1,k-2,\cdots,1,0$ 有

$$F_{j'j''\cdots j(l)}(x)\equiv\sum_{h=0}^{n_{l+1}n_{l+2}\cdots n_k}\phi_{lh}(\xi_{j'},\xi_{j'j''},\cdots,\xi_{j'j''\cdots j(l)})x^h \tag{9.3}$$

($l=0$ 时左边为 $F(x)$,$l=k$ 时 $\sum$ 的上限为1,$\phi_{0h}\in\Delta$,ϕ_{lh} 为域 Δ 上 l 元多项式(当 $l\geqslant 1$)).

特例:$F(x)$ 为域 Δ 上多项式.

证 用归纳法:$l=k$ 时,由式(9.2)知式(9.3)成立;

设式(9.3)对 l 成立,则

$$F_{j'j''\cdots j(l-1)}(x)\equiv\prod_{j^{(l')}=0}^{n_l-1}F_{j'j''\cdots j(l-1)j(l')}(x)\equiv$$

$$\prod_{j^{(l')}=0}^{n_l-1}\sum_{h=0}^{n_{l+1}n_{l+2}\cdots n_k}\phi_{lh}(\xi_{j'},\xi_{j'j''},\cdots,\xi_{j'j''\cdots j(l')})x^h$$

上式展开合并 x 的同次项,易见 x 的各次幂系数为所有 $\xi_{j'j''\cdots j(l-1)j(l')}(j^{(l')}=0,1,2,\cdots,n_l-1)$ 的对称多项式(系数为 $\xi_{j'}$, $\xi_{j'j''},\cdots,\varepsilon_{j'j''\cdots j(l-1)}$ 的多项式;下同),这些 $\xi_{j'j''\cdots j(l'-1)j(l')}$ 为下述方程的所有根

$$y^{n_l}=x_{j'j''\cdots j(l-1)}$$

故上述对称多项式为上述方程的所有根的基本对称多项式(据 Vieta 定理,其值为 $0,0,\cdots,0,(-1)^{n_l+1}x_{j'j''\cdots j(l-1)}$,最后一个数为

$(-1)^{n_l+1}\varphi_l(\xi_{j'},\xi_{j'j''},\cdots,\xi_{j'j''\cdots j(l-1)}))$ 的多项式，从而 $F_{j'j''\cdots j(l-1)}(x)$ 形如（ψ 为 Δ 上 l 元多项式）

$$F_{j'j''\cdots j(l-1)}(x) \equiv \sum_{h=0}^{n_l n_{l+1} n_{l+2}\cdots n_k} \psi(\xi_{j'},\xi_{j''},\cdots,\xi_{j'j''\cdots j(l-1)},(-1)^{l+1})\cdot \varphi_l(\xi_{j'},\xi_{j'j''},\cdots,\xi_{j'j''\cdots j(l-1)})x^h \triangleq \sum_{h=0}^{n_l n_{l+1}\cdots n_k} \phi_{l-1,h}(\xi_{j'},\xi_{j'j''},\cdots,\xi_{j'j''\cdots j(l-1)})x^h$$

得证式(9.3)对 $l-1$ 成立，证毕.

定理9.1[①] 设 $f(x)$ 在域 Δ 上不可约，且有一根 x^* 可用域 Δ 上多层根式解出，则 $f(x)$ 的任一根为 $F(x)$ 的某一根 $x_{j'j''\cdots j(k)}$（即 $f(x)$ 所有根可用与 x^* 同型的多层根式解出）.

证 由域 Δ 上两多项式 $f(x)$ 与 $F(x)$ 有公共根 x^*（即 $x_{00\cdots0}$），且 $f(x)$ 在 Δ 上不可约，故由引理3.1知 $f(x)$ 的任一根为 $F(x)$ 的某一根 $x_{j'j''\cdots j(k)}$. 证毕.

反之（不必假定多项式 $f(x)$ 在域 Δ 不可约）可证多项式 $F(x)$ 的任一根必为多项式 $f(x)$ 的根（但多项式 $F(x)$ 可有重根）. 当多项式 $f(x)$ 在域 Δ 不可约，更可证 $f(x)$ 与 $F(x)$ 的所有（互不相同的）根集相同，且 $f(x)$ 的所有（单重）根在 $F(x)$ 的重数相同.

从第六章四次方程两个解法所得求根公式可见，根 x^* 用多层根式的表示式并非唯一. 又从上述根 x^* 用多层根式解出的定义，未见得各方根 $\xi^{(1)},\xi^{(2)},\cdots,\xi^{(k)}$ 为 $f(x)$ 所有根在域

$$Q^* = Q(\omega_{n_1},\omega_{n_2},\cdots,\omega_{n_k})$$

上的有理式.

但对二次方程 $x^2+px+q=0$ 的根的公式中的 $\sqrt{p^2-4q}$，易见其为方程二根之差，属二根在域 $\mathbf{Q}$ 上的有理式.

在三次方程的根的公式中，从式(6.2)知公式中两个立方根是方程三根 x_1,x_2,x_3 在域 $Q(\omega_3)$ 上的有理式，而

$$\sqrt{\frac{q^2}{4}+\frac{p^3}{27}} = \left(\frac{x_1+\omega_3x_2+\omega_3^2x_3}{3}\right)^3+\frac{q}{2} = \left(\frac{x_1+\omega_3x_2+\omega_3^2x_3}{3}\right)^3-\frac{x_1x_2x_3}{2}$$

①如解三次方程的 Cardan 公式中的 $f(x)$ 为3次多项式，但 $F(x)$ 为($n_1n_2=2\times3$)6次多项式，易验证 $f(x)$ 的每个根都是 $F(x)$ 的2重根.

(据 Vieta 定理知 $q=-x_1x_2x_3$) 亦属三根在 $Q(\omega_3)$ 上的有理式.

在四次方程的第一解法(Euler) 中,从式(6.10) 知 α,β,γ(三次方程(6.9) 的三根之平方根) 属四次方程四根 x_1,x_2,x_3,x_4 在域 $\mathbf{Q}$ 的有理式;又求方程(6.9) 之根时所要求的立方根、平方根为其三根 $\alpha^2,\beta^2,\gamma^2$ 在域 $Q(\omega_3)$ 的有理式,从而属 x_1,x_2,x_3,x_4 在 $Q(\omega_3)$ 的有理式.

在四次方程解法二(Ferrari) 中,三次方程(6.12) 的三根 $\alpha,\beta,\gamma(\alpha=x_1x_3+x_2x_4,\cdots)$ 属四次方程四根在域 $\mathbf{Q}$ 的有理式;同样求方程(6.12) 的根时所要求的平方根、立方根属 x_1,x_2,x_3,x_4 在域 $Q(\omega_3)$ 的有理式;又在方程(6.18) 中的常数项(据 Vieta 定理)

$$\frac{t^*}{2}+\sqrt{\frac{t^{*2}}{4}-r}=x_1x_2$$

$$\sqrt{\frac{t^{*2}}{4}-r}=x_1x_2-\frac{t^*}{2}$$

而

$$t^*=r=x_4x_3+x_2x_1$$

故 $\sqrt{\frac{t^{*2}}{4}-r}$ 亦属 x_1,x_2,x_3,x_4 在域 $\mathbf{Q}$ 的有理式,类似知

$$\mp\sqrt{\frac{a^2}{4}+t^*-p}=-(x_1+x_2)-\frac{a}{2}=$$

$$-(x_1+x_2)+\frac{1}{2}(x_1+x_2+x_3+x_4)$$

亦为 x_1,x_2,x_3,x_4 在域 $\mathbf{Q}$ 上的有理式;而解方程(6.18)、方程(6.19) 时所要求的(二次方程的) 判别式平方根,实为二次方程二根之差,故亦属 x_1,x_2,x_3,x_4 在域 $\mathbf{Q}$ 上有理式.

又如域 $\mathbf{Q}$ 上方程

$$x^3=4$$

之一根 $x^*=\sqrt[3]{4}$,即

$$x^*=\xi^{(1)}=\sqrt[3]{4}\notin\mathbf{Q}$$
$$(\xi^{(1)})^3=4\in\mathbf{Q}$$

这时 $\xi^{(1)}$ 为方程各根在 $Q(\omega_3)$ 的有理式.

但 x^* 可化成

$$\sqrt[3]{4}=\sqrt[3]{(\sqrt{2})^3\sqrt{2}}=\sqrt[3]{(\sqrt{2})^3}\sqrt[3]{\sqrt{2}}=\sqrt{2}\sqrt[3]{\sqrt{2}}$$

即

$$x^* = \xi^{(1)}\xi^{(2)}$$

其中

$$\xi^{(1)} = \sqrt{2} \notin \mathbf{Q}$$

$$(\xi^{(1)})^2 = 2 \in \mathbf{Q}$$

$$\xi^{(2)} = \sqrt[3]{\sqrt{2}} \notin Q(\sqrt{2})$$

$$(\xi^{(2)})^3 = \sqrt{2} = \xi^{(1)}$$

这时 $\xi^{(1)},\xi^{(2)}(=\sqrt[6]{2})$ 不是方程各根 $\sqrt[3]{4},\sqrt[3]{4}\omega_3,\sqrt[3]{4}\omega_3^2$ 在 $Q(\omega_3)$ 的有理式.

但一般有如下 Abel(阿贝尔)定理.

定理9.2(Abel) 设域 Δ 上多项式 $f(x)$ 的一根 x^* 可用域 Δ 上多层根式表示如前述,其中根指数序列为 $n_1,n_2,\cdots,n_k$,又

$$\omega_{n_1},\omega_{n_2},\cdots,\omega_{n_k} \in \Delta$$

则此根指数序列有子序列

$$n_{a_\alpha},\cdots,n_{a_{k-2}},n_{a_{k-1}},n_{a_k}(=n_k)$$

$$(1 \leqslant a_\alpha < \cdots < a_{k-2} < a_{k-1} < a_k = k)$$

及相应方根列

$$\xi'^{(a_\alpha)},\cdots,\xi'^{(a_{k-2})},\xi'^{(a_{k-1})},\xi'^{(a_k)}(=\xi'^{(k)})$$①

它们是 x^* 在域 Δ 的一个多层根式表示式中的根指数列及相应方根列,且此方根列中各数均可表示为 $f(x)$ 各根在域

$$Q^* = Q(\omega_{n_{a_\alpha}},\cdots,\omega_{n_{a_{k-2}}},\omega_{n_{a_{k-1}}},\omega_{n_k})$$

的有理式.(即在 x^* 的多层根式表示式中不妨设各方根均可表示为 $f(x)$ 诸根在域 Q^* 的有理式.)

先证:

引理9.2 设 $u \notin \Delta,\omega_p \in \Delta,p$ 为素数,$u^p = a \in \Delta$,则对任何 $v \in \Delta(u)\backslash\Delta$ 有:

① 存在 $u' \in \Delta(u)\backslash\Delta$,使 $u'^p \in \Delta,\Delta(u') = \Delta(u)$ 且

$$v = \sum_{j=0}^{p-1} v'_j u'^j,v'_1 = 1,v'_0,v'_2,v'_3,\cdots,v'_{p-1} \in \Delta \tag{9.4}$$

(即在扩域 $\Delta(u') = \Delta(u)$ 上,v 用 u' 的多项式表示式中不妨设一次项系数为1).

①定理证明过程是先确定 $\xi'^{(k)},n_{a_{k-1}},\xi'^{(a_{k-1})},\cdots$.

②设v为域Δ上多项式$P(x)$的根，则u'及式(9.4)中的$v'_0,v'_2,v'_3,\cdots,v'_{p-1}$均可表为$P(x)$的各根在域$Q(\omega_p)$上的有理式.

证①：由引理7.2推论知域Δ上方程$t^p=a$不可约，据定理3.3知v可唯一地表示为

$$v=\sum_{j=0}^{p-1}v_{j'}u^{j'},v_{j'}\in\Delta \tag{9.5}$$

由于$v\notin\Delta$，故有某个$v_k\neq 0(1\leqslant k\leqslant p-1$，因若$v_1=v_2=\cdots=v_{p-1}=0$，则$v=v_0\in\Delta$与$v\notin\Delta$矛盾). 令

$$u'=v_ku^k\in\Delta(u)$$

则

$$u^k=u'v_k^{-1}$$

$$(u')^p=v_k^pu^{kp}=v_k^pa^k\in\Delta$$

若$u'\in\Delta$，由k,p互素，取$k',p'\in\mathbf{Z}$使

$$kk'+pp'=1$$

于是

$$u=u^{kk'}\cdot u^{pp'}=(u'v_k^{-1})^{k'}a^{p'}\in\Delta$$

与假设矛盾，得证$u'\notin\Delta,u'\in\Delta(u)\setminus\Delta$.

对$j=0,1,2,\cdots,p-1$作带余除法(kj除以p)得

$$kj=pq_j+j'\quad(q_j\in\mathbf{Z},j'\in\{0,1,2,\cdots,p-1\})$$

j'为j所唯一确定，因k,p互素，由附录Ⅱ定理1知kj对p取完全剩余类，相应j'遍取$0,1,2,\cdots,p$

$$v_{j'}u^{j'}=v_{j'}u^{kj}u^{-pq_j}=v_{j'}(u'v_k^{-1})^ja^{-q_j}=$$
$$v_{j'}v_k^{-j}a^{-q_j}\cdot(u')^j=v'_j(u')^j$$

其中

$$v'_j=v_{j'}v_k^{-j}a^{-q_j}\in\Delta$$

于是式(9.5)各项分别改为u'的$0\sim p-1$次项(次序不一定相同)，其中已知$v_ku^k=u'$，即$v_1'=1$，易见$v_0'=v_0$不变，故式(9.4)成立.

对任何$\zeta\in\Delta(u)$，同样有

$$\zeta=\sum_{j'=0}^{p-1}\zeta'_j(u')^j\quad(\zeta'_j\in\Delta)$$

且可化成

$$\zeta=\sum_{j=0}^{p-1}\zeta'_j(u')^j\quad(\zeta'_j\in\Delta)$$

得证$\Delta(u)\subset\Delta(u')$；又从$u'=v_ku^k\in\Delta(u)$知$\Delta(u')\subset\Delta(u)$，于

是$\Delta(u')=\Delta(u)$.

证②:由$u'\notin\Delta,(u')^p=v_k^p a^k\in\Delta$,按引理7.2推论知域$\Delta$上方程

$$t^p=v_k^p a^k \tag{9.6}$$

不可约.因

$$P(\sum_{j=0}^{p-1}v_j' u'^j)=P(v)=0$$

故域Δ上t的方程

$$P(\sum_{j=0}^{p-1}v_j' t^j)=0 \tag{9.7}$$

与方程(9.6)有公共根u',方程(9.6)在Δ上不可约,故由引理3.1知其任一根

$$t_s=u'\omega_p^s\quad(s=0,1,2,\cdots,p-1)$$

均为方程(9.7)的根,即$P(x)$有根

$$x_s=\sum_{j=0}^{p-1}v_j'(\omega_p^s u')^j$$

又对$j'=0,1,2,\cdots,p-1$取

$$\sum_{s=0}^{p-1}\omega_p^{-j's}x_s=\sum_{j=0}^{p-1}\sum_{s=0}^{p-1}v_j'(u')^j\omega_p^{s(j-j')}=pv_{j'}'(u')^{j'} \tag{9.8}$$

(因$j\neq j'$时$\sum_{s=0}^{p-1}v_j'(u')^j\omega_p^{s(j-j')}=v_j'(u')^j\sum_{s=0}^{p-1}\omega_p^{(j-j')s}=0$,由于$j-j'$与$p$互素,$(j-j')s$除以$p$所得余数遍取$0,1,2,\cdots,p-1$;而$j=j'$时$\omega_p^{s(j-j')}=\omega_p^0=1$,$\sum_{s=0}^{p-1}\omega_p^{(j-j')s}=\sum_{s=0}^{p-1}1=p$)特别当$j'=1$时因$v_1'=1$,故

$$u'=\frac{1}{p}\sum_{s=0}^{p-1}\omega_p^{-s}x_s$$

代入式(9.8)得

$$v_{j'}'=\frac{1}{p}(u')^{-j'}\sum_{s=0}^{p-1}\omega_p^{-j's}x_s=$$

$$\frac{1}{p}(\frac{1}{p}\sum_{s=0}^{p-1}\omega_p^{-s}x_s)^{-j'}\cdot\sum_{s=0}^{p-1}\omega_p^{-j's}x_s$$

②得证.

证定理9.2 记域

$$\Delta_j=\Delta(\xi^{(1)},\xi^{(2)},\cdots,\xi^{(j)})\quad(j=1,2,\cdots,k)$$

因

$$x^*\in\Delta_k\backslash\Delta_{k-1}=\Delta_{k-1}(\xi^{(k)})\backslash\Delta_{k-1}$$

(不妨设$x^* \notin \Delta_{k-1}$,否则若$x^* \in \Delta_{k-1}$,则x^*的表示式中可省去$\xi^{(k)}$),有

$$(\xi^{(k)})^{n_k} = x^{(k-1)} \in \Delta_{k-1}$$
$$\xi^{(k)} \notin \Delta_{k-1}$$

由引理9.2(其中取u,Δ,p,a,v,P分别为这里的$\xi^{(k)},\Delta_{k-1},n_k,x^{(k-1)},x^*,f$)知存在

$$\xi'^{(k)} \in \Delta_{k-1}(\xi^{(k)}) \backslash \Delta_{k-1}$$

使

$$(\xi'^{(k)})^{n_k} = x'^{(k-1)} \in \Delta_{k-1}$$
$$\Delta_{k-1}(\xi'^{(k)}) = \Delta_{k-1}(\xi^{(k)})$$

且

$$x^* = \sum_{j=0}^{n_k-1} v_j(\xi'^{(k)})^j \quad (v_1 = 1, v_0, v_2, v_3, \cdots, v_{n_k-1} \in \Delta_{k-1}) \tag{9.9}$$

且(又因x^*为域Δ上多项式$f(x)$之根)$\xi'^{(k)}$及$v_0,v_2,v_3,\cdots,v_{n_k-1}$均为$f(x)$诸根在域$Q(\omega_{n_k})$的有理式,于是$x'^{(k-1)} = (\xi'^{(k)})^{n_k}$亦然.

若$x'^{(k-1)}$及$v_0,v_2,v_3,\cdots,v_{n_k-1}$在域$\Delta_{k-1}$的表示式中不含$\xi^{(k-1)}$,$\xi^{(k-2)},\cdots,\xi^{(1)}$,即它们属于域$\Delta$,则所要证的结论成立($x^*$的表示式只含一层根式——$x'^{(k-1)}$的$n_k$次方根$\xi'^{(k)}$,它适合定理结论的要求);若这些表示式不含$\xi^{(k-1)},\xi^{(k-2)},\cdots,\xi^{(k-r)}$,但至少有一数(记为$\beta$)的表示式含$\xi^{(k-r-1)}(r \leqslant k-2)$,则$x^*$的表示式不含$\xi^{(k-1)},\xi^{(k-2)},\cdots,\xi^{(k-r)}$,即(其中$\Delta_0 = \Delta$)

$$x'^{(k-1)}, v_0, v_2, v_3, \cdots, v_{n_k-1} \in \Delta_{k-r-1} = \Delta_{k-r-2}(\xi^{(k-r-1)}) \tag{9.10}$$

其中一数

$$\beta \in \Delta_{k-r-2}(\xi^{(k-r-1)}) \backslash \Delta_{k-r-2}$$

又

$$(\xi^{(k-r-1)})^{n_{k-r-1}} = x^{(k-r-2)} \in \Delta_{k-r-2}$$
$$\xi^{(k-r-1)} \notin \Delta_{k-r-2}$$

再由上引理的①(其中u,Δ,p,a,v分别取这里的$\xi^{(k-r-1)},\Delta_{k-r-2},n_{k-r-1},x^{(k-r-2)},\beta$)知存在

$$\xi'^{(k-r-1)} \in \Delta_{k-r-2}(\xi^{(k-r-1)}) \backslash \Delta_{k-r-2} \tag{9.11}$$

使

$$(\xi'^{(k-r-1)})^{n_{k-r-1}} = x'^{(k-r-2)} \in \Delta_{k-r-2} \tag{9.12}$$

$$\Delta_{k-r-2}(\xi'^{(k-r-1)}) = \Delta_{k-r-2}(\xi^{(k-r-1)}) \tag{9.13}$$

$$\beta = \sum_{j=0}^{n_{k-r-1}-1} v_j'(\xi'^{(k-r-1)})^j \quad (v_1' = 1, v_0', v_2', v_3', \cdots, v'_{n_{k-r-1}-1} \in \Delta_{k-r-2}) \tag{9.14}$$

由式(9.9),式(9.10),式(9.13)知

$$x^* \in \Delta_{k-r-1}(\xi'^{(k)}) = \Delta_{k-r-2}(\xi^{(k-r-1)}, \xi'^{(k)}) = \Delta_{k-r-2}(\xi'^{(k-r-1)}, \xi'^{(k)})$$

因已证 $v_0, v_2, v_3, \cdots, v_{n_k-1}$ 及 $x'^{(k-1)}$ 属于 $f(x)$ 诸根在域 $Q(\omega_{n_k})$ 的有理式,故它们中的 β 亦然.在 β 的上述有理表示式中对 $f(x)$ 的 n 个根进行所有 n 次置换($n!$ 个),设分别得出 $\beta_1=\beta$ 及 $\beta_2, \beta_3, \cdots, \beta_{n!}$,取多项式

$$g(t) = \prod_{s=1}^{n!}(t - \beta_s)$$

易见其中 t 的各次幂的系数为 $f(x)$ 诸根的对称式,可表为 $f(x)$ 诸根的基本对称多项式的(由 Vieta 定理即 $f(x)$ 的各系数(属域 Δ)的)有理式.故 $g(t)$ 为域 Δ 上多项式,β 为其一根.再由上引理的②知式(9.14)中的 $\xi'^{(k-r-1)}$ 及 $v_0', v_2', v_3', \cdots, v'_{n_{k-r-1}-1}$ 均可表为 $g(t)$ 的诸根在 $Q(\omega_{n_{k-r-1}})$ 有理式.又易见 $g(t)$ 的诸根如同 β 可表示为 $f(x)$ 诸根在域 $Q(\omega_{n_k})$ 的有理式,于是 $\xi'^{(k-r-1)}$ 及 $v_0', v_2', v_3', \cdots, v'_{n_{k-r-1}-1}$ 可表示为 $f(x)$ 诸根在域 $Q(\omega_{n_{k-r-1}}, \omega_{n_k})$ 的有理式,故 $x'^{(k-r-2)} = (\xi'^{(k-r-1)})^{n_{k-r-1}}$ 亦然.

对 $x'^{(k-1)}$ 及 $v_0, v_2, v_3, \cdots, v_{n_k-1}$ 中除 β 以外各数(记为 $\widetilde{\beta}$),因它们属于(据式(9.10),式(9.13))域

$$\Delta_{k-r-2}(\xi^{(k-r-1)}) = \Delta_{k-r-2}(\xi'^{(k-r-1)})$$

故每个 $\widetilde{\beta}$ 可表示为

$$\widetilde{\beta} = \sum_{j=0}^{n_{k-r-1}-1} \widetilde{v}_j(\xi'^{(k-r-1)})^j \quad (\widetilde{v}_j \in \Delta_{k-r-2}) \tag{9.15}$$

据前述每个 $\widetilde{\beta}$ 表示为 $f(x)$ 的 n 个根在域 $Q(\omega_{n_k})$ 的有理式中,对此 n 个根进行所有 n 次置换得 $\widetilde{\beta}_1 = \widetilde{\beta}, \widetilde{\beta}_2, \widetilde{\beta}_3, \cdots, \widetilde{\beta}_{n!}$.取多项式

$$\widetilde{g}(t) = \prod_{s=1}^{n!}(t - \widetilde{\beta}_s)$$

同样 $\widetilde{g}(t)$ 为域 Δ 上多项式,它有根 $\widetilde{\beta}$,故

$$\widetilde{g}\left(\sum_{j=0}^{n_{k-r-1}-1} \widetilde{v}_j(\xi'^{(k-r-1)})^j\right) = \widetilde{g}(\widetilde{\beta}) = 0$$

记其左边为 $G(\xi'^{(k-r-1)})$,由$\widetilde{g}(t)$ 为域 Δ 上多项式,$\widetilde{v}_j \in \Delta_{k-r-2}$,故 $G(\xi'^{(k-r-1)})$ 是 $\xi'^{(k-r-1)}$ 在域 Δ_{k-r-2} 的多项式,又按方程(9.12)知 $\xi'^{(k-r-1)}$ 是域 Δ_{k-r-2} 方程

$$t^{n_{k-r-1}} = x'^{(k-r-2)} \tag{9.16}$$

之一根,而 $\xi'^{(k-r-1)} \notin \Delta_{k-r-2}$,$\omega_{n_{k-r-1}} \in \Delta \subset \Delta_{k-r-2}$,故方程(9.16)在域 Δ_{k-r-2} 不可约(据引理7.2推论),它与域 Δ_{k-r-2} 多项式 $G(t)$ 有公共根 $\xi'^{(k-r-1)}$,于是由引理3.1知方程(9.16)的任一根

$$\xi'^{(k-r-1)}\omega^u_{n_{k-r-1}} \quad (u = 0,1,2,\cdots,n_{k-r-1}-1)$$

亦为 $G(t)$ 的根,即

$$G(\xi'^{(k-r-1)}\omega^u_{n_{k-r-1}}) = 0$$

$$\widetilde{g}\Big(\sum_{j=0}^{n_{k-r-1}-1} \widetilde{v}_j(\xi'^{(k-r-1)}\omega^u_{n_{k-r-1}})^j\Big) = 0$$

故得$\widetilde{g}(t)$ 的根

$$\sum_{j=0}^{n_{k-r-1}-1} \widetilde{v}_j(\xi'^{(k-r-1)}\omega^u_{n_{k-r-1}})^j \triangleq \widetilde{\beta}_{m_u} \tag{9.17}$$

$$\widetilde{\beta}_{m_u} \in \{\widetilde{\beta}_s : s = 1,2,\cdots,n!\}$$
$$(u = 0,1,2,\cdots,n_{k-r-1}-1)$$

但易见$\widetilde{g}(t)$ 的每个根$\widetilde{\beta}_{m_u}$ 如同$\widetilde{\beta}$ 为 $f(x)$ 的 n 个根在域 $Q(\omega_{n_k})$ 上的有理式,从式(9.17)得

$$\sum_{u=0}^{n_{k-r-1}-1} \omega^{-j'u}_{n_{k-r-1}}\Big[\sum_{j=0}^{n_{k-r-1}-1} \widetilde{v}_j(\xi'^{(k-r-1)}\omega^u_{n_{k-r-1}})^j\Big] = \sum_{u=0}^{n_{k-r-1}-1} \omega^{-j'u}_{n_{k-r-1}}\widetilde{\beta}_{m_u}$$
$$(j' = 0,1,2,\cdots,n_{k-r-1}-1)$$

$$\text{左边} = \sum_{j=0}^{n_{k-r-1}-1} \widetilde{v}_j(\xi'^{(k-r-1)})^j\Big(\sum_{u=0}^{n_{k-r-1}-1} \omega^{(j-j')u}_{n_{k-r-1}}\Big) = n_{k-r-1}\widetilde{v}_{j'}(\xi'^{(k-r-1)})^{j'}$$

于是

$$\widetilde{v}_{j'} = \frac{1}{n_{k-r-1}}(\xi'^{(k-r-1)})^{-j'}\sum_{u=0}^{n_{k-r-1}-1} \omega^{-j'u}_{n_{k-r-1}}\widetilde{\beta}_{m_u}$$

由已证 $\xi'^{(k-r-1)}$ 为 $f(x)$ 的诸根在域 $Q(\omega_{n_{k-r-1}},\omega_{n_k})$ 上的有理式,每个$\widetilde{\beta}_{m_u}$(即某个$\widetilde{\beta}_s$)为 $f(x)$ 的诸根在域 $Q(\omega_{n_k})$ 上的有理式,故每个$\widetilde{v}_j$ 为 $f(x)$ 的诸根在域 $Q(\omega_{n_{k-r-1}},\omega_{n_k})$ 上的有理式.

上述论证说明,定理结论所述根指数列的子序列最后第二项 $n_{a_{k-1}} = n_{k-r-1}$,相应的方根 $\xi'^{(a_{k-1})} = \xi'^{(k-r-1)}$.

若$k-r-1>1$,同样继续论证一讨论$x'^{(k-r-2)}$及上述两情况所得v'_j及$\widetilde{v}_j(j=0,1,2,\cdots,n_{k-r-1}-1$(即$n_{a_{k-1}}-1$))在域$\Delta_{k-r-2}$的表示式,可得相应的$\xi'^{(a_{k-2})}$表示为$f(x)$诸根在域$Q(\omega_{n_{a_{k-2}}},\omega_{n_{a_{k-1}}},\omega_{n_k})$上的有理式,且上述$v_j'$,$\widetilde{v}_j$的表示式中各项系数亦然;又$\xi^{(1)}$,$\xi^{(2)},\cdots,\xi^{(a_{k-2}-1)},\xi'^{(a_{k-2})},\xi'^{(a_{k-1})},\xi'^{(k)}$仍适合"$x^*$可用域$\Delta$上多层根式表示"的定义的要求……最后可得$\xi'^{(a_\alpha)},\cdots,\xi'^{(a_{k-2})},\xi'^{(a_{k-1})}$,$\xi'^{(k)}$适合上述定义要求,且它们都可表示为$f(x)$诸根在域$Q^*$上的有理式.

第十章 判定代数方程可用多层二次根式解出的准则

在探讨圆规直尺作图可能性时需判定代数方程能否用多层二次根式解出,这是上节的特殊情况,虽然这可据下文第十二章Galois理论基本定理用Galois群进行判定,但由于Galois群一般难以求出,然而由二次根式特殊性,可以不涉及Galois群而据方程根域维数判定.这比用Galois群较为不那么抽象且较易检验.由此容易论证古代几何三大作图难题的不可能性及等分圆的可能性.

先证明几个引理.

引理10.1 设域Δ的扩域Ω,$(\Omega:\Delta)=2$,则存在$\xi\in\Omega\backslash\Delta$,使$\xi^2\in\Delta,\Omega=\Delta(\xi)$.

证 任取$u\in\Omega\backslash\Delta$,由$(\Omega:\Delta)=2$知$1,u,u^2$在$\Delta$上线性相关,即存在不全为0的$c_0,c_1,c_2\in\Delta$,使

$$c_0+c_1u+c_2u^2=0$$

若$c_2=0$,则

$$c_0+c_1u=0$$

$$u=-\frac{c_0}{c_1}\in\Delta\subset\Omega$$

与$u\in\Omega\backslash\Delta$矛盾.故证得$c_2\neq0$,从而

$$c_2(u+\frac{c_1}{2c_2})^2+\frac{4c_0c_2-c_1^2}{4c_2}=0 \qquad (10.1)$$

令

$$\xi=u+\frac{c_1}{2c_2}\in\Omega$$

则 $\xi \notin \Delta$(否则 $u=\xi-\dfrac{c_1}{2c_2}\in\Delta\subset\Omega$,与 $u\in\Omega\backslash\Delta$ 矛盾),故 $\xi\in\Omega\backslash\Delta$. 又由式(10.1)知

$$\xi^2=\frac{c_1^2-4c_0c_2}{4c_2^2}\in\Delta$$

对任何 $u'\in\Omega$,由 $(\Omega:\Delta)=2$ 知 $1,\xi,u'$ 在 Δ 上线性相关,即存在不全为 0 的 $a,b,c\in\Delta$ 使

$$a+b\xi+cu'=0$$

现证 $c\neq0$:否则 $c=0$,从而 $a+b\xi=0$. 这时若 $b=0$,则 $a=0$,与 a,b,c 不全为 0 矛盾;若 $b\neq0$,则 $\xi=-\dfrac{a}{b}\in\Delta\subset\Omega$,与上述 $\xi\in\Omega\backslash\Delta$ 矛盾. 于是得证 $c\neq0$,从而

$$u'=-\frac{a}{c}-\frac{b}{c}\xi\in\Delta(\xi)$$

因 u' 为 Ω 中任何数,故

$$\Omega\subset\Delta(\xi)$$

反之由 $\xi\in\Omega,\Delta\subset\Omega$ 知

$$\Delta(\xi)\subset\Omega(\xi)=\Omega$$

于是

$$\Omega=\Delta(\xi)$$

引理 10.2　设域 Δ 上多项式 $f(x)$ 的根域为 Ω, $(\Omega:\Delta)=2^m(m\in\mathbf{Z}^+)$,则存在域 $\Delta_1,\Delta\subset\Delta_1\subset\Omega,(\Delta_1:\Delta)=2$.

证　$f(x)$ 在 Δ 上的不可约因式不可能全为一次因式(否则 $f(x)$ 所有根属于 $\Delta,\Omega=\Delta,(\Omega:\Delta)=1$ 与假设矛盾). 若 $f(x)$ 有在 Δ 不可约的部分因式为一次因式,则约去这些一次因式(即除去在 Δ 上的所有根)后所得多项式的根域仍与在 Δ 上 $f(x)$ 的根域 Ω 相同,故不妨设 $f(x)$ 所有根不属 Δ. 取 $f(x)$ 在 Δ 上的一个不可约因式 $p(x)$,不妨设其最高次项系数为 1,由定理 4.2 推论 2 知 $p(x)$ 的次数 n' 整除 $(\Omega:\Delta)=2^m$,故 $n'=2^k(k\in\{1,2,\cdots,m\})$. 对 k 用归纳法证引理结论成立:

当 $k=1$ 时,$p(x)$ 为 Δ 上二次不可约多项式,设其一根为 α,据定理 3.3 推论知 $(\Delta(\alpha_1):\Delta)=2,\Delta\subset\Delta(\alpha)\subset\Omega$,可取 $\Delta_1=\Delta(\alpha)$,故引理结论成立;

当 $k>1$ 时,设引理结论对任何正整数 $k'<k$ 成立,证引理结论对 k 成立:设 $p(x)$ 之根为 $\alpha_1,\alpha_2,\cdots,\alpha_{n'}$,先证,可取 $c\in\Delta$,使对任何

$i_1, i_2 \in \{1, 2, \cdots, n'\}$ 有

$$\alpha_{i_1}\alpha_{i_2} + c(\alpha_{i_1} + \alpha_{i_2}) \notin \Delta \tag{10.2}$$

否则对任何 $c \in \Delta$,存在相应的 $\alpha_{i_1}, \alpha_{i_2}$ 使

$$\alpha_{i_1}\alpha_{i_2} + c(\alpha_{i_1} + \alpha_{i_2}) \in \Delta$$

但 Δ 中的 c 有无限个,而组合 $\alpha_{i_1}, \alpha_{i_2}$ 只有有限个,故必有 c 的两值 $c_1 \neq c_2$,使相应的 $\alpha_{i_1}, \alpha_{i_2}$ 相同,于是得

$$\alpha_{i_1}\alpha_{i_2} + c_1(\alpha_{i_1} + \alpha_{i_2}) = \delta_1 \in \Delta$$

$$\alpha_{i_1}\alpha_{i_2} + c_2(\alpha_{i_1} + \alpha_{i_2}) = \delta_2 \in \Delta$$

因 $c_1 \neq c_2$ 可解得

$$\alpha_{i_1} + \alpha_{i_2} = \frac{\delta_1 - \delta_2}{c_1 - c_2} \triangleq a \in \Delta$$

$$\alpha_{i_1}\alpha_{i_2} = \delta_1 - c_1\frac{\delta_1 - \delta_2}{c_1 - c_2} \triangleq b \in \Delta$$

从而 $\alpha_{i_1}, \alpha_{i_2}$ 为 Δ 上二次方程

$$x^2 - ax + b = 0$$

之二根,而

$$x^2 - ax + b = (x - \alpha_{i_1})(x - \alpha_{i_2})$$

为 $p(x)$ 在 Δ 上的二次因式,但由 $k > 1$ 知 $p(x)$ 次数 $n' = 2^k > 2$,与 $p(x)$ 在 Δ 上不可约矛盾,取适合方程(10.1)的 c 作多项式

$$f_1(x) = \prod_{1 \leq i_1 < i_2 \leq n'} [x - \alpha_{i_1}\alpha_{i_2} - c(\alpha_{i_1} + \alpha_{i_2})]$$

其任一根 $\alpha_{i_1}\alpha_{i_2} + c(\alpha_{i_1} + \alpha_{i_2}) \notin \Delta$,又易见其 x 的各次幂的系数对 $\alpha_1, \alpha_2, \cdots, \alpha_n$ 对称,故为 $\alpha_1, \alpha_2, \cdots, \alpha_n$ 的基本对称多项式(据 Vieta 定理,即 $p(x)$ 的各项系数乘 ± 1)的有理式,$p(x)$ 的各项系数属 Δ,故 $f_1(x)$ 的 x 的各次幂之系数属 Δ,即 $f_1(x)$ 为域 Δ 上多项式. 因其各根属 Ω,故其在 Δ 上的根域 Ω_1 含于 Ω. 又 $f_1(x)$ 的次数(即其根个数)

$$C_n^2 = \frac{n'(n'-1)}{2} = \frac{2^k(2^k-1)}{2} = 2^{k-1}(2^k-1) = 2^{k-1}q$$

其中

$$q = 2^k - 1(\text{奇数})$$

取 $f_1(x)$ 的一个在 Δ 上不可约因式 $p_1(x)$,因 $(\Omega:\Delta) = 2^m$,$\Omega_1 \subset \Omega$,由定理 3.2 知 $(\Omega_1:\Delta)$ 也是 2 的正整数次幂,于是如同 $p(x)$ 一样,$p_1(x)$ 的次数

$$n_1 = 2^{k_1} \quad (k_1 \in \mathbf{Z}^+)$$

但它不大于$f_1(x)$的次数$2^{k-1}q$(q为奇数),故$k_1 \leqslant k-1$,于是对$f_1(x)$按归纳假设知,存在域Δ'_1,$\Delta \subset \Delta'_1 \subset \Omega_1$,$(\Delta'_1:\Delta)=2$,但$\Omega_1 \subset \Omega$,故$\Delta \subset \Delta'_1 \subset \Omega$,引理结论对$k$成立.证毕.

现引述本章标题所述准则.

定理10.1 设域Δ上不可约方程$f(x)=0$的根域为Ω,其一根x^*可用上节所述(特殊情况)多层二次根式表示(由定理9.1即$f(x)$的所有根可用同型的多层二次根式表示),其充要条件为

$$(\Omega:\Delta)=2^{k'} \quad (k' \in \mathbf{Z}^+)$$

证 $\Leftarrow$:上节中x^*可用域Δ上多层根式表示定义中$n_1=n_2=\cdots=n_k=2$,故对$r=1,2,\cdots,k$有

$$(\xi^{(r)})^2=x^{(r-1)} \in \Delta(\xi^{(1)},\xi^{(2)},\cdots,\xi^{(r-1)})$$
$$\xi^{(r)} \notin \Delta(\xi^{(1)},\xi^{(2)},\cdots,\xi^{(r-1)})$$

又因$\omega_2=-1 \in \mathbf{Q} \subset \Delta$,据引理7.2推论(③⇔①)知$\Delta(\xi^{(1)},\xi^{(2)},\cdots,\xi^{(r-1)})$上方程

$$y^2=x^{(r-1)}$$

不可约,再据定理3.1及定理3.3推论知

$$(\Delta(\xi^{(1)},\xi^{(2)},\cdots,\xi^{(k)}):\Delta)=$$
$$\prod_{r=1}^{k}(\Delta(\xi^{(1)},\xi^{(2)},\cdots,\xi^{(r-1)},\xi^{(r)}):\Delta(\xi^{(1)},\xi^{(2)},\cdots,\xi^{(r-1)}))=$$
$$\prod_{r=1}^{k}2=2^k$$

(其中$r=1$时$\Delta(\xi^{(1)},\xi^{(2)},\cdots,\xi^{(r-1)})$表$\Delta$).

因x^*的在域Δ上的多层二次根式表示式中根指数列全由2组成,其在定理9.2(Abel)所述子序列亦然,又$\omega_2=-1 \in \mathbf{Q}$,故Abel定理中的域$Q^*=\mathbf{Q}$.于是不妨设二次方根$\xi^{(1)},\xi^{(2)},\cdots,\xi^{(k)}$可表示为多项式$f(x)$诸根$x_1,x_2,\cdots,x_n$在域$\mathbf{Q}$的有理式,又因$\mathbf{Q} \subset \Delta$,故

$$\xi^{(1)},\xi^{(2)},\cdots,\xi^{(k)} \in Q(x_1,x_2,\cdots,x_n) \subset \Delta(x_1,x_2,\cdots,x_n)=\Omega$$

从而

$$x^* \in \Delta(\xi^{(1)},\xi^{(2)},\cdots,\xi^{(k)}) \subset \Delta(x_1,x_2,\cdots,x_n)=\Omega$$

但域$\Delta(\xi^{(1)},\xi^{(2)},\cdots,\xi^{(k)})$未包含多项式$f(x)$的全部根.设$f(x)$尚有根$x^{**}$不属此域,$x^{**}$在域$\Delta$上的多层二次根式表示式可视为在域$\partial=\Delta(\xi^{(1)},\xi^{(2)},\cdots,\xi^{(k)})$上多层二次根式表示式.设其中二次方

根列为$\xi'^{(1)},\xi'^{(2)},\cdots,\xi'^{(m)}$①,同样可证

$$(\partial(\xi'^{(1)},\xi'^{(2)},\cdots,\xi'^{(m)}):\partial)=2^m$$

于是域$\partial(\xi'^{(1)},\xi'^{(2)},\cdots,\xi'^{(m)})(\supset\Delta(\xi^{(1)},\xi^{(2)},\cdots,\xi^{(k)})$所含$f(x)$的根数有增加,而

$$\begin{aligned}&(\partial(\xi'^{(1)},\xi'^{(2)},\cdots,\xi'^{(m)}):\Delta)=\\&(\partial(\xi'^{(1)},\xi'^{(2)},\cdots,\xi'^{(m)}):\partial)\cdot(\Delta(\xi^{(1)},\xi^{(2)},\cdots,\xi^{(k)}):\Delta)=\\&2^m\cdot 2^k=2^{m+k}=2^{m+k}\end{aligned}$$

同样有$\xi'^{(1)},\xi'^{(2)},\cdots,\xi'^{(m)}\in\Omega$. 于是

$$(\Delta(\xi^{(1)},\xi^{(2)},\cdots,\xi^{(k)},\xi'^{(1)},\xi'^{(2)},\cdots,\xi'^{(m)}):\Delta)=2^{m+k}$$

$$\xi^{(1)},\xi^{(2)},\cdots,\xi^{(k)},\xi'^{(1)},\xi'^{(2)},\cdots,\xi'^{(m)}\in\Omega$$

继续同样论证,最后可得一域

$$\Delta(\xi^{(1)},\xi^{(2)},\cdots,\xi^{(k)},\xi'^{(1)},\xi'^{(2)},\cdots,\xi'^{(m)},\xi''^{(1)},\xi''^{(2)},\cdots,\xi''^{(m')},\cdots)$$

含有多项式$f(x)$的所有根$x_1,x_2,\cdots,x_n$,且

$$\begin{aligned}&(\Delta(\xi^{(1)},\xi^{(2)},\cdots,\xi^{(k)},\xi'^{(1)},\xi'^{(2)},\cdots,\xi'^{(m)},\xi''^{(1)},\xi''^{(2)},\cdots,\\&\xi''^{(m')},\cdots):\Delta)=2^{k'}\end{aligned}\tag{10.3}$$

又所有平方根

$$\xi^{(1)},\xi^{(2)},\cdots,\xi^{(k)},\xi'^{(1)},\xi'^{(2)},\cdots,\xi'^{(m)},\xi''^{(1)},\xi''^{(2)},\cdots,\xi''^{(m')},\cdots\in\Omega$$

于是

$$\begin{aligned}\Omega=\Delta(x_1,x_2,\cdots,x_n)\subset&\\&\Delta(\xi^{(1)},\cdots,\xi^{(k)},\xi'^{(1)},\cdots,\xi'^{(m)},\xi''^{(1)},\cdots,\xi''^{(m')},\cdots)\end{aligned}$$

又因$\Delta\subset\Omega$,故

$$\begin{aligned}&\Delta(\xi^{(1)},\cdots,\xi^{(k)},\xi'^{(1)},\cdots,\xi'^{(m)},\xi''^{(1)},\cdots,\xi''^{(m')},\cdots)\subset\\&\Omega(\xi^{(1)},\cdots,\xi^{(k)},\xi'^{(1)},\cdots,\xi'^{(m)},\xi''^{(1)},\cdots,\xi''^{(m')},\cdots)=\Omega\end{aligned}$$

从而

$$\Omega=\Delta(\xi^{(1)},\cdots,\xi^{(k)},\xi'^{(1)},\cdots,\xi'^{(m)},\xi''^{(1)},\cdots,\xi''^{(m')},\cdots)$$

再由式(10.3)得

$$(\Omega:\Delta)=2^{k'}$$

$\Rightarrow$:当$(\Omega:\Delta)=2^{k'}$,由引理10.2取域Δ_1,使$\Delta\subset\Delta_1\subset\Omega$,$(\Delta_1:\Delta)=2$. 由定理3.1得

①如域$\mathbf{Q}$上多项式x^4-3的一根$\sqrt[4]{3}$可表示为:$(\xi^{(1)})^2=3,\xi^{(1)}=\sqrt{3},(\xi^{(2)})^2=\xi^{(1)},\xi^{(2)}=\sqrt{\xi^{(1)}}=\sqrt[4]{3}$,易见域$Q(\xi^{(1)},\xi^{(2)})=Q(\sqrt[4]{3})$为实域,而$x^4-3$尚有虚根$\pm\sqrt[4]{3}\mathrm{i}\notin Q(\xi^{(1)},\xi^{(2)})$,但$(\pm\sqrt[4]{3}\mathrm{i})^2=-\sqrt{3}\in Q(\xi^{(1)},\xi^{(2)})$,故它们可在域$Q(\xi^{(1)},\xi^{(2)})$表示为$(\xi'^{(1)})^2=-\sqrt{3},\xi'^{(1)}=\pm\sqrt[4]{-3}=\pm\sqrt[4]{3}\mathrm{i}$.

$$(\Omega:\Delta)=(\Omega:\Delta_1)\cdot(\Delta_1:\Delta)$$

即

$$2^{k'}=(\Omega:\Delta_1)\cdot 2$$

从而

$$(\Omega:\Delta_1)=\frac{2^{k'}}{2}=2^{k'-1}$$

以Δ_1代替Δ,同样取域Δ_2,使$\Delta_1\subset\Delta_2\subset\Omega$,$(\Delta_2:\Delta_1)=2$;继续取域$\Delta_3$,使$\Delta_2\subset\Delta_3\subset\Omega$,$(\Delta_3:\Delta_2)=2$,…,最后得

$$\Delta\subset\Delta_1\subset\Delta_2\subset\Delta_3\subset\cdots\subset\Delta_{k'-1}\subset\Omega$$

$$(\Delta_1:\Delta)=(\Delta_2:\Delta_1)=(\Delta_3:\Delta_2)=\cdots=(\Delta_{k'-1}:\Delta_{k'-2})=(\Omega:\Delta_{k'-1})=2$$

由引理10.1取$\xi^{(1)}\in\Delta_1\backslash\Delta$使

$$\Delta_1=\Delta(\xi^{(1)}),(\xi^{(1)})^2\triangleq x^{(0)}\in\Delta,\xi^{(1)}\notin\Delta$$

再由引理10.1取$\xi^{(2)}\in\Delta_2\backslash\Delta_1$,使

$$\Delta_2=\Delta_1(\xi^{(2)})=\Delta(\xi^{(1)},\xi^{(2)})$$

$$(\xi^{(2)})^2\triangleq x^{(1)}\in\Delta_1=\Delta(\xi^{(1)})$$

$$\xi^{(2)}\notin\Delta_1=\Delta(\xi^{(1)})$$

再由引理10.1取$\xi^{(3)}\in\Delta_3\backslash\Delta_2$,使

$$\Delta_3=\Delta_2(\xi^{(1)})=\Delta(\xi^{(1)},\xi^{(2)},\xi^{(3)})$$

$$(\xi^{(3)})^2\triangleq x^{(2)}\in\Delta_2=\Delta(\xi^{(1)},\xi^{(2)})$$

$$\xi^{(3)}\notin\Delta_2=\Delta(\xi^{(1)},\xi^{(2)})$$

$$\vdots$$

最后由引理10.1取$\xi^{(k')}\in\Omega\backslash\Delta_{k'-1}$,使

$$\Omega=\Delta_{k'-1}(\xi^{(k')})=\Delta(\xi^{(1)},\xi^{(2)},\cdots,\xi^{(k')})$$

$$(\xi^{(k')})^2\triangleq x^{(k'-1)}\in\Delta_{k'-1}=\Delta(\xi^{(1)},\xi^{(2)},\cdots,\xi^{(k'-1)})$$

$$\xi^{(k')}\notin\Delta_{k'-1}=\Delta(\xi^{(1)},\xi^{(2)},\cdots,\xi^{(k'-1)})$$

因$f(x)$所有根$x_j(j=1,2,\cdots,n)$属于Ω,故

$$x_j\in\Delta(\xi^{(1)},\xi^{(2)},\cdots,\xi^{(k')})$$

可用Δ上多层二次根式表示.

推论 若域Δ上不可约多项式$f(x)$的次数含奇素数约数p,则$f(x)$任一根不能用Δ上多层二次根式表示.

(否则$(\Omega:\Delta)=2^k,k\in\mathbf{Z}^+$,而据定理4.2推论2知$2^k$可被奇素数$p$整除,得矛盾).

考察域Δ上的三次方程$f(x)=0$.

若它在域Δ不可约,则由上推论知此方程的任一根不能用域Δ上多层二次根式表示.

若方程$f(x)=0$在域Δ上可约,则$f(x)$可分解为域Δ上三个一次因式之积,或分解为域Δ上一个一次因式与域Δ上一个不可约二次因式之积,易见此时方程$f(x)=0$的三根属于域Δ或可用域Δ上二次根式表示.

于是得:

定理 10.2 域Δ上三次方程的一根属域Δ或可用域Δ上多层二次根式表示的充要条件为:此方程在域Δ至少有一根.

对域$\mathbf{Q}$上的三次方程,通过去分母,不妨设各项系数为整数,易用综合除法检验上述充要条件.

对域Δ上一般四次方程,可有简单准则判定其所有根能否用Δ上多层二次根式表示或属于Δ.

定理 10.3 要域Δ上四次方程

$$x^4+ax^3+px^2+qx+r=0 \tag{10.4}$$

的所有根x_1,x_2,x_3,x_4属于域Δ或能用Δ上多层二次根式表示,其充要条件为:方程(10.4)的 Lagrange 预解方程(方程(6.12))

$$t^3-pt^2+(aq-4r)t+(4rp-q^2-ra^2)=0$$

在Δ上至少有一根.

证 $\Rightarrow$:当方程(6.12)在Δ上有一根t^*,则从方程(6.18)、方程(6.19)分别解出x_1,x_2与x_3,x_4,知此四根为(当$\sqrt{\frac{t^{*2}}{4}-r}\neq 0$,注意式(6.20)、式(6.21))$\sqrt{\frac{t^{*2}}{4}-r}$在$\Delta$的有理式的平方根的有理式;或(当$\sqrt{\frac{t^{*2}}{4}-r}=0$)为$\sqrt{\frac{a^2}{4}+t^*-p}$在$\Delta$上的有理式的平方根的有理式.故四根可用$\Delta$上的多层二次根式表示或属于$\Delta$(当特殊情况——上述两平方根属于$\Delta$,即被开方数为$\Delta$上的数的平方).

$\Leftarrow$:若x_1,x_2,x_3,x_4属于Δ或可用Δ上多层二次根式表示,则从方程(6.18),方程(6.19)分别为x_1,x_2与x_3,x_4所适合的二次方程,据 Vieta 定理知

$$x_1x_2+x_3x_4=\left(\frac{t^*}{2}+\sqrt{\frac{t^{*2}}{4}-r}\right)+\left(\frac{t^*}{2}-\sqrt{\frac{t^{*2}}{4}-r}\right)=t^*$$

从而方程(6.12)的一根x^*属于域Δ或可用域Δ多层二次根式表

示. 这时若方程(6.12)在域Δ不可约,则按定理10.1推论知其所有根不能用域Δ多层二次根式表示,更不属于域Δ,与上结论矛盾,故方程(6.12)必在域Δ可约,从而必在域Δ有一次因式,即在域Δ有根.

第十一章　直尺、圆规作图的可能性

先把几何作图问题化为解代数方程的问题.

首先建立坐标系[①],确定题目已知图形的已知量(非具体数值,而是用图中无向线段(长)、有向线段(代数值)表示,但角除外),如圆的半径,已知线段,两已知点(特定点,如两直线交点、线段端点,非任意点……)的距离,已知点与已知直线的距离,已知点的坐标(易用圆规、直尺作出表示点与直线距离的线段,作出表示点的坐标的两有向线段),已知直线在坐标轴的截距(两有向线段)……取含上述所有已知量的最小数域Δ,则易见图中所有圆及直线的方程(两点式方程(当直线上有二特定点)或截距式方程(含特殊情况:$x=a$或$y=b$——与坐标轴平行的直线的方程))的各项系数亦属于Δ.

在用直尺、圆规作图过程中,不外是:① 过两已知点作直线,易见所得直线各项系数亦属Δ,从而其在坐标轴上的截距亦然;② 以一已知点为圆心以上述无向线段之长或有向线段之模为半径作圆,所得圆之方程各项系数亦属Δ;③ 取任二线(直线与圆,原有的线或作得之线)交点,则两直线之交点坐标仍属Δ(从解两直线方程所构成的方程组时只有加、减、乘、除运算得知),直线与圆交点坐标(解二元一次方程及二元二次方程构成的方程组所得)为含有Δ中之数的平方根(实根)之有理式;两圆交点坐标(解两圆方程构成的方程组,消去两方程的二次项实际亦得二元一次方程与二元二次方程构成的方程组)亦然,从而这些交点及原已知点中任两点之距离亦然,继续进行作图——作直线,圆,取交点,……,可见所得直线,圆的方程的各项系数,直线在坐标轴上的截距,所得交点及圆心之坐标亦必属于Δ或可用Δ上多层(实)二次根式表示,所

①因实际作图与坐标系选择无关,故为检验下文所述(可作图)的充要条件方便,应选坐标系,使尽可能多的点、线落在坐标轴上.

得各点与原已知点中任两点之距离亦然.如作图过程中可取任意长(如当作线段中垂线时,取大于已知线段长的一半之长为半径作圆).因所取之长必与作图所得结果(如上述中垂线)无关,故不妨设此长属于Δ;或可取平面任一点,不妨设此点两坐标属Δ;或可取直线,圆上任一点,不妨设此点为直线或圆与坐标轴平行线(在另一坐标轴的截距属Δ,所得此平行线方程最简单,亦易用直尺、圆规作出)的交点.于是易见此点的坐标亦属Δ或可用Δ上多层二次(实)根式表示.

故所求作的图形可作(用直尺、圆规,下同)的必要条件为:要求作出的最后图形的一切有(无)向线段的代数值(长)属于Δ或可用Δ上多层二次(实)根式表示.

反之证明,若一有向线段之代数值,或无向线段之长属于Δ或可用Δ上的多层二次(实)根式表示,则此线段可作.

因这有(无)向线段用已知线段$a,b,c,d,e,\cdots$的表示式的单位必为长度单位,故此表示式的各项必为一次式,即不能为ab,$\frac{a}{b}$(其单位分别为长度单位之平方,纯数1)$\cdots$,可以为例如$\frac{2ab}{\sqrt{2}c}$,$\frac{abc}{de}$(分子次数比分母次数大1),$\sqrt{a^2+bc-\frac{ade}{\sqrt{bc}}}$(被开方数每项为二次).这些式子可以看成下列基本运算的复合

$$a\pm b,\alpha a,\sqrt{\alpha}a,\sqrt{ab},\sqrt{a^2\pm b^2},\frac{ab}{c}\quad(\alpha\in\mathbf{Q}^+)$$

如

$$\frac{2ab}{\sqrt{2}c}=\frac{(2a)b}{(\sqrt{2}c)}$$

$$\frac{abc}{de}=\frac{\frac{ab}{d}\cdot c}{e}$$

$$\sqrt{a^2+bc-\frac{ade}{\sqrt{bc}}}=\sqrt{(\sqrt{a^2+\sqrt{bc}^2})^2-(\sqrt{\frac{ad}{\sqrt{bc}}\cdot e})^2}$$

又如

$$\sqrt[4]{abcd}=\sqrt{\sqrt{ab}\sqrt{cd}}$$

$$\sqrt{2+\sqrt{3+\sqrt{5}}}a=\sqrt{(2+\sqrt{3+\sqrt{5}})a\cdot a}=$$

$$\sqrt{(2a+\sqrt{(3a+\sqrt{5}a)a})a}$$

上述已知线段的基本运算式所表线段显然可用圆规,直尺作出.

若能确定所求作图形的一切有(无)向线段①如上述,则所求作图形可作出.

综合上述结论得判定准则.

定理11.1 几何作图题可作的充要条件为:能确定所求作图形的一切有向线段(无向线段)的代数值(长)均属于Δ或可用Δ上多层二次(实)根式表示,其中Δ为含有题目中已知图形的所有已知量的最小数域.

推论 若题目已知图形中所有已知量(有(无)向线段)与其中一个已知量之比a为有理数或可表示为域$\mathbf{Q}$上多层二次实根式,则作图题可作的充要条件为:能确定所求作图形的一切有(无)向线段与a之比为有理数或可表示为域$\mathbf{Q}$上多层二次实根式.实际上不妨设$a=1$(自设长度单位),则上述充要条件为:能确定所求作图形的一切有(无)向线段的代数值(长)为有理数或可表示为域$\mathbf{Q}$上多层二次实根式.

例1 倍立方问题:作一立方体(实际是作等于其棱长之线段),使其体积等于已知立方体的2倍.

解 因所有已知线段:已知立方体对角线长,各面对角线长分别为此立方体棱长的$\sqrt{3}$,$\sqrt{2}$倍,故不妨设此立方体棱长为1.设所求立方体棱长为x,则

$$x^3=2$$

它在域$\mathbf{Q}$不可约,据定理10.1推论知此方程之根$\sqrt[3]{2}$不属于域$\mathbf{Q}$也不能用域$\mathbf{Q}$上多层二次根式表示,于是据定理11.1推论知,长为$\sqrt[3]{2}$的线段及所求立方体不可作.

类似知,若有理数a不是任何有理数之立方,则不可能作一立方体,使其体积为已知立方体的a倍.

例2 化圆为方:作一正方形,使与已知圆面积相等.

---

①如所作之一点用其坐标确定,若已作出坐标轴上表此两坐标之有向线段,显然可作出此点;如所作之一点用其到两已知点之距离确定,若已作出表示此二距离的两线段,易见可作出此点(两圆之交点);作直线,由作图时所连两点之坐标确定,或由其在坐标轴上的截距(有向线段)确定;作圆,由其圆心坐标及半径确定;作一角,可归结为先作其两边所在的直线,或当已确定其一边时再由此角之三角函数(可用单位圆的有向线段,即正弦线、余弦线……表示)所确定,易见已作出单位圆的正弦线、余弦线……时可作出此角的另一边.

先证引理:

引理11.1 代数数α(有理数域$\mathbf{Q}$上代数方程的根)的m次方根($m\in\mathbf{Z}^+$)β亦为代数数;所有代数数构成一个数域.

证 由α适合域$\mathbf{Q}$上代数方程$F(x)=0$,即$F(\alpha)=0$,而$\alpha=\beta^m$,故$F(\beta^m)=0$,即β为域$\mathbf{Q}$上代数方程$F(x^m)=0$之根,故β为代数数.

再证第二结论,只要证对任何两代数数α_1,α_2,则$\alpha_1\pm\alpha_2$,$\alpha_1\alpha_2,\frac{\alpha_1}{\alpha_2}(\alpha_2\neq 0)$亦为代数数.

因α_1为代数数,故α_1为域$\mathbf{Q}$上代数方程$F_1(x)=0$之根,不妨设$F_1(x)$在$\mathbf{Q}$上不可约(否则取$F_1(x)$的以α_1为一根且在$\mathbf{Q}$上不可约的因式代之).设此不可约多项式$F_1(x)$的次数为m_1,又因β为代数数,故β为域$\mathbf{Q}$上代数方程$F_2(x)=0$之根,此方程亦可视为域$Q(\alpha_1)$上的代数方程,亦不妨设$F_2(x)$在域$Q'=Q(\alpha_1)$上不可约,设此在域$Q(\alpha_1)$上不可约的多项式$F_2(x)$的次数为m_2,则由定理3.1及定理3.3推论知

$$(Q(\alpha_1,\alpha_2):\mathbf{Q})=(Q(\alpha_1,\alpha_2):Q(\alpha_1))\cdot(Q(\alpha_1):\mathbf{Q})=(Q'(\alpha_2):Q')\cdot(Q(\alpha_1):\mathbf{Q})=m_2\cdot m_1$$

于是对域$Q(\alpha_1,\alpha_2)$中每一数ξ,m_2m_1+1个数

$$1,\xi,\xi^2,\cdots,\xi^{m_2m_1+1}$$

在域$\mathbf{Q}$上线性相关,存在不全为0的

$$c_0,c_1,c_2,\cdots,c_{m_2m_1+1}\in\mathbf{Q}$$

使

$$c_0+c_1\xi+c_2\xi^2+\cdots+c_{m_2m_1+1}\xi^{m_2m_1+1}=0$$

故ξ为域$\mathbf{Q}$上下述代数方程之根

$$c_0+c_1x+c_2x^2+\cdots+c_{m_2m_1+1}x^{m_2m_1+1}=0$$

即域$Q(\alpha_1,\alpha_2)$的任何数ξ(含$\alpha_1\pm\alpha_2,\alpha_1\alpha_2,\frac{\alpha_1}{\alpha_2}(\alpha_2\neq 0)$)为代数数.

推论 域$\mathbf{Q}$上的多层根式必为代数数.

证 易见任何有理数c(域$\mathbf{Q}$上方程$x-c=0$之根)为代数数.于是第六章中域$\mathbf{Q}$上多层根式中定义中的$x^{(0)}$为代数数,按上引理知$x^{(0)}$的n_1次方根$\xi^{(1)}$为代数数,从而$\xi^{(1)}$在$\mathbf{Q}$上的有理式亦为代数数,其n_2次方根$\xi^{(2)}$亦为代数数,$x^{(2)}$(即$\xi^{(1)}$,$\xi^{(2)}$在$\mathbf{Q}$上的有理式)亦为代数数……最后知x^*亦为代数数.

解例2 问题归结为作等于所求正方形边长的线段x,因已知

线段只是圆的半径,不妨设已知圆半径长为1,则

$$x^2 = \pi \cdot 1^2 = \pi$$

$$x = \sqrt{\pi}$$

但Lindemann(林德曼)在1882年证明π非代数数,从而$\sqrt{\pi}$非代数数(否则$\pi = (\sqrt{\pi})^2$为代数数),故不是域**Q**上多层二次实根式,故线段x不能作出.

反之也不能"化方为圆":作一圆使与已知正方形面积相等.

(与例1类似知可设正方形边长为1,圆半径为r,则$\pi r^2 = 1, r = \sqrt{\frac{1}{\pi}}$,$r$非代数数(否则$\pi = \frac{1}{r^2}$为代数数),线段$r$不可作出.)

下面讨论等分圆周可能性,先证四引理.

引理11.2 设p为素数,$m \in \mathbf{Z}^+$,则分圆方程

$$x^{p^{m-1}(p-1)} + x^{p^{m-1}(p-2)} + x^{p^{m-1}(p-3)} + \cdots + x^{p^{m-1}} + 1 = 0$$

$$\left(\text{左边为 } \Phi(x) = \frac{x^{p^m} - 1}{x^{p^{m-1}} - 1}, \text{有根 } x = \mathrm{e}^{2\pi \mathrm{i}/p^m}\right)$$

在有理数域**Q**不可约.

证 $m = 1$时,由定理7.4知上方程在有理数域**Q**不可约.

$m \geqslant 2$时,作代换$x = y + 1$得

$$\Phi(x) = \Phi(y+1) = \frac{(y+1)^{p^m} - 1}{(y+1)^{p^{m-1}} - 1} =$$

$$\frac{y^{p^m} + \mathrm{C}_{p^m}^1 y^{p^m - 1} + \mathrm{C}_{p^m}^2 y^{p^m - 2} + \cdots + \mathrm{C}_{p^m}^{p^m - 1} y}{y^{p^{m-1}} + \mathrm{C}_{p^{m-1}}^1 y^{p^{m-1} - 1} + \mathrm{C}_{p^{m-1}}^2 y^{p^{m-1} - 2} + \cdots + \mathrm{C}_{p^{m-1}}^{p^{m-1} - 1} y} =$$

$$\frac{y^{p^m - 1} + \mathrm{C}_{p^m}^1 y^{p^m - 2} + \mathrm{C}_{p^m}^2 y^{p^m - 3} + \cdots + \mathrm{C}_{p^m}^{p^m - 1}}{y^{p^{m-1} - 1} + \mathrm{C}_{p^{m-1}}^1 y^{p^{m-1} - 2} + \mathrm{C}_{p^{m-1}}^2 y^{p^{m-1} - 3} + \cdots + \mathrm{C}_{p^{m-1}}^{p^{m-1} - 1}}$$

由附录Ⅱ引理1知分子(分母)除最高次项系数1不能被p^m(不能被p^{m-1})整除外,其余各项可被p^m(被p^{m-1})整除,故可被p整除,实际进行竖式除法易见商式(即$\Phi(y+1)$)除最高次项系数1不能被p整除外,其余各项系数均可被p整除①,再检验商的常数项(为分子常数项除以分母常数项所得之商)

$$\frac{\mathrm{C}_{p^m}^{p^m - 1}}{\mathrm{C}_{p^{m-1}}^{p^{m-1} - 1}} = \frac{\mathrm{C}_{p^m}^1}{\mathrm{C}_{p^{m-1}}^1} = \frac{p^m}{p^{m-1}} = p$$

---

①易见商的第一项系数为1,而被除式减去除式与商之第一项之积所得之差$r_1(y)$的各项系数均为p之倍数,于是商的第二项系数为p之倍数,而$r_1(y)$减去除式与商第二项之积所得的差各项均为p之倍数,如此继续论证知商的其余各项系数均为p之倍数.

不能被 p^2 整除,故由 Eisentein 判别法知 y 的多项式 $\Phi(y+1)$ 在域 $\mathbf{Q}$ 不可约,易见 $\Phi(x)$ 亦然.

注 $m=1$ 时同样作代换 $x=y+1$,(直接)按 Eisentein 判别法亦可证 $\Phi(x)$ 在域 $\mathbf{Q}$ 不可约.

引理 11.3 设域 Δ 上偶次对称多项式

$$p_{2n}(x) \equiv \sum_{k=0}^{n} a_k(x^{2n-k}+x^k) \quad (a_k \in \Delta)$$

(x^n 的系数为 $2a_n$). 则当 $x \neq 0$ 时

$$p_{2n}(x) \equiv x^n \sum_{k=0}^{n} c_k(x+x^{-1})^k \quad (c_k \in \Delta) \tag{11.1}$$

证 对 n 用归纳法:

$n=0$ 时,取 $c_0=2a_0$,易见式(11.1)成立(两边分别为 $2a_0, c_0$);

设方程(11.1)对 n 成立,则

$$\begin{aligned} p_{2(n+1)}(x) &\equiv a_0(x^{2n+2}+1)+\cdots①\equiv \\ &a_0x^{n+1}(x^{n+1}+x^{-n-1})+\cdots\equiv \\ &a_0x^{n+1}[(x+x^{-1})^{n+1}-\cdots]+\cdots \end{aligned}$$

方括号内省略号含 $x^0, x^k, x^{-k}(k=1,2,\cdots,n)$ 项的多项式,其中 x^k 与 x^{-k} 的系数相等,乘 a_0x^{n+1} 后亦得含 $x^1 \sim x^{2n+1}$ 项的对称多项式. 于是

$$p_{2(n+1)}(x)=a_0x^{n+1}(x+x^{-1})^{n+1}+x\cdots②=$$

$$a_0x^{n+1}(x+x^{-1})^{n+1}+x\cdot x^n\sum_{k=0}^{n}c_k(x+x^{-1})^k(\text{按归纳假设})=$$

$$x^{n+1}\sum_{k=0}^{n+1}c_k(x+x^{-1})^k \quad (c_{n+1}=a_0)$$

即式(11.1)对 $n+1$ 成立. 证毕.

引理 11.4 若式(11.1)中 $p_{2n}(x)$ 在域 Δ 不可约,则

$$p_n^*(y) \equiv \sum_{k=0}^{n} c_k y^k$$

在 Δ 上不可约.

证 令 $y=x+x^{-1}$,则由式(11.1)得

$$p_{2n}(x)=x^n\sum_{k=0}^{n}c_ky^k=x^np_n^*(y)$$

若 $p_n^*(y)$ 在 Δ 可约,则

---

①省略号为含 $x^1 \sim x^{2n+1}$ 项的对称多项式.

②省略号为含 $x^0 \sim x^{2n}$ 项的对称多项式.

$$p_n^*(y) \equiv \widetilde{p}_{n_1}(y) \cdot \widetilde{p}_{n_2}(y)$$

$\widetilde{p}_{n_1}, \widetilde{p}_{n_2}$ 分别为 n_1, n_2 次多项式, $n_1, n_2 \in \mathbf{Z}^+$, $n_1 + n_2 = n$, 于是

$$p_{2n}(x) = x^n p_n^*(y) = x^{n_1} \widetilde{p}_{n_1}(y) \cdot x^{n_2} \widetilde{p}_{n_2}(y) =$$

$$x^{n_1} \widetilde{p}_{n_1}(x + x^{-1}) \cdot x^{n_2} \widetilde{p}_{n_2}(x + x^{-1})$$

而易见 $x^{n_1} \widetilde{p}_{n_1}(x + x^{-1})$, $x^{n_2} \widetilde{p}_{n_2}(x + x^{-1})$ 分别为 Δ 上 $2n_1(\geqslant 2)$ 次与 $2n_2(\geqslant 2)$ 次对称多项式, 与假设 $p_{2n}(x)$ 在 Δ 上不可约矛盾. 故得证 $p_n^*(y)$ 在 Δ 上不可约.

引理 11.5 设 $n \in \mathbf{Z}^+$, 则 $\cos n\alpha$ 可用 $\cos \alpha$ 的 n 次多项式表示, 其中各项系数为整数.

证 由 De Moivre 公式

$$\cos n\alpha + \mathrm{i}\sin n\alpha = (\cos \alpha + \mathrm{i}\sin \alpha)^n$$

故 $\cos n\alpha$ 等于上式右边按二项式定理展开式中所有实数项之和, 其每一项形如

$$\mathrm{C}_n^{2k} \cos^{n-2k} \alpha \cdot (\mathrm{i}\sin \alpha)^{2k} = (-1)^k \mathrm{C}_n^{2k} \cos^{n-2k} \alpha \cdot (1 - \cos^2 \alpha)^k$$

按 $\cos \alpha$ 的幂展开后, 最高次幂 $\cos^n \alpha$ 的系数为①

$$\sum_{k=0}^{[n/2]} (-1)^k \mathrm{C}_n^{2k} \cdot (-1)^k = \sum_{k=0}^{[n/2]} \mathrm{C}_n^{2k} > 0$$

再把这些项之和合并化简, 得 $\cos \alpha$ 的多项式, 显然 $\cos \alpha$ 的各次幂的系数均为整数. 且最高次幂 $\cos^n \alpha$ 的系数为正整数, 故 $\cos n\alpha$ 可表为 $\cos \alpha$ 的 n 次整系数多项式.

例 3 分圆为 n 等份, 正整数 $n \geqslant 2$.

先论证:

引理 11.6 设两正整数 n_1, n_2 互素, 则可分圆为 $n_1 n_2$ 等份的充要条件为: 可分圆为 n_1 等份及 n_2 等份.

推广 设正整数 $n_1, n_2, \cdots, n_k$ 两两互素, 则可分圆为 $n_1 n_2 \cdots n_k$ 等份的充要条件为: 可分圆为任何 $n_s(s = 1, 2, \cdots, k)$ 等份.

(注意 n_s 与 $n_1 n_2 \cdots n_{s-1}$ 互素, 屡次应用上引理知结论成立).

证引理 11.6.

必要性显然, 下证充分性:

---

① $[n/2]$ 为 $n/2$ 的整数部分, 即 $[n/2] = \begin{cases} n/2 & \text{当 } n \text{ 为偶数} \\ (n-1)/2 & \text{当 } n \text{ 为奇数} \end{cases}$.

由 n_1,n_2 互素,可取 $m_1,m_2\in\mathbf{Z}$ 使

$$n_1m_1+n_2m_2=1$$

从而

$$\frac{m_1}{n_2}+\frac{m_2}{n_1}=\frac{1}{n_1n_2}$$

故把圆分为 n_1 等份及 n_2 等份后,取圆的 $\frac{m_1}{n_2}+\frac{m_2}{n_1}$ 便得圆的 $\frac{1}{n_1n_2}$. 得证充分性.

分解 n 为素因数乘积

$$n=p_1^{m_1}p_2^{m_2}\cdots p_k^{m_k} \tag{11.2}$$

其中 $p_1,p_2,\cdots,p_k$ 为互不相同的素数,k 为正整数,$m_1,m_2,\cdots,m_k$ 为正整数.

由上引理的推广知,当且仅当可把圆分成任何 $p_s^{m_s}(s=1,2,\cdots,k)$ 等份时可分圆为 n 等份,故先考虑特殊情况:分圆为 p^m 等份(p 为素数,$m\in\mathbf{Z}^+$).

$p=2$ 时显然可分圆为 2^m 等份,下设 p 为奇素数.

不妨设圆半径为长度单位 1,问题实质是要在复平面单位圆上作出点 $z_0=\mathrm{e}^{2\pi\mathrm{i}/p^m}$,它为下述分圆方程的一根

$$z^{p^{m-1}(p-1)}+z^{p^{m-1}(p-2)}+z^{p^{m-1}(p-3)}+\cdots+z^{p^{m-1}}+1=0 \tag{11.3}$$

按引理 11.2 知它在有理数域 $\mathbf{Q}$ 不可约,又因 p 为奇素数,$p^{m-1}(p-1)$ 为偶数,故方程(11.3)左边为偶次对称多项式,由引理 11.3 知当 $z\neq0$ 时左边恒等于

$$z^{\frac{p^{m-1}(p-1)}{2}}\sum_{j=0}^{\frac{p^{m-1}(p-1)}{2}}c_j(z+z^{-1})^j\quad(c_j\in\mathbf{Z})$$

因 $z_0\neq0$,故由它为方程(11.3)之一根知

$$z_0^{\frac{p^{m-1}(p-1)}{2}}\sum_{j=0}^{\frac{p^{m-1}(p-1)}{2}}c_j(z_0+z_0^{-1})^j=0$$

$$\sum_{j=0}^{\frac{p^{m-1}(p-1)}{2}}c_j(z_0+z_0^{-1})^j=0$$

但

$$z_0+z_0^{-1}=\mathrm{e}^{2\pi\mathrm{i}/p^m}+(\mathrm{e}^{2\pi\mathrm{i}/p^m})^{-1}=\mathrm{e}^{2\pi\mathrm{i}/p^m}+\mathrm{e}^{-2\pi\mathrm{i}/p^m}=$$

$$\left(\cos\frac{2\pi}{p^m}+\mathrm{i}\sin\frac{2\pi}{p^m}\right)+\left(\cos\frac{2\pi}{p^m}-\mathrm{i}\sin\frac{2\pi}{p^m}\right)=$$

$$2\cos\frac{2\pi}{p^m}$$

故 $2\cos\frac{2\pi}{p^m}$ 为下述方程之根

$$\sum_{j=0}^{\frac{p^{m-1}(p-1)}{2}} c_j y^j = 0 \tag{11.4}$$

由方程(11.3)在域 $\mathbf{Q}$ 不可约，据引理 11.4 知此方程在域 $\mathbf{Q}$ 不可约，易见当且仅当在复平面可作点 $2\cos\frac{2\pi}{p^m}$（即在实轴作代数值为 $2\cos\frac{2\pi}{p^m}$ 的有向线段）时可在单位圆作点 $e^{2\pi i/p^m}$. 据定理 11.1 推论，问题归结为要 $2\cos\frac{2\pi}{p^m}$ 属于域 $\mathbf{Q}$ 或可用域 $\mathbf{Q}$ 上多层二次实根式表示. 又由于 $2\cos\frac{2\pi}{p^m}$ 为方程(11.4)之根，据定理 10.1，问题又归结为：对方程(11.4)在域 $\mathbf{Q}$ 上的根域 Ω，有

$$(\Omega:\mathbf{Q}) = 2^k \quad (k \in \mathbf{N}) \tag{11.5}$$

（$k=0$ 时，即方程(11.4)之根属于 $\mathbf{Q}$）.

现求 $(\Omega:\mathbf{Q})$，先证 $2\cos\frac{2\pi}{p^m}$ 为域 $\mathbf{Q}$ 上方程(11.4)的本原数：

由方程(11.3)左边等于

$$\frac{z^{p^m}-1}{z^{p^{m-1}}-1}$$

知方程(11.3)的所有 $p^{m-1}(p-1)$ 个根是所有单位质根（即 1 的 p^m 次方根而非 1 的 p^{m-1} 次方根），它们成对共轭，可表为

$$e^{\pm a_j 2\pi i/p^m} \quad \left(j=1,2,\cdots,\frac{p^{m-1}(p-1)}{2}; a_j \in \mathbf{Z}^+, p^{m-1} \nmid a_j, a_j < \frac{p^m}{2}\right)$$

如同方程(11.3)的根 $z_1 = e^{2\pi i/p^m}$，相应方程(11.4)有根 $z_1 + z_1^{-1} = 2\cos\frac{2\pi}{p^m}$，方程(11.3)的每对根 $e^{\pm a_j 2\pi i/p^m}$（互为例数），相应方程(11.4)有根

$$e^{a_j 2\pi i/p^m} + (e^{a_j 2\pi i/p^m})^{-1} = 2\cos\frac{a_j 2\pi}{p^m}$$

因 $0 < \frac{a_j 2\pi}{p^m} < \pi$，故这些根互不相同，共 $\frac{p^m(p-1)}{2}$ 个，为方程

(11.4) 的所有根. 再据引理 11.5 知 $\cos\frac{a_j 2\pi}{p^m}$ 为 $\cos\frac{2\pi}{p^m}$ 的整系数多项式,从而易见方程(11.4) 的所有根 $2\cos\frac{a_j 2\pi}{p^m}$ 为 $2\cos\frac{2\pi}{p^m}$ 的有理系数多项式,故根域

$$\Omega = Q(2\cos\frac{2\pi}{p^m})$$

又以 $2\cos\frac{2\pi}{p^m}$ 为一根且在域 $\mathbf{Q}$ 上不可约方程(11.4) 次数为 $\frac{p^{m-1}(p-1)}{2}$,据定理 3.3 推论知

$$(\Omega:\mathbf{Q}) = \frac{p^{m-1}(p-1)}{2}$$

故条件(11.5) 为

$$\frac{p^{m-1}(p-1)}{2} = 2^k \quad (k \in \mathbf{N})$$

即

$$p^{m-1}(p-1) = 2^{k'} \quad (k' = k+1 \in \mathbf{Z}^+) \tag{11.6}$$

因 p 为奇素数,故必须

$$m = 1, p = 2^{k'} + 1 \tag{11.7}$$

反之易见(11.7)$\Rightarrow$(11.6).

又因 p 为素数,故必须 k' 不含奇素数约数(否则有 $k' = p'q'$,p' 为奇素数,$q' \in \mathbf{Z}^+$),则

$$p = 2^{p'q'} + 1 = (2^{q'})^{p'} + 1 = (2^{q'} + 1)(2^{q'(p'-1)} - 2^{q'(p'-2)} + \cdots - 2^{q'} + 1)$$

p 有约数 $2^{q'} + 1 \in \mathbf{Z}^+$ 且易见 $q' < k'$,$2^{q'} + 1 \neq p$,与 p 为素数矛盾).

于是

$$k' = 2^q \quad (q \in \mathbf{N})$$

$$p = 2^{2^q} + 1$$

即 p 为 Fermat 素数(注意 $2^{2^q} + 1$ 型整数未必为素数).

故要把圆 n 等分,对式(11.2) 中每一奇素数 $p_s(s = 1,2,\cdots,k)$,必须且只需相应因式 $p_s^{m_s}$ 为 $2^{2^{q_s}} = 1$ 型的 Fermat 素数,故有:

定理 11.2(Gauss(高斯)) 可把圆 n 等分的充要条件为

$$n = 2^m p_1 p_2 \cdots p_k \quad (m \in \mathbf{N}, k \in \mathbf{N})$$

($k = 0$ 时,即 $n = 2^m$) 又对 $s = 1,2,\cdots,k$,$p_s = 2^{2^{q_s}} + 1(q_s \in \mathbf{N})$ 且为

素数.

例 4 (用直尺、圆规)可作之角.

① $\frac{m}{n}\cdot 2\pi$(m,n 为互素的正整数)型.

易见:可作角$\frac{m}{n}\cdot 2\pi \Leftrightarrow$可作角$\frac{1}{n}\cdot 2\pi$(即把圆 n 等分).

(因 m,n 互素,可取 $a,b\in \mathbf{Z}$)使

$$am+bn=1$$

从而

$$\frac{am}{n}\cdot 2\pi+b\cdot 2\pi=\frac{1}{n}\cdot 2\pi$$

故若可作角$\frac{m}{n}\cdot 2\pi$,则可作角$\frac{1}{n}\cdot 2\pi$(反之更显然).

② 反三角函数型. $\arcsin a(|a|\leqslant 1)$, $\arccos a(|a|\leqslant 1)$, $\arctan a$, $\operatorname{arccot} a$.

显然,当且仅当有向线段 a 可作时,(据定理 11.1)即 a 属于域 $\mathbf{Q}$ 或可用域 $\mathbf{Q}$ 上多层二次实根式表示时,上述角可作.

例 5 把已知角 α(用具体式子表示)分成奇素数 q 等份.

①$\alpha=\frac{m}{n}\cdot 2\pi$($m,n\in \mathbf{Z}^+$ 且互素).

这时

$$\frac{1}{q}\alpha=\frac{m}{qn}\cdot 2\pi$$

当且仅当可把圆 qn 等分时可把角 α 分为 q 等份.

②$\alpha=\cos^{-1}a$($a\in \mathbf{Q}$, $|a|\leqslant 1$).

设 $\beta=\frac{1}{q}\alpha$,则

$$a=\cos\alpha=\cos q\beta$$

按引理 11.5, $\cos q\beta$ 可用域 $\mathbf{Q}$ 上 q 次整系数多项式 $p(\cos\beta)$ 表示,从而有

$$p(\cos\beta)=a\quad(a\in \mathbf{Q})$$

因 q 为奇素数,故当方程 $p(y)=a$ 在域 $\mathbf{Q}$ 不可约时,由定理 10.1 推论知,此方程之根 $\cos\beta$ 不属域 $\mathbf{Q}$ 亦不能用域 $\mathbf{Q}$ 上多层二次实根式表示,从而据定理 11.1 推论知(视为单位圆的有向线段) $\cos\beta$ 及角 $\beta=\frac{1}{q}\alpha$ 不可作. 而当方程 $p(y)=a$ 在域 $\mathbf{Q}$ 可约时再考虑以 $\cos\beta$

为一根且在域$\mathbf{Q}$不可约的方程$\tilde{p}(y)=0$($\tilde{p}(y)$为$p(y)-a$在域$\mathbf{Q}$的不可约因式). 若$\tilde{p}(y)$的次数含奇素数因数,则$\cos\beta$及角$\beta=\frac{1}{q}\alpha$亦不可作;若$\tilde{p}(y)$的次数为1,2(为$2^n,n\geqslant 2$),则$\cos\beta$及角$\beta=\frac{1}{q}\alpha$可作(则难以判定).

特别当$q=3$时,易判定三次方程$p(y)=a$在域$\mathbf{Q}$上是否可约,当不可约(可约)时,$\cos\beta$及角$\beta=\frac{1}{q}\alpha$不可作(由定理10.2及11.1知$\cos\beta$及角$\beta=\frac{1}{q}\alpha$可作).

由上述可知,与可把任意角α分成$2^s(s\in\mathbf{N})$等份不同,对$q\neq 2^s(s\in\mathbf{N})$及随意作出的任意角$\alpha$,因不能确定此角的量值$\alpha$,故不能把它$q$等分.

例6 矩形$ABCD$,$AB=2AD$,过点A作直线l与两直线CD,CB分别交于M,N,使$MN=5AD$.

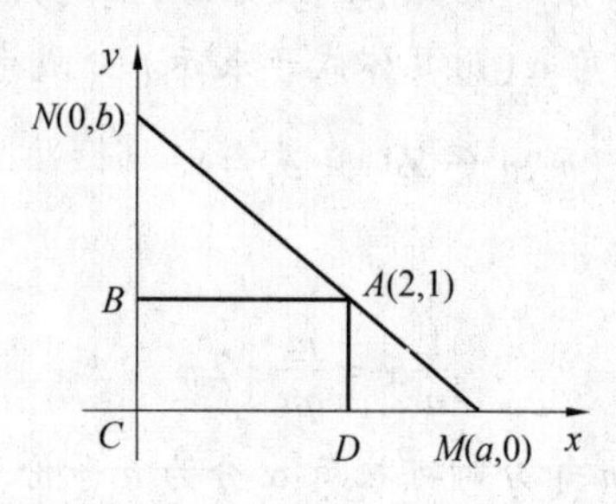

图11.1

解 因所有已知线段,$BC=AD$,$AB=DC=2AD$,$AC=BD=\sqrt{5}AD$,故不妨设$AD=1$(坐标轴单位线段),以C为原点,射线CD,CB分别为正半x,y轴作坐标系,则得坐标$A(2,1)$. 设坐标$M(a,0)$,$N(0,b)$,则由$MN=5AD=5$得

$$a^2+b^2=25 \tag{11.8}$$

又直线MN方程为

$$\frac{x}{a}+\frac{y}{b}=1$$

它过点$A(2,1)$,故有

$$\frac{2}{a}+\frac{1}{b}=1$$

$$2b + a = ab$$

$$b = \frac{a}{a-2}$$

代入方程(11.8)得

$$a^2 + \frac{a^2}{(a-2)^2} = 25$$

$$a^4 - 4a^3 + 4a^2 + a^2 = 25a^2 - 100a + 100$$

$$a^4 - 4a^3 - 20a^2 + 100a - 100 = 0 \qquad (11.9)$$

求得其 Lagrange 预解方程(6.12)为

$$t^3 + 20t^2 - 400 = 0$$

经检验知它无有理数根,由定理10.3知对 a 的四次方程(11.9)不存在多层二次实根式解,再由定理11.1知本作图题不能用圆规直尺作图.

第十二章　Galois 理论基本定理
——代数方程可用根式解的判定准则

Galois 理论基本定理,是在多项式 $f(x)$ 的系数域 Δ 逐步添加循环方程之根后所得 Galois 群序列来判定 $f(x)$ 的根可用域 Δ 上多层根式表示.

设域 Δ 上多项式 $f(x)$ 如同第四章所述,其 Galois 群为 G,在域 Δ 上添加在 Δ 上不可约素数次循环方程之根 ξ 后所得域 $\Delta(\xi)$,$f(x)$ 又可看成域 $\Delta(\xi)$ 上的多项式,先考虑域 $\Delta(\xi)$ 上 $f(x)$ 的 Galois 群 G^* 与 G 之关系.

引理 12.1　设在数域 F 上不可约多项式 $\psi(y)$ 的两根为 α,β,又有域 F 上二元多项式 $\Phi(x,y)$,且对 x 有

$$\Phi(x,\alpha) \equiv 0 \qquad (12.1)$$

则

$$\Phi(x,\beta) \equiv 0$$

证　因任一数域 $F \supset \mathbf{Q}$(有理数域),取任一有理数 $r \in F$,对域 F 上 y 的多项式 $\Phi(r,y)$,由式(12.1)知

$$\Phi(r,\alpha) = 0$$

即域 F 上 y 的多项式 $\Phi(r,y)$ 与 $\psi(y)$(在 F 上不可约)有公共根 α,故由引理3.1知 $\psi(y)$ 的根 β 是 $\Phi(r,y)$ 的根,即

$$\Phi(r,\beta)=0$$

于是域 F 上 x 的多项式 $\Phi(x,\beta)$ 有无数个(有理数)根,故

$$\Phi(x,\beta)\equiv 0$$

定理 12.1 设域 Δ 上多项式 $f(x)$ 的 Galois 群为 G,又域 Δ 上素数 m 次不可约循环方程 $\psi(t)=0$ 的一根为 ξ,域 $\Delta(\xi)$ 上多项式 $f(x)$ 的 Galois 群为 G^*,则 G^* 为 G 的不变子群,且 $[G:G^*]$ 等于 m 或 1.

证 ① 证 G^* 为 G 的子群.

设域 Δ 上多项式 $f(x)$ 根域 Ω 的一个本原数为 α. 先证 α 亦为域 $\Delta(\xi)$ 上多项式 $f(x)$ 的根域 Ω^* 的本原数,由定理 4.3 知这只要证

$$\Omega^*=\Delta(\xi,\alpha) \tag{12.2}$$

由根域 Ω^* 之意义知

$$\Omega^*=\Delta(\xi,x_1,x_2,\cdots,x_n)$$

($x_1,x_2,\cdots,x_n$ 为 $f(x)$ 的全部(单)根).

由定理 4.3 知 $x_1,x_2,\cdots,x_n$ 均为 α 在域 Δ 上的多项式,从而

$$\Omega^*=\Delta(\xi,x_1,x_2,\cdots,x_n)\subset\Delta(\xi,\alpha)$$

反之,因 $\alpha\in\Omega$,故 α 可表为域 Δ 上 $x_1,x_2,\cdots,x_n$ 的有理式,从而

$$\Delta(\xi,\alpha)\subset\Delta(\xi,x_1,x_2,\cdots,x_n)=\Omega^*$$

得证式(12.2).

设以域 Δ(域 $\Delta(\xi)$)上多项式 $f(x)$ 的根域 $\Omega(\Omega^*)$ 的本原数 α 为一根在域 Δ(域 $\Delta(\xi)$)上不可约多项式为 $g(t)$(为 $g^*(t)$),其各根为根域 $\Omega(\Omega^*)$ 的所有本原数. $g(t)$ 亦可看成域 $\Delta(\xi)$ 上的多项式,它与域 $\Delta(\xi)$ 上不可约多项式 $g^*(t)$ 有公共根 α,由引理 3.1 知 $g^*(t)$ 的根集 S 每一根均为 $g(t)$ 的根. 群 G^* 在域 Ω^* 相应的本原数变换集为

$$\{\alpha\to\alpha':\alpha'\in S\}$$

但集 S 之任一根 α' 亦为 $g(t)$ 之根,故这些变换集组成群 G 在域 Ω 相应的本原数变换集之子集,由 α 变成每个 α' 在 $f(x)$ 的根集所导致的(Galois 群 G^* 的)置换可视为 Galois 群 G 的置换,因此 G^* 为 G 之子群.

② 证 $[G:G^*]$ 等于 m 或 1(即证 $[G:G^*]\mid m$(素数)).

对上述 α,Ω,Ω^* 及 $g(t),g^*(t)$,因 Galois 群 $G(G^*)$ 的阶数等于 $g(t)(g^*(t))$ 的次数 $q(q^*)$,故

$$[G:G^*]=\frac{q}{q^*}(\text{正整数})$$

于是只要证

$$\frac{q}{q^*} \mid m$$

由①之证,在域$\Delta(\xi)$上$g^*(t)$是$g(t)$的因式,可设

$$g(t)=g^*(t)h(t)$$

$g^*(t)$,$h(t)$为域$\Delta(\xi)$上多项式,它们的各项系数为ξ的有理式,而ξ为域Δ上多项式$\psi(t)$之根,由定理3.3可设上述系数为域Δ上ξ的多项式.于是$g^*(t)$,$h(t)$分别等于域Δ上二元多项式$\Gamma(t,\xi)$,$H(t,\xi)$,故

$$g(t)=\Gamma(t,\xi)H(t,\xi) \tag{12.3}$$

又设$\xi_j(j=1,2,\cdots,m)$为域Δ上不可约m次多项式$\psi(t)$之根,其中$\xi_1=\xi$,由引理12.1(令其中$\Phi(t,y)=g(t)-\Gamma(t,y)H(t,y)$)从上式得

$$g(t)=\Gamma(t,\xi_j)H(t,\xi_j)\quad(j=1,2,\cdots,m) \tag{12.4}$$

从而

$$(g(t))^m=\prod_{j=1}^{m}\Gamma(t,\xi_j)\cdot\prod_{j=1}^{m}H(t,\xi_j)$$

易见右边两连乘积中t的各次幂系数为$\xi_1,\xi_2,\cdots,\xi_m$的对称式,可表为它们的基本对称多项式的有理式,即域Δ上多项式$\psi(t)$各项系数的有理式(按Vieta定理),故属于域Δ,于是两连乘积实为域Δ上多项式.从而域Δ上t的多项式

$$\prod_{j=1}^{m}\Gamma(t,\xi_j)$$

为$(g(t))^m$的因式,但$g(t)$在域Δ上不可约,故存在正整数$m'\leqslant m$,使

$$\prod_{j=1}^{m}\Gamma(t,\xi_j)=(g(t))^{m'} \tag{12.5}$$

比较两边t的次数得

$$mq^*=m'q$$

$$m=m'\frac{q}{q^*}$$

即$\frac{q}{q^*}$整除m.

③证群G^*为群G之不变子群.

对上述$\alpha,\Omega,\Omega^*,g(t),g^*(t)=\Gamma(t,\xi)$,设$\tau^*,\tau$分别为群$G^*$,

G 的任何置换,在①中已证 α 亦为域 $\Delta(\xi)$ 上多项式 $f(x)$ 根域 Ω^* 的本原数. 设根域 Ω^* 相应于 Galois 群 G^* 的置换 τ^* 的本原数变换为 α 变成本原数 $\alpha^*=\alpha\tau^*$,α^* 为 $g^*(t)$ 的一根亦为 $g(t)$ 的一根(②中已证 $g^*(t)$ 的根都是 $g(t)$ 的根). 因 α,α^* 都是 $g(t)$ 的根,故 α,α^* 亦为域 Δ 上多项式 $f(x)$ 根域 Ω 的本原数. 设根域 Ω 相应于 Galois 群 G 的置换 τ 的本原数变换把 α,α^* 分别变成本原数

$$\alpha'=\alpha\tau\in\Omega\subset\Omega^* \tag{12.6}$$

$$\alpha^{*\prime}=\alpha^*\tau\in\Omega\subset\Omega^* \tag{12.7}$$

则

$$\alpha=\alpha'\tau^{-1}$$

$$\alpha^*=\alpha^{*\prime}\tau^{-1}$$

$$\alpha'(\tau^{-1}\tau^*\tau)=(\alpha'\tau^{-1})(\tau^*\tau)=\alpha(\tau^*\tau)=(\alpha\tau^*)\tau=$$

$$\alpha^*\tau=\alpha^{*\prime}$$

置换 $\tau^{-1}\tau^*\tau$ 把 α' 变成 $\alpha^{*\prime}$,我们证明这置换也是 G^* 的一置换,从而由于 τ,τ^* 分别是 G,G^* 的任意置换知 G^* 为 G 之不变子群.

与 α,α^* 不同,$\alpha',\alpha^{*\prime}$ 不再是域 $\Delta(\xi)$ 上(q^* 次)不可约多项式 $g^*(t)$ 即 $\Gamma(t,\xi)$ 的根,但我们可证明它们是另一个在域 $\Delta(\xi)$ 上不可约 q^* 次多项式的根,由此因 q^* 是多项式 $g^*(t)$ 的次数,$g^*(t)$ 为域 $\Delta(\xi)$ 上多项式 $f(x)$ 根域 Ω^* 上本原数所适合的不可约方程,故 $g^*(t)$ 的次数即域 $\Delta(\xi)$ 上多项式 $f(x)$ 根域 Ω^* 的维数,又因 α',$\alpha^{*\prime}\in\Omega^*$,从而由 $\alpha',\alpha^{*\prime}$ 是域 $\Delta(\xi)$ 上不可约 q^* 次多项式的根,据第四章本原数定义可知 $\alpha',\alpha^{*\prime}$ 亦为域 $\Delta(\xi)$ 上多项式 $f(x)$ 根域 Ω^* 的本原数,把 α' 变成 $\alpha^{*\prime}$ 的置换 $\tau^{-1}\tau^*\tau$ 为 Galois 群 G^* 的一置换.

因 α' 为 $g(t)$ 的根,由式(12.5)知 α' 为某个 $\Gamma(t,\xi_{j'})$ 之根,下面证明 $\Gamma(t,\xi_{j'})$ 正是上段所述域 $\Delta(\xi)$ 上不可约 q^* 次多项式即可.

先证 $\Gamma(t,\xi_{j'})$ 作为域 $\Delta(\xi_{j'})$ 上 t 的多项式,也是域 $\Delta(\xi)$ 上 t 的多项式,即证 $\Delta(\xi_{j'})=\Delta(\xi)$:因 $\xi,\xi_{j'}$ 为域 Δ 上不可约循环方程 $\psi(t)$ 之根,据定理7.2推论知其所有根为域 Δ 上多项式 $\psi(t)$ 根域 K 的本原数,故 $K=\Delta(\xi_{j'})=\Delta(\xi)$(按定理4.3).

又证 $\Gamma(t,\xi_{j'})$ 在域 $\Delta(\xi_{j'})$ 即 $\Delta(\xi)$ 上不可约:否则有

$$\Gamma(t,\xi_{j'})=\Gamma_1(t,\xi_{j'})\cdot\Gamma_2(t,\xi_{j'})$$

(Γ_1,Γ_2 为域 Δ 上二元多项式,对 t 至少1次). 又 $\xi_{j'},\xi$ 为域 Δ 上不可约多项式 $\psi(t)$ 之根,由引理12.1从上式得

$$\Gamma(t,\xi)=\Gamma_1(t,\xi)\cdot\Gamma_2(t,\xi)$$

即 $g^*(t)=\Gamma(t,\xi)$ 在域 $\Delta(\xi)$ 上可约,与前述矛盾.

已知 $g(t)$ 的根 α' 为 $\Gamma(t,\xi_{j'})$(对 t)之根,最后证 $\alpha^{*\prime}$ 亦为 $\Gamma(t,\xi_{j'})$(对 t)之根:因 α,α^* 都是域 Δ 上多项式 $f(x)$ 根域 Ω 的本原数,由定理 4.3 知

$$\alpha^*\in\Omega=\Delta(\alpha)$$

再由定理 3.3 知

$$\alpha^*=\varphi(\alpha)\text{—— 域 }\Delta\text{ 上多项式}\tag{12.8}$$

又因 α 与 α^* 为 $g^*(t)$ 即 $\Gamma(t,\xi)$ 的根,故

$$\Gamma(\varphi(\alpha),\xi)=\Gamma(\alpha^*,\xi)=0$$

$$\Gamma(\alpha,\xi)=0$$

域 $\Delta(\xi)$ 上 t 的多项式 $\Gamma(\varphi(t),\xi)$ 与 $\Gamma(t,\xi)$ 有公共根 α,而 $\Gamma(t,\xi)=g^*(t)$ 在 $\Delta(\xi)$ 不可约,故由引理 3.1 知 $\Gamma(t,\xi)$ 为 $\Gamma(\varphi(t),\xi)$ 的因式,于是

$$\Gamma(\varphi(t),\xi)=\Gamma(t,\xi)B(t,\xi)$$

($B(t,\xi)$ 为域 Δ 上二元多项式). 而 $\xi,\xi_{j'}$ 为域 Δ 上不可约多项式 $\psi(t)$ 之根,由引理 12.1 从上式得

$$\Gamma(\varphi(t),\xi_{j'})=\Gamma(t,\xi_{j'})B(t,\xi_{j'})\tag{12.9}$$

但 α' 是 $\Gamma(t,\xi_{j'})$ 之根,故

$$\Gamma(\alpha',\xi_{j'})=0$$

又从(12.9)得

$$\Gamma(\varphi(\alpha'),\xi_{j'})=\Gamma(\alpha',\xi_{j'})B(\alpha',\xi_{j'})=0\tag{12.10}$$

α,α^* 均为域 $\Delta(\xi)$ 上多项式 $f(x)$ 根域 Ω^* 的本原数,对式(12.8)两边用 $f(x)$ 诸根在域 Δ 上的有理式表示式进行域 Δ 上多项式 $f(x)$ 的 Galois 群 G 的置换 τ(按定理 5.1 推论)得

$$\alpha^*\tau=\varphi(\alpha\tau)$$

再按式(12.6),式(12.7)得

$$\alpha^{*\prime}=\varphi(\alpha')$$

从而又再(12.10)得

$$\Gamma(\alpha^{*\prime},\xi_{j'})=0$$

即 $\alpha^{*\prime}$ 为 $\Gamma(t,\xi_{j'})$(对 t)的根.

例 1 有理数域 **Q** 上不可约二次方程

$$x^2-2=0$$

的两根 $\pm\sqrt{2}$ 互为相反数,即可在域 **Q** 相互有理表示,又此方程在域

$\mathbf{Q}$ 无根,据定理 7.3 知此为域 $\mathbf{Q}$ 上不可约循环方程,其 Galois(循环) 群 G 的生成元为其所有两根的轮换,故

$$G=\{I,(\sqrt{2},-\sqrt{2})\}$$

$\sqrt{3}$ 不属于域 $\mathbf{Q}$ 上此方程根域,则在域 $Q(\sqrt{3})$ 上此方程仍不可约:否则它有$a+b\sqrt{3}(a,b\in\mathbf{Q})$ 型的根,即

$$a+b\sqrt{3}=\pm\sqrt{2}$$

两边平方得

$$a^2+3b^2+2\sqrt{3}ab=2$$

显然 $a,b\neq 0$,于是

$$\sqrt{3}=\frac{2-a^2-3b^2}{2ab}$$

为有理数,得矛盾.故在域 $Q(\sqrt{3})$ 上此方程仍是不可约方程,相应 Galois 群 $G'=G$ 显然为 G 本身的不变子群

$$[G:G']=1$$

例 2 考虑域 $\mathbf{Q}$ 上三次方程(见第八章例 1)

$$t^3+t^2-4t+1=0$$

其三根为

$$\eta_1=\omega_{13}^6+\omega_{13}^9+\omega_{13}^7+\omega_{13}^4$$

$$\eta_2=\omega_{13}^{10}+\omega_{13}^2+\omega_{13}^3+\omega_{13}^{11}$$

$$\eta_3=\omega_{13}^8+\omega_{13}^{12}+\omega_{13}^5+\omega_{13}$$

由定理 7.7 知此方程为域 $\mathbf{Q}$ 上不可约循环方程,其 Galois 群为以此三根的轮换(η_1,η_2,η_3) 为生成元的循环群(3 阶)A_3.

取不属方程根域的 $\xi=\omega_{13}$,因 η_1,η_2,η_3 均为域 $\mathbf{Q}$ 上 ξ 的有理式,属于域 $Q(\xi)$,故在域 $Q(\xi)$ 上方程的 Galois 群 G' 为恒等置换群 $\{I\}$(1 阶),它是 A_3 的不变子群

$$[A_3:\{I\}]=\frac{3}{1}=3$$

正是定理所述循环方程次数 m.

例 3 考虑有理数域 $\mathbf{Q}$ 上多项式

$$x^3-4=0$$

其三根为

$$x_1=\sqrt[3]{4}$$

$$x_2=\sqrt[3]{4}\,\omega_3$$

$$x_3=\sqrt[3]{4}\,\omega_3^2$$

取

$$\rho=\sqrt[3]{4}\sqrt{3}\,\mathrm{i}=\sqrt[3]{4}(\omega_3-\omega_3^2)$$

属域 **Q** 上此多项式根域 Ω. 又三根

$$\sqrt[3]{4}=\frac{1}{36}\rho^4$$

$$\sqrt[3]{4}\,\omega_3=\frac{1}{36}\rho^4\left(-\frac{1}{2}-\frac{1}{24}\rho^3\right)$$

$$\sqrt[3]{4}\,\omega_3^2=\frac{1}{36}\rho^4\left(-\frac{1}{2}+\frac{1}{24}\rho^3\right)$$

故 $\Omega=Q(\rho)$. 由定理4.3知 ρ 为根域 Ω 的本原数. 现求以 ρ 为一根在域 **Q** 上不可约的多项式: 易求得

$$\rho^6=-432$$

$$\rho^6+432=0$$

ρ 为域 **Q** 上6次多项式

$$t^6+432$$

之根, 它是域 **Q** 上不可约多项式, 因它在复数域的分解式为

$$\prod_{k=0}^{5}\left[\rho-\sqrt[6]{432}\,\mathrm{e}^{\left(\frac{\pi}{6}+\frac{2\pi k}{6}\right)\mathrm{i}}\right]$$

其中任何 $j(j=1,2,\cdots,5)$ 个一次因式之积的常数项之模 $(\sqrt[6]{432})^j$ 非有理数, 从而此常数项非有理数(有理数之模为有理数). 于是共轭本原数有6个, 从而域 **Q** 上此多项式的 Galois 群 G 为3次6(=3!)阶群(唯有)S_3——3次对称群.

若视此多项式为域 $Q^*=Q(\omega_3)$ 上多项式, 则因它的一根 $\sqrt[3]{4}\notin Q^*$①, 由定理7.5知方程

$$x^3-4=0$$

为域 Q^* 上不可约循环方程, 相应 Galois 群 G 为3阶群 A_3, A_3 是 S_3 的不变子群

$$[S_3:A_3]=\frac{6}{3}=2$$

为 ω_3 所适合的在域 Q 上素数次不可约循环方程 $t^2+t+1=0$ 的次数.

---

①否则若 $\sqrt[3]{4}\in Q(\omega_3)$, 因 ω_3 适合域 **Q** 上不可约方程 $t^2+t+1=0$. 据定理3.3知 $\sqrt[3]{4}=a\omega_3+b=(-a/2+b)+\frac{\sqrt{3}}{2}a\mathrm{i}(a,b\in\mathbf{Q})$, 从而 $\sqrt[3]{4}=-a/2+b\in\mathbf{Q}$, $0=\sqrt{3}/2a$, 得矛盾.

而在域 $Q^*(\sqrt[3]{4})=Q(\omega_3,\sqrt[3]{4})$ 上多项式 x^3-4 可分解为一次因式之积

$$(x-\sqrt[3]{4})(x-\sqrt[3]{4}\omega_3)(x-\sqrt[3]{4}\omega_3^2)$$

相应 Galois 群为恒等置换群 $\{I\}$（1 阶群）（因为此多项式之根全属域 $Q^*(\sqrt[3]{4})$，由定理5.2知Galois群中所有置换不改变这些根），$\{I\}$ 为 A_3 的不变子群

$$[A_3:\{I\}]=\frac{3}{1}=3$$

正是 $\sqrt[3]{4}$ 所适合的在域 $Q^*=Q(\omega_3)$ 上素数3次不可约循环方程 $x^3-4=0$ 的次数 m.

这里 $\omega_3=\dfrac{x_2}{x_1}$ 属域 $\mathbf{Q}$ 上多项式 x^3-4 的根域，$\sqrt[3]{4}$ 属域 Q^* 上多项式 x^3-4 的根域. 实际上若定理12.1中 ξ 属于域 Δ 上多项式 $f(x)$ 的根域，则可证必有 $[G:G']=m$.

现表述用 Galois 群判定可用多层根式解出多项式 $f(x)$ 一根的条件.

定理 12.2 设域 Δ 上多项式 $f(x)$ 的一根 x^* 可用域 Δ 上多层根式表示，则（必要条件）域 Δ 上多项式 $f(x)$ 的 Galois 群 G 有组合因数为素数的不变子群列

$$G,G_1,G_2,\cdots,G^*,G_1^*,G_2^*,\cdots,G_{k-1}^*,G_k^*$$

其中相继两项的指标 $[G:G_1]$，$[G_1:G_2]$，$\cdots$ 是素数，且 x^* 经群 G 的任一置换不变，但不能经 G_{k-1}^* 的所有置换不变.

证 设域 Δ 上多层根式表示式中根指数列为

$$n_1,n_2,\cdots,n_k$$

先把单位质根列

$$\omega_{n_1},\omega_{n_2},\cdots,\omega_{n_k}$$

中属于域 Δ 者去掉，再把有重复者只留其一，得互不相同的单位质根列

$$\omega_{n_1'},\omega_{n_2'},\cdots,\omega_{n_{k'}'}$$

先考察域 $\Delta(\omega_{n_1'})$ 上多项式 $f(x)$ 的 Galois 群：按定理7.7，分解 n'_1-1 为素因数连乘积

$$n_1'-1=p'p''\cdots p^{(l)}q^{(l)}$$

则 $\omega_{n_1'}$ 必为该定理所述的某个 $\eta_{j'j''\cdots j^{(l)}j^{(l+1)}}$. 因由定理7.2知循环方程的各根可相互有理表示，用 $\eta_{j'j''}$ 在域 $\Delta(\eta_{j'})$ 上的有理式可表示 $\eta_{j'1}$，

$\eta_{j'2},\cdots,\eta_{j'p''}$,从而表示其和 $\eta_{j'}$,故

$$\Delta(\eta_{j'},\eta_{j'j''})=\Delta(\eta_{j'j''})$$

类似知

$$\Delta(\eta_{j'j''},\eta_{j'j''j'''})=\Delta(\eta_{j'j''j'''})$$

$$\vdots$$

由上段结论及定理 12.1 知:

域 $\Delta(\eta_{j'})$ 上多项式 $f(x)$ 的 Galois 群 Γ_1 为域 Δ 上多项式 $f(x)$ 的 Galois 群 G 之不变子群

$$[G:\Gamma_1]=p' \text{ 或 } 1$$

域 $\Delta(\eta_{j'j''})$ 上多项式 $f(x)$ 的 Galois 群 Γ_2 为域 $\Delta(\eta_{j'})$ 上多项式 $f(x)$ 的 Galois 群 Γ_1 之不变子群

$$[\Gamma_1:\Gamma_2]=p'' \text{ 或 } 1$$

域 $\Delta(\eta_{j'j''j'''})$ 上多项式 $f(x)$ 的 Galois 群 Γ_3 为域 $\Delta(\eta_{j'j''})$ 上多项式 $f(x)$ 的 Galois 群 Γ_2 之不变子群

$$[\Gamma_3:\Gamma_2]=p''' \text{ 或 } 1$$

$$\vdots$$

域 $\Delta(\omega_{n_1'})=\Delta(\eta_{j'j''\cdots j^{(l)}j^{(l+1)}})$ 上多项式 $f(x)$ 的 Galois 群 Γ_{l+1} 为域 $\Delta(\eta_{j'j''\cdots j^{(l)}})$ 上多项式 $f(x)$ 的 Galois 群 Γ_l 的不变子群

$$[\Gamma_{l+1}:\Gamma_l]=q^{(l)} \text{ 或 } 1$$

于是得不变子群列

$$G,\Gamma_1,\Gamma_2,\Gamma_3,\cdots,\Gamma_{l+1}$$

记 $\Gamma_{l+1}=G_1$,若接连两项之指标 $[\Gamma_j:\Gamma_{j+1}]=1$,显然 $\Gamma_j=\Gamma_{j+1}$,这时省去 Γ_j 或 Γ_{j+1},于是得组合因数为素数的不变子群列

$$G,\cdots,G_1$$

G_1 为域 $\Delta(\omega_{n_1'})$ 上多项式 $f(x)$ 的 Galois 群,以 G_1 取代 G,以域 $\Delta(\omega_{n'_1})$ 取代域 Δ,同样可证 G_1 有组合因数为素数的不变子群列

$$G_1,\cdots,G_2$$

G_2 为域 $\Delta(\omega_{n_1'},\omega_{n_2'})$ 上多项式 $f(x)$ 的 Galois 群,同样继续论证,最后得群 G 的组合因数为素数的不变子群列

$$G,\cdots,G_1,\cdots,G_2,\cdots,G_{k'}$$

记 $G_{k'}=G^*$,它是在域 $\Delta(\omega_{n_1'},\omega_{n_2'},\cdots,\omega_{n_{k'}'})$(记为 Δ^*)上多项式 $f(x)$ 的 Galois 群.

把 x^* 用域 Δ 上多层根式表示式中的域 Δ 上多元有理式 $\varphi_j(j=1,2,\cdots,k)$ 看成为域 Δ^* 上多元有理式,$x^{(0)}$ 看成域 Δ^* 的数. 若有某些

$$\xi^{(j)} \in \Delta^*(\xi^{(1)},\xi^{(2)},\cdots,\xi^{(j-1)})$$

同样可省去$\xi^{(j)}$,以下不妨设所有

$$\xi^{(j)} \notin \Delta^*(\xi^{(1)},\xi^{(2)},\cdots,\xi^{(j-1)}) \quad (j=1,2,\cdots,k)$$

因

$$\omega_{n_j} \in \Delta^* \subset \Delta^*(\xi^{(1)},\xi^{(2)},\cdots,\xi^{(j-1)})$$

再由$(\xi^{(j)})^{n_j}=x^{(j-1)}$及定理7.5知,方程

$$t^{n_j}=x^{(j-1)}$$

为域$\Delta^*(\xi^{(1)},\xi^{(2)},\cdots,\xi^{(j-1)})$上素数$n_j$次不可约循环方程.再由定理12.1可得$G^*$的不变子群列

$$G^*,G_1^*,G_2^*,G_3^*,\cdots,G_k^*$$

其中域$\Delta^*(\xi^{(1)})$上多项式$f(x)$的Galois群G_1^*为域Δ^*上多项式$f(x)$的Galois群G的不变子群

$$[G^*:G_1^*]=n_1 \text{ 或 } 1$$

对$j=2,3,\cdots,k$,域$\Delta^*(\xi^{(1)},\xi^{(2)},\cdots,\xi^{(j)})$上多项式$f(x)$的Galois群$G_j^*$为域$\Delta^*(\xi^{(1)},\xi^{(2)},\cdots,\xi^{(j-1)})$上多项式$f(x)$的Galois群$G_{j-1}^*$之不变子群

$$[G_{j-1}^*:G_j^*]=n_j \text{ 或 } 1$$

省去使$[G_{j-1}^*:G_j^*]=1$的G_{j-1}^*或G_j^*,连同前面已得的不变子群列,可得G的组合因数为素数的不变子群列

$$G,\cdots,G_1,\cdots,G_2,\cdots,G_{k'}=G^*,\cdots,G_k^*$$

因$x^* \in \Delta^*(\xi^{(1)},\xi^{(2)},\cdots,\xi^{(k)})$域$\Delta^*(\xi^{(1)},\xi^{(2)},\cdots,\xi^{(k)})$上多项式$f(x)$的Galois群为$G_k^*$,故由定理5.2知群$G_k^*$的所有置换不改变$x^*$.又因$x^* \notin \Delta^*(\xi^{(1)},\xi^{(2)},\cdots,\xi^{(k-1)})$,而域$\Delta^*(\xi^{(1)},\xi^{(2)},\cdots,\xi^{(k-1)})$上多项式$f(x)$的Galois群为$G_{k-1}^*$,据定理5.2(用反证法)知$x^*$不能经$G_{k-1}^*$的所有置换不变.

若第九章中所述x^*的域Δ上多层根式适合定理9.2(Abel)的结论,则可证必有$[G^*:G_1^*]=n_1,[G_{j-1}^*:G_j^*]=n_j(j=2,3,\cdots,k)$.

又在近世代数教材证明了,上述必要条件也是充分条件(下文定理12.3亦然).

Galois Theory of Algebraic Equations[①] 需补充颇多预备知识证其充分性(本书作者未拜读).文献 *Galois Theory*[②](研究生教材)对

---

① JEAN - PIERRE TIGNOL. Galois Theory of Algebraic Equations[M]. Word Scientific, 2001.

② JEAN - PIERRE ESCOFIER. Galois Theory[M]. Springer, 2000.

Δ 增加条件"域 Δ 含有 q 次单位质根 ω_q(q 为 Galois 群 G 的阶数)"证其充分性. 因实际应用一般只用到条件的必要性,故在此不论述充分性.

在二、三、四次方程的求根公式中似未见要先在有理数域 $\mathbf{Q}$ 求单位质根的多层根式表示式(在三次方程的 Cartan 公式中的 $\sqrt[3]{-\frac{q}{2}+\sqrt{\frac{q^2}{4}+\frac{p^3}{27}}\omega_3}$ 及 $\sqrt[3]{-\frac{q}{2}+\sqrt{\frac{q^2}{4}+\frac{p^3}{27}}\omega_3^2}$ 实为 $-\frac{q}{2}+\sqrt{\frac{q^2}{4}+\frac{p^3}{27}}$ 的另两个立方根,……),似乎在域 Δ 解方程与在域 Δ^* 解方程无区别,即前者并不比后者要先求单位质根. 下列例子说明此语不对.

例 4 解方程

$$x^4+x^3+x^2+x+1=0$$

(即求 ω_5 的分圆方程,见第八章例 2).

在域 $Q(\omega_5)$ 解此方程,可立即写出其四根

$$\omega_5,\omega_5^2,\omega_5^3,\omega_5^4$$

它们全属于 $Q(\omega_5)$,相应 Galois 群为 $\{I\}$,其组合因数为素数的不变子群列只有一项 $\{I\}$.

在域 $\mathbf{Q}$ 解此方程,第八章例 2 中要先解方程

$$t^2+t-1=0$$

求 η_1,η_2(即 $2\cos\frac{4\pi}{5},2\cos\frac{2\pi}{5}$),其值为 $\frac{-1\pm\sqrt{5}}{2}$,再按

$$\omega_5^2+\omega_5^3=\eta_1$$
$$\omega_5^2\omega_5^3=1$$
$$\omega_5+\omega_5^4=\eta_2$$
$$\omega_5\omega_5^4=1$$

分别解方程

$$t^2-\eta_1t+1=0$$
$$t^2-\eta_2t+1=0$$

求得 ω_5^2,ω_5^3 及 ω_5,ω_5^4.

现求域 $\mathbf{Q}$ 上此分圆方程的 Galois 群 G 的组合因数为素数的不变子群列.

在第八章例 2 中已求得 Galois 群 G 为以

$$s=(\omega_5^2,\omega_5^4,\omega_5^3,\omega_5)$$

为生成元的循环群

$$G=\{I,s,s^2,s^3,s^4\}=$$
$$\{I,(\omega_5^2,\omega_5^4,\omega_5^3,\omega_5),(\omega_5^2,\omega_5^3)(\omega_5^4,\omega_5),(\omega_5^2,\omega_5,\omega_5^3,\omega_5^4)\}$$

再求域 $Q(\eta_1)$ 上此分圆方程的 Galois 群 Γ_1：此时在域 $Q(\eta_1)$ 上以本原数 ω_5 为一根的不可约方程为

$$t^2-\eta_1 t+1=0 \tag{12.11}$$

其根为 ω_5,ω_5^4，即域 $Q(\eta_1)$ 上此分圆方程的共轭本原数. 由共轭本原数变换 $\omega_5\to\omega_5$（不变）导致此分圆方程各根的恒等置换 I；由共轭本原数变换 $\omega_5\to\omega_5^4$，使此分圆方程四根

$$\omega_5^2,\omega_5^4,\omega_5^3,\omega_5$$

分别变成

$$(\omega_5^4)^2=\omega_5^3,(\omega_5^4)^4=\omega_5,(\omega_5^4)^3=\omega_5^2,\omega_5^4$$

即导致各根的置换为

$$(\omega_5^2,\omega_5^3)(\omega_5^4,\omega_5)$$

于是得在域 $Q(\eta_1)$ 上此分圆方程的 Galois 群为

$$\Gamma_1=\{I,(\omega_5^2,\omega_5^3)(\omega_5^4,\omega_5)\}$$

按定理 2.7 易验证 Γ_1 为 G 的不变子群

$$[G:\Gamma_1]=\frac{4}{2}=2$$

正是方程(12.11)的次数.

求在域 $Q(\eta_1,\omega_5)=Q(\omega_5)$ 上此分圆方程的 Galois 群 Γ_2：此时在域 $Q(\omega_5)$ 上以本原数 ω_5 为一根之不可约方程为

$$t-\omega_5=0$$

共轭本原数集只有 ω_5 一数，共轭本原数变换只有 $\omega_5\to\omega_5$（不变），故导致此分圆方程四根的置换为 I，所述 Galois 群

$$\Gamma_2=\{I\}$$

为群 Γ_1 的不变子群

$$[\Gamma:\Gamma_2]=\frac{2}{1}=2$$

于是所求的不变子群列为

$$\{I,(\omega_5^2,\omega_5^4,\omega_5^3,\omega_5),(\omega_5^2,\omega_5^3)(\omega_5^4,\omega_5),(\omega_5^2,\omega_5,\omega_5^3,\omega_5^4)\},$$
$$\{I,(\omega_5^2,\omega_5^3)(\omega_5^4,\omega_5)\},\{I\}$$

例 5　考虑多项式

$$x^3-4$$

其三根为

$$x_1=\sqrt[3]{4}$$
$$x_2=\sqrt[3]{4}\omega_3$$
$$x_3=\sqrt[3]{4}\omega_3^2$$

若视此多项式为域 $Q^*=Q(\omega_3)$ 上多项式,则只需求一次立方根$\sqrt[3]{4}$,使可在域 Q^* 上写出它的三根. 在定理12.1例3中已求得在域 Q^* 中此多项式的Galois群为 A_3;而在域 $Q^*(\sqrt[3]{4})$ 上此多项式的Galois群为单位置换群$\{I\}$. 即在域 Q^* 上此多项式的Galois群的不变子群列为

$$A_3,\{I\}$$

若认为此多项式为域 $\mathbf{Q}$ 上的多项式,则要先求

$$\omega_3=\frac{-1+\sqrt{3}\mathrm{i}}{2}=\frac{-1+\sqrt{-3}}{2}$$

即先求平方根$\sqrt{-3}$,再求$\sqrt[3]{4}$,共两次求方根才得 x_1,x_2,x_3 的表示式. 在定理12.1例中已求得在域 $\mathbf{Q}$ 上此多项式的Galois群为 S_3;在域 $Q(\omega_3)$ 上此多项式的Galois群为 A_3;而在域 $Q(\omega_3,\sqrt[3]{4})=Q^*(\sqrt[3]{4})$ 上此多项式的Galois群为$\{I\}$. 故在域 $\mathbf{Q}$ 上此多项式的Galois群的不变子群列的

$$S_3,A_3,\{I\}$$

例6 考察三次方程

$$x^3-7x+7=0$$

它在有理数域 $\mathbf{Q}$ 上不可约(由易见它无有理数根可知).

按第六章中Cardan公式可得其三根

$$x_1=\sqrt[3]{-\frac{7}{2}+\sqrt{\frac{-49}{108}}}+\sqrt[3]{-\frac{7}{2}-\sqrt{\frac{-49}{108}}}=\alpha+\frac{\frac{7}{3}}{\alpha}$$

其中

$$\alpha=\sqrt[3]{-\frac{7}{2}+\frac{7}{18}\sqrt{3}\,\mathrm{i}}$$

(注意要使两立方根之积为 $-\frac{p}{3}=\frac{7}{3}$)

$$x_2=\alpha\omega_3+\frac{\frac{7}{3}}{\alpha\omega_3}$$

$$x_3 = \alpha\omega_3^2 + \frac{\frac{7}{3}}{\alpha\omega_3^2}$$

即在域 **Q** 表示根要用两层根式.

但在域 $Q(\omega_3) = Q^*$ 表示根，因

$$\omega_3 = \frac{-1-\sqrt{3}\mathrm{i}}{2}$$

$$\sqrt{3}\mathrm{i} = -1-2\omega_3$$

实际上只用 $Q(\omega_3)$ 的数 $-\frac{7}{2}+\frac{7}{18}\sqrt{3}\mathrm{i}$ 的立方根一次即可表示 x_1, x_2, x_3.

现表述用 Galois 群判定可用多层根式解出 $f(x)$ 所有根的准则.

定理12.3 设域 Δ 上多项式 $f(x)$ 可用域 Δ 上多层根式表示其所有根，则(必要条件)域 Δ 上多项式 $f(x)$ 的 Galois 群 G 有组合因数的素数的不变子群列

$$G, \cdots, \{I\}$$

证 设多项式 $f(x)$ 的根 x_1 在域 Δ 的多层根式表示式的根指数列为

$$n_1, n_2, \cdots, n_k$$

则由定理12.2知相应 Galois 群有组合因数为素数的不变子群列的最后一项 G_k^* 为在该定理证明所述的域 $\Delta^*(\xi^{(1)}, \xi^{(2)}, \cdots, \xi^{(k)})$ 上多项式 $f(x)$ 的 Galois 群

$$x_1 \in \Delta^*(\xi^{(1)}, \xi^{(2)}, \cdots, \xi^{(k)})$$

多项式 $f(x)$ 的属于此域的根(设有 m 个)如同 x_1 不被 G_k^* 的任一置换改变(据定理5.2). 这些根可用上述同型多层根式表示.

若多项式 $f(x)$ 尚有上述 m 个根以外的根，因这些根在域 Δ 上的多层根式表示式亦可看成在该定理证明所述的域 $\Delta^*(\xi^{(1)}, \xi^{(2)}, \cdots, \xi^{(k)})$ 上的多层根式表示式，在这些根中任取 x^{**} 代替 x^*，用域 $\Delta^*(\xi^{(1)}, \xi^{(2)}, \cdots, \xi^{(k)})$ 代替域 Δ，同样知子群 G_k^* 有组合因数为素数的不变子群列

$$G_k^*, \cdots, G^{**}$$

其中 G^{**} 为在域 $\Delta^*(\xi^{(1)}, \xi^{(2)}, \cdots, \xi^{(k)})$ 的一扩域上多项式 $f(x)$ 的 Galois 群，x^{**} 属于此扩域. 多项式 $f(x)$ 属于此扩域的根(含前述 m 个根)数有增加，即被 G^{**} 所有置换不变的根数有增加.

(如此继续论证)最后得出组合因数为素数的不变子群列

$$\cdots,G_k^*,\cdots,G^{**},\cdots,G^{***},\cdots,\widetilde{G}$$

$\widetilde{G}$ 为域 Δ 的某扩域上多项式 $f(x)$ 的 Galois 群,多项式 $f(x)$ 的所有根属于此扩域,被群 $\widetilde{G}$ 的所有置换不变. 故 $\widetilde{G}=\{I\}$.

例7 取域 $\mathbf{Q}$ 上多项式

$$f(x)=(x^2-2)(x^2-3)$$

其四根为

$$\sqrt{2},-\sqrt{2},\sqrt{3},-\sqrt{3}$$

在第四章的例中求得根域的共轭本原数为

$$\sqrt{2}+\sqrt{3},\sqrt{2}-\sqrt{3},-\sqrt{2}+\sqrt{3},-\sqrt{2}-\sqrt{3}$$

取

$$\rho=\sqrt{2}+\sqrt{3}$$

在第五章中已求得相应的 Galois 群为

$$G=\{I,(\sqrt{2},-\sqrt{2}),(\sqrt{3},-\sqrt{3}),(\sqrt{2},-\sqrt{2})(\sqrt{3},-\sqrt{3})\}$$

取根

$$x^*=\xi^{(1)}=\sqrt{2}\notin\mathbf{Q}$$

$$(\xi^{(1)})^2=2\in\mathbf{Q}$$

在域 $Q(\sqrt{2})$ 上因式(4.3)中两因式之积

$$(x-\sqrt{2}-\sqrt{3})(x-\sqrt{2}+\sqrt{3})=x^2-2\sqrt{2}x-1$$

已是域 $Q(\sqrt{2})$ 上不可约多项式,故共轭本原数只是

$$\sqrt{2}+\sqrt{3},\sqrt{2}-\sqrt{3}$$

在域 $Q(\sqrt{2})$ 上多项式 $f(x)$ 的 Galois 群 G' 只是由于 $\sqrt{2}+\sqrt{3}$ 不变及 $\sqrt{2}+\sqrt{3}\to\sqrt{2}-\sqrt{3}$ 两变换在多项式 $f(x)$ 的四根产生的两置换所组成

$$G'=\{I,(\sqrt{3},-\sqrt{3})\}$$

易见 G' 为 G 的不变子群(按定理2.7易见对任何 $\sigma\in G,\tau\in G'$ 有 $\sigma^{-1}\tau\sigma\in G'$),$[G:G']=\frac{4}{2}=2$. 它的所有置换不改变两根 $\sqrt{2}$,$-\sqrt{2}$,但 G' 非恒等置换群 $\{I\}$.

若改取余下一根

$$x^*=\xi^{(1)}=\sqrt{3}$$

$$(\xi^{(1)})^2 = 3 \in \mathbf{Q}$$

类似可知,在域 $Q(\sqrt{3})$ 上多项式 $f(x)$ 根域的共轭本原数为

$$\sqrt{2}+\sqrt{3}, -\sqrt{2}+\sqrt{3}$$

在域 $Q(\sqrt{3})$ 上多项式 $f(x)$ 的 Galois 群为

$$G' = \{I, (\sqrt{2}, -\sqrt{2})\}$$

其所有置换不改变二根 $\sqrt{3}, -\sqrt{3}$,但非恒等置换群 $\{I\}$.

若在上述两域分别再取

$$x^{**} = \sqrt{3}$$

$$x^{**} = \sqrt{2}$$

则在域 $Q(\sqrt{2},\sqrt{3})$ 上多项式 $f(x)$ 分解为4个一次因式之积,故显然相应 Galois 群为恒等置换群 $\{I\}$,它为上述两个 G' 的不变子群. $[G':\{I\}] = \frac{2}{1} = 2$. 即按两不同次序解出多项式 $f(x)$ 的全部根,G 的组合因数为素数(2,2)的不变子群列分别为

$$G, \{I, (\sqrt{3}, -\sqrt{3})\}, \{I\}$$

$$G, \{I, (\sqrt{2}, -\sqrt{2})\}, \{I\}$$

实际上任何可约方程都会出现上述情况,由于其各个不可约因式的根的多层根式表示式不同型,故按其中一个不可约因式的根 x^* 的多层根式表示式按定理 12.2 从 x^* 的表示式中的 $\xi^{(1)}$, $\xi^{(2)}$, $\cdots$ 求Galois群 G 的不变子群列时一般与其他不可约因式无关,故得到不改变 x^* 的不变子群时未能使其任一置换不改变其他不可约因式的根,按不同次序解出多项式 $f(x)$ 全部根所得出的不变子群列不会完全相同.

在定理 12.2 中的各例都是不可约方程,在求得的不变子群列的(其所有置换)不改变一根 x^* 的群都是恒等置换群 $\{I\}$,即未出现如同本例(可约方程)的不变子群列的情况. 但实际上当域 Δ 上多项式 $f(x)$ 不可约时也可能出现上述如同本例的情况,只因在扩展多项式 $f(x)$ 的系数域时可能在某次扩展到域 $\Delta(\xi^{(1)}, \xi^{(2)}, \cdots, \xi^{(j)})(1 \leqslant j < k)$,多项式在此域变成可约,从而在此域上多项式 $f(x)$ 的 Galois 群的不变子群列便出现上述情况.

如群($G_8^{(4)}$ 型)

$$\{I, (x_1, x_2)(x_3, x_4), (x_1, x_3)(x_2, x_4), (x_1, x_4)(x_2, x_3),$$

$$(x_1,x_2),(x_3,x_4),(x_1,x_3,x_2,x_4),(x_1,x_4,x_2,x_3)\}$$

按定理 2.7 易验证它有不变子群($G_4^{(4)}$ Ⅲ 型)

$$\{I,(x_1,x_2),(x_3,x_4),(x_1,x_2)(x_3,x_4)\}$$

$$[G_{(8)}^4 : G_{(4)}^4 \text{Ⅲ}] = \frac{8}{4} = 2$$

取在域 **Q** 上多项式(暂未求得具体例子)$f(x)$,使相应 Galois 群为 $G_8^{(4)}$,则按定理 12.2 知,在根域存在数 ξ,使在域 $Q(\xi)$ 上此多项式 $f(x)$ 的 Galois 群为 G_4^4Ⅲ(非可迁群). 由引理 7.1 及其推论知,在域 $Q(\xi)$ 上多项式 $f(x)$ 必定可约,且可分解为两个(域 $Q(\xi)$ 上的)二次因式之积,此两个二次固式分别为

$$(x-x_1)(x-x_2)$$

$$(x-x_3)(x-x_4)$$

(因 $G_4^{(4)}$Ⅲ的所有置换把 x_1,x_2 及 x_3,x_4 两组根分别仍变为此两组根本身). 于是继续扩展系数域时会得出 $G_4^{(4)}$Ⅱ的不变子群列如同上例.

第十三章 至少五次的代数方程不存在用多层根式表示的求根公式(卢芬尼 - 阿贝尔(Ruffini-Abel) 定理)

Galois 理论基本定理 12.2,12.3 虽是代数方程的根可用多层根式表示的必要条件,但方程的根未求出时何以知道方程的 Galois 群是否存在所述的不变子群列? 此二定理确极抽象似无什么意义. 但实际上按这二定理解决了历史上长期未能解决的几个难题——证明了其不可能性. 第十三章和第十四章就是论述 Galois 理论基本定理的应用.

一、二、三、四次方程有求根公式,且适用于复系数方程. 并非所有至少 5 次的复系数方程都不能用其系数的多层根式表示其根(例如方程 $x^5-2=0$ 的根就是方根). 而是几乎所有(即下文所述“一般的”)至少五次方程的根不能用多层根式表示,当然就不存在用多层根式表示的一般求根公式(由 Ruffini 所提出,但其证明有漏洞. 由 Abel 最先给予了严密证明).

现先表述“一般”复系数 n 次方程之定义,并证明相应 Galois 群为 n 次对称群 S_n.

定义 13.1　n 个复数 $x_1,x_2,\cdots,x_n$ 称为代数独立的，如果不存在有理数域 $\mathbf{Q}$ 上的 n 元多项式 P，使

$$P(x_1,x_2,\cdots,x_n)=0$$

现证所述 n 复数组 $(x_1,x_2,\cdots,x_n)$ 存在且是几乎所有的 n 复数组：取超越数 x_1（非“代数数”，即不是任何有理系数代数方程之根）；取复数 x_2 不是域 $Q(x_1)$ 上任何代数方程之根；取复数 x_3 不是域 $Q(x_1,x_2)$ 上任何代数方程之根；……；取复数 x_n 不是域 $Q(x_1,x_2,\cdots,x_{n-1})$ 上任何代数方程之根. 则易见 $x_1,x_2,\cdots,x_n$ 为代数独立的. 由于所有有理系数方程组成之集为可列集（或称可数集），它们的所有根即所有代数数为可列集，而（所有）复数集为不可列集，于是上述所有（超越数）x_1 组成不可列集；取定 x_1，则域 $Q(x_1)$ 的所有数（x_1 的两有理系数多项式之商）组成可列集，故域 $Q(x_1)$ 上所有代数方程及其根都组成可列集，从而上述所有 x_2 组成不可列集，而余下的数组成可列集；同样知上述所有 $x_3,x_4,\cdots,x_n$ 都组成不可列集，余下的数都组成可列集. 即上述所有 n 复数组 $(x_1,x_2,\cdots,x_n)$——代数独立的数组是几乎所有的 n 复数组.

定义 13.2　设方程

$$x^n+\sum_{j=1}^{n}a_jx^{n-j}=0 \tag{13.1}$$

的 n 个根 $x_1,x_2,\cdots,x_n$ 为代数独立的，则称此方程为 n 次“一般”方程，其系数的有理域为

$$Q(a_1,a_2,\cdots,a_n)$$

这时所有 a_j 必非有理数（否则据 Vieta 定理知 $x_1,x_2,\cdots,x_n$ 非代数独立）.

易见系数列

$$a_1,a_2,\cdots,a_n$$

与根集

$$\{x_1,x_2,\cdots,x_n\}$$

有一一对应关系，故由代数独立的所有 n 复数组是几乎所有的 n 复数组，知所有 n 次“一般”复系数方程是几乎所有 n 次复系数方程. 但常见的有理系数方程必然不是“一般”方程.

定理 13.1　设方程（13.1）为 n 次“一般”复系数方程，视其为域 $Q(a_1,a_2,\cdots,a_n)$ 上的代数方程，则在域 $Q(a_1,a_2,\cdots,a_n)$ 上此方程的 Galois 群为 n 次对称群 S_n.

证　设域 $Q(a_1,a_2,\cdots,a_n)$ 上方程（13.1）的 Galois 群为 G.

对任何取定的置换

$$\tau \in G$$

易见置换集

$$\{\sigma\tau : \sigma \in G\} = G$$

(由置换群之定义知左边任一置换属于群 G;反之对任何置换 $\sigma' \in G$,由群之定义有逆置换 $\tau^{-1} \in G$,且(记)$\sigma = \sigma'\tau^{-1} \in G$,于是 $\sigma' = \sigma\tau$ 属左边置换集).

取有理数域 $\mathbf{Q}$ 上 n 元多项式

$$M(x_1, x_2, \cdots, x_n) = \sum_{\sigma \in G} (x_1 x_2^2 x_3^3 \cdots x_n^n)\sigma$$

对上述 τ 有

$$M(x_1, x_2, \cdots, x_n)\tau = \sum_{\sigma \in G} (x_1 x_2^2 x_3^3 \cdots x_n^n)\sigma\tau = \sum_{\sigma' \in G} (x_1 x_2^2 x_3^3 \cdots x_n^n)\sigma' = M(x_1, x_2, \cdots, x_n)$$

故对任何 $\tau \in G$ 有

$$M(x_1, x_2, \cdots, x_n)\tau = M(x_1, x_2, \cdots, x_n)$$

因 τ 为域 $Q(a_1, a_2, \cdots, a_n)$ 上方程(13.1)的 Galois 群 G 的任一置换,$M(x_1, x_2, \cdots, x_n)$ 经置换 τ 不变,故据定理 5.2 知

$$M(x_1, x_2, \cdots, x_n) \in Q(a_1, a_2, \cdots, a_n)$$

再据 Vieta 定理知 $a_1, a_2, \cdots, a_n$ 都为方程(13.1)的 n 个根 $x_1, x_2, \cdots, x_n$ 的对称式,从而群 S_n 的任一置换不改变 $M(x_1, x_2, \cdots, x_n)$.

若群 $G \neq S_n$,任取

$$\tau' \in S_n \backslash G$$

则 $(x_1 x_2^2 x_3^3 \cdots x_n^n)\tau'$ 属

$$M(x_1, x_2, x_3, \cdots, x_n)\tau' = \sum_{\sigma \in G} (x_1 x_2^2 x_3^3 \cdots x_n^n)\sigma\tau'$$

与

$$\sigma = I \in G$$

相应的一项;但因 $\tau \notin G$,$(x_1 x_2^2 x_3^3 \cdots x_n^n)\tau'$ 不属 $M(x_1, x_2, \cdots, x_n)$ 表示式的项,视 $x_1, x_2, \cdots, x_n$ 为任意变元时 $(x_1 x_2^2 x_3^3 \cdots x_n^n)\tau'$ 不会与 $M(x_1, x_2, \cdots, x_n)$ 的表示式中任一项恒等(因易见 $x_1 x_2^2 x_3^3 \cdots x_n^n$ 经群 S_n 的不同置换(τ' 及 G 内任一置换)得出不恒等的项),故

$$M(x_1, x_2, \cdots, x_n) \neq M(x_1, x_2, \cdots, x_n)\tau'$$

于是 $M(x_1, x_2, \cdots, x_n) - M(x_1, x_2, \cdots, x_n)\tau'$ 为域 $\mathbf{Q}$ 上 n 元多项式

(不恒等于0). 又由于 $x_1, x_2, \cdots, x_n$ 代数独立,故

$$M(x_1, x_2, \cdots, x_n) - M(x_1, x_2, \cdots, x_n)\tau' \neq 0$$

即

$$M(x_1, x_2, \cdots, x_n) \neq M(x_1, x_2, \cdots, x_n)\tau'$$

这与上段最后结论矛盾,得证 $G = S_n$.

再考察 $n \geqslant 5$ 时对称群 S_n 及交代群 A_n 之不变子群,以推出上述 Galois 群 S_n 无适合 Galois 理论基本定理的不变子群列(从而知至少5次的复系数"一般"方程无多层根式解).

定理13.2 $n \geqslant 5$ 时 n 个元素 $1, 2, \cdots, n$ 的 n 次交代群 A_n 的不变子群只有 A_n 本身及恒等置换群 $\{I\}$.

证 在第二章中已论述 A_n 及 $\{I\}$ 为 A_n 的不变子群,设

$$H \neq \{I\}$$

为群 A_n 之不变子群. 群 H 含非恒等置换 I,把群 H 的所有非恒等置换表示为若干个无共同元素之轮换的积. 设改变元素个数最小者为置换 h.

先证:

①h 的上述轮换积表示式中各轮换因式所含元素个数相同:否则设各轮换因式所含的元素个数最小为 r,则 h 尚有超过 r 个元素的轮换因式. 因群 H 必含置换 h',但置换 h' 使含 r 个元素的所有轮换因式(也要 r 次幂,因含互不相同的元素的轮换之积适合交换律)中所有元素不变,于是群 H 的置换 h' 所改变的元素个数小于置换 h 所改变的元素个数,而其余轮换因式因所含元素超过 r 个,其 r 次幂仍使其中各元素改变,故 h' 非恒等置换,这与置换 h 的定义矛盾. 所述得证.

再证:

②h 的各轮换因式(所含元素个数相同)不能含至少4个元素:否则设 h 含轮换因式(其中 $\alpha_1, \alpha_2, \alpha_3, \alpha_4, \cdots$ 为 $1, 2, 3, 4, \cdots, n$ 中任意不同元素)

$$(\alpha_1, \alpha_2, \alpha_3, \alpha_4, \cdots)$$

取群 A_n 的(偶)置换

$$g = (\alpha_2, \alpha_3, \alpha_4)$$

由置换 $h \in H$ 及群之定义知

$$h^{-1} \in H$$

又因群 H 为群 A_n 之不变子群, $g \in A_n$, $h^{-1} \in H$ 知

$$g^{-1}h^{-1}g \in H$$

于是

$$hg^{-1}h^{-1}g = h(g^{-1}h^{-1}g) \in H$$

除轮换$(\alpha_1,\alpha_2,\alpha_3,\alpha_4,\cdots)$中的元素$\alpha_1,\alpha_2,\alpha_3,\alpha_4,\cdots$外其余元素经置换$h$会有改变,再经置换$g^{-1}$(只与$\alpha_2,\alpha_3,\alpha_4$有关)不变,又经置换$h^{-1}$变回原来的元素,最后经置换$g$又不变,故这些元素经置换$hg^{-1}h^{-1}g$都不变.计算$hg^{1}h^{-1}g$只需考虑轮换$(\alpha_1,\alpha_2,\alpha_3,\alpha_4,\cdots)$

$$\begin{aligned} hg^{-1}h^{-1}g = &(\alpha_2,\alpha_3,\alpha_4,\cdots,\alpha_1)(\alpha_4,\alpha_3,\alpha_2)(\alpha_4,\alpha_3,\alpha_2,\alpha_1,\cdots)\cdot \\ &(\alpha_2,\alpha_3,\alpha_4)^{①} = \\ &(\alpha_2,\alpha_3,\alpha_4,\cdots,\alpha_1)(\alpha_2,\alpha_4,\alpha_3,\alpha_1,\cdots) = \\ &(\alpha_2,\alpha_1,\alpha_4) \end{aligned}$$

(上式第一轮换因式中省略号内各元素及α_1在第二轮换因式的次序正好与第一因式相反,故省略号内各元素经两置换后不变;又从第一轮换因式知α_4要变为其后第一元素,此第一元素在第二轮换因式处最后位置,故再变成α_2).

群H的置换$hg^{-1}h^{-1}g$只改变3个元素,而置换h至少改变4个元素,又与置换h之定义矛盾.得证所述.

于是置换h为若干个3阶轮换之积或若干个2阶轮换之积.下证轮换因式个数为1:

若

$$h = (\alpha_1,\alpha_2,\alpha_3)(\alpha_4,\alpha_5,\alpha_6)\cdots$$

取群A_n的(偶)置换(据引理1.5)

$$g = (\alpha_1,\alpha_2,\alpha_5)$$

因置换$h \in H$,故置换$h^{-1} \in H, g^{-1}hg \in H$,于是

$$h^{-1}g^{-1}hg \in H$$

但(同样不必考虑省略号内的轮换的元素)

$$\begin{aligned} h^{-1}g^{-1}hg = &(\alpha_6,\alpha_5,\alpha_4)(\alpha_3,\alpha_2,\alpha_1)(\alpha_5,\alpha_2,\alpha_1)(\alpha_1,\alpha_2,\alpha_3)\cdot \\ &(\alpha_4,\alpha_5,\alpha_6)(\alpha_1,\alpha_2,\alpha_5)^{②} = \\ &(\alpha_6,\alpha_5,\alpha_4)(\alpha_3,\alpha_2,\alpha_1)(\alpha_2,\alpha_5,\alpha_3)(\alpha_4,\alpha_1,\alpha_6) = \\ &(\alpha_6,\alpha_3,\alpha_5,\alpha_1,\alpha_2) \end{aligned}$$

群H之置换$h^{-1}g^{-1}hg$只改变5个元素,而置换h至少改变6个元素,

---

①按定理2.7求后三因式之积.

②按定理2.7求后四轮换因式之积.

亦与置换 h 之定义矛盾.

若

$$h=(\alpha_1,\alpha_2)(\alpha_3,\alpha_4)(\alpha_5,\alpha_6)\cdots$$

取群 A_n 之(偶) 置换

$$g=(\alpha_1,\alpha_2,\alpha_5)$$

则群 H 之置换

$$\begin{aligned}hg^{-1}h^{-1}g=&(\alpha_1,\alpha_2)(\alpha_3,\alpha_4)(\alpha_5,\alpha_6)(\alpha_5,\alpha_2,\alpha_1)\cdot\\&(\alpha_2,\alpha_1)(\alpha_4,\alpha_3)\cdot\\&(\alpha_6,\alpha_5)(\alpha_1,\alpha_2,\alpha_5)=\\&(\alpha_1,\alpha_2)(\alpha_3,\alpha_4)(\alpha_5,\alpha_6)(\alpha_5,\alpha_2)\cdot\\&(\alpha_4,\alpha_3)(\alpha_6,\alpha_1)=\\&(\alpha_1,\alpha_5)(\alpha_2,\alpha_6)\end{aligned}$$

只改变 4 个元素,而置换 h 改变至少 6 个元素,又与 h 之定义矛盾.

若

$$h=(\alpha_1,\alpha_2)(\alpha_3,\alpha_4)$$

仍取

$$g=(\alpha_1,\alpha_2,\alpha_5)$$

则 H 之置换

$$\begin{aligned}hg^{-1}h^{-1}g=&(\alpha_1,\alpha_2)(\alpha_3,\alpha_4)(\alpha_5,\alpha_2,\alpha_1)(\alpha_2,\alpha_1)\cdot\\&(\alpha_4,\alpha_3)(\alpha_1,\alpha_2,\alpha_5)=\\&(\alpha_1,\alpha_2)(\alpha_3,\alpha_4)(\alpha_5,\alpha_2)(\alpha_4,\alpha_3)=\\&(\alpha_1,\alpha_5,\alpha_2)\end{aligned}$$

只改变 3 个元素,而 h 改变 4 个元素,亦与 h 之定义矛盾.

于是 h 只能是一个 3 阶轮换或一个 2 阶轮换,但后者是奇置换(引理 1.5),由(全部) 偶置换组成的群 A_n 的不变子群 H 只含偶置换,于是 h 不可能是一个 2 阶轮换,只能是一个 3 阶轮换

$$h=(\alpha_1,\alpha_2,\alpha_3)$$

再证 n 个元素 $1,2,\cdots,n$ 的任何 3 阶轮换 $(\beta_1,\beta_2,\beta_3)\in H$:把排列 $\alpha_1,\alpha_2,\alpha_3$ 及 β_1,β_2,β_3 都补成 $1,2,3,\cdots,n(n\geqslant 5)$ 的排列

$$\alpha_1,\alpha_2,\alpha_3,\alpha_4,\alpha_5,\cdots$$

$$\beta_1,\beta_2,\beta_3,\beta_4,\beta_5,\cdots$$

若两排列的奇偶性不同,则对调 (β_4,β_5) 可使它们的奇偶性相同(据引理 1.5 及引理 1.4 推论 1),不妨设第二排列已作上述处理,得

群 A_n 中的偶置换

$$g=\begin{pmatrix}\alpha_1,\alpha_2,\alpha_3,\cdots,\alpha_n\\ \beta_1,\beta_2,\beta_3,\cdots,\beta_n\end{pmatrix}$$

于是(按定理 2.7 算)

$$g^{-1}hg=(\beta_1,\beta_2,\beta_3)$$

但 $h\in H, g\in A_n, H$ 为 A_n 的不变子群,故 $g^{-1}hg=(\beta_1,\beta_2,\beta_3)\in H$.

于是群 H 含有所有三阶轮换,从而含它们之积.

现证群 A_n 中任一非恒等置换可表示为若干个 3 阶轮换之积:由引理 1.4 知群 A_n 的任一非恒等置换(偶置换)g 可表示为偶数个对换之积,故可设

$$g=t_1t_2t_3t_4\cdots t_{2m-1}t_{2m}\quad (m\in\mathbf{Z}^+)$$

其中每个 t_j 为二元素对换. 若 t_1,t_2 是同一对元素对换,则易见 $t_1t_2=I$(可省去);若 t_1,t_2 对换的两对元素中有一元素相同,设

$$t_1=(\alpha,\beta),\ t_2=(\alpha,\gamma)$$

则

$$t_1t_2=(\alpha,\beta,\gamma)$$

若 t_1,t_2 对换的两对元素无相同元素,设

$$t_1=(\alpha,\beta),\ t_2=(\gamma,\delta)$$

则

$$\begin{aligned}t_1t_2&=(\alpha,\beta)(\gamma,\delta)=\\&(\alpha,\beta)(\alpha,\gamma)(\gamma,\alpha)(\gamma,\delta)=\\&(\alpha,\beta,\gamma)(\gamma,\alpha,\delta)\end{aligned}$$

于是 t_1t_2 可分别表示为 0,1,2 个 3 阶轮换之积. $t_3t_4,t_5t_6,\cdots$ 亦然,所述得证.

由于群 A_n 任一置换可表示为若干个 3 阶轮换之积,前面已证 3 阶轮换之积属于群 H,故

$$A_n\subset H$$

反之因群 H 为 A_n 的子群,故

$$H\subset A_n$$

从而

$$H=A_n$$

于是 A_n 的不变子群只有 $\{I\}$ 及 A_n 定理证毕.

定理 13.3 当 $n\geqslant 5$ 时 n 次对称群 S_n 的不变子群只有 S_n,A_n

及$\{I\}$.

证 第二章中已论述群S_n,A_n及$\{I\}$是群S_n的不变子群,设群S_n尚有不变子群

$$S' \neq S_n, A_n, \{I\}$$

设群S'与A_n之交集$S' \cap A_n$中任二置换σ_1,σ_2. 因

$$\sigma_1, \sigma_2 \in S' \cap A_n$$

故

$$\sigma_1, \sigma_2 \in S'$$

从而

$$\sigma_1 \sigma_2 \in S'$$

同理有

$$\sigma_1 \sigma_2 \in A_n$$

于是

$$\sigma_1 \sigma_2 \in S' \cap A_n$$

即置换集$S' \cap A_n$内可进行乘法运算,据定理2.4知$S' \cap A_n$为置换群,显然为群A_n的子群.

现证

$$S' \cap A_n \neq A_n$$

否则$S' \cap A_n = A_n$,从而$A_n \subset S'$,又因$S' \neq A_n, S_n$,故A_n为S'的真子集,S'为S_n的真子集,于是

$$[A_n:A_n] < [S':A_n] < [S_n:A_n]$$

即

$$1 < [S':A_n] < 2$$

从而$[S':A_n]$非整数,与定理2.5矛盾. 所述得证.

再证

$$S' \cap A_n \neq \{I\} \tag{13.2}$$

否则$S' \cap A_n = \{I\}$. 因群A_n由所有n次偶置换所组成,从而$S' \cap A_n$由群S'中所有偶置换组成,于是$S' \cap A_n = \{I\}$即群S'的偶置换只有恒等置换I. 但$S' \neq \{I\}$,故群S'应至少还有一个奇置换μ,从而有置换μ^2. 据引理1.4推论2,μ^2为偶置换,而S'中的偶置换只有I,故

$$\mu^2 = I$$

μ的无共同元素的轮换积表示式不能含至少有3个元素的轮换(否则因μ^2使轮换也要平方,但轮换的平方未能使其中各元素不变,从

而$\mu^2 \neq I$),故奇置换μ为无共同元素的2阶轮换(即对换)之积. 据引理1.4知其中2阶轮换个数为奇数,故可设

$$\mu = (a_1, a_2)(a_3, a_4)(a_5, a_6)\cdots(a_{4n+1}, a_{4n+2}) \quad (n \geqslant 0)$$

取置换

$$(a_1, a_3) \in S_n$$

由$\mu \in S'$及S'为S_n的不变子群,故

$$(a_1, a_3)^{-1}\mu(a_1, a_3) \in S' \tag{13.3}$$

即当$n \geqslant 1$时为

$$(a_1, a_3)^{-1}(a_1, a_2)(a_3, a_4)(a_5, a_6)\cdots(a_{4n+1}, a_{4n+2})(a_1, a_3) \in S'$$

按定理2.7算得

$$(a_3, a_2)(a_1, a_4)(a_5, a_6)\cdots(a_{4n+1}, a_{4n+2}) \in S'$$

且为奇置换,它与群S'的奇置换μ之积为

$$(a_3, a_2)(a_1, a_4)(a_1, a_2)(a_3, a_4) = (a_3, a_1)(a_2, a_4) \in S'$$

于是群S'尚有偶置换$(a_1, a_3)(a_2, a_4)$(据引理1.5推论),与前述S'中只有偶置换I矛盾;而当$n = 0$时$\mu = (a_1, a_2)$,式(13.3)为

$$(a_1, a_3)^{-1}(a_1, a_2)(a_1, a_3) \in S'$$

即

$$(a_3, a_2) \in S'$$

它与μ之积

$$(a_3, a_2)(a_1, a_2) = (a_3, a_1, a_2) \in S'$$

据引理1.5,它也是偶置换,亦与前述矛盾. 于是式(13.2)得证.

现证群$S' \cap A_n$为群A_n之不变子群:

对任何置换

$$\alpha \in S' \cap A_n \subset S'$$

及任何置换

$$\beta \in A_n$$

因群S'为群A_n之不变子群知

$$\beta^{-1}\alpha\beta \in S' \tag{13.4}$$

又因置换$\alpha \in S' \cap A_n \subset A_n$,$\alpha$为偶置换,置换$\beta \in A_n$,$\beta$为偶置换,据引理2.1知$\beta^{-1}$亦为偶置换,再据引理1.4推论3知$\beta^{-1}\alpha\beta$亦为偶置换,于是

$$\beta^{-1}\alpha\beta \in A_n$$

又从式(13.4)知

$$\beta^{-1}\alpha\beta \in S' \cap A_n$$

由于 α,β 分别为 $S' \cap A_n$ 及 A_n 的任意置换,从而(据不变子群的第二个定义)知 $S' \cap A_n$ 为 A_n 之不变子,但上面已证 $S' \cap A_n \neq A_n$,$\{I\}$,于是群 A_n 除有不变子群 A_n,$\{I\}$ 外尚有不变子群 $S' \cap A_n$,这与定理 13.2 矛盾.即前面所述 S' 不存在.本定理得证.

定理 13.4 整数 $n \geqslant 5$ 时"一般"复系数 n 次方程(13.1)的任一根 x^* 不能表示为系数域 $Q(a_1,a_2,\cdots,a_n)$ 上的多层根式.

证 由定理 13.3 知群 S_n 的不变子群只有 S_n,A_n 及 $\{I\}$,而

$$[S_n:S_n]=1$$

$$[S_n:A_n]=\frac{n!}{\frac{n!}{2}}=2$$

$$[S_n:\{1\}]=\frac{n!}{1}=n!\ =2\cdot 3\cdot 4\cdot 5\cdot \cdots \cdot n$$

(当 $n \geqslant 5$ 时非素数).故以 S_n 为首项的组合因数为素数的下一项只能为 A_n.

由定理 13.2 知 A_n 的不变子群只有 A_n 及 $\{I\}$,而

$$[A_n:A_n]=1$$

$$[A_n:\{1\}]=\frac{\frac{n!}{2}}{1}=\frac{n!}{2}=3\cdot 4\cdot 5\cdot \cdots \cdot n$$

(当 $n \geqslant 5$ 时非素数).故没有以 A_n 为首项的组合因数为素数的不变子群列.

由定理 13.1 知域 $Q(a_1,a_2,\cdots,a_n)$ 上复系数"一般"方程(13.1)的 Galois 群为 S_n,但据上面证明,此 Galois 群 S_n 的组合因数为素数的不变子群列只有

$$S_n, A_n$$

而方程(13.1)的任一根 x^* 不能被 A_n 的所有置换不变(因当 n 为奇数,由引理 1.5 可取偶置换 $(x_1,x_2,\cdots,x_n) \in A_n$,此置换改变所有根 $x_1,x_2,\cdots,x_n$;当 n 为偶数,取偶置换 $(x_1,x_2,\cdots,x_{n-1}) \in A_n$,此置换改变除 x_n 外的其余各根,再取 A_n 的(偶)置换 (x_1,x_2,x_n) $(n \geqslant 5$,x_1,x_2,x_n 互不相同),则此置换改变 x_n).于是据定理 12.2,方程(13.1)的任一根 x^* 不能用域 $Q(a_1,a_2,\cdots,a_n)$ 上多层根式表示.

附注.

《苏俄教育科学院初等数学全书(代数)》[1] 和 *Galois Theory of Algebraic Equations*[2] 未用 Galois 的定理证定理 13.3[3],又(未用 Galois 的定理)直接证 Ruffini-Abel 定理:至少五次代数方程不存在用多层根式表示的求根公式[4]. 证定理 13.3 时实际上"证明"了域 Δ 上任何 $n \geqslant 5$ 次代数方程的根都不能用域 Δ 上多层根式表示. 又把《苏俄教育科学院初等数学全书(代数)》中对 Ruffini-Abel 定理的证明最后的置换 $t = (3,4,5)$, $u = (1,2,3)$ 分别改为 $t = (2,3)$, $u = (1,2)$,则限制"$n \geqslant 5$"可改为"$n \geqslant 3$",同样可"证"域 Δ 上至少三次方程都不存在用域 Δ 上多层根式表示的求根公式. 实际上,在《苏俄教育科学院初等数学全书(代数)》和 *Galois Theory of Algebraic Equations* 中证这二定理时都有一个同样的错误:在根域中一数用各根 $x_1, x_2, \cdots, x_n$ 的两不同形式的表示式中取相应 Galois 群以外的置换 $\sigma = (x_1, x_2, x_3)$,认为结果仍相等. 但如同第五章中所述,对根域的数用各根表示的不同表示式进行相应 Galois 群以外的置换会得出不同的数.

一至四次方程都有求根公式,三、四次方程的求根公式本是在设域 Δ 上方程的 Galois 群是对称群的情况下推导出来的(实际上从推导过程知所得公式适用于任何三、四次方程). 不同于 $n \geqslant 5$ 的情况,$n = 1,2,3,4$ 时对称群 S_n 都有(适合定理 12.3 的)组合因数为素数的不变子群列.

注意(见第二章):A_n 是 S_n 的不变子群,恒等置换群 $\{I\}$ 是 A_n 的不变子群.

因

$$S_1 = \{I\}$$

故 S_1 的上述不变子群列为 $\{I\}$.

因

$$A_2 = \{I\} \quad (\text{唯一二次偶群})$$

故 S_2 的上述不变子群列为

---

① 乌兹科夫,等. 苏俄教育科学院初等数学全书(代数)[M]. 丁寿田,译. 北京:商务印书馆,1954.

② JEAN – PIERRE TIGNOL. Galois Theory of Algebraic Equations[M]. Word Scientific,2001.

③ JEAN – PIERRE TIGNOL. Galois Theory of Algebraic Equations[M]. Word Scientific,2001.

④ JEAN – PIERRE TIGNOL. Galois Theory of Algebraic Equations[M]. Word Scientific,2001.

$$S_2, \{I\}$$

(组合因数为2).

S_3 的上述不变子群列为(从 A_3 起为所有三次偶群,见附录 Ⅰ 的三次偶群表)

$$S_3, A_3, \{I\}$$

(组合因数为:2,3).

S_4 的上述不变子群列为(从 A_4 起为所有四次偶群,见附录 Ⅰ 中的四次偶群表)

$$S_4, A_4, G_4^{(4)} \text{Ⅱ}, G_2^{(4)}, I$$①

(组合因数列为:2,3,2,2).

上述各组合因数列正好为求根公式中的多层根式的根指数列.

一至四次偶群,即 A_n 的所有子群已全部列入上述不变子群列,故它们都有组合因数为素数的不变子群列,最后一项为 $\{I\}$. 又易见任一非偶群 G 的偶部群 G_+ 为 G 之不变子群(任取 $\sigma \in G_+, \tau \in G$,由引理2.1知 τ^{-1} 与 τ 有相同奇偶性,于是无论 τ 为奇、偶置换,据引理1.4推论3易见 $\tau^{-1}\sigma\tau$ 为 G_+ 的置换,据不变子群第二定义知 G_+ 为 G 之不变子群),且由附录 Ⅰ 中引理1之推论知 $[G:G_+]=2$ 为素数,于是当 $n=1,2,3,4$,在 G 后接上 G_+,再在 G_+ 接上前述组合因数为素数的不变子群列便得 G 的组合因数为素数的不变子群列,最后一项为 $\{I\}$.

故任何一至四次群都有组合因数为素数的不变子群列,且最后一项为 $\{I\}$,再次阐明任何一至四次方程所有根都可用多层根式表示.

第十四章　实域上素数次不可约方程无多层根式解的充分条件

域 Δ 上至少五次的"一般"方程的根不能表示为域 Δ 的多层根式."一般"方程虽然包含了几乎所有代数方程,但却无具体方法判定一个方程是否为"一般"方程(极少数明显可见有多层根式解的方程当然不是"一般"方程,但除此以外却无法判定).但对域 Δ 上

---

① 易按定理2.7验证 A_4 与 $G_4^{(4)}$ Ⅱ, $G_4^{(4)}$(Ⅱ)与 $G_2^{(4)}$ 适合定义2.3.

素数$p(\geqslant 5)$次不可约方程若正好有一对共轭虚根(即正好有$p-2$个实根),则其任一根不能用域Δ上多层根式表示.这个充分条件还是较易检验的——用附录Ⅲ的Sturm(斯图姆)定理可求出任何实系数方程的实根个数;附录Ⅳ提供了判定任何五次以内有理系数多项式可约性的方法.本节论述这个充分条件.

引理14.1　设n次可迁群G有一置换为(单独一个)对换,则所有元素可分为$l\geqslant 1$组(各组元素个数相同)

$$C_1=\{a_{11},a_{12},\cdots,a_{1m}\}$$
$$C_2=\{a_{21},a_{22},\cdots,a_{2m}\}$$
$$\vdots$$
$$C_l=\{a_{l1},a_{l2},\cdots,a_{lm}\}$$
$$(m\geqslant 2,lm=n)$$

使同(异)组两元素的对换属于(不属于)群G;又对任何两组C_{i_1},$C_{i_2}(i_1\neq i_2)$,群G必有置换s使C_{i_1}的所有元素经置换s分别变成C_{i_2}的各个(不同)元素.

证　设群G有对换(a_0,b_0),对任一元素a,因G是可迁群故必存在一置换σ,它把元素a_0换成a,设置换σ把元素b_0换成元素b,由于逆置换$\sigma^{-1}\in G$,故群G含置换

$$\sigma^{-1}(a_0,b_0)\sigma=(a,b)$$

(据定理2.7)即对任一元素a,群G必有a与某元素b的对换(a,b).又若群G含两对换(a,b),(b,c),则群G中必有对换

$$(b,c)^{-1}(a,b)(b,c)=(a,c)$$

即群G中可对换的元素有传递性,于是可把n个元素分成l组:C_1,C_2,$\cdots$,C_l,使同(异)组两元素的对换属于(不属于)群G,但其中各组最后的元素先设为$a_{1m_1},a_{2m_2},\cdots,a_{lm_l}(m_1+m_2+\cdots+m_l=n)$.

若对$i_1,i_2=1,2,\cdots,l$,有$m_{i_1}\neq m_{i_2}$(即两组C_{i_1},C_{i_2}所含元素个数不同),不妨设$m_{i_1}>m_{i_2}$,因G为可迁群,存在置换$s\in G$使

$$a_{i_11}s=a_{i_21}$$

设

$$a_{i_1j}s=b_j\quad(j=2,3,\cdots,m_{i_1})$$

因a_{i_11}与a_{i_1j}属同一组C_i,故

$$(a_{i_11},a_{i_1j})\in G$$

群G含有置换

$$s^{-1}(a_{i_11},a_{i_1j})s=(a_{i_21},b_j)$$

于是

$$b_j \in C_{i_2}$$

即

$$\{a_{i_2 1}, b_2, b_3, \cdots, b_{m_{i_1}}\} \in C_{i_2}$$

又易见(其中 $b_1 = a_{i_2 1}$)

$$b_{j'} \neq b_j (当 j' \neq j, j', j \in \{1, 2, \cdots, m_{i_1}\})$$

(否则有 $b_{j'} = b_j$,从而 $a_{i_1 j'} = b_{j'} s^{-1} = b_j s^{-1} = a_{i_1 j}$ 得矛盾).

但 S_{i_2} 中只有 m_{i_2} 个元素,$m_{i_2} < m_{i_1}$,这与上述结果矛盾.故知任两组 C_{i_1}, C_{i_2} 所含元素个数相同(设为 m).且

$$a_{i_1 j} s \quad (j = 1, 2, \cdots, m)$$

取尽 C_{i_2} 的所有元素.

若 $m = 1$,则任二元素在不同组,此二元素之对换不属于群 G,与假设群 G 有对换矛盾,故 $m \geqslant 2$.

推论 素数 m 次可迁群 G 必为对称群 S_n.

证 因素数

$$n = lm$$

其中 l, m 为正整数,$m \geqslant 2$,故必有 $l = 1, m = n$.即 n 个元素在同一组 C_1,群 G 含有任二元素的对换,又从引理 1.4 知 S_n 的任一置换可分解为若干对换之积.故此置换属于群 G,于是 $S_n \subset G$,又显然 $G \subset S_n$ 故 $G = S_n$.

定理 14.1 设素数 $p \geqslant 5$,(部分)实域 Δ 上 p 次不可约方程 $f(x) = 0$ 正好有一对共轭虚根 x_1, x_2,则其所有根不能用域 Δ 上多层根式表示.

证 由引理 7.1 推论知,域 Δ 上不可约方程 $f(x) = 0$ 的 Galois 群 G 为素数 p 次可迁群.下证,对换 (x_1, x_2) 为 Galois 群 G 的置换.先证此对换导致域 Δ 上方程 $f(x) = 0$ 根域 Ω 在域 Δ 上的自同构映射.

对 $f(x)$ 各根 $x_1, x_2, \cdots, x_n$ 在域 Δ 上的有理式 $M(x_1, x_2, \cdots, x_n)$,若其值

$$M(x_1, x_2, x_3, x_4, \cdots, x_n) = \mu \in \Delta \subset R$$

上式两边取共轭复数,正好使 x_1, x_2 对换,实根 $x_3, x_4, \cdots, x_n$ 不变,μ 不变,故得

$$M(x_2, x_1, x_3, x_4, \cdots, x_n) = \mu$$

即置换 (x_1, x_2) 导致根域 Ω 的一个使域 Δ 的数不变的映射;又从前述知,对换 (x_1, x_2) 使根域 Ω 的每个数 α 变成其共轭复数 $\overline{\alpha}$,反

之对根域 Ω 的任一数

$$\beta = M'(x_1, x_2, x_3, x_4, \cdots, x_n) \text{—— 域 } \Delta \text{ 上有理式}$$

可取

$$\alpha = M'(x_2, x_1, x_3, x_4, \cdots, x_n) \in \Omega$$

使

$$\alpha(x_1, x_2) = M'(x_2, x_1, x_3, x_4, \cdots, x_n)(x_1, x_2) = M'(x_1, x_2, x_3, x_4, \cdots, x_n) = \beta$$

又因二复数之和、积的共轭复数分别等于二复数的共轭复数之和、积. 故对换 (x_1, x_2) 导致根域 Ω 在域 Δ 上的自同构映射.

于是由定理5.1的5°知导致此自同构映射的置换 (x_1, x_2) 为域 Δ 上多项式 $f(x)$ 的Galois群 G 的置换，再由引理14.1推论知群 $G = S_n$. 又由定理13.4的证明知 $f(x)$ 的任一根不能用域 Δ 的多层根式表示 —— 因该定理的证明实际是证明当多项式 $f(x)$ 适合上述结论对它的任一根不能用域 Δ 上多层根式表示.

例1 设多项式

$$x^5 - pqx + p$$

其中 p 为(正)素数，q 为正整数，则当 $q \geqslant 2$ 或 $p \geqslant 13$ 时此多项式的所有根不能用有理数域 $\mathbf{Q}$ 上多层根式表示.

解 此多项式除最高次项系数为1外其余各项系数可被素数 p 整除，又常数项不能被 p^2 整除，由Eisenstein(爱森斯坦)定理知此多项式在域 $\mathbf{Q}$ 不可约，又本例条件较附录Ⅲ中例的条件为强，附录Ⅲ例中已证这时此多项式正好有3个实根，即正好有一对共轭虚根，故由定理14.1知此多项式的所有根不能表示为域 $\mathbf{Q}$ 上的多层根式.

附录Ⅰ 构造三、四次偶群表及三、四次对称群 S_n 的真子群(指标小于 n)

只含(全部或部分)偶置换的群称偶群(如全部 n 次偶置换组成的交代群 A_n 及 $\{I\}$)，否则称非偶群(如 $n \geqslant 2$ 时的对称群 S_n).

本附录论述如何求出全部三、四次偶群，又为了用Lagrange(拉格朗日)预解方程推导求三、四次方程的根的公式的需要，求 $n = 3, 4$ 时对称群 S_n 的真子群 $G(G \neq S_n)$，使指标 $[S_n : G] < n$.

引理1 设非偶群 G 的所有奇(偶) 置换组成的集为G_- (G_+,它显然为一偶群,称为 G 的偶部群),在 G_- 中任取置换 τ_1,则

$$G_- = \{\sigma\tau_1 : \sigma \in G_+\}$$

这时称由偶群 G_+ 乘 τ_1 产生(非偶) 群 G.

证 因 τ_1 为奇置换,$\sigma \in G_+$ 为偶置换,由引理 1.4 推论 2 知上式右边集每一置换 $\sigma\tau_1$ 为奇置换;又

$$\tau_1 \in G_- \subset G$$

$$\sigma \in G_+ \subset G$$

故 $\sigma\tau_1 \in G$,于是 $\sigma\tau_1$ 属于 G 的奇置换集 G_-,得证

$$G_- \supset \{\sigma\tau_1 : \sigma \in G_+\}$$

反之,对任何 $\sigma' \in G_-$,因由引理2.1知 τ_1^{-1} 为群 G 中的奇置换,由引理1.4 推论 2 知

$$\sigma'\tau_1^{-1} \triangleq \sigma_0 \in G_+$$

于是

$$\sigma' = \sigma_0\tau_1 \in \{\sigma\tau_1 : \sigma \in G_+\}$$

从而

$$G_- \subset \{\sigma\tau_1 : \sigma \in G_+\}$$

得证

$$G_- = \{\sigma\tau_1 : \sigma \in G_+\}$$

推论 非偶群的阶数为其偶部群阶数的 2 倍.

例 1 三次交代群(所有偶置换,据引理 1.5)

$A_3 = S_3^+ = \{I, (1,2,3), (1,3,2)\}$

$S_3 = A_3 \cup \{(1,2), (2,3), (1,3)\}$(记此置换集为 S_3^-)

在 S_3^- 取 $\tau_1 = (1,2)$,易验证 $A_3 = S_3^+$ 各置换乘 τ_1 分别得出 S_3^- 各置换.

但对一个奇置换 τ_1 及一个偶群 G',$\{\sigma\tau_1 : \sigma \in G'\} \cup G'$ 不一定是非偶群. 如取四次偶群 $G' = \{I\}$,奇置换 $\tau_1 = (1,2,3,4)$,上述并集为 $\{(1,2,3,4), I\}$,它不是一个群,因它不含 $(1,2,3,4)^2 = (1,3)(2,4)$.

一个 n 次群,若其所有置换都只使其中某 $k(k < n)$ 个元素改变,而其余元素不变,则称此群为退化群. 它可看成由一个 k 次群的所有元素添上 $n-k$ 个不变元素所成的 n 次群,称之为由这 k 次群导出的 n 次群.

① 求所有三次偶群及 S_3 的真子群 G,使$[S_3 : G] < 3$.

从 $A_3=S_3^+$ 的上述表示式易见三次偶群只有

$$\{I\},A_3$$

因 G 为 S_3 的真子群,易见$[S_3:G]>1$,于是唯有$[S_3:G]=2$,因 S_3 为 6 阶群,故 G 为 3 阶群,又因按上推论知非偶群 G 的阶数为偶数,故 G 只能为偶群,而 3 阶偶群只有 A_3,故所求 $G=A_3$.

② 求全部四次偶群及 S_4 的真子群 G,使$[S_4:G]<4$.

因一次群只有$\{I\}$(偶群),易见二次群只有$\{I\}$(偶群)及 S_2(非偶群),于是四次退化偶群只有$\{I\}$及由 S_3 导出的四次退化群. 今只求四次非退化偶群:由 S_4 不是偶群,故偶群为 S_4 的真子群,其阶数为 S_4 的阶数4! 的比其本身小的约数,故 A_4(阶数4! /2)为阶数最大的四次偶群.

现求所有其余四次非退化偶群:易见

$$A_4=S_4^+=\{I,\tau_{12},\tau_{13},\tau_{14},\sigma_1,\sigma_1^{-1},\sigma_2,\sigma_2^{-1},\sigma_3,\sigma_3^{-1}\}$$

其中(τ 类)

$$\tau_{12}=(1,2)(3,4)$$
$$\tau_{13}=(1,3)(2,4)$$
$$\tau_{14}=(1,4)(2,3)$$

(σ 类)

$$\sigma_1=(2,3,4)$$
$$\sigma_2=(3,4,1)$$
$$\sigma_3=(4,1,2)$$
$$\sigma_4=(1,2,3)$$

(σ^{-1} 类)

$$\sigma_1^{-1}=(4,3,2)$$
$$\sigma_2^{-1}=(1,4,3)$$
$$\sigma_3^{-1}=(2,1,4)$$
$$\sigma_4^{-1}=(3,2,1)$$
$$S_4=A_4\cup S_4^-$$

$$S_4^-=\{(1,2),(1,3),(1,4),(2,3),(2,4),(3,4),$$
$$(1,2,3,4),(1,2,4,3),(1,3,2,4),$$
$$(1,3,4,2),(1,4,2,3),(1,4,3,2)\}$$

易见

$$\tau_{12}^2=\tau_{13}^2=\tau_{14}^2=I$$

$$\tau_{12}\tau_{13}=\tau_{13}\tau_{12}=\tau_{14}$$

$$\tau_{12}\tau_{14} = \tau_{14}\tau_{12} = \tau_{13}$$

$$\tau_{13}\tau_{14} = \tau_{14}\tau_{13} = \tau_{12}$$

故

$$\{I,\tau_{12}\},\{I,\tau_{13}\},\{I,\tau_{14}\}$$

都是偶群(只含 I 及 τ 类一个转换,称之为 $G_2^{(4)}$ 类);

$$\{I,\tau_{12},\tau_{13},\tau_{14}\}$$

也是偶群(只含 I 及 τ 类至少两转换,记为 $G_4^{(4)}$ Ⅱ).

下证其余四次非退化偶群 G' 必为 A_4:G' 必含有 σ,σ^{-1} 类至少一对互逆的置换,但不能只含一对此互逆置换(否则 $G'=\{I,\sigma_i,\sigma_i^{-1}\}$ 为退化群).故 G' 必含 σ,σ^{-} 类两对互逆置换.不妨设至少含 $\sigma_1,\sigma_1^{-1},\sigma_2,\sigma_2^{-1}$,或至少含 $\sigma_1,\sigma_1^{-1},\sigma_3,\sigma_3^{-1}$(其余情况与上述之一类似).

若 G' 至少含 $\sigma_1,\sigma_1^{-1},\sigma_3,\sigma_3^{-1}$,则 G' 含

$$\sigma_1\sigma_3 = \sigma_4$$

$$\sigma_3\sigma_1 = \sigma_2$$

$$\sigma_1\sigma_2 = \tau_{13}$$

$$\sigma_2\sigma_1 = \tau_{14}$$

又含

$$\tau_{13}\tau_{14} = \tau_{12}$$

即 G' 含所有偶置换,从而

$$G' = A_4$$

(阶数最大的四次偶置换群).

若 G' 至少含 $\sigma_1,\sigma_1^{-1},\sigma_2,\sigma_2^{-1}$,则它含

$$\sigma_2\sigma_1^{-1} = \sigma_3$$

从上述情况知 G' 含所有四次偶置换,亦有 $G'=A_4$.

求得所有四次非退化偶群分类为(上(下)标表次(阶)数)

$G_2^{(4)}$:$\{I,(1,2)(3,4)\},\{I,(1,3),(2,4)\},\{I,(1,4)(2,3)\}$

$G_4^{(4)}$ Ⅱ 型:①$\{I,(1,2)(3,4),(1,3)(2,4),(1,4)(2,3)\}$

A_4(4! /2 = 12 阶)

求 S_4 的真子群 G,使

$$[S_4:G] = 2,3$$

①还有六个 4 阶的 4 次非退化(非偶)群.

当$[S_4:G]=2$. 因S_4为$4!=24$阶群，故G为$24/2=12$阶群，于是G为偶群A_4(因若为非偶群，则其偶部群为$12/2=6$阶群，但无四次的6阶偶群).

当$[S_4:G]=3$. 则G为$24/3=8$阶群，只能为非偶群，其偶部群为$8/2=4$阶群(只有)$G_4^{(4)}$ Ⅱ. 据上引理知G为$G_4^{(4)}$ Ⅱ 各置换乘G中一奇置换τ_1所产生. 下证$G_4^{(4)}$ Ⅱ 各置换乘S_4^-任一(奇)置换τ_1所得置换集连同$G_4^{(4)}$ Ⅱ 各置换均能组成一群.

因$G_4^{(4)}$ Ⅱ 所有置换对元素1,2,3,4对称，故不妨设S_4^-中的$\tau_1=(1,3)$或$\tau_1=(1,2,3,4)$.

取$\tau_1=(1,3)$. $G_4^{(4)}$ Ⅱ 各置换乘$(1,3)$分别得

$$(1,3),(1,2,3,4),(2,4),(1,4,3,2)$$

易验证它们连同$G_4^{(4)}$ Ⅱ 各置换组成一群(称$G_8^{(4)}$)

$$G=\{I,(1,2)(3,4),(1,3)(2,4),(1,4)(2,3),\\(1,3),(1,2,3,4),(2,4),(1,4,3,2)\}$$

取$\tau_1=(1,2,3,4)$. 因它属上述八阶群G，故按上述引理知，$G^+=G_4^{(4)}$ Ⅱ 中所有置换乘τ_1得出G^-中各置换，它们连同A_4各置换仍得此群G.

把上述群G中元素1,3;2,4分别改成1,2;3,4或1,4;2,3，便得出$G_8^{(4)}$型的其余两群.

附录 Ⅱ　数论预备知识

① 正整数整除性

正整数整除性有与多项式整除性完全类似的性质(因两正整数同样有带余除法). 如用辗转相除法求最大公约数，二正整数的最大公约数可表示为此二数分别乘某二整数之积的和，素因数分解的唯一性等.

② 同余式

设整数$a,b,b>0$，则存在唯一整数q与r，使

$$a=qb+r,0\leqslant r<b$$

(或记为$\frac{a}{b}=q+\frac{r}{b}$).

称a除以b得不完全商q及余数为r(即“带余除法”)，

如

$$\frac{7}{3}=2+\frac{1}{3}①$$

$$7=2\times3+1$$

$$\frac{-7}{3}=(-3)+\frac{2}{3}$$

$$-7=(-3)\times3+2$$

若两整数 a,b 除以正整数 m 所得余数相同，即等价于 $m\mid a-b$（表示 m 整除 $a-b$，即 $a-b$ 为 m 的（整数）倍数），则称 a 与 b 对模 m 同余，或称 a 与 b 属对模 m 的同一剩余类，记为

$$a\equiv b(\bmod m)$$

当 a 与 b 对模 m 不同余，即 $a-b$ 非 m 的倍数（$m\nmid a-b$），记为 $a\not\equiv b(\bmod m)$，显然对模 m 的剩余类只有 m 个，即除以 m 所得余数分别为 $0,1,2,\cdots,m-1$ 的剩余类.

如（m,k,d 为整数，$m>0$）

$$km\equiv0(\bmod m)$$

$$km+d\equiv d(\bmod m)$$

易见（各小写字母均表整数，$m>0$）

$$a\equiv a(\bmod m)$$

$$a\equiv b(\bmod m)\Leftrightarrow b\equiv a(\bmod m)$$

若 $a\equiv b(\bmod m)$，$b\equiv c(\bmod m)$，则

$$a\equiv c(\bmod m)$$

（简记为 $a\equiv b\equiv c(\bmod m)$）.

若 $a_1\equiv a_2(\bmod m)$，$b_1\equiv b_2(\bmod m)$，则

$$a_1+b_1\equiv a_2+b_2(\bmod m)$$

$$a_1-b_1\equiv a_2-b_2(\bmod m)$$

$$a_1b_1\equiv a_2b_2(\bmod p)$$

（特例）

$$ka_1\equiv ka_2(\bmod p)$$

$$a_1^r\equiv a_2^r(\bmod p)\quad(r\geqslant0)$$

即同余式及其加、减、乘法运算有如同整数等式及其加、减、乘法运算同样的性质.

③（正）素数模 p 的完全剩余类

①不要按小学课本写成 $7\div3=2\cdots1$，因 $7\div3$ 结果是一个分数，$2\cdots1$ 是什么数？

定理1 设p为素数,(正、负)整数m非p的(整数)倍数(即$m \not\equiv 0 (\bmod p)$,亦即$p \nmid m$),则

$$jm \quad (j = 0, 1, 2, \cdots, p-1)$$

对p互不同余——它们除以p所得余数互不相同.

即上述p个数分别属对模p的所有p个不同剩余类,称它们取对模p的完全剩余系.

证 设

$$0 \leqslant j_1 < j_2 \leqslant p-1$$

因p为素数,$p \nmid m$,故p与m互素,又易见$0 < j_2 - j_1 \leqslant p-1$,故$p$亦与$j_2 - j_1$互素,于是$(j_2 - j_1)m$非$p$的倍数,从而$j_2 m \not\equiv j_1 m (\bmod p)$.

例如9非素数7的倍数

$$0 \times 9, 1 \times 9, 2 \times 9, 3 \times 9, 4 \times 9, 5 \times 9, 6 \times 9$$

分别等于

$$0, 9, 18, 27, 36, 45, 54$$

它们除以7所得余数分别为

$$0, 2, 4, 6, 1, 3, 5$$

互不相同,遍取除数为7时的所有余数0 ~ 6.

④Fermat(费马)定理

引理1 设p为素数,$j = 1, 2, \cdots, p-1$,则组合数

$$C_p^j \equiv 0 (\bmod p)$$

(即C_p^j为p之倍数).

证 由组合数之意义知

$$C_p^j = \frac{p(p-1)(p-2)\cdots(p-j+1)}{1 \cdot 2 \cdot 3 \cdot \cdots \cdot j}$$

为整数,分子、分母相约后最后得整数,分母的因数2可整除分子,但$2 < p_1$,2与素数p互素,故2能整除分子中除p外的各数之积,此积约去分母的因数2后所得之商同理可被3整除,此商再约去3之后同理又可被4整除,继续同样论证最后知上述积约去分母各因数后得整数,即C_p^j为p与整数之积.

定理2(Fermat) 设p为素数,正整数a非p的倍数,则

$$a^{p-1} \equiv 1 (\bmod p)$$

证 设b为任何正整数,由二项式定理,上引理及同余式性质得

$$(b+1)^p = b^p + C_p^1 b^{p-1} + C_p^2 b^{p-2} + \cdots + C_p^{p-1} b + 1 \equiv$$
$$b^p + 0 + 0 + \cdots + 0 + 1 \equiv$$

$$b^p + 1 \pmod p$$

现用归纳法证

$$b^p \equiv b \pmod p \qquad (1)$$

当 $b=1$ 时式(1) 显然成立;

设式(1) 对 b 成立,则对 $b+1$ 按前述结论及归纳假设有

$$(b+1)^p \equiv b^p + 1 \equiv b + 1 \pmod p$$

即式(1) 亦对 $b+1$ 成立,所述得证.

于是

$$a^p \equiv a \pmod p$$

$$a^p - a \equiv 0 \pmod p$$

$$a(a^{p-1} - 1) \equiv 0 \pmod p$$

即 p 整除 $a(a^{p-1}-1)$,但素数 p 不整除 a,于是 p 整除 $a^{p-1}-1$,即

$$a^{p-1} - 1 \equiv 0 \pmod p$$

$$a^{p-1} \equiv 1 \pmod p$$

⑤ 正整数对素数模的指数与素数模的原根.

定义　设 p 为素数,正整数 a 非 p 的倍数,适合

$$a^{d'} \equiv 1 \pmod p$$

的正整数 d'(必存在,因据 Fermat 定理 d' 可取 $p-1$) 中最小者 $d(\leqslant p-1)$ 称为 a 对模 p 的指数; 若此指数为 $p-1$,则称 a 为素数模 p 的原根.

下证素数模 p 的原根的存在性.

引理 2　设 p,a,d 如上述,则有等价关系

$$a^m \equiv 1 \pmod p \Leftrightarrow d \mid m$$

特别是

$$d \mid p-1$$

证　设 $d \mid m$,则

$$m = kd \quad (k \in \mathbf{Z}^+)$$

由

$$a^d \equiv 1 \pmod p$$

得

$$(a^d)^k \equiv 1^k \equiv 1 \pmod p$$

即

$$a^m = a^{kd} = (a^d)^k \equiv 1 \pmod p$$

反之,设 $a^m \equiv 1 \pmod p$,证 $d \mid m$.

否则 $d \nmid m$,作带余除法得

$$m = kd + r \quad (k \in \mathbf{N}, r \in \mathbf{Z}^+, r < d)$$

$$a^m = (a^d)^k \cdot a^r \equiv 1^k \cdot a^r \equiv a^r (\bmod p)$$

据 $r \in \mathbf{Z}^+, r < d$ 及指数 d 之定义知

$$a^r \not\equiv 1 (\bmod p)$$

但又有

$$a^r \equiv a^m \equiv 1 (\bmod p)$$

得矛盾. 故必有 $d \mid m$.

引理3　设 p 为素数,则对模 p 的同余式方程

$$a_0 x^n + a_1 x^{n-1} + a_2 x^{n-2} + \cdots + a_{n-1} x + a_n \equiv 0 (\bmod p) \qquad (2)$$

$$(\text{其中 } a_0, a_1, a_2, \cdots, a_n \in \mathbf{N}, p \nmid a_0, n \in \mathbf{Z}^+)$$

对 x 至多有 n 个对模 p 的剩余类解①.

证　用归纳法.

当 $n = 1$ 时,式(2) 为

$$a_0 x + a_1 \equiv 0 \ (\bmod p)$$

由定理1知存在唯一 $j \in \{0,1,2,\cdots,p-1\}$ 使

$$a_0 j \equiv - a_1 (\bmod p)$$

即

$$a_0 j + a_1 \equiv 0 \ (\bmod p)$$

从而式(2) 只有一个对模 p 的剩余类解

$$x \equiv j \ (\bmod p)$$

当 $n \geqslant 2$ 时,设 $n-1$ 次对模 p 的同余式方程最多有 $n-1$ 个对模 p 的剩余类解,而设式(2) 有 m 个对模 p 的剩余类解,证 $m \leqslant n$ 即可:当 $m = 0,1$ 时结论成立; 当 $m \geqslant 2$ 时设这些解为

$$x \equiv x_1, x_2, \cdots, x_m (\bmod p)$$

(其中 $x_1, x_2, \cdots, x_m$ 中任二者不对 p 同余). 作变换

$$x = y + x_1$$

得方程

$$a_0 y^n + a'_1 y^{n-1} + a'_2 y^{n-2} + \cdots + a'_{n-1} y + a'_n \equiv 0 (\bmod p)$$

它的所有解为(因 $y = x - x_1$)

$$y \equiv 0, x_2 - x_1, x_3 - x_1, \cdots, x_m - x_1 (\bmod p) \qquad (3)$$

---

①设 $x = x^*$ 适合式(2),易见与 x^* 对 p 属同一剩余类的任何数也适合式(2),即 $x \equiv x^* (\bmod p)$ 也适合式(2),故式(2) 的解为若干个对模 p 的剩余类.

由 $y \equiv 0(\bmod p)$ 适合上方程,代入得

$$a'_n \equiv 0(\bmod p)$$

于是上方程为

$$a_0 y^n + a'_1 y^{n-1} + a'_2 y^{n-2} + \cdots + a'_{n-1} y \equiv 0(\bmod p)$$

即

$$(a_0 y^{n-1} + a'_1 y^{n-2} + a'_2 y^{n-3} + \cdots + a'_{n-1}) y \equiv 0(\bmod p) \quad (4)$$

因 $x_2, x_3, \cdots, x_m$ 都与 x_1 不对模 p 同余,故 $x_2 - x_1, x_3 - x_1, \cdots, x_m - x_1$ 都非 p 之倍数.

而

$$y \equiv x_2 - x_1, x_3 - x_1, \cdots, x_m - x_1(\bmod p) \quad (5)$$

时,方程(3) 左边二因式之积为素数 p 之倍数,但第二因式非 p 之倍数,故第一因式为 p 之倍数,即方程(4) 为下述方程之解

$$a_0 y^{n-1} + a'_1 y^{n-2} + \cdots + a'_{n-1} \equiv 0(\bmod p) \quad (6)$$

又由 $x_2, x_3, \cdots, x_m$ 中任二数不对模 p 同余(即其差非 p 之倍数),知 $x_2 - x_1, x_3 - x_1, \cdots, x_m - x_1$ 中任二数不对模 p 同余,为(6) 的互异剩余类解. 当 $a'_{n-1} \not\equiv 0(\bmod p)$ 时, $y \equiv 0(\bmod p)$ 不是(6) 的解,这时(6) 无(5) 以外的解(因(6) 的解必须适合(4),(4) 的所有解为(3)),即(6) 有 $m - 1$ 个解,由归纳假设 $m - 1 \leqslant n - 1$,即 $m \leqslant n$;当 $a'_{n-1} \equiv 0(\bmod p)$ 时(6) 的所有解为(3),共 m 个解,由归纳假设 $m \leqslant n - 1 < n$. 即两情形都有 $m \leqslant n$.

引理 4 设 p, q 为素数, m 为正整数, $q^m \mid p - 1$,则存在正整数 a 非 p 之倍数,使 a 对模 p 的指数为 q^m.

证 由上引理知同余式方程(易见 $\frac{p-1}{q}$ 为正整数)

$$x^{p-1/q} \equiv 1(\bmod p)$$

至多有 $\frac{p-1}{q}$ 个对 p 的剩余类解,且这些解不与 0 对 p 同余(用 $x \equiv 0(\bmod p)$ 代入上方程左边得 0),即不是 p 的倍数,又因素数 $q \geqslant 2$, $\frac{p-1}{q} \leqslant \frac{p-1}{2} < p - 1$,故在与 0 不对 p 同余的 $p - 1$ 个剩余类中可求得一剩余类 $x^* \not\equiv 0(\bmod p)$,使 $x \equiv x^*(\bmod p)$ 不适合上方程,(因由引理 3 知上方程至多只有 $\frac{p-1}{2}$ 个对模 p 的剩余类解),即

$$x^{*(p-1)/q} \not\equiv 1(\bmod p)$$

取

$$a = x^{*(p-1)/q^m} \tag{6}$$

因 x^* 非素数 p 之倍数,故 a 亦然,从 Fermat 定理知

$$x^{*p-1} \equiv 1 (\bmod p)$$

又从方程(6) 两边取 q^m 次幂得

$$a^{q^m} = x^{*p-1} \equiv 1 (\bmod p)$$

故由引理2知 a 对模 p 的指数整除 q^m. 但由方程(6) 两边取 q^{m-1} 次幂得

$$a^{q^{m-1}} = x^{*(p-1)/q} \not\equiv 1 (\bmod p)$$

故再由引理2(用反证法)a 对模 p 的指数不整除 q^{m-1},因 q 为素数,能整除 q^m 而不能整除 q^{m-1} 的正整数只有 q^m,于是得证所述指数为 q^m.

定理3(素数模 p 之原根存在定理)　设 p 为素数,则对模 p 的原根必存在.

证　分解素因数

$$p - 1 = q_1^{m_1} q_2^{m_2} \cdots q_r^{m_r} \tag{7}$$

其中 $q_1, q_2, \cdots, q_r$ 为互不相同之素数,$m_1, m_2, \cdots, m_r$ 为正整数. 由上引理求出非(素数)p 倍数之正整数 $a_1, a_2, \cdots, a_r$,使其对模 p 之指数分别为 $q_1^{m_1}, q_2^{m_2}, \cdots, q_r^{m_r}$. 下证 $a_1 a_2 \cdots a_r$(亦非 p 之倍数)对模 p 之指数 d 为 $p-1$ 即可.

由引理2知 $d \mid p-1$,若 $d \neq p-1$,则 d 较 $p-1$ 至少失去一个素因数,不妨设为 q_1,于是 $d \mid \frac{p-1}{q_1}$,再由引理2知

$$(a_1 a_2 \cdots a_r)^{\frac{p-1}{q_1}} \equiv 1 (\bmod p) \tag{8}$$

对 $i = 2, 3, \cdots, r$,因

$$q_i^{m_i}(q_1^{m_1-1} q_2^{m_2} \cdots q_{i-1}^{m_i-1} q_{i+1}^{m_{i+1}} \cdots q_r^{m_r}) = \frac{p-1}{q_1}$$

即 $q_i^{m_i} \left| \frac{p-1}{q} \right.$,故又由引理2知

$$a_i^{(p-1)/q_1} \equiv 1 (\bmod p)$$

由此式及式(8) 按同余式性质得

$$a_1^{(p-1)/q_1} \equiv a_1^{(p-1)/q_1} \prod_{i=2}^{r} a_i^{(p-1)/q_1} \equiv (a_1 a_2 \cdots a_r)^{(p-1)/q_1} \equiv 1 (\bmod p)$$

再由引理2知 a_1 对模 p 之指数 $q_1^{m_1} \left| \frac{p-1}{q_1} \right.$,即

$$\frac{p-1}{q_1} = k q_1^{m_1} \quad (k \in \mathbf{Z}^+)$$

$$p-1=kq_1^{m_1+1}$$

这与前述 $p-1$ 的(唯一)素因素分解式(7)中 q_1 的指数矛盾. 故反证假设 $d \neq p-1$ 不成立,必应有 $d=p-1$.

例如,设素数 $p=7$,取 $a=3$,则

$$3^1=3,3^2=9,3^3=27,3^4=81,3^5=243,3^6=729$$

它们除以 7 分别得余数

$$3,2,6,4,5,1(\equiv 3^6 (\mathrm{mod}\ 7))$$ ①

于是 3 对模 7 的指数为 $6=7-1$,3 为模 7 的原根;

改取 $a=2$,则

$$2^1=2,2^2=4,2^3=8$$

它们除以 7 分别得余数

$$2,4,1(\equiv 2^3 (\mathrm{mod}\ 7))$$

故 2 对模 7 的指数为 $3 \neq 7-1$,2 非模 7 的原根,但 $3 \mid 7-1$(若再求 $2^4,2^5,2^6$ 除以 7 所得余数,则得 2,4,1 与前重复).

再论述素数模的原根性质.

定理 4　设 p 为素数,g 为模 p 之原根,则 $g,g^2,\cdots,g^{p-1}$ 分别与(次序不相同)$1,2,\cdots,p-1$ 对模 p 同余(称 $g,g^2,\cdots,g^{p-1}$ 为模 p 之简化剩余系).

证　由于 g 是模 p 之原根,故 $g^{p-1}\equiv 1(\mathrm{mod}\ p)$,又对 $j=1,2,\cdots,p-2$ 有 $g^j \not\equiv 1(\mathrm{mod}\ p)$,又易证 $1\leqslant j_1<j_2\leqslant p-2$ 时

$$g^{j_1} \not\equiv g^{j_2} (\mathrm{mod}\ p)$$

(否则 $g^{j_1}\equiv g^{j_2}(\mathrm{mod}\ p)$,$1\equiv g^{j_2-j_1}(\mathrm{mod}\ p)$,但 $1\leqslant j_2-j_1<p-2$ 与前述矛盾). 从而 $p-1$ 个数 $g,g^2,g^3,\cdots,g^{p-2},g^{p-1}$(它们都不与 0 对 p 同余)分别与 $1,2,\cdots,p-1$(次序不相同)对模 p 同余.

例　取 $p=7$,$g=3$ 见前述例.

附录 Ⅲ　求实系数多项式的实根个数

先介绍实数列变号次数及多项式的 Sturm(斯图姆)序列两概念.

---

①实际求这些余数可不必先求 3 的各次幂. 只要从第一数 3 起逐次乘 3 后取其除以 7 所得余数,再用此余数乘 3 后取其除以 7 所得余数……$3\times3=9$,9 除以 7 得余数 2;$2\times3=6$,6 除以 7 得余数 6;$6\times3=18$,18 除以 7 得余数 4;$4\times3=12$,12 除以 7 得余数 5;$5\times3=15$,15 除以 7 得余数 1.

一个实数列除去其中等于 0 的项后,若接连两项的符号相反,就认为此数列于此处有一次变号,称数列中所有接连两项变号次数之和为此数列的变号次数.

如数列

$$5, -8, 0, -7, 1, -3$$

除去其中的 0 后得

$$5, -8, -7, 1, -3$$

在 5, -8 处, -7,1 处; 1, -3 处有变号,此数列变号次数为 3.

对无重根的实系数多项式 $f(x)$,令

$$f_0(x)=f(x)$$

$$f_1(x)=f'(x) \quad (\text{导函数})$$

作带余除法:$f_0(x)$ 除以 $f_1(x)$,设得商式(多项式)为 $q_1(x)$,余式(多项式,次数较 $f_1(x)$ 低)为 $r(x)$,令

$$f_2(x)=-r(x)$$

$$r(x)=-f_2(x)$$

则

$$f_0(x)=f_1(x)q_1(x)+r(x)=f_1(x)q_1(x)-f_2(x)$$

又 $f_1(x)$ 除以 $f_2(x)$,设商式为 $q_2(x)$,余式为 $-f_3(x)$,则

$$f_1(x)=f_2(x)q_2(x)-f_3(x)$$

如此继续辗转相除,一般有

$$f_{j-1}(x)=f_j(x)q_j(x)-f_{j+1}(x) \quad j=0,1,2,\cdots,m-1 \qquad (1_j)$$

其中 f_{j-1}, f_j, q_j 均为多项式,f_{j+1} 的次数低于 f_j 的次数,与 Euclid(欧几里得)序列不同,每次取 f_{j+1} 实为将余式变号. 直至最后余式为 0

$$f_{m-1}(x)=f_m(x)q_m(x)$$

即得出多项式 $f(x)$ 的 Sturm 序列

$$f_0(x), f_1(x), f_2(x), f_3(x), \cdots, f_{m-1}(x), f_m(x)$$

与 Euclid 序列最后一项为 $f_1(x), f_2(x)$ 的最高公因式同样,知 $f_m(x)$ 为 $f(x)$ 及 $f'(x)$ 的最高公因式①. 因 $f(x)$ 无重根,$f(x)$ 与 $f'(x)$ 互素(互质),故它们的最高公因式 $f_m(x)$ 为 0 次多项式(非 0 常数).

Sturm 序列性质:

① 相邻两项(多项式)无公共根.

①或先证,Sturm 序列最后一项等于 Euclid 序列最后一项与 ±1 之积.

证 否则设 $f_j(x)$ 与 $f_{j+1}(x)$ 有公共根 α

$$f_j(\alpha)=f_{j+1}(\alpha)=0$$

从式(1_j) 得

$$f_{j-1}(\alpha)=0$$

再从式(1_{j-1})(即 $f_{j-2}(x)=f_{j-1}(x)q_{j-1}(x)-f_j(x)$) 得

$$f_{j-2}(\alpha)=0$$

如此继续,最后知 $f_1(\alpha)=f_0(\alpha)=0$,即 $f_1(x)=f'(x)$ 与 $f_0(x)=f(x)$ 有公共根 α,从而 $f(x)$ 有重根 α,与假设矛盾.

② 设 α 为序列中间项 $f_j(x)(j\neq 0,m)$ 之根,则 $f_{j-1}(\alpha)$ 与 $f_{j+1}(\alpha)$ 都不为 0(由性质 ① 知) 且符号相反,又当 x 增加通过点 α 时 Sturm 序列的变号次数不变.

证 设符号函数

$$\operatorname{sgn} f_{j-1}(\alpha)=s\neq 0$$①

则由 $f_j(\alpha)=0$ 及式(1_j) 知

$$\operatorname{sgn} f_{j+1}(\alpha)=\operatorname{sgn}(-f_{j-1}(\alpha))=-s$$

得证第一结论.

再由多项式函数连续性知,在 α 的充分小邻域内

$$\operatorname{sgn} f_{j-1}(x)=s$$

$$\operatorname{sgn} f_{j+1}(x)=-s$$

于是在此邻域内无论 $f_j(x)$ 取什么符号,即无论 $\operatorname{sgn} f_j(x)$ 等于 s,$-s$ 或 0,易见序列

$$f_{j-1}(x),f_j(x),f_{j+1}(x)$$

在此邻域内的变号次数都是 1,即 x 增加通过点 α 时 Sturm 序列在此处的变号次数不变.

若 α 是不止一个中间项之根(由性质 ① 知这些项不相连),则在每个这样的中间项前后,序列都有一次变号,同样 x 通过点 α 时在其接连两项处序列变号次数亦不变,故 x 增加通过点 α 时,整个序列变号次数不变.

③ 当 x 增加通过 $f(x)$ 的实根 α,则 $f(x)$ 与 $f'(\alpha)$ 间的变号次数减少 1.

---

① $\operatorname{sgn} t=\begin{cases}1 & \text{当 } t>0 \text{ 时}\\ 0 & \text{当 } t=0 \text{ 时.}\\ -1 & \text{当 } t<0 \text{ 时}\end{cases}$

证 这时$f(\alpha)=0, f'(\alpha)\neq 0$(否则$\alpha$为$f(x)$的重根). 若$f'(\alpha)>0$,则$x$增加通过点$\alpha$时$f(x)$由负变正,$f(x)$,$f'(x)$的符号由“$-,+$”变成“$+,+$”,故变号次数减少1; 若$f'(\alpha)<0$,则$x$增加通过点$\alpha$时$f(x)$由正变负,$f(x)$,$f'(x)$的符号由“$+,-$”变成“$-,-$”,变号次数亦减少1.

Sturm 定理 设$a<b$,a,b都不是多项式$f(x)$的根,则$f(x)$在区间$[a,b]$的实根个数等于其Sturm序列在a点的变号次数减在b点的变号次数所得之差.

证 因Sturm序列所有项的符号只有在x通过各项之根时才会改变,而在通过中间项之根时Sturm序列变号次数不改变; 末项为非0常数无根; 而x通过$f(x)$的每一根时序列变号次数减少1. 故易见定理成立.

求实系数多项式(无重根)实根个数,可取充分大的正数M,使在区间$[M,+\infty)$及$(-\infty,-M]$内$f(x)$无根,故只需求$f(x)$在区间$(-M,M)$内实根个数. 又因正数M充分大时各项$f_j(x)$在M点的符号函数值

$$\operatorname{sgn} f_j(M)=\operatorname{sgn} a_{j0}$$

(a_{j0}为多项式$f_j(x)$最高次项系数). 而

$$\operatorname{sgn} f_j(-M)=(-1)^{n_j}\operatorname{sgn} a_{j0}$$

(n_j为多项式$f_j(x)$的次数). 故实际上只需计算两序列

$$(-1)^{n_0}\operatorname{sgn} a_{00},(-1)^{n_1}\operatorname{sgn} a_{10},(-1)^{n_2}\operatorname{sgn} a_{20},\cdots,(-1)^{n_m}\operatorname{sgn} a_{m0}$$

$$\operatorname{sgn} a_{00},\operatorname{sgn} a_{10},\operatorname{sgn} a_{20},\cdots,\operatorname{sgn} a_{m0}$$

的变号次数差,即得$f(x)$的实根个数.

例1 求多项式$f(x)=x^5-pqx+p$(“$p\geqslant 1,q\geqslant 2$”或“$p\geqslant 13,q\geqslant 1$”)的实根个数.

解 求得$f(x)$的Sturm序列

$$x^5-pqx+p$$

$$5x^4-pq$$

$$\frac{4}{5}pqx-p$$

$$pq-\frac{3\ 125}{256q^4}$$

当“$p\geqslant 1,q\geqslant 2$”或“$p\geqslant 13,q\geqslant 1$”时

$$\frac{4}{5}pq>0$$

$$pq - \frac{3\ 125}{256q^4} = \frac{1}{256q^4}(256pq^5 - 3\ 125) > 0$$

故上述两序列为

$$-1,1,-1,1$$

$$1,1,1,1$$

两序列变号次数差 $3-0=3$,故实根个数为 3.

附录 Ⅳ 检验有理系数多项式的可约性

Eisenstein(爱森斯坦)定理只是在颇特殊情况下判定有理系数多项式不可约.现论述检验不超过五次的一般有理系数多项式的可约性.

检验有理系数二、三次多项式可约性.因当且仅当此多项式有理系数一次因式,即有有理数根时此多项式在有理数域 **Q** 可约,故可用综合除法检验.

检验有理系数四次多项式 $f(x)$ 可约性.因四次多项式可约有两种情况:① 它可分解为一次与三次有理系数多项式之积,即有有理数根,可用综合除法检验;② 它可分解为两个有理系数二次多项式之积即其各根为有理数或有理数域二次根式,这可按定理 10.3 判定,即检验此四次方程 $f(x)=0$ 的 Lagrange 预解方程(6.12)在域 **Q** 上至少有一根(按综合除法).

但当情况 ① 不成立时,还有一方法可检验情况 ② 是否成立,从而知此四次多项式是否可约.设有理系数四次方程

$$a_0x^4 + a_1x^3 + a_2x^2 + a_3x + a_4 = 0$$

无有理数根,检验其左边的四次多项式的可约性.作未知数的变换

$$x = x' - \frac{a_1}{4a_0}$$

得 x' 的无三次项的四次方程

$$a_0x'^4 + a_2'x'^2 + a_3'x' + a_4' = 0$$

再去分母(两边乘所有系数的分母的最小公倍数)得整系数四次方程

$$a''_0x'^4 + a''_2x'^2 + a''_3x' + a''_4 = 0$$

再作变换

$$x' = \frac{1}{a''_0}t$$

后去分母(各项乘 a''^3_0)得最高次项系数为1的整系数方程

$$t^4+p_2t^2+p_3t+p_4=0\quad (p_2,p_3,p_4\in \mathbf{Z})\tag{1}$$

因每次作变换不改变方程左边的多项式是否有有理系数二次因式,故考虑最后的方程即可. 设其左边有(据 Gauss(高斯)引理推论,可设)整系数二次因式

$$t^2+\alpha t+\beta\quad (\alpha,\beta\in \mathbf{Z},\beta \text{ 为 } p_4 \text{ 的约数})\tag{2}$$

作带余除法,$t^4+p_2t^2+p_3t+p_4$ 除以 $t^2+\alpha t+\beta$,得商式为

$$t^2-\alpha t+(p_2-\beta+\alpha^2)$$

余式为

$$[p_3+\beta\alpha-\alpha(p_2-\beta+\alpha^2)]t+[p_4-\beta(p_2-\beta+\alpha^2)]$$

易见:方程(1)左边多项式有因式(2)⇔上余式恒等于0,即

$$p_3+\beta\alpha-\alpha(p_2-\beta+\alpha^2)=0\tag{3}$$

$$p_4-\beta(p_2-\beta+\alpha^2)=0\tag{4}$$

即

$$p_3+\beta\alpha=\alpha(p_2-\beta+\alpha^2)$$
$$p_4=\beta(p_2-\beta+\alpha^2)$$

二式相除得

$$\frac{p_3+\beta\alpha}{p_4}=\frac{\alpha}{\beta}$$

解得

$$\alpha=\frac{p_3\beta}{p_4-\beta^2}\tag{5}$$

代入式(3)或式(4),去分母得 β 的整式方程,取所有可作为 p_4 之(正、负)约数的根,再把每个这样的根代入式(5),检验所得的 α 是否整数. 若有一个这样的根代入式(5)所得的 α 为整数,则原四次多项式在域 $\mathbf{Q}$ 可约;若 p_4 的所有(正、负)约数均非所得 β 的方程之根或其所有可作为 p_4 之约数之根代入式(5)所得的 α 均非整数,则原四次多项式在域 $\mathbf{Q}$ 不可约,或用 p_4 的所有(正,负)约数作为 β 代入式(5)检验所得 α 是否为整数,当 α 为整数时再检验 α 与 β 是否适合式(3)或式(4),当 p_4 有一约数适合上述要求时,则原四次多项式在域 $\mathbf{Q}$ 可约,若 p_4 的一切约数作为 β 都不能适合上述要求,则原四次多项式在域 $\mathbf{Q}$ 不可约.

可用类似(但较复杂)方法检验有理系数五次多项式可约性. 当它可约时亦有两种情况:

① 可分解为有理系数一次与四次多项式之积,亦可用综合除法检验;

② 可分解为有理系数二次与三次多项式之积,可用类似上述方法检验,设有理系数五次方程

$$a_0x^5+a_1x^4+a_2x^3+a_3x^2+a_4x+a_5=0$$

无有理数根,检验其左边五次多项式的可约性. 作未知数的变换

$$x=x'-\frac{a_1}{5a_0}$$

得 x' 的无四次项的五次方程

$$a_0x'^5+a_2'x'^3+a_3'x'^2+a_4'x'+a_5'=0$$

再去分母得整系数五次方程

$$a''_0x'^5+a''_2x'^3+a''_3x'^2+a''_4x'+a''_5=0$$

作变换

$$x'=\frac{1}{a''_0}t$$

后去分母得最高次项系数为 1 的整系数方程

$$t^5+p_2t^3+p_3t^2+p_4t+p_5=0$$

同样考虑最后的方程即可. 取二次多项式

$$t^2+\alpha t+\beta \quad (\alpha,\beta\in\mathbf{Z},\beta \text{ 为 } p_5 \text{ 的约数})$$

作带余除法,$t^5+p_2t^3+p_3t^2+p_4t+p_5$ 除以 $t^2+\alpha t+\beta$,得商式

$$t^3-\alpha t^2+(p_2-\beta+\alpha^2)t+(p_3+2\alpha\beta-p_2\alpha-\alpha^3)$$

余式为

$$[p_4+\beta^2-p_2\beta-\alpha^2\beta-\alpha(p_3+2\alpha\beta-p_2\alpha-\alpha^3)]t+$$
$$[p_5-\beta(p_3+2\alpha\beta-p_2\alpha-\alpha^3)]$$

类似检验此余式是否恒等于 0,即

$$p_4+\beta^2-p_2\beta-\alpha^2\beta=\alpha(p_3+2\alpha\beta-p_2\alpha-\alpha^3) \qquad (6)$$

$$p_5=\beta(p_3+2\alpha\beta-p_2\alpha-\alpha^3) \qquad (7)$$

两式相除再去分母得

$$\beta^2\alpha^2+p_5\alpha+(p_2\beta^2-p_4\beta-\beta^3)=0$$

解得

$$\alpha=\frac{1}{2}\left(-\frac{p_5}{\beta^2}\pm\sqrt{\frac{p_5^2}{\beta^4}+4\left(\frac{p_4}{\beta}+\beta-p_2\right)}\right)$$

用 p_5 的一切(正,负)约数作为 β 代入上式检验所得 α 是否为整数,若 α 为整数时再检验 α 与 β 是否适合式(6) 或式(7),当 p_5 有一约

数适合上述要求时,原五次多项式在域 $\mathbf{Q}$ 可约,否则不可约.

检验一般有理系数多项式可约性,可用如下 Kronecker 算法(著者译自一本综述多项式理论研究的外文书):

对于计算多项式 $f \in \mathbf{Z}[x]$①的不可约因式分解式,Kronecker 提出如下算法(Kronecker 算法).设 $\deg f = n$②而 $r = [n/2]$.如果多项式 $f(x)$ 可约,则它有不高于 r 次的因式 $g(x)$.为了求出这个 $g(x)$,考察数 $c_j = f(j), j = 0,1,\cdots,r$.如果 $c_j = 0$,则 $x - j$ 为 $f(x)$ 的因式③;如果 $c_j \neq 0$,则 $g(j)$ 为数 c_j 的因数.数 $c_0,\cdots,c_r$ 的每一套因数 $d_0,\cdots,d_r$ 正好对应于一个不高于 r 次的多项式 $g(x)$,使 $g(j) = d_j, j = 0,1,\cdots,r$.这就是④

$$g(x) = \sum_{j=0}^{r} d_j g_j(x), g_j(x) = \prod_{\substack{D \leqslant k \leqslant r \\ k \neq j}} \left(\frac{x-k}{j-k}\right)$$

对每个这样的多项式 $g(x)$ 要检验,它的系数是整数且它整除 $f(x)$.

日本著名作家、诺贝尔文学奖获得者大江健三郎在回答记者问其:请您按关注的程度排列美术、音乐和科学这三个选项时说(如果按虽然并不真正理解,却是十分着迷的因素排)是科学、音乐、美术;(就经常自信地认为非常了解这一点而论)是美术、音乐、科学([日]大江健三郎著,许金龙译,《大江健三郎口述自传》,北京:新世界出版社,2008).的确科学特别是数学都很让人向往,但要理解是很困难的,特别是以抽象著称的伽罗瓦理论,不仅是对普通人,即使是数学专业人士也是很难懂的.以上是一位数学系的退休教授对这一理论的学习体会.

按照厄布迪尔的理论,"象征资本的积累"必是反经济的,也就是说一个人编什么书,最初只是因为他想编:他考虑的只是他自己,而不是读者.一般的老年人退休之后就是打打麻将、遛遛弯,但数学人不一样,他们总想弄点费脑筋的东西.用谢教授的话来说是为了预防老年痴呆.

黑格尔有一句话说:这个世界上没有什么理想是完全实现的,但人之所以高贵,就是看他死的时候,他的头是不是还朝着他理想的方向(死时候的

---

① 不妨设 $f(x)$ 为整系数多项式 $\mathbf{Z}[x]$.

② $\deg f$ 表 f 的次数.

③ 数 $j = 0,1,\cdots,r$ 可改为任意 $r+1$ 个不同的整数.若对每个 $c_j = 0$,则 f 可约;若 $c_0,\cdots c_r$ 都不为 0,再按下述过程检验.

④ 按 Lagrange 插值公式.

情形谁都难预测,但我们感到自豪的是起码到现在为止,我们的头还是朝着理想的方向,那就是做中国的斯普林格!).

谢先生年轻时曾对伽罗瓦理论很向往,但没时间搞懂.执教大学后又搞微分方程,属分析类,所以退休后的这一举动也算完成了他年轻时代的一个心愿.

美国作家霍姆斯(Oliver Wendell Holmes)在《用晨餐时的诗人》(*The poet at the Breakfast Table*)第十一章中写道:一本坏书,就像一艘有漏洞的船,在智慧的大海中航行,总会有些智慧从漏洞中流进去的.古人云:开卷有益,再差的书读之都会有收获,况且是严谨的数学书,只不过有点难读罢了.

谢先生虽不是代数界人士,也远不能算作行业先锋,以此身份写一本抽象代数的书难免会有人说三道四,但笔者认为这本书的一个最大看点是高度原创性,完全是一位老人自己对一门自己感兴趣的理论的解悟.音乐界中苏姗大妈和“忐忑”一曲之所以能迅速走红,一个主要原因是原创加草根,这也是社会进入后现代的一个特征.人人都有权搞自己所喜爱的学术,只要你遵守学术共同体的规则,别自成一派即可.

出一本属于自己的著作是所有自认为是文化人的光荣与梦想.笔者认为中国即将进入老年化时代,选择著书养老也不失一种既高雅又另类的方式.有人提出疑问为什么谢先生在长达几十年的高校教学生涯中没写一部著作而非要到退休后才写了这样一本著作.在我们现有的大学体制和学术团体中,有些人找到了一种途径,像海明威小说《老人与海》中的大鱼一样把他们自己与海上老人紧紧绑在一起,控制着整个学术界……他们鼓吹有抱负的精神,为他们的专著寻找细小的主题,训练他们从事单调而无用的学科,热衷于脚注和参考书目,扼杀了有生命力的火花.

谢先生似乎是那种忠实于自己内心感受的人,乐于为自己感兴趣的东西投入时间和精力而不是屈服于利益.目前大学教师很难愿意静下心来写一本书.而写书这个活是好人不愿干,孬人干不了.大学教师这一群体的集体放弃造成目前图书市场良莠不齐、鱼龙混杂之势.

据梁文道先生在《读者》一书中所说:研究印刷史和书籍史的学者有个共识,认为古登堡(Gutenberg)印刷术的发明,是人类两种阅读取向的分水岭.在印刷术普及之前,读者追求的是“精读”(intensive reading),犹如古人注经,务求一字一句都要看出个道理,往往一本书能耗上一辈子的生命……而印刷术的年代是个“泛读”(extensive reading)为王的时代,读书首要是求多求广.

以上内容是谢先生精读的产物,但不幸遇到了这个泛读的时代.笔者在审读过程中发现,要想快速读完这本书几乎是不可能的,需要极大的耐心.笔者曾编发过德国数学家阿廷的《伽罗瓦理论》这本小册子,由北大老前辈李同孚先生从德文版翻译出.如果说阿廷用的是文言文,那么谢先生这本就是白话文.

上面这段长文的写作与出版都是十年前的事了.由于谢先生与笔者在数学界都属于"非著名",所以几乎没有收到什么反响.十年弹指一挥间,末了,想起了义山的一首《无题》,其中有句"万里风波一叶舟,忆归初罢更夷犹.碧江地没元相引,黄鹤沙边亦少留",所说的,乃是人生的诸般寥落,与渐行渐远的个体的孤单.这自然不能简单地看作一种感伤,细想此理,或许年少时会易于伤怀,步入中年时亦多愁苦,但真正到了必须面对的年纪,还是坦然以对.就像心焕吾师,唯有一弯明月,一缕清风而已.故义山之诗是最好的慰藉:"人生岂得长无谓,怀古思乡共白头."这一切,非经历而不能解也.

刘培杰

2021年8月12日

于哈工大

刘培杰数学工作室
已出版(即将出版)图书目录——原版影印

书　　名	出版时间	定　价	编号
数学物理大百科全书.第1卷	2016－01	418.00	508
数学物理大百科全书.第2卷	2016－01	408.00	509
数学物理大百科全书.第3卷	2016－01	396.00	510
数学物理大百科全书.第4卷	2016－01	408.00	511
数学物理大百科全书.第5卷	2016－01	368.00	512
zeta函数,q-zeta函数,相伴级数与积分	2015－08	88.00	513
微分形式:理论与练习	2015－08	58.00	514
离散与微分包含的逼近和优化	2015－08	58.00	515
艾伦·图灵:他的工作与影响	2016－01	98.00	560
测度理论概率导论,第2版	2016－01	88.00	561
带有潜在故障恢复系统的半马尔柯夫模型控制	2016－01	98.00	562
数学分析原理	2016－01	88.00	563
随机偏微分方程的有效动力学	2016－01	88.00	564
图的谱半径	2016－01	58.00	565
量子机器学习中数据挖掘的量子计算方法	2016－01	98.00	566
量子物理的非常规方法	2016－01	118.00	567
运输过程的统一非局部理论:广义波尔兹曼物理动力学,第2版	2016－01	198.00	568
量子力学与经典力学之间的联系在原子、分子及电动力学系统建模中的应用	2016－01	58.00	569
算术域	2018－01	158.00	821
高等数学竞赛:1962—1991年的米洛克斯·史怀哲竞赛	2018－01	128.00	822
用数学奥林匹克精神解决数论问题	2018－01	108.00	823
代数几何(德文)	2018－04	68.00	824
丢番图逼近论	2018－01	78.00	825
代数几何学基础教程	2018－01	98.00	826
解析数论入门课程	2018－01	78.00	827
数论中的丢番图问题	2018－01	78.00	829
数论(梦幻之旅):第五届中日数论研讨会演讲集	2018－01	68.00	830
数论新应用	2018－01	68.00	831
数论	2018－01	78.00	832

刘培杰数学工作室

已出版(即将出版)图书目录——原版影印

书　　名	出版时间	定　价	编号
湍流十讲	2018－04	108.00	886
无穷维李代数:第3版	2018－04	98.00	887
等值、不变量和对称性:英文	2018－04	78.00	888
解析数论	2018－09	78.00	889
《数学原理》的演化:伯特兰·罗素撰写第二版时的手稿与笔记	2018－04	108.00	890
哈密尔顿数学论文集(第4卷):几何学、分析学、天文学、概率和有限差分等	2019－05	108.00	891
偏微分方程全局吸引子的特性:英文	2018－09	108.00	979
整函数与下调和函数:英文	2018－09	118.00	980
幂等分析:英文	2018－09	118.00	981
李群,离散子群与不变量理论:英文	2018－09	108.00	982
动力系统与统计力学:英文	2018－09	118.00	983
表示论与动力系统:英文	2018－09	118.00	984
分析学练习.第1部分	2021－01	88.00	1247
分析学练习.第2部分,非线性分析	2021－01	88.00	1248
初级统计学:循序渐进的方法:第10版	2019－05	68.00	1067
工程师与科学家微分方程用书:第4版	2019－07	58.00	1068
大学代数与三角学	2019－06	78.00	1069
培养数学能力的途径	2019－07	38.00	1070
工程师与科学家统计学:第4版	2019－06	58.00	1071
贸易与经济中的应用统计学:第6版	2019－06	58.00	1072
傅立叶级数和边值问题:第8版	2019－05	48.00	1073
通往天文学的途径:第5版	2019－05	58.00	1074
拉马努金笔记.第1卷	2019－06	165.00	1078
拉马努金笔记.第2卷	2019－06	165.00	1079
拉马努金笔记.第3卷	2019－06	165.00	1080
拉马努金笔记.第4卷	2019－06	165.00	1081
拉马努金笔记.第5卷	2019－06	165.00	1082
拉马努金遗失笔记.第1卷	2019－06	109.00	1083
拉马努金遗失笔记.第2卷	2019－06	109.00	1084
拉马努金遗失笔记.第3卷	2019－06	109.00	1085
拉马努金遗失笔记.第4卷	2019－06	109.00	1086
数论:1976年纽约洛克菲勒大学数论会议记录	2020－06	68.00	1145
数论:卡本代尔1979:1979年在南伊利诺伊卡本代尔大学举行的数论会议记录	2020－06	78.00	1146
数论:诺德韦克豪特1983:1983年在诺德韦克豪特举行的Journees Arithmetiques数论大会会议记录	2020－06	68.00	1147
数论:1985－1988年在纽约城市大学研究生院和大学中心举办的研讨会	2020－06	68.00	1148

刘培杰数学工作室
已出版(即将出版)图书目录——原版影印

书　　名	出版时间	定　价	编号
数论:1987年在乌尔姆举行的 Journees Arithmetiques 数论大会会议记录	2020－06	68.00	1149
数论:马德拉斯 1987:1987年在马德拉斯安娜大学举行的国际拉马努金百年纪念大会会议记录	2020－06	68.00	1150
解析数论:1988年在东京举行的日法研讨会会议记录	2020－06	68.00	1151
解析数论:2002年在意大利切特拉罗举行的 C. I. M. E. 暑期班演讲集	2020－06	68.00	1152
量子世界中的蝴蝶:最迷人的量子分形故事	2020－06	118.00	1157
走进量子力学	2020－06	118.00	1158
计算物理学概论	2020－06	48.00	1159
物质,空间和时间的理论:量子理论	2020－10	48.00	1160
物质,空间和时间的理论:经典理论	2020－10	48.00	1161
量子场理论:解释世界的神秘背景	2020－07	38.00	1162
计算物理学概论	2020－06	48.00	1163
行星状星云	2020－10	38.00	1164
基本宇宙学:从亚里士多德的宇宙到大爆炸	2020－08	58.00	1165
数学磁流体力学	2020－07	58.00	1166
计算科学:第1卷,计算的科学(日文)	2020－07	88.00	1167
计算科学:第2卷,计算与宇宙(日文)	2020－07	88.00	1168
计算科学:第3卷,计算与物质(日文)	2020－07	88.00	1169
计算科学:第4卷,计算与生命(日文)	2020－07	88.00	1170
计算科学:第5卷,计算与地球环境(日文)	2020－07	88.00	1171
计算科学:第6卷,计算与社会(日文)	2020－07	88.00	1172
计算科学.别卷,超级计算机(日文)	2020－07	88.00	1173
代数与数论:综合方法	2020－10	78.00	1185
复分析:现代函数理论第一课	2020－07	58.00	1186
斐波那契数列和卡特兰数:导论	2020－10	68.00	1187
组合推理:计数艺术介绍	2020－07	88.00	1188
二次互反律的傅里叶分析证明	2020－07	48.00	1189
旋瓦兹分布的希尔伯特变换与应用	2020－07	58.00	1190
泛函分析:巴拿赫空间理论入门	2020－07	48.00	1191
卡塔兰数入门	2019－05	68.00	1060
测度与积分	2019－04	68.00	1059
组合学手册.第一卷	2020－06	128.00	1153
*一代数、局部紧群和巴拿赫*一代数丛的表示.第一卷,群和代数的基本表示理论	2020－05	148.00	1154
电磁理论	2020－08	48.00	1193
连续介质力学中的非线性问题	2020－09	78.00	1195
多变量数学入门(英文)	2021－05	68.00	1317
偏微分方程入门(英文)	2021－05	88.00	1318
若尔当典范性:理论与实践(英文)	2021－07	68.00	1366
伽罗瓦理论.第4版(英文)	2021－08	98.00	1408

 刘培杰数学工作室 已出版(即将出版)图书目录——原版影印 

书　　名	出版时间	定　价	编号
典型群,错排与素数	2020—11	58.00	1204
李代数的表示:通过 gln 进行介绍	2020—10	38.00	1205
实分析演讲集	2020—10	38.00	1206
现代分析及其应用的课程	2020—10	58.00	1207
运动中的抛射物数学	2020—10	38.00	1208
2—纽结与它们的群	2020—10	38.00	1209
概率,策略和选择:博弈与选举中的数学	2020—11	58.00	1210
分析学引论	2020—11	58.00	1211
量子群:通往流代数的路径	2020—11	38.00	1212
集合论入门	2020—10	48.00	1213
酉反射群	2020—11	58.00	1214
探索数学:吸引人的证明方式	2020—11	58.00	1215
微分拓扑短期课程	2020—10	48.00	1216
抽象凸分析	2020—11	68.00	1222
费马大定理笔记	2021—03	48.00	1223
高斯与雅可比和	2021—03	78.00	1224
π与算术几何平均:关于解析数论和计算复杂性的研究	2021—01	58.00	1225
复分析入门	2021—03	48.00	1226
爱德华·卢卡斯与素性测定	2021—03	78.00	1227
通往凸分析及其应用的简单路径	2021—01	68.00	1229
微分几何的各个方面.第一卷	2021—01	58.00	1230
微分几何的各个方面.第二卷	2020—12	58.00	1231
微分几何的各个方面.第三卷	2020—12	58.00	1232
沃克流形几何学	2020—11	58.00	1233
仿射和韦尔几何应用	2020—12	58.00	1234
双曲几何学的旋转向量空间方法	2021—02	58.00	1235
积分:分析学的关键	2020—12	48.00	1236
为有天分的新生准备的分析学基础教材	2020—11	48.00	1237
数学不等式.第一卷.对称多项式不等式	2021—03	108.00	1273
数学不等式.第二卷.对称有理不等式与对称无理不等式	2021—03	108.00	1274
数学不等式.第三卷.循环不等式与非循环不等式	2021—03	108.00	1275
数学不等式.第四卷.Jensen 不等式的扩展与加细	2021—03	108.00	1276
数学不等式.第五卷.创建不等式与解不等式的其他方法	2021—04	108.00	1277

刘培杰数学工作室

已出版(即将出版)图书目录——原版影印

书　名	出版时间	定　价	编号
冯·诺依曼代数中的谱位移函数:半有限冯·诺依曼代数中的谱位移函数与谱流(英文)	2021—06	98.00	1308
链接结构:关于嵌入完全图的直线中链接单形的组合结构(英文)	2021—05	58.00	1309
代数几何方法. 第 1 卷(英文)	2021—06	68.00	1310
代数几何方法. 第 2 卷(英文)	2021—06	68.00	1311
代数几何方法. 第 3 卷(英文)	2021—06	58.00	1312

书　名	出版时间	定　价	编号
代数、生物信息和机器人技术的算法问题. 第四卷,独立恒等式系统(俄文)	2020—08	118.00	1199
代数、生物信息和机器人技术的算法问题. 第五卷,相对覆盖性和独立可拆分恒等式系统(俄文)	2020—08	118.00	1200
代数、生物信息和机器人技术的算法问题. 第六卷,恒等式和准恒等式的相等 问题、可推导性和可实现性(俄文)	2020—08	128.00	1201
分数阶微积分的应用:非局部动态过程,分数阶导热系数(俄文)	2021—01	68.00	1241
泛函分析问题与练习:第 2 版(俄文)	2021—01	98.00	1242
集合论、数学逻辑和算法论问题:第 5 版(俄文)	2021—01	98.00	1243
微分几何和拓扑短期课程(俄文)	2021—01	98.00	1244
素数规律(俄文)	2021—01	88.00	1245
无穷边值问题解的递减:无界域中的拟线性椭圆和抛物方程(俄文)	2021—01	48.00	1246
微分几何讲义(俄文)	2020—12	98.00	1253
二次型和矩阵(俄文)	2021—01	98.00	1255
积分和级数. 第 2 卷,特殊函数(俄文)	2021—01	168.00	1258
积分和级数. 第 3 卷,特殊函数补充:第 2 版(俄文)	2021—01	178.00	1264
几何图上的微分方程(俄文)	2021—01	138.00	1259
数论教程:第 2 版(俄文)	2021—01	98.00	1260
非阿基米德分析及其应用(俄文)	2021—03	98.00	1261
古典群和量子群的压缩(俄文)	2021—03	98.00	1263
数学分析习题集. 第 3 卷,多元函数:第 3 版(俄文)	2021—03	98.00	1266
数学习题:乌拉尔国立大学数学力学系大学生奥林匹克(俄文)	2021—03	98.00	1267
柯西定理和微分方程的特解(俄文)	2021—03	98.00	1268
组合极值问题及其应用:第 3 版(俄文)	2021—03	98.00	1269
数学词典(俄文)	2021—01	98.00	1271
确定性混沌分析模型(俄文)	2021—06	168.00	1307
精选初等数学习题和定理. 立体几何. 第 3 版(俄文)	2021—03	68.00	1316
微分几何习题:第 3 版(俄文)	2021—05	98.00	1336
精选初等数学习题和定理. 平面几何. 第 4 版(俄文)	2021—05	68.00	1335

刘培杰数学工作室

已出版(即将出版)图书目录——原版影印

书名	出版时间	定价	编号
狭义相对论与广义相对论:时空与引力导论(英文)	2021—07	88.00	1319
束流物理学和粒子加速器的实践介绍:第2版(英文)	2021—07	88.00	1320
凝聚态物理中的拓扑和微分几何简介(英文)	2021—05	88.00	1321
混沌映射:动力学、分形学和快速涨落(英文)	2021—05	128.00	1322
广义相对论:黑洞、引力波和宇宙学介绍(英文)	2021—06	68.00	1323
现代分析电磁均质化(英文)	2021—06	68.00	1324
为科学家提供的基本流体动力学(英文)	2021—06	88.00	1325
视觉天文学:理解夜空的指南(英文)	2021—06	68.00	1326
物理学中的计算方法(英文)	2021—06	68.00	1327
单星的结构与演化:导论(英文)	2021—06	108.00	1328
超越居里:1903年至1963年物理界四位女性及其著名发现(英文)	2021—06	68.00	1329
范德瓦尔斯流体热力学的进展(英文)	2021—06	68.00	1330
先进的托卡马克稳定性理论(英文)	2021—06	88.00	1331
经典场论导论:基本相互作用的过程(英文)	2021—07	88.00	1332
光致电离量子动力学方法原理(英文)	2021—07	108.00	1333
经典域论和应力:能量张量(英文)	2021—05	88.00	1334
非线性太赫兹光谱的概念与应用(英文)	2021—06	68.00	1337
电磁学中的无穷空间并矢格林函数(英文)	2021—06	88.00	1338
物理科学基础数学.第1卷,齐次边值问题、傅里叶方法和特殊函数(英文)	2021—07	108.00	1339
离散量子力学(英文)	2021—07	68.00	1340
核磁共振的物理学和数学(英文)	2021—07	108.00	1341
分子水平的静电学(英文)	2021—08	68.00	1342
非线性波:理论、计算机模拟、实验(英文)	2021—06	108.00	1343
石墨烯光学:经典问题的电解解决方案(英文)	2021—06	68.00	1344
超材料多元宇宙(英文)	2021—07	68.00	1345
银河系外的天体物理学(英文)	2021—07	68.00	1346
原子物理学(英文)	2021—07	68.00	1347
将光打结:将拓扑学应用于光学(英文)	2021—07	68.00	1348
电磁学:问题与解法(英文)	2021—07	88.00	1364
海浪的原理:介绍量子力学的技巧与应用(英文)	2021—07	108.00	1365
多孔介质中的流体:输运与相变(英文)	2021—07	68.00	1372
洛伦兹群的物理学(英文)	2021—08	68.00	1373
物理导论的数学方法和解决方法手册(英文)	2021—08	68.00	1374
非线性波数学物理学入门(英文)	2021—08	88.00	1376
波:基本原理和动力学(英文)	2021—07	68.00	1377
光电子量子计量学.第1卷,基础(英文)	2021—07	88.00	1383
光电子量子计量学.第2卷,应用与进展(英文)	2021—07	68.00	1384
复杂流的格子玻尔兹曼建模的工程应用(英文)	2021—08	68.00	1393
电偶极矩挑战(英文)	2021—08	108.00	1394
电动力学:问题与解法(英文)	2021—09	68.00	1395
自由电子激光的经典理论(英文)	2021—08	68.00	1397

刘培杰数学工作室

已出版(即将出版)图书目录——原版影印

书　　名	出版时间	定　价	编号
曼哈顿计划——核武器物理学简介(英文)	2021—09	68.00	1401
粒子物理学(英文)	2021—09	68.00	1402
引力场中的量子信息(英文)	2021—09	128.00	1403
器件物理学的基本经典力学(英文)	2021—09	68.00	1404
等离子体物理及其空间应用导论.第1卷,基本原理和初步过程(英文)	2021—09	68.00	1405
拓扑与超弦理论焦点问题(英文)	2021—07	58.00	1349
应用数学:理论、方法与实践(英文)	2021—07	78.00	1350
非线性特征值问题:牛顿型方法与非线性瑞利函数(英文)	2021—07	58.00	1351
广义膨胀和齐性:利用齐性构造齐次系统的李雅普诺夫函数和控制律(英文)	2021—06	48.00	1352
解析数论焦点问题(英文)	2021—07	58.00	1353
随机微分方程:动态系统方法(英文)	2021—07	58.00	1354
经典力学与微分几何(英文)	2021—07	58.00	1355
负定相交形式流形上的瞬子模空间几何(英文)	2021—07	68.00	1356
广义卡塔兰轨道分析:广义卡塔兰轨道计算数字的方法(英文)	2021—07	48.00	1367
洛伦兹方法的变分:二维与三维洛伦兹方法(英文)	2021—08	38.00	1378
几何、分析和数论精编(英文)	2021—08	68.00	1380
从一个新角度看数论:通过遗传方法引入现实的概念(英文)	2021—07	58.00	1387
动力系统:短期课程(英文)	2021—08	68.00	1382
几何路径:理论与实践(英文)	2021—08	48.00	1385
论天体力学中某些问题的不可积性(英文)	2021—07	88.00	1396
广义斐波那契数列及其性质(英文)	2021—08	38.00	1386
对称函数和麦克唐纳多项式:余代数结构与Kawanaka恒等式	2021—09	38.00	1400
杰弗里·英格拉姆·泰勒科学论文集:第1卷.固体力学(英文)	2021—05	78.00	1360
杰弗里·英格拉姆·泰勒科学论文集:第2卷.气象学、海洋学和湍流(英文)	2021—05	68.00	1361
杰弗里·英格拉姆·泰勒科学论文集:第3卷.空气动力学以及落弹数和爆炸的力学(英文)	2021—05	68.00	1362
杰弗里·英格拉姆·泰勒科学论文集:第4卷.有关流体力学(英文)	2021—05	58.00	1363

刘培杰数学工作室
已出版(即将出版)图书目录——原版影印

书　名	出版时间	定　价	编号
非局域泛函演化方程:积分与分数阶(英文)	2021—08	48.00	1390
理论工作者的高等微分几何:纤维丛、射流流形和拉格朗日理论(英文)	2021—08	68.00	1391
半线性退化椭圆微分方程:局部定理与整体定理(英文)	2021—07	48.00	1392
非交换几何、规范理论和重整化:一般简介与非交换量子场论的重整化(英文)	2021—09	78.00	1406
数论论文集:拉普拉斯变换和带有数论系数的幂级数(俄文)	2021—09	48.00	1407
挠理论专题:相对极大值,单射与扩充模(英文)	2021—09	88.00	1410
强正则图与欧几里得若尔当代数:非通常关系中的启示(英文)	2021—10	48.00	1411
拉格朗日几何和哈密顿几何:力学的应用	2021—10	48.00	1412

联系地址:哈尔滨市南岗区复华四道街 10 号　哈尔滨工业大学出版社刘培杰数学工作室
网　　址:http://lpj.hit.edu.cn/
邮　　编:150006
联系电话:0451—86281378　　13904613167
E-mail:lpj1378@163.com